地球卫星遥感

卷1：理论与仪器

Earth Science Satellite Remote Sensing

Science and Instruments

Volume 1

John J.Qu · Wei Gao · Menas Kafatos
Robert E. Murphy · Vincent V. Salomonson
EDITORS

TSINGHUA UNIVERSITY PRESS
Beijing

Springer

Heng

内 容 简 介

本书共有两卷。此为第1卷，包含20章，主要提供了有关中分辨率成像光谱仪(MODIS)产品的信息和科学应用，介绍了美国国家极轨环境卫星系统 (NPOESS) 和 NPOESS 预备计划 (NPP)，还探讨了其他卫星遥感装备和应用。有关数据格式、数据处理、数据查询和订购等方面的内容在第 2 卷中给出。

本卷论及美国宇航局(NASA)用于监测和探测地球变化的主要卫星系统——地球观测系统(EOS)，EOS 包括的卫星 Terra、Aqua 和 Aura 及其装载的 MODIS、AIRS、AMSU、AMSR-E、OMI 等遥感仪器，并讨论了 NPP 将携带的 4 个 NPOESS 系统重要部件：可见光红外成像辐射组件(VIIRS)，航线交叉红外探测器(CrIS)，先进技术微波探测器(ATMS)以及臭氧成图和廓线仪装置(OMPS)，可用于研究地球气候环境和天气变化。

本书作者均为相关领域具有权威性的专家与学者。图书内容既包括现代遥感技术的基础知识，又涉及卫星遥感的前沿领域，有广泛的实用性，可作为遥感、地学、环境、空间信息等地球科学领域的专业参考书。

图书在版编目(CIP)数据

地球卫星遥感.卷 1，理论与仪器 = Earth Science Satellite Remote Sensing Volume Ⅰ: Science and Instruments/(美)曲(Qu, J. J.)等主编. —北京：清华大学出版社，2006.9
ISBN 7-302-12844-8

Ⅰ.地…　Ⅱ.曲…　Ⅲ.①卫星遥感 – 理论 – 英文 ②卫星遥感 – 遥感仪 – 英文 ③卫星遥感 – 遥感设备 – 英文　Ⅳ.TP72

中国版本图书馆 CIP 数据核字(2006)第 034365 号

出 版 者：清华大学出版社　　地　址：北京清华大学学研大厦
http://www.tup.com.cn　　邮　编：100084
社 总 机：010-62770175　　客户服务：010-62776969
组稿编辑：张兆琪
文稿编辑：赵彤伟
印 装 者：北京华联印刷有限公司
发 行 者：新华书店总店北京发行所
开　本：153×235　印张：27.5　字数：639 千字
版　次：2006 年 9 月第 1 版　2006 年 9 月第 1 次印刷
书　号：ISBN 7-302-12844-8/TP・8243
ISBN 3-540-35606-1
印　数：1～1500
定　价：128.00 元

This book is dedicated to Dr. Vincent V. Salomonson

Few individuals have had such profound impact on the development of Earth remote sensing as Dr. Vincent V. Salomonson. We, his co-editors of this volume, have been chosen to dedicate it to him in recognition of his many contributions to our field. There is not a topic discussed in the book that has not been strongly influenced either by his personal research or his leadership.

After completing his undergraduate studies, he began his career as a weather officer in the US Air Force. He then returned to graduate school, earning a PhD in Atmospheric Science from Colorado State University in 1968. The bulk of his career was spent at the NASA Goddard Space Flight Center (1968 – 2005) where he conducted research and served as a branch head, laboratory chief, and, for 11 years, as the Director of Earth Sciences. He was deeply engaged in mission development, serving as the Project Scientist for Landsat-4 and -5 (1977 – 1989), and as the team leader for the Moderate Resolution Imaging Spectroradiometer (MODIS) from 1989 to the present. Under his leadership as a laboratory chief and as the Director of Earth Sciences, the men and women of the NASA Goddard Space Flight Center built the foundation for the study of global climate and environmental change using space-based systems and theoretical modeling.

He has served as the president of the American Society for Photogrammetry and Remote Sensing (ASPRS), and is a fellow of the ASPRS, as well as of the Institute for Electrical and Electronics Engineers (IEEE). He has served as an associate editor of several journals and has twice received the NASA Exceptional Scientific Achievement Medal. He has been recognized for career achievements twice, first with the William T Pecora award for his work on Landsat and the NASA Outstanding Leadership Medal for his role in establishing the Earth Sciences Directorate as an internationally recognized entity performing interdisciplinary Earth System Science.

Dr. Salomonson is now a Research Professor at the University of Utah, and the Director of Earth Sciences (Emeritus) at NASA Goddard Space Flight Center.

Foreword

From the late 1990's to the present there has been a truly spectacular series of missions brought to fruition by NASA in the context of the Earth Observing System (EOS). Beginning with the launch and operation in 1997 of the Orbview-2 and the SeaWiFS sensor followed soon by the Tropical Rainfall Monitoring Mission (TRMM) and in 1999, followed by the Quicksat mission and Landsat-7 missions, NASA embarked on a truly spectacular set of missions and sensors that have and are enabling very marked increases in understanding of Earth-atmosphere system processes and trends related to climate change potentially related to future habitability of the earth and the management of the earth's natural resources. The first "flagship" mission launched in very late 1999 of the EOS was the Terra mission devoted to providing a suite of observations that emphasized land processes, but also provided extensive observations of ocean and atmosphere features by utilizing a rich array of sensors with multi-band, multi-angle observations at several spatial resolutions ranging from 15 meters to several kilometers and observing extensive portions of the globe each day. The Terra mission was followed by its sister mission, Aqua, with another formidable array of sensors primarily devoted to observations of the atmosphere, but also key components of the hydrological cycle. On both the Terra and Aqua mission, a sensor often referred to as the "keystone" instrument, was the Moderate Resolution Imaging Spectroradiometer (MODIS). MODIS has provided global, daily observations in 26 spectral bands with spatial resolution ranging from 250 to 1,000 meters that observe many land, ocean, and atmospheric features (e.g. cloud microphysical properties) that serve to not only offer exciting new data for scientific studies, but also provide substantial context around and within which the other instruments on the Terra and Aqua missions are also providing very insightful and useful geophysical products for use by the science and applications communities.

Within the background and context provided in the previous paragraph, nearly the first half (Chapters 1 – 9) of this book offers the reader some in-depth explanatory material describing the MODIS instruments, the background and information included in representative products from MODIS, and the data product development, processing and validation efforts that go into these products. The Chapters 10 – 15 that follow describe the succeeding operational, environmental missions called the National Polar-Orbiting Environmental Satellite Series (NPOESS) and the NPOESS Preparatory Project (NPP) and the attendant sensors including the Visible and Infrared Imaging Radiometer Suite (VIIRS) that is essentially a follow-on instrument to the MODIS with very similar capabilities. The NPP bridges the gap between the Terra and Aqua missions and the beginning of the NPOESS series that should occur early in the next decade; i.e. after 2010. Chapters 16 – 20 describe other various instruments and applications of remote sensing from satellites that give some indication of not only what other missions like the TRMM mission and its envisioned follow-on, the Global Precipitation Mission (GPM) are doing or will do with regard to precipitation, but also what some of the applications of spaceborne observations might be as well as the plans for meteorological satellite development and operation in China.

It is hoped by the authors of this volume that readers will benefit in terms of understanding how to utilize spaceborne observations of the Earth, its processes, and changes over time that affect everyone. Certainly improved information derived from these spaceborne observations of the Earth on a daily, synoptic, high observation basis is valuable and necessary if everyone is to work more closely and harmoniously together to sustain this Earth that is home to humankind.

Vincent V. Salomonson

Senior Scientist and MODIS Science Team Leader
NASA, Goddard Space Flight Center Greenbelt, MD

Preface

Earth science satellite remote sensing has seen rapid expansion during the last decade. NASA's Earth Observing System (EOS) program is providing data for in-depth scientific understanding of the functioning of the Earth as a system through a constellation of satellites that have been launched in recent years or will be launched in the near future. The National Polar-orbiting Operational Environmental Satellite System (NPOESS) Preparatory Project (NPP) mission is the latest in a series transitioning from research to operational satellite status. NPOESS will provide NASA with a continuation of global change observations following EOS Terra and Aqua. NPP will provide NPOESS with risk reduction demonstration and validation for the four critical NPOESS sensors, algorithms, and processing. The NPOESS mission will provide a national, operational, polar-orbiting remote-sensing capability by converging Department of Defense (DoD) and National Oceanic and Atmospheric Administration (NOAA) satellite programs while incorporating new technologies from NASA. Scientists and students have expressed great interest in these missions. However, there is currently no textbook for graduate students to learn about the EOS, NPP and NPOESS missions, or the current and potential applications of the resulting data.

The core of this book arose from the Workshop for Earth Science Satellite Remote Sensing held at George Mason University (GMU) from October 15 to 22, 2002. Updated information is included in this book. This book is designed to give readers having limited remote sensing background a thorough introduction to current and future NASA, NOAA and other Earth science remote sensing. It covers missions/sensors, such as Tropical Rainfall Measuring Mission (TRMM), Atmospheric Infrared Sounder (AIRS), and Advanced Microwave Sounding Unit (AMSU). It also discusses the NPOESS and NPP missions. Emphasis is placed on the recently launched Moderate Resolution Imaging Spectroradiometer (MODIS) on board both of the satellites Terra and Aqua. Key MODIS science team members were invited to contribute several chapters. The editors acknowledge support by the Center of Earth Observing and Space Research (CEOSR) at GMU and NASA/GSFC MODIS and NPP projects.

The goals of this volume are to: (1) provide information on the MODIS products and data processing; (2) give an introduction to the NPOESS and NPP missions; and (3) explore other satellite remote sensing instruments and applications. Detailed information about data formats, data searching and ordering, remote sensing and GIS products, Web GIS applications and tools can be found in volume 2 of Earth Science Satellite Sensing.

There are many people who assisted with this book. First, the editorial team would like to thank all authors involved in contributing chapters for the Earth Science Satellite Remote Sensing. Each author has spent extra hours in addition to existing workloads and ongoing commitments. Second, we would like to thank over eighty anonymous reviewers for their constructive comments and suggestions. The most chapters in this book were originally presented at the Second Workshop of the Earth Science Satellite Remote Sensing at George Mason University (GMU). We would also like to thank many of the Moderate Resolution Imaging Spectroradiometer (MODIS) and the NPOESS Preparatory Project (NPP) science team members who contributed their MODIS and NPOESS/NPP chapters. Much appreciation also goes to the Center for Earth Observation and Space Research (CEOSR) at GMU for supporting the workshop and this book. Special thanks and appreciation go to Mr. Manny Smith for providing editing assistance and tracking chapter status with leading authors. We would like to acknowledge Ms. Lingli Wang, Ms. Bockhwa Kim and Ms. Wanting Wang of the School of Computational Sciences (SCS) at GMU spending tremendous effort working on templates, tables and figures for this book. The efforts of many individuals including Prof. George Taylor, Dr. William Sommers, Prof. Ruixin Yang, Dr. Xianjun Hao and Mr. Hank Wolf at GMU and Dr. Xiaoxiong Xiong at NASA/GSFC, who supported this book, are highly appreciated.

John J. Qu
George Mason University
Fairfax, VA
October 25, 2005

Contents

List of Contributors

Prof. John J. Qu	EastFIRE Lab George Mason University 4400 University Drive Fairfax, VA 22030, USA Phone: 703-993-3958 Fax: 703-993-1993 E-mail: jqu@scs.gmu.edu
Dr. Wei Gao	USDA UV-B Monitoring and Research Program Natural Resource Ecology Laboratory Colorado State University Fort Collins, CO 80523, USA Phone: 970-491-3609 Fax: 970-491-3601 E-mail: wgao@uvb.nrel.colostate.edu
Prof. Menas Kafatos	School of Computational Sciences George Mason University 4400 University Drive, Fairfax, VA 22030, USA Phone: (703)993-3616 Fax: (703)993-1993 E-mail: mkafatos@gmu.edu
Dr. Robert E. Murphy	Earth Sciences Directorate NASA/Goddard Space Flight Center Greenbelt, MD 20771, USA Phone: (301)713-4875 E-mail: robert.murphy@noaa.gov
Dr. Vincent V. Salomonson	University of Utah, Salt Lake City UT 84112, Eearth-Sun Division, Code 600 NASA/Goddard Space Flight Center Greenbelt, MD 20771, USA Phone: (301)614-5631 Fax: (301)614-5808 E-mail: vincent.v.salomonson@nasa.gov

Dr. Alice Isaacman	Science Applications International Corp. 7501 Forbes Blvd, Lanham, MD20706, USA
Dr. Bryan A. Baum	Atmospheric Sciences Directorate, NASA Langley Research Center, Hampton, VA, USA E-mail: bryan.baum@ssec.wisc.edu
Carl F. Schueler	Raytheon Santa Barbara Remote Sensing, 75 Coromar Drive, Goleta CA 93117, USA
Ms. Chaohua Dong	National Satellite Meteorological Center (NSMC), China Meteorological Administration (CMA), Beijing 100081, P. R. China E-mail: dchua@nsmc.cma.gov.cn
Dr. Colin J. Seftor	Raytheon, Information Technology and Scientific Services, Upper Marlboro, MD 20774, USA E-mail: colin_j_seftor@raytheon.com
Dr. Allen Chu	Joint Center for Earth Systems Technology, University of Maryland, Baltimore County, Baltimore, Maryland, USA Laboratory for Atmospheres, NASA Goddard Space Flight Center, Greenbelt, Maryland, USA E-mail: achu@climate.gsfc.nasa.gov
Dr. Dong-Bin Shin	Center for Earth Observing and Space Research, School of Computational Sciences, George Mason University, Fairfax Virginia 22030-4444, USA E-mail: dshin@scs.gmu.edu
Dr. Dorothy K. Hall	Cryospheric Sciences Branch, Code 614.1, NASA/Goddard Space Flight Center, Greenbelt, Maryland 20771, USA E-mail: dhall@glacier.gsfc.nasa.gov
Edward J. Masuoka	Earth-Sun Division, NASA/Goddard Space Flight Center, Greenbelt, MD 20771, USA
Dr. Eric F. Vermote	Associate Research Scientist, Department of Geography, University of Maryland, College Park & NASA/GSFC Code 922 E-mail: eric@kratmos.gsfc.nasa.gov

Dr. Erich Stocker	NASA/GSFC, Code 902, TRMM/TSDIS, Greenbelt, MD 20770, USA E-mail: stocker@tsdis.gsfc.nasa.gov
Frank J. DeLuccia	The Aerospace Corporation, P. O. Box 92957, Los Angeles, CA 90009, USA
Dr. Fuzhong Weng	Brach Chief, Sensor Physics Branch, NOAA/NESDIS/Office of Research and Applications, Camp Springs, MD 20746, USA E-mail: fuzhong.weng@noaa.gov
Dr. George A. Riggs	Senior Scientist/Programmer, Science Systems & Applications, Inc. (SSAI) E-mail: griggs@ltpmail.gsfc.nasa.gov
J E Clement	Raytheon Santa Barbara Remote Sensing, 75 Coromar Drive, Goleta CA 93117, USA
Dr. Jack C. Larsen	Raytheon, Information Technology and Scientific Services, Upper Marlboro, MD 20774, USA E-mail: jack_c_larsen@raytheon.com
Mr. Jianmin Xu	National Satellite Meteorological Center (NSMC), China Meteorological Administration (CMA), Beijing 100081, P. R. China E-mail: jxusmc@public.bta.net.cn
Dr. Ji-Zhong Jin	Department of Meteorology & Earth System Science Interdisciplinary Center(ESSIC), University of Maryland, College Park, MD 20742-2465, USA
Dr. Joel Susskind	Senior Scientist, Laboratory for Atmospheres, NASA Goddard Space Flight Center, Greenbelt, MD 20771, USA E-mail: joel.susskind-1@nasa.gov
Dr. John Kwiatkowski	NASA/Goddard Space Flight Center, TRMM Science Data and Information System, Greenbelt MD 20771, USA/George Mason University E-mail: johnk@tsdis.gsfc.nasa.gov

Mr. Jun Yang	National Satellite Meteorological Center (NSMC), China Meteorological Administration (CMA), Beijing 100081, P. R. China
Dr. Lawrence E. Flynn	Staff Scientist, NOAA/NESDIS, Camp Springs, MD 20734, USA E-mail: lawrence.e.flynn@noaa.gov
Dr. Long S. Chiu	Assoc. Proffessor, Center for Earth Observing and Space Research, School of Computational Sciences, George Mason University, Fairfax Virginia 22030-4444 NASA/Goddard Space Flight Center, Data and Information Services Center, Distributed Active Archive Center, Greenbelt, Maryland 20771, USA E-mail: lchiu@gmu.edu
Dr. Lorraine Remer	Laboratory for Atmospheres, NASA Goddard Space Flight Center, Greenbelt, Maryland, USA E-mail: remer@climate.gsfc.nasa.gov
Mitchell D. Goldberg	Chief, Satellite Meteorology and Climatology Division, NOAA/NESDIS/Office of Research and Applications, Camp Springs, MD 20746, USA E-mail: mitch.Goldberg@noaa.gov
Mr. Nazmi Saleous	Raytheon ITSS at NASA Goddard Space Flight Center (GSFC), Code 922, Greenbelt, Maryland, USA E-mail: nazmi@kratmos.gsfc.nasa.gov
Dr. Peng Gong	Professor and Co-Director, Department of Environmental Science, Policy, and Management Center for Assessment & Monitoring of Forest & Environmental Resources, University of California, Berkeley, CA 94720, USA E-mail: gong@nature.berkeley.edu
Phillip Ardanuy	Raytheon Information Solutions, 12220 Sunrise Valley Drive, Reston, VA 20191, USA
Dr. Philippe Xu	IM Systems Group, Inc., Kensington, MD 20895, USA

Mr. Robert E. Wolfe	Raytheon ITSS at NASA Goddard Space Flight Center (GSFC), Code 922, Greenbelt, Maryland, USA E-mail: rwolfe@pop900.gsfc.nasa.gov
Dr. Ruiliang Pu	Department of Environmental Science, Policy, and Management Center for Assessment & Monitoring of Forest & Environmental Resources, University of California, Berkeley, CA 94720, USA E-mail: rpu@nature.berkeley.edu
Dr. Steven Platnick	Deputy Aqua Project Scientist, Laboratory for Atmospheres, NASA Goddard Space Flight Center, Greenbelt, MD 20771, USA E-mail: steven.e.platnick@nasa.gov
Dr. Wenjian Zhang	National Satellite Meteorological Center (NSMC), China Meteorological Administration (CMA), Beijing 100081, P. R. China E-mail: wjzhang@nsmc.cma.gov.cn
William Barnes	University of Maryland, Baltimore County, Baltimore, MD 21250, USA NASA/GSFC E-mail: wbarnes@neptune.gsfc.nasa.gov
Dr. Xiaoxiong (Jack) Xiong	Physical Scientist, Earth Sciences Directorate, NASA/Goddard Space Flight Center, Greenbelt, MD 20771, USA E-mail: jxiong@ltpmail.gsfc.nasa.gov
Dr. Yimin Ji	Research Associate Professor, Center for Earth Observing and Space Research, School of Computational Sciences, George Mason University, Fairfax, VA 22030, USA E-mail: yji@tsdis.gsfc.nasa.gov
Dr. Zhanqing Li	Professor, Department of Meteorology & Earth System Science Interdisciplinary Center, University of Maryland, College Park, MD 20742-2465, USA E-mail: zli@atmos.umd.edu

1 Introduction to Science and Instruments

Menas Kafatos and John J. Qu

The first volume of this book covers the important topics of Earth science remote sensing. The volume methodically covers a variety of Earth science data sets obtained from several remote sensing sensors, and is divided into three main sections, Moderate Resolution Imaging Spectroradiometer (MODIS) data and data processing; National Polar Orbiting Environmental Satellite System (NPOESS) and NPOESS Preparatory Project (NPP) missions; and several other missions. In some ways the division into 3 main components makes a lot of sense: MODIS represents advanced global Earth observing covering all components of the Earth as it presently stands. NPP and NPOESS represent advanced future observing systems and, in the case of NPOESS future operational systems, also rely on heritage sensors such as MODIS. The third category represents equally valuable observing systems. They all help, we believe, to provide the reader with information about the exciting field of Earth Science Satellite Remote Sensing (RS). We provide here a synopsis of the volume, heavily using material provided in the chapters themselves. The material is surely not ours; it is the authors' works, which we have relied on. The reader is guided to references and more detailed discussion in the chapters themselves, written by the experts in the relevant areas.

The first Moderate Resolution Imaging Spectroradiometer (MODIS) (Salomonson et al., 2006) was launched in December 1999 on the polar orbiting National Aeronautics and Space Administration (NASA) Earth Observing System (EOS) Terra satellite and the second MODIS was launched on the polar orbiting Aqua satellite in May, 2002. Each MODIS acquires daily global data in 36 spectral bands—29 with 1 km, 5 with 500 m and 2 with 250 m nadir pixels. MODIS was designed to take measurements in a broad spectral range, at three spatial resolutions and over a wide field of view, being ideal for studying different components of the Earth system. Algorithms developed by the researchers and scientists of the MODIS Science Team are used to generate approximately 40 science products that are used in formulating a wide range of parameters that are needed for the short- and long-term studies of the Earth's land, oceans, and atmosphere. Both Terra and Aqua MODIS are currently operating on-orbit and making continuous global observations.

The MODIS section starts with a chapter by Vincent Salomonson, William Barnes, and Edward J. Masuoka, Introduction to MODIS and an Overview of

Associated Activities. It provides an overview of the MODIS and data products that can be used for a variety of applications. The authors demonstrate the power and versatility of the MODIS instrument and the associated data, algorithm systems and organizational approach that have led to the usefulness of MODIS in a wide variety of land, ocean and atmosphere science areas and applications. The chapter provides the background, history of development and a technical description of the instrument, followed by the MODIS team description and division of responsibilities. This helps us understand the central role of MODIS in the Earth Observing System (EOS) Data and Information System (EOSDIS). Finally, the data archive of MODIS, present status and future development of an instrument onboard of the National Polar-Orbiting Environmental Satellite Series (NPOESS).

The next chapter, Chapter 3 by Xiaoxiong Xiong, Alice Isaacman, and William Barnes, covers MODIS Level-1B Products, specifically describing the MODIS L1B data products and the calibration algorithms used to generate them. The higher level MODIS data processing is done at the MODIS Data Adaptive Processing System (MODAPS). The MODIS products, including L1B products, are distributed through the Goddard Space Flight Center Distributed Active Archive Center (GDAAC). MODIS Level-1B algorithms convert raw data from the sensor's observations (in the L1A format) into radiometrically calibrated and geometrically located data sets. The MODIS L1B algorithms are developed and maintained by the MODIS Characterization Support Team (MCST) under the direction of the MODIS science team leader. In terms of preparing the data for DAAC access, working closely with the MODIS Science Team representatives, MODIS Characterization Support Team (MCST) analysts and L1B code developers perform requisite evaluations before a new version of L1B software is sent to the Goddard DAAC for additional testing and subsequent data production. The emphasis here is on the implementation of the MODIS calibration algorithms in the L1B code. The code standards and associated properties, data processing, and data product retrieval are also discussed.

The MODIS Geolocation chapter is authored by Robert E. Wolfe. As described by Wolfe, a global network of ground control points has been used to improve the geolocation accuracy for MODIS/Terra to better than 45 m and for MODIS/Aqua to better than 60 m. This chapter contains an overview of the geolocation approach, a summary of several combined years of MODIS/Terra and MODIS/Aqua geolocation results, as well as an overview of present work, and a discussion of the applicability to future missions. The MODIS approach allows an operational characterization of the MODIS geolocation errors and enables individual MODIS observations to be geolocated to sub-pixel accuracies.

Introduction to MODIS cloud products is presented in Chapter 5 by Bryan A. Baum and Steven Platnick. MODIS offers a new and perhaps unique perspective for deriving global and regional cloud properties. It is well known that cloud physics and their role in climate and global modeling are both crucial and also challenging. As such, this chapter takes on added value in terms of global climate change and associated climatological data. MODIS provides spectral measurements as several operational weather platforms such as the Advanced Very High Resolution Radiometer (AVHRR), High Resolution Infrared Sounder (HIRS) and Geostationary Operational Environmental Satellite (GOES) and offers new spectral measurements that expand the potential for global cloud retrievals. The authors describe that for the first time, various cloud properties, e.g. cloud top pressure, thermodynamic phase, optical thickness, and particle size can be derived at 1-km spatial resolution.

D. Allen Chu and Lorraine Remer explore the MODIS observations of aerosol loading in the period 2000 – 2004 in Chapter 6. Here, aerosol optical depth is a measure of aerosol loading in an atmospheric column. The satellite-based measurements can cover large areas, providing the way to study aerosol effects at global scales. As Salomonson and co-authors have pointed out, MODIS is one sensor designed to measure aerosol on EOS. The unique set of seven MODIS channels with higher coverage and higher resolution of 250 – 500 m enables scientists to not only derive aerosol optical depth but also other parameters. As such, it is expected to aid significantly in studying aerosol effects on the Earth's energy budget and hydrological cycle. One of the success stories is the correlations between MODIS aerosol optical depth (AOD) and $PM_{2.5}$ mass concentration which showed that MODIS AOD can be used as an indicator for $PM_{2.5}$ mass concentration, pinpointing the pollution source to local, State, and continental origins, generated by urban/industrial pollution, forest fires, or dust storms.

Robert E. Wolfe and Nazmi Saleous examine MODIS land products and associated processing in Chapter 7, summarizing the MODIS Land Science team's products, describing the data processing approach as well as the process for monitoring and improving the product quality. The two MODIS instruments allow key measurements for understanding the Earth's terrestrial ecosystems. Global time-series of terrestrial geophysical parameters have been produced from the two MODIS. These well calibrated instruments, a team of land scientists and a large data production system have allowed for the development of a new suite of high quality land product variables at spatial resolutions as fine as 250 m in support of global change research and natural resource applications. The authors examine a great deal of the history and development of the land products. Although heritage systems such as AVHRR produced long time series, the global resolution has been much lower than MODIS. Processing evolved into a combination of processing the lower level products for the EOS DAAC's and production of the higher level discipline specific products for the MODIS Science Investigator Lead Processing System.

Chapter 8 authored by Eric F. Vermote and Nazmi Z. Saleous describes operational atmospheric correction of MODIS visible to middle infrared land surface data for an infinite Lambertian target. It starts first with a theoretical background section and then shows how the solution of the equation of transfer is implemented in operations; also, it describes the input of the atmospheric corrections, and finally it discusses the error budget and validation.

The atmospheric correction of remote sensing data has always been a concern for the ocean color products, since the signal of interest is almost an order of magnitude smaller than the top of the atmosphere signal. Atmospheric correction over ocean on Sea-viewing Wide Field-of-view Sensor (SeaWiFS), MODIS and Medium Resolution Imaging Spectrometer (MERIS) is now including various corrections. Over land, because of the lesser impact of the atmosphere compared to ocean, and the lack of dedicated missions, the use of standard atmospheric corrections has been slower to establish. The chapter describes the operational procedure for atmospheric correction over Land for an infinite Lambertian target.

Chapter 9, authored by Dorothy K. Hall, George A. Riggs, and Vincent V. Salomonson examine MODIS snow and ice products. As is known, the Earth's snow and sea ice are among the Earth's most dynamic features. Over 40% of the Earth's land surface may be covered by snow cover during the Northern Hemisphere winter, and sea ice itself covers 5% – 8% of the ocean surface at any given time. They both act as an effective insulating layer between the atmosphere and land or ocean surfaces. Numerous studies have shown the importance of accurate measurements of snow and ice extent and albedo, as well as snow depth and water equivalent, and sea ice concentration and thickness, as they relate to the Earth's climate and climate change. Satellites are well suited to the measurement of snow cover and sea ice because the high albedo of snow and sea ice presents a good contrast with most other natural features except clouds.

Weekly snow mapping of the Northern Hemisphere using NOAA data began in 1966 and continues today in the US, with improved spatial resolution. The MODIS snow and sea ice products are available globally and at a variety of different resolutions and projections to serve different user groups. The authors point to the need of validated products and re-processing. There are several different data-product levels starting with Level-2 (L2). MODIS algorithms have been developed to map snow and sea ice and to calculate snow albedo and sea ice surface temperature.

In the next section, R. E. Murphy gives a description of the NPOESS Preparatory Project (NPP). NPP is a mission designed to bridge the gap between the NASA EOS and the National Polar Orbiting Operational Environmental Satellite System (NPOESS), the United States' next generation of operational environmental satellites. The NPOESS is a satellite system used to monitor global environmental conditions, and to collect data related to weather, atmosphere, oceans, land, and near-space environment. NPOESS carries 10 major satellite sensors, and provides the retrievals for 55 environmental data records (EDRs).

NPP, on the other hand, forms a crucial link between primarily scientific missions (EOS, etc.) to primarily operational missions (NPOESS) and the goals are to provide data continuity for global change research and to provide risk reduction. NPP is a partnership between the Integrated Program Office (IPO) (which itself is an activity of Department of Defense (DoD), NOAA and NASA) and NASA. NPOESS will replace the current civilian Polar Operational Environmental Satellite (POES) program and the military Defense Meteorological Satellite Program (DMSP) which operate independently with a joint system. It incorporates advanced sensors patterned, in many cases, after those developed by NASA for its research oriented EOS.

In its history of development, NASA participated strongly in the Visible Infrared Imaging Radiometer Suite (VIIRS) as it became apparent that the proposed VIIRS capabilities were close to those of MODIS. A NASA study conducted with the aid of the MODIS science team concluded that a number of important new capabilities were too preliminary for flight on an operational sensor, while others were consistent with operational use. The article goes into many details of the evolution of NPP as well as key aspects of the two program philosophies, IPO (operational) and NASA (scientific). A novel aspect of NPP is that the sensor vendors not only developed the sensors, they also developed (or adapted) the science algorithms.

The Visible Infrared Imaging Radiometer Suite (VIIRS) is discussed by R. E. Murphy, Phillip Ardanuy, Frank J. DeLuccia, J. E. Clement, and Carl F. Schueler in Chapter 11. As the case with MODIS, VIIRS is used to obtain measurements of the Earth's oceans, land surface and atmosphere, producing a wide range of geophysical data products, and providing global coverage at moderate (better than 1 km) spatial resolution. This combination of spatial and temporal scales has been chosen to provide needed input to operational weather and environmental models while sampling the natural variability of biological processes on the land surface and in the oceans. VIIRS has very high radiometric and geometric fidelity, enabling it to acquire data for global climate change. VIIRS will fly on NPOESS in three polar sun-synchronous orbit planes with equator crossing times of 1730 (Ascending Node), 0930 (Descending Node) and 1330 (Ascending Node). It will also fly on NPP with a 1030 Ascending Node.

VIIRS consists of a single sensor and draws heavily on the experience gained in building and operating MODIS, as well as the Sea-viewing Wide Field-of-view Sensor (SeaWiFS) for measuring ocean color, the Along Track Scanning Radiometer (ATSR) for measuring sea-surface temperature (SST), and the Operational Linescan System (OLS) for terminator imaging.

Fuzhong Weng provides an overview for the Conically-scanned Microwave Imager Sounder (CMIS) on board the future NPOESS in Chapter 12. CMIS is used to image the Earth's surface through clouds, and it alone will measure more than 20 EDRs. The EDRs are especially important for ocean surface wind speed and direction, and for soil moisture measurements—key performance parameters for the Department of Defense (DoD) and useful for civilian agricultural and flood warning applications. Its sounding channels will provide information on atmospheric profiles of temperature and water vapor and clouds that can be directly assimilated into numerical weather prediction models.

The author makes the point that during the past 20 years, there has been a dramatic increase in the use of satellite based—microwave products by both meteorological and oceanographic organizations, attributed to the launch of the Special Sensor Microwave Imager (SSM/I) in 1987 onboard the first of a series of Defense Meteorological Satellite Program (DMSP) satellites. SSM/I contains six channels in window regions with dual polarization, and a seventh channel centered on a water vapor line with vertical polarization.

The experience learned from SSM/I led to the development of more operational products from other microwave sensors. NOAA now generates operational products including cloud liquid, water vapor, rain rate, snow cover and sea ice concentration with the Advanced Microwave Sounding Unit (AMSU) sensor.

The Advanced Technology Microwave Sounder (ATMS) is described by Mitchell D. Goldberg and Fuzhong Weng. The next generation of microwave sounders—the Advanced Technology Microwave Sounder (ATMS)—are scheduled to be flown on the NPOESS starting with the NPP in 2006. The ATMS has nearly the same functional requirements as those from the current operational microwave sensor suite of the Advanced Microwave Sounding Unit (AMSU) – A and – B. However the instrument design of ATMS is very different, characterized by much smaller volume, weight and power consumption.

The assimilation of AMSU observations in numerical weather prediction models have resulted in significant improvements in forecast skills. The authors highlight the differences between ATMS and AMSU instruments and they present some current applications of AMSU measurements.

Chapter 14 authored by Joel Susskind introduces Atmospheric Infrared Spectroradiometer (AIRS) and Cross-Track Infrared Sounder (CrIS). AIRS/AMSU/HSB is a state of the art advanced infra-red microwave sounding system that was launched on the EOS Aqua platform. Besides being part of a climate mission, one of the objectives of AIRS is to provide sounding information of sufficient accuracy such that when assimilated into a general circulation model, significant improvement in forecast skill can arise. CrIS, like AIRS, is an advanced IR sounder and will fly on the NPP mission, as well as on NPOESS. CrIS will be accompanied by ATMS, a microwave sounder with characteristics similar to those of AMSU – A/HSB. CrIS/ATMS is expected to produce products of similar accuracy as can be obtained from AIRS/AMSU/HSB.

Observations in the sounding channels for AIRS and CrIS are sensitive to atmospheric temperature and constituent profiles and can be used to determine these parameters. The spatial resolution of AIRS and CrIS is coarser than MODIS and VIIRS. However, high spectral resolution observations in the infra-red are desirable for a number of reasons. Having a large number of channels with high spectral resolution, allows for absorption features to be isolated. As the author explains, the best channels for sounding purposes are usually either between lines or on line centers.

The primary AIRS products are atmospheric temperature profiles, water vapor profile, and ozone profile; land/ocean surface skin temperature and IR and microwave spectral emissivity; AIRS clear column radiances; cloud top pressure and effective fractional cloud cover for up to two cloud types; and Outgoing Longwave Radiation (OLR) and clear sky OLR (the longwave radiation which would have gone to space if no clouds were present). Atmospheric profiles of CO and CH_4, as well as total integrated atmospheric CO_2 burden are being derived in a research mode.

Both AIRS and CrIS are accompanied by a microwave sounding instrument. IR and microwave observations are very complementary and the strengths of a combined system are greater than the sum of the strengths of the individual components: High resolution IR observations produce higher vertical resolution, and hence better accuracy, of mid-lower tropospheric temperature profile than the microwave observations. In addition, high spectral resolution IR observations provide the best information about surface skin temperature and constituent profiles.

The author describes the basic physics giving rise to the AIRS channel radiances, including how different channels are sensitive to different surface and surface quantities and other aspects of AIRS and CrIS.

Lawrence E. Flynn, Colin J. Seftor, Jack C. Larsen, and Philippe Xu discuss the Ozone Mapping and Profiler Suite (OMPS), the next-generation US ozone monitoring system, designed for NPOESS. The first flight of an OMPS is scheduled for late 2006 on the NPP satellite. The OMPS is designed to replace both the NASA Total Ozone Mapping Spectrometer (TOMS) and NOAA Solar Backscatter Ultraviolet Spectrometer/2 (SBUV/2) systems. OMPS was designed to meet the stringent set of performance requirements for atmospheric ozone products detailed in the original NPOESS system specifications.

In Chapter 16, John Qu estimates estimated the surface irradiances at the Earth surface using multi-satellite remote sensing measurements. The purpose of this study was to simulate UV-B irradiance at the Earth's surface using satellite ozone, cloud and aerosol measurements applied in a two-stream atmospheric radiative transfer model based on a delta-Eddington approximation. The study also evaluated the model's performance compared to high-quality surface measurements of UV-B in remote areas. The satellite measurements used include the TOMS derived total column ozone; the International Satellite Climatology Cloud Project (ISCCP) cloud products; and the Advanced Very High Resolution Radiometer (AVHRR) and Stratospheric Aerosol Gas Experiment (SAGE II) aerosol observations. The Earth surface UV-B measurements used in this study to evaluate the two-stream model were made in North America and Russia by the United States Department of the Interior, Bureau of Land Management (BLM) using broadband Yankee Environmental SUV-1 meters and from a spectroradiometer at a National Science Foundation site in San Diego, California. The evaluation revealed that overall, the two- stream model simulated monthly UV-B radiation at the Earth's surface within ±6% of the measured values 95% of the time. A wide-ranging sensitivity analysis revealed that, aside from the solar zenith angle, the model was sensitive to (in order of most to least) clouds, ozone, Rayleigh scattering, and aerosols. As expected, model performance was best for clear days and poorer for cloudy days. Nevertheless, this research demonstrated that the two-stream model could reliably simulate UV-B radiation at the Earth surface, if provided with satellite remote sensing derived and other model input parameters.

In the last section, Long S. Chiu and Dong-Bin Shin discuss surface rain rates from TRMM Version 5 Algorithms. The Tropical Rainfall Measuring Mission (TRMM), jointly sponsored by NASA and Japan Aerospace Exploration Agency (JAXA, previously known as National Space Development Agency, or NASDA), is the first international satellite mission to monitor and study tropical and subtropical rain systems and patterns. The TRMM rain sensor package includes the first space-borne Precipitation Radar (PR), the TRMM Microwave Imager (TMI) and the Visible and Infrared Scanner (VIRS). Rainfall estimates provided by the TRMM are applicable to climate analysis, data assimilation, water resource management, and agricultural decision support to health issues.

Zhanqing Li, Ji-Zhong Jin, Peng Gong, and Ruiliang Pu discuss how Remote Sensing (RS) data can be modeled for carbon emission from fires, they apply this to North America. They emphasize that accurate accounting of carbon cycling is paramount to understanding and modeling global climate change. It has been argued that the missing carbon may be absorbed in the terrestrial biomes of the Northern Hemisphere and in particular the temperate and boreal forests in North America could account for the bulk of the missing carbon. Fire is a driving factor controlling the carbon dynamics in North America, affecting both the sign and magnitude of the carbon budget. The close correlation between the carbon budget and overall fire activity demonstrates the importance of the accurate estimation of carbon emissions from fires. RS data can be used to provide continental coverage and is capable of providing additional spatial and temporal fire information to improve fire emission estimations. In addition to burned area, RS can provide "snapshots" of fire dynamics information (starting and ending dates and daily spread), and spatial heterogeneity. The authors discuss that the best strategy is to combine conventional and satellite data to maximize their utility for fire emission estimation.

Yimin Ji and Erich Stocker review fire algorithms, products, and applications that have been developed at NASA's TRMM Science Data and Information System (TSDIS) in Chapter 19. Land fires not only are frequent hazards to human lives and property, they also change the state of the vegetation and contribute to the climate forcing by releasing large amounts of aerosols and greenhouse gases. As such, the release of aerosols during fires may also lead to significant changes of cloud microphysics and radiative properties. A recent study of TRMM data in Indonesia showed evidence that smoke from sustained fires may even suppress regional rainfall completely for certain rain types and therefore create a feedback or an even more favorable environment for fire to occur. This chapter reviews the existing and future satellite fire and aerosol products.

In Chapter 20, Wenjian Zhang, Jianmin Xu, Chaohua Dong, and Jun Yang describe China's meteorological satellites, which have become an indispensable tool for weather and environment observations. These satellites are integrated components of China's Earth observation system for meteorological operations, major natural disasters monitoring and improving the efficiency of many sectors of China's national economy.

The meteorological satellite program of China consists of two series: the polar orbiting series and geostationary series. The authors describe the main objectives of the program, which are to establish, with the combination of polar and geostationary orbits, a comprehensive operational meteorological satellite system as well as the associated ground monitoring and application data system, in order to meet the needs of Chinese society, and enhance its ability of contribution to the international community. The China Meteorological Administration (CMA) contracts China Aerospace Science and Technology Corporation (CASC) to develop the space segment including the launchers and the satellites, while the National Satellite Meteorological Center (NSMC), a sub-organization of China Meteorological Administration, is responsible for developing the satellite ground segment.

In China, meteorological satellites are named simply as Feng-Yun series, or abbreviated as FY-series, where the words Feng-Yun stand for "Winds and Clouds". The FY-odd numbers are used for the generations of the polar orbiting satellite series, i.e. FY-1 for the first generation, FY-3 for the second generation, etc., whereas the FY-even numbers are used for the generations of the geostationary series, i.e. FY-2 for the first generation, etc.

This book volume covers EOS and NPOESS/NPP major instruments and products. It provides basic information about EOS and NPOESS/NPP missions. More detailed information regarding remote sensing data, computational processing and tools can be obtained in volume 2.

References

Baum B, Platnick S (2006) Introduction to MODIS Cloud Products. In: Qu, Gao, Kafatos, Murphy and Salomonson (eds) Earth Science Satellite Remote Sensing: Vol 1. Springer-Verlag and Tsinghua University Press (This publicatioin)

Chiu LS, Shin D-B, Kwiatkowski J (2006) Surface Rain Rates from TRMM Satellite Algorithms. In: Qu, Gao, Kafatos, Murphy and Salomonson (eds) Earth Science Satellite Remote Sensing: Vol 1. Springer-Verlag and Tsinghua University Press (This publication)

Chu DA, Remer L (2006) MODIS Observation of Aerosol Loading from 2000 to 2004. In: Qu, Gao, Kafatos, Murphy and Salomonson (eds) Earth Science Satellite Remote Sensing:

Vol 1. Springer-Verlag and Tsinghua University Press (This publication)

Flynn LE, Seftor CJ, Larsen JC, Xu P (2006) The Ozone Mapping and Profiler Suite (OMPS). In: Qu, Gao, Kafatos, Murphy and Salomonson (eds) Earth Science Satellite Remote Sensing: Vol 1. Springer-Verlag and Tsinghua University Press (This publication)

Goldberg MD, Weng F (2006) Advanced Technology Microwave Sounder (ATMS). In: Qu, Gao, Kafatos, Murphy and Salomonson (eds) Earth Science Satellite Remote Sensing: Vol 1. Springer-Verlag and Tsinghua University Press (This publication)

Hall DK, Riggs GA, Salomonson VV (2006) MODIS Snow and Sea Ice Products. In: Qu, Gao, Kafatos, Murphy and Salomonson (eds) Earth Science Satellite Remote Sensing: Vol 1. Springer-Verlag and Tsinghua University Press (This publication)

Ji Y, Stocker E (2006) TRMM Fire Algorithm, Product and Applications. In: Qu, Gao, Kafatos, Murphy and Salomonson (eds) Earth Science Satellite Remote Sensing: Vol 1. Springer-Verlag and Tsinghua University Press (This publication)

Li Z, Jin J-Z, Gong P, Pu R (2006) Use of Satellite Remote Sensing Data for Modeling Carbon Emissions from Fires: A Perspective in North America. In: Qu, Gao, Kafatos, Murphy and Salomonson (eds) Earth Science Satellite Remote Sensing: Vol 1. Springer-Verlag and Tsinghua University Press (This publication)

Murphy RE (2006) The NPOESS Preparatory Project (NPP). In: Qu, Gao, Kafatos, Murphy and Salomonson (eds) Earth Science Satellite Remote Sensing: Vol 1. Springer-Verlag and Tsinghua University Press (This publication)

Murphy RE, Ardanuy P, DeLuccia FJ, Clement JE, Schueler CF (2006). The Visible Infrared Imaging Radiometer Suite (VIIRS). In: Qu, Gao, Kafatos, Murphy and Salomonson (eds) Earth Science Satellite Remote Sensing: Vol 1. Springer-Verlag and Tsinghua University Press (This publication)

Qu JJ (2006), Estimating solar UV-B irradiance at the Earth's surface using multi-satellite remote sensing measurements. In: Qu, Gao, Kafatos, Murphy and Salomonson (eds) Earth Science Satellite Remote Sensing: Vol 1. Springer-Verlag and Tsinghua University Press (This publication)

Salomonson VV, Barnes W, Masuoka EJ (2006) Introduction to MODIS and an Overview of Associated Activities. In: Qu, Gao, Kafatos, Murphy and Salomonson (eds) Earth Science Satellite Remote Sensing: Vol 1. Springer-Verlag and Tsinghua University Press (This publication)

Susskind J (2006) Introduction to AIRS and CrIS. In: Qu, Gao, Kafatos, Murphy and Salomonson (eds) Earth Science Satellite Remote Sensing: Vol 1. Springer-Verlag and Tsinghua University Press (This publication)

Vermote EF, Saleous NZ (2006) Operational Atmospheric Correction of MODIS Visible to Middle Infrared Land Surface Data in the Case of an Infinite Lambertian Target. In: Qu, Gao, Kafatos, Murphy and Salomonson (eds) Earth Science Satellite Remote Sensing: Vol 1. Springer-Verlag and Tsinghua University Press (This publication)

Weng F (2006) Conically – Scanned Microwave Imager Sounder (CMIS). In: Qu, Gao, Kafatos, Murphy and Salomonson (eds) Earth Science Satellite Remote Sensing: Vol 1. Springer-Verlag and Tsinghua University Press (This publication)

Wolfe RE, Saleous NZ (2006) MODIS Land Products and Data Processing. In: Qu, Gao,

Kafatos, Murphy and Salomonson (eds) Earth Science Satellite Remote Sensing: Vol 1. Springer-Verlag and Tsinghua University Press (This publication)

Wolfe RE (2006) MODIS Geolocation. In: Salomonson V, et al. (eds) Earth Science Satellite Remote Sensing: Vol 1. Springer-Verlag and Tsinghua University Press (This publication)

Xiong X, Isaacman A, Barnes W (2006) MODIS Level-1B Products. In: Qu, Gao, Kafatos, Murphy and Salomonson (eds) Earth Science Satellite Remote Sensing: Vol 1. Springer-Verlag and Tsinghua University Press (This publication)

Zhang W, Xu J, Dong C, Yang J (2006) China's Current and Future Meteorological Satellite Systems. In: Qu, Gao, Kafatos, Murphy and Salomonson (eds) Earth Science Satellite Remote Sensing: Vol 1. Springer-Verlag and Tsinghua University Press (This publication)

2 Introduction to MODIS and an Overview of Associated Activities

Vincent V. Salomonson, William Barnes and Edward J. Masuoka

2.1 Introduction

This chapter provides an overview of the Moderate Resolution Imaging Spectroradiometer (MODIS) and associated activities devoted to and resulting in products developed and validated that can be used by the scientific and applications communities. The intent and purpose of this chapter is to enable the reader to understand and appreciate the power of the MODIS instrument and the associated systems and organizational approach that have led to the very considerable impact the observations that are progressively leading to a wide variety of improvements in science associated with land, ocean and atmosphere processes and trends as well as a wide variety of resource and environmental applications. This overview will begin by providing the background (Section 2.2) and history (Section 2.3) for the conception and development of the MODIS instrument followed by a technical description of the instrument (Section 2.4). Then a description of the role and the responsibilities of the MODIS Science Team (MST) and supporting elements will be provided in Section 2.5 so as to enable the reader to understand the roles and responsibilities of the MST and how they relate not only to the development of the instrument, but also to the scope involved in the development of the MODIS algorithms, code, and products that feed into the operational processings systems of the Earth Observing System (EOS) Data and Information System (EOSDIS). In Section 2.6 the present elements that process the MODIS data into products and archive them will be described. The last section (Section 2.7) will provide a summary of the present status of the MODIS instruments on the EOS Terra and Aqua spacecraft and insight to the future when an instrument derived from the MODIS experience will be operated on the National Polar Orbiting Environmental Satellite System (NPOESS).

2.2 Background

In the mid-1970s a number of scientists began to realize that there was a growing need to study the Earth as a system comprised of physical, chemical and biological processes operating over a wide range of temporal and spatial scales.

At about the same time there were significant advances in spaceborne remote sensing technology that enabled the systematic global measurement of a number of key geophysical parameters.

The Landsat series of satellites, beginning in 1972, and several other spaceborne optical sensors including the High Resolution Infrared Radiation Sounder (HIRS-2) and the Advanced Very High Resolution Radiometer (AVHRR) onboard the TIROS-N satellite launched on 13 October, 1978 followed by the Total Ozone Monitoring Spectrometer (TOMS) and Coastal Zone Color Scanner (CZCS) launched onboard the Nimbus-7 satellite eleven days later (24 October, 1978) have proven to be major milestones in the systemic monitoring of the Earth from low Earth orbit (LEO). The common goals of these sensors were: (1) to measure geophysical parameters from the Earth's land, atmosphere and oceans using multiple spectral bands of reflected solar and emitted thermal energy; (2) to measure these parameters globally; and (3) to continue the measurements for several years. These sensors were the culmination of 10 – 15 years of sensor and algorithm development based on data gathered during numerous field experiments using airborne and surface-based systems. Major geophysical parameters that they were designed to measure included cloud cover, snow and sea ice extent and variability, sea surface temperature (AVHRR), atmospheric temperature profiles (HIRS) and chlorophyll-a concentration (CZCS). Post-launch developments included algorithms monitoring land cover conditions using, for example, the normalized-difference vegetation index (NDVI) (AVHRR), water vapor profiles (HIRS), ocean turbidity (CZCS) and numerous others.

In parallel with the development of these passive electro-optical systems, there was an equally rapid development of passive and active (radar) microwave and active optical (lidar) sensors. These advances in spaceborne remote sensing technology were accompanied by increases in computer processing power that enabled regional parametric models to be extended to a global scale. In the early 1980s, these developments fostered an interest for the development of a large LEO platform that would facilitate the study of the Earth as a system. The original concept of a very large spacecraft carrying as many of the new sensor concepts as possible was designated "System Z". However, as the studies progressed, it became evident that a single spacecraft was not feasible and that a multi-mission concept was preferable. This approach was named the Earth Observing System (EOS).

In the spring of 1983, management at the National Aeronautics and Space Administration (NASA) created the EOS Science and Mission Requirements Working Group and tasked it with developing an overall concept for the EOS. The Science and Mission Requirements Working Group Report (Butler et al., 1984) called for developing a fifteen year climate data set and included brief descriptions of seven major "facility" sensors that were deemed necessary for the task. A further extensive description of the need and rationale for studying the Earth as a system along with the necessary observational needs including the

EOS can be found in Bretherton (1988).

2.3 MODIS History

In the spring of 1984, NASA formed instrument panels to develop science requirements and sensor concepts for each of the facility sensors described in the EOS Science and Mission Requirements Working Group Report. The term "facility sensor", as compared to a "Principal Investigator-P.I." instrument, is important because it represents an instrument that is developed by NASA to support the science community and for which, via peer-review, a Science Team is selected (to be discussed in Section 2.5). One of the sensors under consideration was the Moderate Resolution Imaging Spectrometer (MODIS) that was to include many of the attributes of the CZCS, AVHRR, HIRS, and relevant characteristics of the Landsat Thematic Mapper (TM).

The MODIS Instrument Panel, a group of nineteen scientists and remote sensing technologists from government laboratories and academia, examined the then current state of Earth remote sensing science and developed a MODIS concept (Barnes, 1985; Esaias and Barnes, 1986) calling for two sensors, MODIS-N (nadir) and MODIS-T (tilt). MODIS-N was a conventional imaging filter radiometer with 35 spectral bands and MODIS-T was a 64-band imaging spectrometer capable of tilting fore and aft to avoid sun-glint from the ocean's surface (Magner and Salomonson, 1991).

Management/development of the MODIS system was assigned to the Goddard Space Flight Center (GSFC) where the decision was made to develop MODIS-T in-house and MODIS-N via a competetively selected contractor. Subsequent studies (NASA, 1985a; NASA, 1985b; NASA, 1989) provided additional detail for the MODIS concept (Barnes et al., 1986) which resulted in the selection in 1991 of a contractor, Hughes/Santa Barbara Research Center (SBRC)[①], for the MODIS-N development. Soon after the start of the SBRC contract, there was a major restructuring of the EOS program and a decision was made to terminate the MODIS-T development and retain the MODIS-N (to be called the Moderate Resolution Imaging Spectroradiometer (MODIS)) (Salomonson et al., 1989; Barnes and Salomonson, 1992; Barnes et al., 1998). In addition, to partially compensate for the loss of MODIS-T, which was primarily focused on observing ocean color, and to obtain better capability to monitor the temporal variability of clouds and land cover conditions, it was decided that the MODIS would fly in both a mid-morning descending and a mid-afternoon ascending orbit, thereby avoiding sunglint and enabling global coverage without tilting the sensor.

Over the next few years, the Santa Barbara Remote Sensing (SBRS) team

① Subsequently Raytheon/Santa Barbara Remote Sensing.

developed, fabricated and tested two MODIS flight models. The first of these, the Protoflight Model (PFM), was completed in June of 1997 and, after integration and testing with the EOS-AM spacecraft and a year-long launch delay, was placed into a 10:30AM descending orbit on December 18, 1999. The second unit, Flight Model 1 (FM1), was launched onboard the EOS-PM satellite into a 1:30PM ascending orbit on May 4, 2002. After launch, the EOS-PM and EOS-AM satellites were renamed "Terra" and "Aqua" respectively.

2.4 MODIS Sensor

The MODIS design was driven by the scientific community's desire to view the entire globe every 1 – 2 days at a moderate spatial resolution (nominally 1 kilometer) with sufficient spectral bands in the visible (VIS) through the long-wave infrared (LWIR) regions to enable the measurement of numerous (40 – 50) geophysical parameters. Due to the heterogeniety of land scenes, it was soon apparent that a nadir resolution of 1 kilometer was too coarse for a number of the proposed land products. Therefore the nadir resolution of several of the reflected solar bands was increased to 250 meters (2 bands) and 500 meters (5 bands). Nearly 2/3 of the data output by MODIS comes from these seven bands. The radiometric drivers in the MODIS design were the sensitivity of the sea-surface temperature (SST) bands at 11 and 12 micrometers and the high signal to noise ratios (SNRs) required for the ocean color and ocean fluorescence visible-near infrared (VIS/NIR) solar bands.

Table 2.1 MODIS design parameters

Orbit	705 km, 10:30 a.m. descending node or 1:30 p.m. ascending node, sun-synchronous, near-polar, circular
Scan Rate	20.3 rpm, cross track
Swath Dimensions	2,330 km (across track) by 10 km (along track at nadir)
Telescope	17.78 cm diam. off-axis, afocal (collimated), with intermediate field stop
Size	1.0 m × 1.6 m × 1.0 m
Weight	250 kg
Power	225 W (orbital average)
Data Rate	11 Mbps (peak daytime)
Quantization	12 bits
Spatial Resolution	250 m (bands 1 – 2)
(at nadir)	500 m (bands 3 – 7), 1,000 m (bands 8 – 36)
Design Life	5 years

After numerous trades between the scientists and technologists, a 36-band crosstrack scanning radiometer design was agreed on (Barnes and Salomonson, 1992). The optical system includes a double-sided scan mirror, an off-axis afocal telescope, three dichroic beam splitters, and four sets of refractive relay optics in front of four focal plane assemblies (FPAs) with band limiting spectral filters and detector arrays. The scan swath is 2,330 kilometer wide (±55 degrees) and extends 10 kilometers along track at nadir. Therefore, the FPA arrays have 10, 20 and 40 detector elements for the 1,000, 500 and 250 meter bands respectively. An overview of the MODIS design parameters is given in Table 2.1. The 36 spectral bands are listed by spatial resolution and wavelength together with their primary uses in Table 2.2. A cutaway drawing that indicates the location of the major MODIS components is shown in Fig. 2.1.

Table 2.2 MODIS spectral bands

Primary Use	Band	Bandwidth*	Spectral Radiance**	Required SNR***
Land/Cloud Boundaries	1	620 – 670	21.8	128
	2	841 – 876	24.7	201
Land/Cloud Properties	3	459 – 479	35.3	243
	4	545 – 565	29.0	228
	5	1,230 – 1,250	5.4	74
	6	1,628 – 1,652	7.3	275
	7	2,105 – 2,155	1.0	110
Ocean Color/ Phytoplankton/ Bioeochemistry	8	405 – 420	44.9	880
	9	438 – 448	41.9	838
	10	483 – 493	32.1	802
	11	526 – 536	27.9	754
	12	546 – 556	21.0	750
	13	662 – 672	9.5	910
	14	673 – 683	8.7	1,087
	15	743 – 753	10.2	586
	16	862 – 877	6.2	516
Atmospheric Water Vapor	17	890 – 920	10.0	167
	18	931 – 941	3.6	57
	19	915 – 965	15.0	250
				Required NEΔT(K)****
Surface/Cloud Temperature	20	3.660 – 3.840	0.45	0.05
	21	3.929 – 3.989	2.38	2.00
	22	3.929 – 3.989	0.67	0.07
	23	4.020 – 4.080	0.79	0.07

Continued

Primary Use	Band	Bandwidth*	Spectral Radiance**	Required SNR***
Atmospheric Temperature	24	4.433 – 4.498	0.17	0.25
	25	4.482 – 4.549	0.59	0.25
Cirrus Clouds	26	1.360 – 1.390	6.00	150*****
Water Vapor	27	6.535 – 6.895	1.16	0.25
	28	7.175 – 7.475	2.18	0.25
	29	8.400 – 8.700	9.58	0.05
Ozone	30	9.580 – 9.880	3.69	0.25
Surface/Cloud Temperature	31	10.780 – 11.280	9.55	0.05
	32	11.770 – 12.270	8.94	0.05
Cloud Top Altitude	33	13.185 – 13.485	4.52	0.25
	34	13.485 – 13.785	3.76	0.25
	35	13.785 – 14.085	3.11	0.25
	36	14.085 – 14.385	2.08	0.35

* Bands 1 to 19, nm; Bands 20 – 36, μm

** (W/m^2-μm-sr)

*** SNR=Signal-to-noise ratio } Performance goal is 30% – 40% better than required

**** NEΔT=Noise-equivalent temperature difference

***** SNR

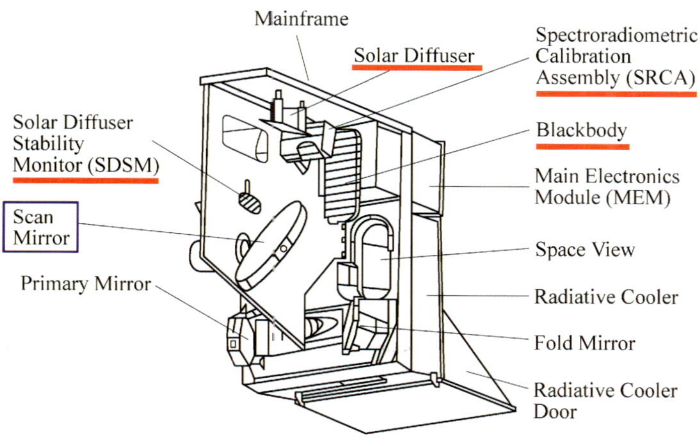

Figure 2.1 The MODIS sensor and its major subsystems

Although there are numerous technological advances incorporated into the MODIS design, at least two are unique. The first of these is simply the number of spectral bands. The incorporation of 36 spectral bands into a single sensor including 490 detector elements, each of which is a distinct electronic channel requiring individual characterization and calibration, is a major advance over the handfull of bands and detectors in sensors such as the AVHRR, CZCS and HIRS.

The need for the data to be seamless across the different copies of the sensors implied that the MODIS system must include a level of on-orbit calibration and characterization heretofore unattainable. In addition to a requirement for long-term stability and accurate radiometric calibration, the system must also be capable of monitoring sensor spatial and spectral charteristics on-orbit. The MODIS design addresses each of these needs.

The MODIS radiometric calibration uses two approaches, one for the 16 thermal emissive bands and a second for the 20 reflected solar bands. The thermal emissive band calibration is essentialy the same as that used by the AVHRR; the sensor's digital counts are related to radiance by viewing a blackbody (BB) of known temperature and deep space (zero input radiance) each scan line. Since a number of the 16 theral emissive bands are non-linear, a quadratic fit to each detector's response vs radiance is utilized. The response curves are determined pre-launch by scanning a variable temperature BB. The non-linear fitting coefficient is updated periodically post-launch by varying the temperature of the onboard BB. The dominant linear coefficient is renewed each scan (Xiong et al., 2002a; Chiang et al., 2003).

The calibrated output from the 20 reflected solar bands is in the form of a reflectance factor, ρ cos (θ), (where ρ is the reflectance and θ is the local solar zenith angle) for each detector at every Earth view pixel. The reflectance factor is readily converted to radiance using the solar spectral irradiance $E(\lambda)$. The reflectance factor is based on a knowledge of the bi-directional reflectance factor (BRF) of the onboard solar diffuser (SD). The 20 solar bands view the sun every 1 – 2 weeks via the SD. Knowledge of the BRF enables a counts-to-radiance gain to be determined for each detector. The BRF is measured pre-launch using reflectance standards tracable to the National Institute of Science and Technology (NIST). The angular dependence of the BRF is checked periodically on-orbit via spacecraft yaw maneuvers (Xiong et al., 2003a). Changes in the reflectance of the diffuser are monitored on-orbit via the solar diffuser stability monitor (SDSM), a miniature radiometer with nine spectral bands that alternately views the sun and the solar illuminated diffuser. Any change in the ratio of these views indicates a spectrally dependent change in system response. Changes in the SD are differentiated from sensor degradation using data from periodic views of the moon (Xiong et al., 2002b).

MODIS's spectral and spatial responses are monitored on-orbit with the spectral radiometric calibration assembly (SRCA). The SRCA includes a lamp-based source and a spectrometer that output wavelength calibrated spectral radiance over the range of the reflected solar bands, thereby enabling measurements of the bands' spectral shape and central wavelengths (Che et al., 2003). It is also possible to add a thermal source, bypass the spectrometer, and, using a set of reticles, measure the relative band-to-band registration (BBR) of all 36 bands (Xiong et al., 2002c).

Additional details of the MODIS design, radiometric calibration algorithms and on-orbit performance can be found in Chapter 4 and at the MODIS Web site: http://modis.gsfc.nasa.gov/.

2.5 MODIS Science Team and Data Products

Participation of the science community in the development of MODIS was initially through membership on the MODIS Instrument Panel. After the Instrument Panel completed their task, an ad hoc science team that included scientists from the Instrument Panel and other volunteers provided guidance during the initial development of sensor requirements. Near the end of 1989, NASA announced the results of a peer-reviewed science team selection process. The initial MODIS Science Team included 24 scientists, 20 from the United States and 4 internationals with Dr. Vincent Salomonson from NASA's Goddard Space Flight Center designated as the Science Team Leader. Each of the Science Team members was selected via peer-review to develop one or more science products derived from MODIS observations that would contribute to a better understanding of global land, ocean or atmospheric processes and trends. This responsibility included algorithm development, computer coding, implementation, and checkout/validation so as to enable delivery of the product(s) to the scientific and applications communities. A second selection added four more members to the team in 1997. These new team members either provided a new product or had the responsibility of providing product validation efforts for selected products. The 28 member team (see Table 2.3) remained essentially unchanged throughout the MODIS development, launches, on-orbit performance verification and product validation. A reconstituted science team composed of previous members along with new members using the same peer-reviewed process, was appointed very late in 2003 to continue the maintenance of the MODIS products along with further validation of the products for scientific and applications use through 2006.

As indicated above, the Science Team members were tasked with developing and testing algorithms to convert the radiometrically calibrated and geometrically located output from the MODIS sensors into geophysical products (see Table 2.4). Approximately 40 principal products are produced from the Terra and Aqua MODIS observations covering a wide range of land, ocean and atmospheric variables needed to study Earth system processes and trends.

Table 2.3 Members of the MODIS Science Team

Name	Occupation
Atmospheres	
Michael King (Group Leader)	NASA/Goddard Space Flight Center
Bo-Cai Gao	NASA/Goddard Space Flight Center
Yoram Kaufman	NASA/Goddard Space Flight Center
W. Paul Menzel	NOAA/University of Wisconsin
Didier Tanre	University des Science et Techniquee de Lille, France

Continued

Name	Occupation
Land	
Christopher Justice (Group Leader)	University of Maryland, College Park
Alfredo Huete	University of Arizona
Jan-Peter Muller	University College London, UK
Ranga Myneni (1997)	Boston University
Vincent Salomonson (Team Leader)	NASA/Goddard Space Flight Center
Steven Running	University of Montana
Alan Strahler	Boston University
John Townshend (1997)	University of Maryland, College Park
Eric Vermote (1997)	NASA/Goddard Space Flight Center
Zhengming Wan	University of California, Santa Barbara
Oceans	
Wayne Esaias (Group Leader)	NASA/Goddard Space Flight Center
Mark Abbott	Oregon State University
Ian Barton	CSIRO, Australia
Otis Brown	University of Miami
Janet Campbell (1997)	University of New Hampshire
Kendall Carder	University of South Florida
Dennis Clark	NOAA/NESDIS, Washington, D.C.
Robert Evans	University of Miami
Howard Gordon	University of Miami
Frank Hoge	NASA/Goddard Space Flight Center
John Parslow	CSIRO, Australia
Calibration	
Phillip Slater (Group Leader)	University of Arizona
Kurt Thome (Group Leader 1999-Present)	University of Arizona
William Barnes	NASA/Goddard Space Flight Center

Table 2.4 MODIS Products

MOD #	Product Name
MOD 01	Level-1A Radiance Counts
MOD 02	Level-1B Calibrated, Geolocated Radiances also Level-1B "subsampled" 5 km × 5 km product
MOD 03	Geolocation Data Set
MOD 04	Aerosol Product
MOD 05	Total Precipitable Water
MOD 06	Cloud Product
MOD 07	Atmospheric Profiles

Continued

MOD #	Product Name
MOD 08	Gridded Atmosphere Products (Level-3)
MOD 09	Atmospherically Corrected Surface Reflectance
MOD 10	Snow Cover
MOD 11	Land Surface Temperature and Emissivity
MOD 12	Land Cover/Land Cover Change
MOD 13	Vegetation Indices
MOD 14	Thermal Anomalies, Fires and Biomass Burning
MOD 15	Leaf Area Index and FPAR
MOD 16	Surface Resistance and Evapotranspiration
MOD 17	Vegetation Production , Net Primary Productivity
MOD 18	Normalized Water Leaving Radiance
MOD 19	Pigment Concentration
MOD 20	Chlorophyll II Fluorescence
MOD 21	Chlorophyll a Pigment Concentration
MOD 22	Photosynthetically Active Radiation (PAR)
MOD 23	Suspended Solids Concentration in Ocean Water
MOD 24	Organic Matter Concentration
MOD 25	Coccolith Concentration
MOD 26	Ocean Water Attenuation Coefficient
MOD 27	Ocean Primary Productivity
MOD 28	Sea Surface Temperature
MOD 29	Sea Ice Cover
MOD 31	Phycoerythrin Concentration
MOD 35	Cloud Mask
MOD 36	Total Absorption Coefficient
MOD 37	Ocean Aerosol Properties
MOD 39	Clear Water Epsilon
MOD 43	Albedo 16-Day Level-3
MOD 44	Vegetation Cover Conversion and Continuous Fields
MODISALB	Snow and Sea Ice Albedo

The Team Leader, in addition to furnishing a set of algorithms for his science products, served as the science point-of-contact with the EOS Projects developing the EOS-AM/"Terra" and EOS-PM/"Aqua" missions. As such, he, with the cognizance and concurrence of the MST, insured that the sensor design met the requirements generated by the team members' algorithms. He was also responsible for the algorithms to convert the raw MODIS data into its radiometrically calibrated and geometrically located form (known as Level-1B or L1B), on-orbit operation and characterization of the sensors, verification and integration of the algorithms delivered by the science team members, and providing administrative support for the team members. These tasks obviously call for a staff of support personnel and an efficient organizational structure. The Science Team Leader organization is shown in Fig. 2.2; major components and their interactions are described above.

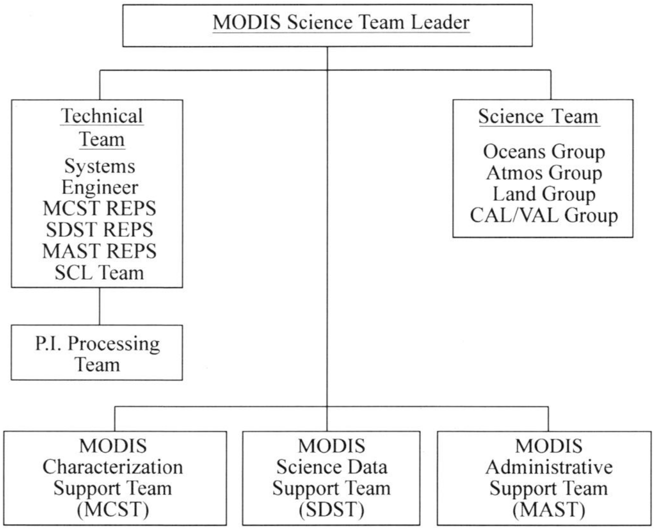

Figure 2.2 MODIS science team organization

The Team Leader meets periodically (~every 6 months) with the Science Team and the support staff to go over progress, issues, and goals. In the early pre-launch period the principal emphasis was on the developement of the instrument relative to its requirements. This is very necessary if the instrument is to achieve the originally projected goals. There is a natural "tension" between sensor requirements and scientific goals verus cost and schedule. Therefore, there has to be a frequent and continuous dialogue between those representing science and the broader user community and those representing cost and schedule concerns. When projected or existing budgets cannot sustain the requirements, a substantive dialogue with the Science Team or, at least, the Science Team Leader (representing the Science Team) needs to occur frequently enough to ensure that nothing is incorported into or taken from the instrument capability that will jepordize mission success. Pursuant to advising the Team Leader and the Science Team an active, expert instrument calibration and characterization group (here called the "MODIS Calibration Support Team"—MCST) was formed. The MCST worked closely with the MODIS contractor and reviewed and performed independent calculations regarding the calibration and performance of the instrument. Experience had shown and was corroborated in the MODIS case that an active and team-like coupling of the MODIS Science Team Leader supported by MCST and with periodic review by the entire MST and the instrument contractor was and is very beneficial, even essential, to the eventual scientific and applications success of the instrument.

Nearer to launch, when instrument performance had been largely determined, the bulk of attention shifted to the coding of algorithms and their injection into the data processing system with good checkout and quality control. This was and is being done by the "Science Data Support Team (SDST)". Given the complexity of Science Team support and interactions, along with the necessity of administering contracts for the Science Team members, a MODIS Administrative Support Team (MAST) was created. Representatives of these support teams meet weekly with the Science Team Leader to go over progress and issues.

Post-launch and during subsequent operations, due diligence and oversight has to be provided by the Science Team Leader, the support groups and the Science Team members to address issues associated with deviances or anomalies that occur with the operations of the MODIS instrument. In addition, challenges and complexities associated with the product algorithms (all closely coupled with the instrument performance) have to be addressed. These include their coding, their performance in the data processing systems (to be discussed later) and the validation and checkout of the resulting data products.

No matter how exhaustive the pre-launch testing and simulation was, all algorithms must be updated post-launch. This is due to differences in on-orbit instrument operations as compared to pre-launch testing and the performance of the data processing systems under a full load. The post-launch process of product validation and checkout for use by the community is a continuing one over the life of the mission. This is mainly due to the fact that the performance of the data product algorithms needs to be examined over the whole data collection period to eliminate artifacts or blemishes caused by everything from unanticipated variances in instrument performance and calibration to mistakes that occur in the coding of the algorithms.

To maximize the utility of the products as quickly as possible it is desirable to keep the community informed and involved in the examination of the data products as much as possible. In the case of the MODIS instruments, the fact that the MODIS sensor data have been and are available via X-band direct broadcast (DB) to regional stations (well over 50 stations world-wide as of late 2003) has been advantageous in this regard. In addition, communication with the community at-large has also been greatly enhanced via use of the Worldwide Web/Internet. MODIS has established a web page (http://modis.gsfc.nasa.gov/) with many component parts to accomplish the sharing of information regarding instrument performance, the theoretical basis of the algorithms (called Algorithm Theoretical Basic Documents—ATBDs) and user guides, Science Team efforts and results, recent and past publications, etc.

After three-plus years of Terra/MODIS on-orbit operation, the data products have achieved early validation status for all of the products and, therefore, can be used for a wide variety of scientific studies and applications. Additional work is needed to extend the validation of all products to include global variability. As stated above, this level of validation cannot be expected to occur until late in the mission when trends over several years can be examined to ensure that instrument and data processing anomalies have been largely eliminated.

2.6 MODIS Data Processing

NASA's Earth Observing System Data and Information System (EOSDIS) produce archives and distribute MODIS products. Figure 2.3 shows the MODIS processing flow in the EOSDIS and the volume of products produced by each system in the processing chain.

The EOS Data and Operations System (EDOS) receives data from ground stations at White Sands, New Mexico (for the EOS Terra spacecraft) and Svaalbard, Norway and Fairbanks, Alaska (for the EOS Aqua spacecraft). The data stream from the ground stations is made up of the Consultative Committee for Space Data Systems (CCSDS) packets from all sensors on the spacecraft.

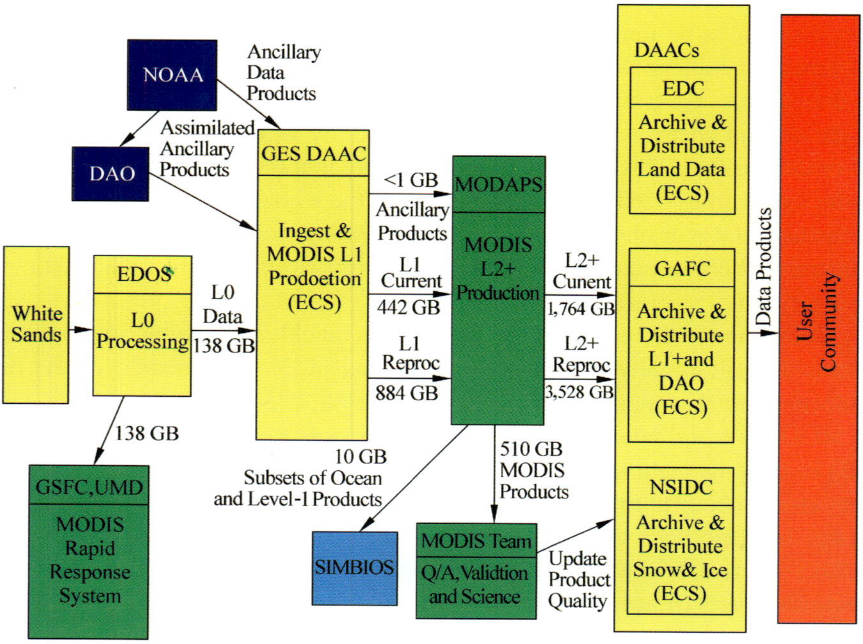

Figure 2.3 Flow of MODIS data products and volumes produced per day during processing of current data from the EOS Terra and Aqua spacecrafts and during reprocessing campaigns

EDOS separates the combined data stream into files for each instrument, time orders packets eliminating duplicates and correcting errors and packaging the raw data into the time interval requested by each instrument team. The raw data from MODIS is sent to the Goddard Earth Sciences Distributed Active Archive Center (GES DAAC) in two 2-hour binary files, 12 sets per day from each spacecraft. One file contains engineering and telemetry data from the instrument and is used by MCST and the other contains raw sensor counts from each detector and is used in downstream processing.

The GES DAAC receives raw data files from EDOS, predicts orbit and ephermis from the Flight Dynamics Facility and ancillary data (such as ozone, and Reynolds weekly sea surface temperature) from NOAA and produces Level-1 products (calibrated radiance and geolocation fields, cloud/land-sea mask and atmospheric profiles). The Level-1 products are in EOS-HDF (National Center for Supercomputing Applications—NCSA's Hierarchical Data Format version 4.5 with EOS extensions for satellite data) swath format. Each product file contains 5 minutes of data that covers an area on the earth 2,330 km by 2,000 km. 5-minute granules were chosen to keep files under 2 GB in size to facilitate data transfer and not burden user resources. Product files also include metadata fields that provide key information about the product, such as quality measures, data field specifications and measurement units.

In most cases, the Level-1 products are produced and available to be ordered within 12 hours of the time MODIS images an area. Four Silicon Graphics Origin supercomputers are used to produce the Level-1 products at the GES DAAC. These systems and their online and nearline storage handle processing, data archive and data distribution. Roughly 70 commercial software packages, over a million lines of custom software control these activities and the hardware and software comprise the EOSDIS Core System (ECS), which is installed at all DAAC sites. Many DAACs have also developed custom software to provide features not included in the ECS or to replace features which were expensive to scale. At the GES DAAC custom software development, sustaining engineering, processing, user support and product distribution are handled by a team of 120 full time staff members who are supported by a group of 250 developers and engineers who support the ECS at all DAACs.

During routine processing for each MODIS instrument, 221 GB of Level-1 and Level-2 products are shipped to the MODIS Adaptive Processing System (MODAPS) to use in the generation of MODIS science products. In addition to routine processing for each MODIS instrument, the GES DAAC has the capacity to reprocess MODIS products from a single MODIS instrument at the rate of 4X (4 times real time) and send the reprocessed data products to MODAPS.

MODAPS was developed by the MODIS science team members and the Science Data Support Team (SDST) to generate the higher level (Level-2 and -3) MODIS science products. Producing higher level products in a system operated by the science team has streamlined the process of getting science software into production systems, provided scientists with the opportunity to tailor a system to meet their requirements and optimize the system for the production of their science products. Since the MODAPS is a separate processing system from the DAACs and relies on only two commercial software packages (Sybase which holds the database used to control production, archiving and distribution of products and StorNext which manages product storage and retrieval from the near-line tape archive), adding new technologies like a Storage Area Network and Linux clusters has been relatively simple and cost effective.

Each processing node in the MODAPS system is comprised of Silicon Graphics supercomputers and clusters of personal computers running Linux. In the present MODAPS there are four separate processing nodes, one node produces the current data products from both MODIS instruments, two are used for reprocessing past data once improved calibration or science algorithms are available and one node is used to conduct large scale tests of improved science software before using it to produce products that will be distributed to the public. The elements of a typical processing node are illustrated in Fig. 2.4. In developing MODAPS, the design philosophy was to keep the processing software relatively simple and to achieve gains in processing rates through the incorporation of ever faster computing systems and disk storage.

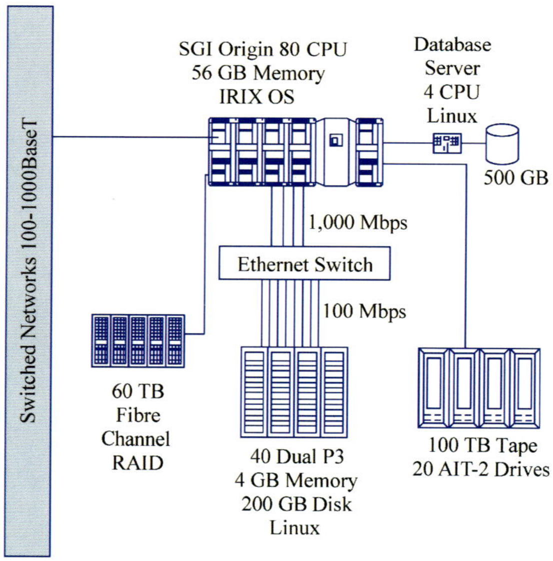

Figure 2.4 Elements of the MODAPS processing node used to produce current data products

In parallel with production of MODIS data products there is an ongoing process of integrating software delivered by the MODIS Team into GES DAAC and MODAPS processing systems and assessing the quality of products being generated. Figure 2.5 illustrates the steps software changes go through from delivery by the science team members to integration in the MODAPS production system. Out of 170 changes to science software, roughly half passed the first stage of testing in which the software is integrated into the production system and a unit test with data from a small number of files is run and the results are examined for correct metadata and product contents. The changes that pass unit testing are put through extensive science tests in which up to 32 days of products are produced and the impact of changes in each product on downstream products is evaluated. Roughly two thirds of the changes that pass unit testing also pass science tests and are placed into production.

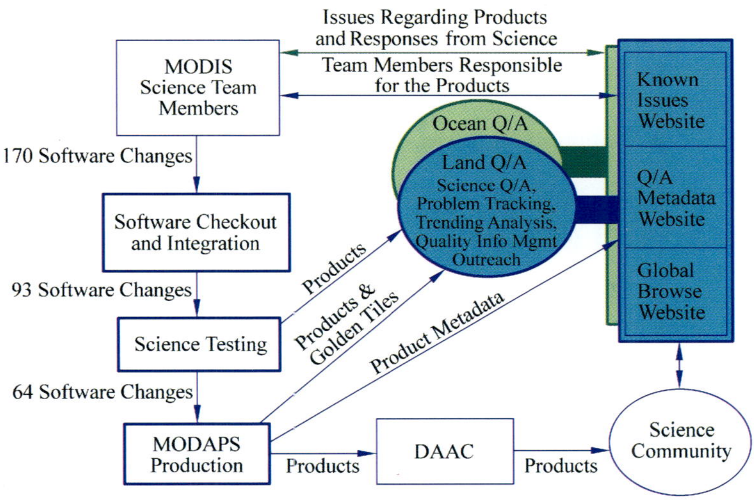

Figure 2.5 Flow of software changes through software testing and product quality assessment. Number of changes in each stage of the process is from a6-month period in 2003

As part of the daily production of MODIS science products, Level-3 products are shipped to quality assessment (Q/A) systems for ocean and land products. Product metadata from all jobs running in the system include quality fields and software is run daily which scans the metadata and generates a list of any products where problems were flagged in the Q/A metadata. These products are then flagged for further analysis by the Q/A staff. For Level-3 land products the world is divided into 10 degree× 10 degree tiles and a set of 9 "golden" tiles distributed across major biomes is pushed to the Land Q/A system. Automated software generates time series statistics for parameters from each product each tile and stores the statistics by date and tile in a Land Q/A database. The Q/A database is in turn used to generate time series plots that analysts, science team members and interested scientists can examine for temporal trends and possible problems with products. Known issues for each product are posted at the Land Q/A Web site, http://landdb1.nascom.nasa.gov/ QA_WWW/newPage.cgi, along with global browse images and time series plots. When a known issue is posted at the site, the science team member responsible for the product is notified and asked to examine the problem further and to develop a software patch to resolve the issue. The Land Q/A approach is described in detail in Roy et al. (2002). The Ocean Q/A process focuses on Level-3 global maps received daily from MODAPS. The maps include measures of product quality and enable Q/A staff and members of the Ocean Science Team at the University of Miami to focus their attention on areas of lower than expected quality. The Web site, http://modis-ocean.gsfc.nasa.gov/qa/, includes plots of statistics from individual products as well as time series plots over the years covered by the mission and the known issues identified for a product. The Q/A of MODIS Atmosphere products are through examination of Level-3 product images and plots of time series statistics. The examination of products and the posting of known issues are handled by the science team member responsible for each product or his staff and are performed on a weekly basis.

2.7 Status and Follow-On Systems

2.7.1 Status

Since their launch in December, 1999 and May, 2002 onboard the EOS Terra and Aqua satellites (aka EOS-AM and EOS-PM), the MODIS sensors have been operating essentially continuously. The ground data system (Section 2.5) has processed and archived nearly two petabytes of MODIS data products (Section 2.4) which are being distributed globally to science and application users via the EOS Distributed Active Archive Centers (DAACs).

Details of sensor performance are in Chapter 4 and the literature (Barnes et al., 1998; Barnes et al., 2003; Xiong et al., 2003b). Overall performance of both sensors has been at or better than the design specifications. There have been failures of two of the Terra MODIS subsystems; a power supply and a formatter. Both of these have been replaced by backup units and there has been no perceptible impact on the science data. In addition, the Terra sensor's solar diffuser (SD) attenuation screen failed to open during a solar calibration event (May 2003) and the screen has been left in front of the SD and the SD door has been left open. This has resulted in a 0.1% – 0.2% decrease in the radiometric accuracy of the reflected solar bands that were previously calibrated with the SD screen open. The Aqua MODIS has had no failures to date. Both sensors were designed for five years of on-orbit operation; this period will be complete in December 2004 and May 2007. Present performance is such that it is anticipated that both sensors will at least meet (and hopefully exceed) these dates.

2.7.2 Follow-On Systems

As noted previously, the EOS was originally planned (Butler et al., 1984) as a set of missions that would generate a seamless fifteen year data set for the study of global climate change. The evolution of MODIS (Section 2.4) into a single 36-band sensor on both EOS-AM and EOS-PM together with its design life of five years resulted in a requirement for six MODIS flight models.

During the mid-1990s, there was a series of programatic decisions that NASA missions were to be essentially for proof-of-concept, that the development and long-term archiving of decadal (or longer) data sets were primarily the responsibility of the operational agencies such as National Oceanic and Atmospheric Administration (NOAA), and that NOAA was to be the lead agency in climate studies. Consequently, the EOS program was shortened to a single copy of each mission with the intention of transfering successful climate measurements to the NOAA systems. At nearly the same time, the United States government decided to combine the low Earth orbiting (LEO) weather satellite systems belonging to NOAA and the Department of Defense (DOD) into a National Polar-Orbiting Environmental Satellite System (NPOESS). The responsibility for developing and operating the NPOESS was assigned to the newly created Integrated Program Office (IPO). The IPO is a tri-agency program managed by personnel from NOAA, DOD and NASA.

Development of the NPOESS requirements as supplied by the three governing agencies resulted in the formulation of a concept for an imaging radiometer that was designated the Visible and Infrared Imaging Radiometer Suite (VIIRS). Studies conducted during the design phase indicated that nearly all of the MODIS attributes could be included in the VIIRS with minimal impact (Murphy et al., 2001).

In 1997 the IPO initiated a three year set of competitive studies to develop the VIIRS design (Scheuler and Barnes, 1998; Ardanuy et al., 2001). These studies resulted in the selection of Raytheon/SBRS, the same group that developed the MODIS, as the VIIRS developer.

The selected design (Scheuler et al., 2001) is a 22-band imaging radiometer that will generate nearly all of the MODIS science products. Externally the VIIRS has the appearance of a scaled down MODIS (see Fig. 2.6), but there are several major differences. The VIIRS scanner is a rotating telescope assembly similar to that used on the Sea Wide-Field Scanner (SeaWiFS) (also developed by SBRS); although the SD/SDSM, blackbody (BB) and space view have been retained, the SRCA has been eliminated; crosstrack pixel growth has been restricted to less than 2X (compared to MODIS' 6X) and the aft optics are all reflective. Retention of the MODIS science parameters with fewer spectral bands was enabled via the use of dual gains in a number of the VIIRS bands.

Figure 2.6 VIIRS (top) and MODIS (bottom). VIIRS mass power and volume are 162 kg, 140 W, and 1.2 m^3. For MODIS, these are 230 kg, 147 W and 2.0 m^3

The first NPOESS spacecraft is scheduled for launch no earlier than 2008. Since, as mentioned above, the design life of the last MODIS sensor (Aqua) ends in May 2007, there is the possibility of a data gap between the EOS and NPOESS sensors. Therefore, NASA and the IPO are developing a joint mission, the NPOESS Preparatory Project (NPP), for launch in 2005 that will include the first VIIRS flight model. This mission will serve as a gap-filler between the EOS and NPOESS data sets and as a risk reduction mission for the IPO.

In total, the development of MODIS providing observations since early 2000, and its evolution into an operational instrument, the VIIRS, that will continue observations for the foreseeable future is a very exciting development. Heretofore, the "workhorse" operational instrument on the NOAA meteorological satellites has been the AVHRR. With the implementation of the VIIRS the scientific and applications community can look forward to a considerable increase in the breadth and power of observations for a long time to come.

References

Ardanuy PE, Schueler CF, Miller SW, Kealy PS, Cota SA, Haas JM, Welsch C (2001) NPOESS VIIRS design process. In: Barnes WL (ed) Proceedings Earth Observing Systems Ⅵ, SPIE 4483: 24 – 34

Barnes WL (1985) Science requirements for a Moderate Resolution Imaging Spectrometer (MODIS) for EOS. Proc AIAA/NASA Earth Observing System (EOS) Conference AIAA-85-2085

Barnes WL, Salomonson VV (1992) MODIS: A global imaging spectroradiometer for the Earth Observing System. Proc Critical Review of Optical Technologies for Aerospace Sensing CR 47: 1 – 23

Barnes WL, Ostrow H, Salomonson VV (1986) Preliminary system concepts for MODIS: Moderate Resolution Imaging Spectrometer for EOS. Proc SPIE Tech Symp Southeast on Optics and Optoelectronic Systems, Orlando, FL

Barnes WL, Pagano TS, Salomonson VV (1998) Pre-launch characteristics of the Moderate Resolution Imaging Spectroradiometer (MODIS) on EOS AM-1. IEEE Transactions on Geoscience and Remote Sensing 36(4): 1,088 – 1,100

Barnes WL, Xiong X, Salomonson VV (2003) Status of Terra MODIS and Aqua MODIS. In press J Advances in Space Research

Bretherton F (1988) Earth system science: A closer view, Report of the Earth Systems Sciences Committee to the NASA Advisory Council. National Aeronautics and Space Administration, Washington, D. C.

Butler DM et al. (1984) Earth Observing System; science and mission requirements working group report. NASA Technical Memorandum 86129

Che N, Xiong X, Barnes WL (2003) On-orbit spectral characterization results for Terra MODIS reflective solar bands, In: Barnes WL (ed) Proceedings Earth Observing Systems Ⅷ. SPIE 5151

Chiang K, Wu A, Barnes WL, Guenther B, Solamonson VV, Xiong X (2003) On-orbit radiometric calibration uncertainty of the Terra MODIS thermal emissive bands. In: Proceedings CALCON '03, Logan, Utah

Esaias WE, Barnes W (1986) Moderate Resolution Imaging Spectrometer (MODIS). MODIS instrument panel report, Earth Observing System Reports, vol. Ⅱb

Magner TJ, Salomonson VV (1991) Moderate Resolution Imaging Spectrometer-Tilt (MODIS-T). International J Imaging Systems and Tech 3: 121 – 130

Murphy RE, Barnes WL, Lyapustin AI, Privettte J, Welsch C, DeLuccia F, Swenson H, Schueler CF, Ardanuy P, Kealy P (2001) Using VIIRS to provide data continuity with MODIS. Proceedings of IGARSS 2001, Sydney, Austrailia

NASA (1985a) Phase-A study for a Moderate Resolution Imaging Spectrometer-Nadir (MODIS-N). Final Report, NAS5-27145

NASA (1985b) Phase-A study for a Moderate Resolution Imaging Spectrometer-Tiltable (MODIS-T). Final Report, NAS5-27147

NASA (1989) Moderate Resolution Imaging Spectrometer-Nadir (MODIS-N). Phase-B Final Report, NAS5-30149

Roy DP, Borak JS, Devadiga S, Wolfe RE, Zheng M, Descloitres J (2002) The MODIS Land product quality assessment approach, Remote Sensing of Environment, Volume 83: 77 – 96

Salomonson VV, Barnes WL, Maymon PW, Montgomery HE, Ostrow H (1989) MODIS: Advanced facility instrument for studies of the Earth. IEEE Transactions on Geoscience and Remote Sensing 27: 145 – 153

Schueler CF, Barnes WL (1998) Next-generation MODIS for polar operational environmental satellites. J Atmos Oce Technol 15(2): 430 – 439

Schueler CF, Clement JE, Ardanuy PE, Welsch C, DeLuccia F, Swenson H (2001) NPOESS VIIRS sensor design overview, In Barnes WL (ed) Proceedings Earth Observing Systems VI, SPIE 4483: 11 – 23

Taubes. G (1993) Earth Scientists Look NASA's Gift Horse in the Mouth, Science Vol. 259: 912 – 914

Xiong X, Chiang K, Guenther B, Barnes WL (2002a) MODIS thermal emissive bands calibration algorithm and on-orbit performance. In: Proceedings Asia-Pacific Symposium on Remote Sensing of the Atmosphere and Clouds III, SPIE 4891: 392 – 401

Xiong X, Sun J, Barnes WL (2002b) MODIS on-orbit characterization using the moon. In: Fujisada H (ed) Proceedings Sensors, Systems, and Next Generation Satellites VIII. SPIE 4881: 299 – 307

Xiong X, Che N, Adimi F, Barnes WL (2002c) On-orbit spatial characterizations for Terra MODIS, In: Barnes WL (ed) Proceedings Earth Observing Systems VII. SPIE 4814: 347 – 357

Xiong X, Sun J, Esposito J, Liu X, Barnes WL, Guenther B (2003a) On-orbit characterization of a solar diffuser's bi-directional reflectance factor using spacecraft maneuvers, In: Barnes WL (ed) Proceedings Earth Observing Systems VIII. SPIE 5151

Xiong X, Chiang K, Esposito J, Guenther B, Barnes WL (2003b) MODIS on-orbit calibration and characterization. Metrologia 40: 89 – 92

3 MODIS Level-1B Products

Xiaoxiong Xiong, Alice Isaacman and William Barnes

3.1 Introduction

A key instrument of the NASA's Earth Observing System (EOS) Terra and Aqua missions, the Moderate Resolution Imaging Spectroradiometer (MODIS) was designed to take measurements in a broad spectral range at three spatial resolutions and with a wide field of view. In addition to many new features, MODIS extends a number of heritage sensors' data sets that are essential for understanding global environmental changes (Salomonson et al., 1989; Barnes and Salomonson, 1992). Algorithms developed by the researchers and scientists of the MODIS Science Team are used to generate approximately 40 science products that are used in formulating a wide range of parameters that are needed for the short- and long-term studies of the Earth's land, oceans, and atmosphere. Both Terra and Aqua MODIS are currently operating on-orbit and making continuous global observations (Salomonson et al., 2002; Barnes et al., 2003).

As illustrated in Fig. 3.1, the sensors' raw data are transmitted to the ground stations, such as the one at the White Sands in New Mexico, through the Tracking Data Relay Satellite System (TDRSS) and then sent to the EOS Data and Operations System (EDOS). At the Goddard Space Flight Center Distributed Active Archive Center (GDAAC), the sensor's original binary data files (Level-0)

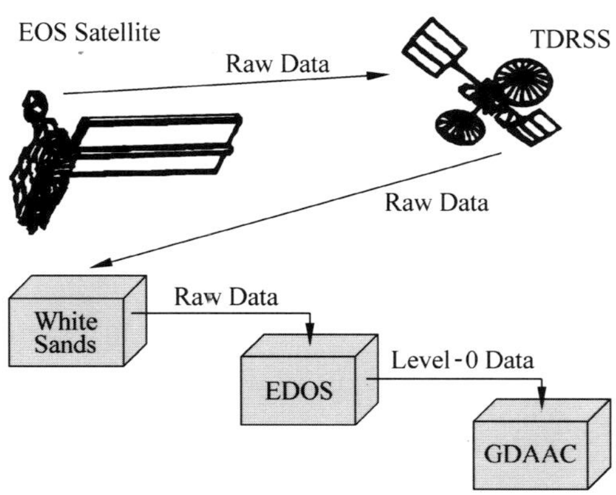

Figure 3.1 Data path from satellite to GDAAC

from the EDOS are first reformatted and separated into Level-1A (L1A) data sets in Hierachical Data Format (HDF). Each L1A data set (referred to as a "granule") contains 5 minutes' worth of data, consisting of the sensor's response in digital numbers (DNs) as well as other engineering and telemetry information. These are the principal inputs to the Level-1B (L1B) process. Other inputs include a geolocation file and a set of L1B Look-Up Tables (LUTs), which provide many calibration parameters determined from pre-launch and on-orbit calibration and characterization. The higher level MODIS data processing is done at the MODIS Data Adaptive Processing System (MODAPS). The MODIS products, including L1B products, are distributed through the GDAAC.

MODIS Level-1B algorithms convert raw data from the sensor's observations (in the L1A format) into radiometrically calibrated and geometrically located data sets. The MODIS L1B algorithms are developed and maintained by the MODIS Characterization Support Team (MCST) under the direction of the MODIS Science Team Leader. Because of the differences in the Terra MODIS and Aqua MODIS sensors' characteristics, operational configurations, and response changes, two separate versions of L1B software and associated LUTs are maintained and updated as necessary based on the instrument calibration and characterization performed by the MCST. Working closely with the MODIS Science Team representatives, MCST analysts and L1B code developers perform initial evaluations before a new version of L1B software or LUTs is sent to the GDAAC for additional testing and subsequent data production.

This chapter describes the MODIS L1B data products and the calibration algorithms used to generate them. The emphasis is on the implementation of the MODIS calibration algorithms in the L1B code. The code standards and associated properties, data processing, and data product retrieval are also discussed. The instrument characteristics and discussions of the radiometric calibration algorithms are presented in Chapter 5 MODIS Calibration and Characterization of Vol.2 and in other documents about MODIS calibration and characterization (Guenther et al., 1998; Xiong et al., 2002a; Xiong et al., 2002b; Xiong et al., 2003). Additional instrument background information is given in Chapter 2 (Introduction to MODIS and an Overview of Associated Activities). The MODIS instrument scan cavity and its on-board calibrators are illustrated in Fig. 2.1.

3.2 L1B Data Product Description

L1B output consists of calibrated earth view (EV) data of all 36 spectral bands, organized in three HDF files corresponding to MODIS' three spatial resolutions, and associated metadata files. These files subsequently serve as the common input for many higher-level science algorithms. A separate file containing on-board calibrator data sets and key telemetry and geolocation data is also produced by the L1B process. It does not contain the EV sector data and is primarily used by

MCST analysts to perform instrument on-orbit calibration and to monitor the instrument's health status (Isaacman et al., 2003; MCST, 2003c).

The MODIS has 36 spectral bands. Bands 1 and 2 have a nadir spatial resolution of 250 m with 40 detectors in each band, bands 3 – 7 have a nadir spatial resolution of 500 m with 20 detectors in each band, and all other bands have a spatial resolution of 1 km with 10 detectors in each band. Bands 13 and 14, each with two columns of detectors, are measured at both low and high gains through the use of time-delay and integration (TDI) technique. Thus MODIS has a total of 490 detectors that require calibration and are used to generate L1B products (Barnes et al., 1998; Guenther et al., 1998). Bands 1 – 19 and 26 are the Reflective Solar Bands (RSBs) and bands 20 – 25 and 27 – 31 are the Thermal Emissive Bands (TEBs). These bands/detectors are located on four different focal plane assemblies (FPAs) as shown in Fig. 4.1.

Level-1B calibrated data products include top of the atmosphere (TOA) reflectance factors for the RSBs, radiances for both the RSBs and TEBs, and associated uncertainty indices and data quality flags. To save space, the calibrated Earth view data are stored as 16-bit unsigned integers coupled with associated scale and offset terms in three separate HDF files, allowing the user to reconstruct calibrated radiance values for all MODIS bands as well as calibrated reflectance values for the Reflective Solar Bands. To assist in user evaluation of data quality, associated uncertainty values are included as unsigned 8-bit integers with coefficients provided to reconstruct the percent uncertainty. The uncertainty is based on band and detector-dependent lookup table values in conjunction with correction terms calculated during the calibration process, e.g. the temperature correction value derived in the reflective calibration process. A scan-by-scan 32-bit data quality flag and a granule level 8-bit detector quality flag for each of the 490 detectors are included as part of each file's metadata.

Level-1B output file names reflect the contents of each type of file. MODIS Terra and Aqua names begin with "MOD02" or "MYD02" respectively where "02" denotes the MODIS process number. The three EV data files correspond to different data resolutions: a 250 meter resolution file consisting of Band 1 and Band 2 data at their native resolution (designated the "MOD02QKM" or quarter-km file), a 500 meter resolution file consisting of Band 3 – 7 data at their native resolution plus Band 1 and 2 data aggregated to 500 meter resolution (the "MOD02HKM" or half-km file), and a 1 km resolution file consisting of all remaining bands at their native resolution plus Band 1 – 7 data aggregated to 1-km resolution (the "MOD021KM" file). There are no TEB data products in the sub-kilometer resolution files. A fourth HDF output file (the "MOD02OBC" file) consists of on-board calibration and telemetry information taken during each scan plus various quality assurance values designed by the MCST to track the sensors' performance in certain areas.

See Tables 3.1 and 3.2 for a complete list of the principal data products contained in each type of L1B output file and the meaning of the data quality flags. All data fields in the output products are specified in the MODIS Level-1B Products

Data Dictionary (MCST, 2003c). The data quality flags provide information on the instrument operational conditions.

Earth View Level-1B output files also contain geolocation information. The 250 m resolution output files contain geodetic latitudes and longitudes for all detectors and all pixels in each scan; the 500 m resolution output files contain latitude/longitude information for every other detector and pixel, and 1 km resolution output files contain latitude/longitude and altitude information for every fifth detector and pixel. 1 km resolution output files also contain Solar and sensor zenith and azimuth angle information.

When the MODIS instrument is viewing the dark side of the Earth ("night mode"), data from the Reflective Solar Bands (with the exception of Band 26) are not telemetered to the ground. To save space, production of high resolution output files is suspended when this happens. The instrument is said to be operating in "day mode" otherwise. A special command may be sent to the instrument in

Table 3.1 Principle MODIS data products

HDF File Information	Data Products
Short Name: MOD021KM MYD021KM Resolution: 1 km Day mode size: 327 mb Night mode size: 136 mb[*]	Earth View Data Products EV_1KM_RefSB EV_1KM_Emissive EV_Band26 EV_250_Aggr1KM_RefSB[*] EV_500_Aggr1KM_RefSB[*] Associated uncertainties
Short Name: MOD02HKM MYD02HKM Resolution: 500 m Day mode size: 262 mb Not produced in night mode	0 Earth View Data Products EV_500_RefSB EV_250_Aggr500_RefSB Associated uncertainties
Short Name: MOD02QKM MYD02QKM Resolution: 250 m Day mode size: 273 mb Not produced in night mode	1 Earth View Data Products EV_250_RefSB Associated uncertainties
Short Name: MOD02OBC MYD02OBC Resolution: 1 km, 500 m, 250 m Day mode size: 56 mb Night mode size: 56 mb	2 On-Board Calibration and Engineering Information Space View, Solar Diffuser, Blackbody data (as digital numbers) Spacecraft Engineering data. Quality Assurance Data.

* Reflective Solar Bands, with the exception of the Band 26 data set, are populated with fill values when the instrument operates in night mode, resulting in a smaller data set size.

Table 3.2 Interpretation of Level-1B EV scan-by-scan data quality flags

<table>
<tr><th>Bit QA
Flags
Bit #</th><th>Description (condition that causes bit to be set to 1)</th></tr>
<tr><td>Bit 0</td><td>Moon within defined limits of SV</td></tr>
<tr><td>Bit 1</td><td>Spacecraft maneuver</td></tr>
<tr><td>Bit 2</td><td>Sector rotation</td></tr>
<tr><td>Bit 3</td><td>Negative radiance beyond noise level</td></tr>
<tr><td>Bit 4</td><td>PC bands Ecal on</td></tr>
<tr><td>Bit 5</td><td>PV bands Ecal on</td></tr>
<tr><td>Bit 6</td><td>SD door open</td></tr>
<tr><td>Bit 7</td><td>SD screen down</td></tr>
<tr><td>Bit 8</td><td>NAD closed</td></tr>
<tr><td>Bit 9</td><td>SDSM on</td></tr>
<tr><td>Bit 10</td><td>Radiative cooler heaters on</td></tr>
<tr><td>Bit 11</td><td>Day mode bands telemetered at night</td></tr>
<tr><td>Bit 12</td><td>Linear emissive calibration (unused)</td></tr>
<tr><td>Bit 13</td><td>DC restore change</td></tr>
<tr><td>Bit 14</td><td>(unused)</td></tr>
<tr><td>Bit 15</td><td>BB heater on</td></tr>
<tr><td>Bit 16</td><td>Missing previous granule</td></tr>
<tr><td>Bit 17</td><td>Missing subsequent granule</td></tr>
<tr><td>Bits 18 – 19</td><td>SRCA calibration mode, determined from telemetry
<table>
<tr><th>Bit 18</th><th>Bit 19</th><th>Meaning</th></tr>
<tr><td>0</td><td>0</td><td>Radiometric (0)</td></tr>
<tr><td>0</td><td>1</td><td>Spatial (1)</td></tr>
<tr><td>1</td><td>0</td><td>Spectral (2)</td></tr>
<tr><td>1</td><td>1</td><td>Undetermined (3)</td></tr>
</table></td></tr>
<tr><td>Bit 20</td><td>Moon within the SV keep-out box for Reflective Solar bands</td></tr>
<tr><td>Bit 21</td><td>Moon within the SV keep-out box for Emissive bands</td></tr>
<tr><td>Bit 22</td><td>All space-view data bad for any Reflective Solar band</td></tr>
<tr><td>Bit 23</td><td>All blackbody data bad for any Reflective Solar band</td></tr>
<tr><td>Bit 24</td><td>Dropped scan(s) between previous granule and granule being calibrated</td></tr>
<tr><td>Bit 25</td><td>Dropped scan(s) between granule being calibrated and subsequent granule</td></tr>
<tr><td>Bit 26</td><td>SCI_ABNORMAL flag</td></tr>
<tr><td>Bits 27, 31</td><td>Reserved for future use</td></tr>
</table>

order to collect RSB EV data when the spacecraft is operating in "night mode". These data are collected occasionally to meet special characterization requirements. The regular calibration data sets from the sensors' view of the on-board calibrators are produced continuously in both "day mode" and "night mode".

MODIS is a scanning radiometer with a two-sided paddle wheel mirror, making observations with both sides of the scan mirror. As the scan mirror sweeps across the Earth scene in a ±55° scanning range relative to instrument nadir, 1,354 frames of data are recorded from each detector. For the MODIS bands at 250 m resolution, 4 subframes are recorded for each (1-km) frame, and 2 subframes per frame are recorded for the 500-m resolution bands. The data products are organized on a scan-by-scan basis, additionally indexed by band, detector, frame, and in the case of Bands 1 – 7, subframe. Thus for a typical 203-scan 5-minute data granule taken in day mode operation, a total of slightly more than 217 million measurements is produced (see Table 3.3), resulting in a data volume of approximately 860 megabytes for the three HDF files comprising the Earth view L1B data output.

Table 3.3 Number of measurements processed from a typical 203-scan 5-minute MODIS day mode granule

		Resolution		
		250 (m)	500 (m)	1 (km)
Per Scan	Number of Detectors Per Band	40	20	10
	Number of Subframes Per Band	5,416	2,708	1,354
	Total Measurements Per Band	216,640	54,160	13,540
	Number of Bands at Resolution	2	5	31*
	Total Measurements	433,280	270,800	419,740
Per 203-Scan Granule	Total Measurements Per Resolution	87,955,840	43,977,920	85,207,220
	Total for All Resolutions	217,140,980		

* Bands 13 and 14 are counted twice because their measurements are collected at both low and high gains using time-delay and integration (TDI).

3.3 L1B Calibration Algorithm

MODIS L1B provides calibration for both TEB and RSB detectors. Calibration is performed for each band, detector, sub-sample (for the sub-kilometer resolution bands 1 – 7), and mirror side (BDSM), and thus is on a pixel-by-pixel basis for all MODIS detectors (Guenther et al., 1998; Guenther et al., 2002; Xiong et al., 2002a; Xiong et al., 2002b; Xiong et al., 2003). The radiometric calibration algorithm is divided into two major modules in the L1B software: one for the thermal emissive bands and the other for the reflective solar bands. The thermal emissive band calibration uses the on-board calibrator (OBC) blackbody (BB) and the key calibration coefficient for each detector is determined on a scan-by-scan basis from its response to the OBC BB. The reflective solar bands are calibrated using a solar diffuser (SD) panel with key calibration coefficients determined offline and periodically updated through the L1B LUTs. The RSB calibration uses some processed data from TEB calibration and is therefore performed after the TEB calibration. Both TEB and RSB calibration modules in the L1B process are executed after L1B pre-processing in which the calibration coefficients are either calculated from on-orbit data sets or determined from the associated LUT values. A detailed description of the MODIS TEB and RSB radiometric calibration algorithms is also provided in Chapter 5 MODIS Calibration and Characterization of Vol.2).

3.3.1 Thermal Emissive Bands Algorithm

The primary data product of the MODIS thermal emissive bands (Bands 20 – 25 and 27 – 36) is TOA radiance. In the current calibration algorithm, the EV radiance of each pixel, L_{EV}, is a function of the sensor's EV digital number (DN_{EV}), the space view DN average ($< DN_{SV} >$), quadratic calibration coefficients (a_0, b_1, and a_2), and the thermal emissive radiance from the scan mirror (L_{SM}). In addition, the L_{EV} also depends on the sensor's response versus scan angle at the earth view as well as the space view (RVS_{EV} and RVS_{SV}):

$$L_{EV} = f\{DN_{EV}, < DN_{SV} >, a_0, b_1, a_2, L_{SM}, RVS_{EV}, RVS_{SV}\} \tag{3.1}$$

Most of the parameters are either loaded from the TEB LUTs or determined from the pre-processed L1A data, including parameters used for the uncertainty assessment. Table 3.4 lists some of the most significant TEB LUTs with their names and simple descriptions.

The dominant linear calibration coefficient b_1, the background term $< DN_{SV} >$, and the scan mirror radiance L_{SM} in Equation (3.1) are pre-processed and passed to the TEB calibration module on a scan-by-scan basis. The offset term a_0 and the nonlinear term a_2 in the quadratic approach are provided by the LUTs. DN_{EV} and RVS_{EV} in Equation (3.1) are the only scan angle (or data frame number) dependent parameters.

Figure 3.2 is a simplified flow diagram that illustrates the core of the TEB calibration algorithm: computing the EV radiance, converting radiance to a scaled integer (SI), and computing the uncertainty index (UI). For each scan of EV data, this is looped through bands, detectors, and frames. The data quality flags are assigned for each scan of the data. An optical crosstalk (PCX) correction algorithm, which applies only to Bands 32 – 36, is included in the TEB calibration. Based on the sensors' calibration and characterization results, the PCX switch is set ON for Terra MODIS L1B and OFF for Aqua MODIS L1B. The PCX coefficients are provided by the LUTs. MODIS Band 21, which is primarily used for fire detection with extremely low gain, uses calibration coefficients from the LUTs. For Aqua MODIS, LUTs also provide calibration coefficients for Bands 33, 35, and 36 when the on-board calibrator blackbody is above these Bands' saturation limits (Xiong et al., 2002a).

Table 3.4 Most significant thermal emissive bands LUTs

LUT Name	Description
A0	Quadratic coefficients for calculating a_0
A2	Quadratic coefficients for calculating a_2
Band_21_b1	The value of b_1 for each Band 21 detector
BB_T_sat_aqua	Saturation temperature for bands 33, 35, and 36 (MODIS Aqua only)
BB_T_sat_default_b1_aqua	Default b_1 for bands 33, 35, and 36 to use if saturated on BB warmup (MODIS Aqua only)
BB_T_sat_switch_aqua	Flag to switch to default b_1 for bands 33, 35, 36 when BB temperature is above saturation temperature (MODIS Aqua only)
epsilon_bb	Black-body emissivity
epsilon_cav	Effective cavity emissivity
PC_XT	PC bands cross-talk correction parameters
PCX_correction_switch	Switch (0 = OFF, 1 = ON) for the PC crosstalk correction
RSR	Relative spectral responses
RVS_BB_SV_Frame_No	Frame number for calculating the BB and SV RVS
RVS_TEB	Quadratic coefficients for calculating EV RVS for TEBs
Sigma_TEB_ADC	ADC uncertainty
Sigma_TEB_PV_resid_elec	PV bands residual electronic cross-talk uncertainty
TEB_specified_uncertainty	Factor used in computing uncertainty index
TEB_UI_scaling_factor	Factor used in computing uncertainty index
Ucoeff	Coefficients of polynomial fit of uncertainty weight vs. DN
Ucoeff_Calibr_resid	Residual uncertainty to the calibration polynomial fit

3.3.2 Reflective Solar Bands Algorithm

The MODIS reflective solar bands use a simple linear calibration algorithm. The earth view's TOA reflectance factor, $\rho_{\text{EV}} \cos(\theta_{\text{EV}})$, is the primary L1B product. It is determined from each pixel's digital number (DN_{EV}), the space view DN average ($< DN_{\text{SV}} >$), two parameters (k_{INST} and T_{INST}) used for the instrument temperature dependent correction, a linear calibration coefficient (m_1), a normalized Earth-Sun distance factor (d_{ES}), and the response versus scan angle at the earth view pixel (RVS_{EV}),

$$\rho_{\text{EV}} \cos(\theta_{\text{EV}}) = f\{DN_{\text{EV}}, < DN_{\text{SV}} >, k_{\text{INST}}, T_{\text{INST}}, m_1, d_{\text{ES}}, RVS_{\text{EV}}\} \tag{3.2}$$

Like the TEB algorithm, these parameters are either loaded from the LUTs or determined from pre-processed data. In the RSB calibration algorithm, determined offline and periodically updated through the L1B LUTs. The RSB calibration uses some processed data from TEB calibration and is therefore performed after the TEB calibration. Both TEB and RSB calibration modules in the L1B process are executed after L1B pre-processing in which the calibration coefficients are either calculated from on-orbit data sets or determined from the associated LUT values. A detailed description of the MODIS TEB and RSB radiometric calibration algorithms is also provided in Chapter 5 MODIS Calibration and Characterization of Vol.2).

3.3.1 Thermal Emissive Bands Algorithm

The primary data product of the MODIS thermal emissive bands (Bands 20 – 25 and 27 – 36) is TOA radiance. In the current calibration algorithm, the EV radiance of each pixel, L_{EV}, is a function of the sensor's EV digital number (DN_{EV}), the space view DN average ($< DN_{SV} >$), quadratic calibration coefficients (a_0, b_1, and a_2), and the thermal emissive radiance from the scan mirror (L_{SM}). In addition, the L_{EV} also depends on the sensor's response versus scan angle at the earth view as well as the space view (RVS_{EV} and RVS_{SV}):

$$L_{EV} = f\{DN_{EV}, < DN_{SV} >, a_0, b_1, a_2, L_{SM}, RVS_{EV}, RVS_{SV}\} \tag{3.1}$$

Most of the parameters are either loaded from the TEB LUTs or determined from the pre-processed L1A data, including parameters used for the uncertainty assessment. Table 3.4 lists some of the most significant TEB LUTs with their names and simple descriptions.

The dominant linear calibration coefficient b_1, the background term $< DN_{SV} >$, and the scan mirror radiance L_{SM} in Equation (3.1) are pre-processed and passed to the TEB calibration module on a scan-by-scan basis. The offset term a_0 and the nonlinear term a_2 in the quadratic approach are provided by the LUTs. DN_{EV} and RVS_{EV} in Equation (3.1) are the only scan angle (or data frame number) dependent parameters.

Figure 3.2 is a simplified flow diagram that illustrates the core of the TEB calibration algorithm: computing the EV radiance, converting radiance to a scaled integer (SI), and computing the uncertainty index (UI). For each scan of EV data, this is looped through bands, detectors, and frames. The data quality flags are assigned for each scan of the data. An optical crosstalk (PCX) correction algorithm, which applies only to Bands 32 – 36, is included in the TEB calibration. Based on the sensors' calibration and characterization results, the PCX switch is set ON for Terra MODIS L1B and OFF for Aqua MODIS L1B. The PCX coefficients are provided by the LUTs. MODIS Band 21, which is primarily used for fire detection with extremely low gain, uses calibration coefficients from the LUTs. For Aqua MODIS, LUTs also provide calibration coefficients for Bands 33, 35, and 36 when the on-board calibrator blackbody is above these Bands' saturation limits (Xiong et al., 2002a).

Table 3.4 Most significant thermal emissive bands LUTs

LUT Name	Description
A0	Quadratic coefficients for calculating a_0
A2	Quadratic coefficients for calculating a_2
Band_21_b1	The value of b_1 for each Band 21 detector
BB_T_sat_aqua	Saturation temperature for bands 33, 35, and 36 (MODIS Aqua only)
BB_T_sat_default_b1_aqua	Default b_1 for bands 33, 35, and 36 to use if saturated on BB warmup (MODIS Aqua only)
BB_T_sat_switch_aqua	Flag to switch to default b_1 for bands 33, 35, 36 when BB temperature is above saturation temperature (MODIS Aqua only)
epsilon_bb	Black-body emissivity
epsilon_cav	Effective cavity emissivity
PC_XT	PC bands cross-talk correction parameters
PCX_correction_switch	Switch (0 = OFF, 1 = ON) for the PC crosstalk correction
RSR	Relative spectral responses
RVS_BB_SV_Frame_No	Frame number for calculating the BB and SV RVS
RVS_TEB	Quadratic coefficients for calculating EV RVS for TEBs
Sigma_TEB_ADC	ADC uncertainty
Sigma_TEB_PV_resid_elec	PV bands residual electronic cross-talk uncertainty
TEB_specified_uncertainty	Factor used in computing uncertainty index
TEB_UI_scaling_factor	Factor used in computing uncertainty index
Ucoeff	Coefficients of polynomial fit of uncertainty weight vs. DN
Ucoeff_Calibr_resid	Residual uncertainty to the calibration polynomial fit

3.3.2 Reflective Solar Bands Algorithm

The MODIS reflective solar bands use a simple linear calibration algorithm. The earth view's TOA reflectance factor, $\rho_{\mathrm{EV}} \cos(\theta_{\mathrm{EV}})$, is the primary L1B product. It is determined from each pixel's digital number (DN_{EV}), the space view DN average ($< DN_{\mathrm{SV}} >$), two parameters (k_{INST} and T_{INST}) used for the instrument temperature dependent correction, a linear calibration coefficient (m_1), a normalized Earth-Sun distance factor (d_{ES}), and the response versus scan angle at the earth view pixel (RVS_{EV}),

$$\rho_{\mathrm{EV}} \cos(\theta_{\mathrm{EV}}) = f\{DN_{\mathrm{EV}}, < DN_{\mathrm{SV}} >, k_{\mathrm{INST}}, T_{\mathrm{INST}}, m_1, d_{\mathrm{ES}}, RVS_{\mathrm{EV}}\} \tag{3.2}$$

Like the TEB algorithm, these parameters are either loaded from the LUTs or determined from pre-processed data. In the RSB calibration algorithm,

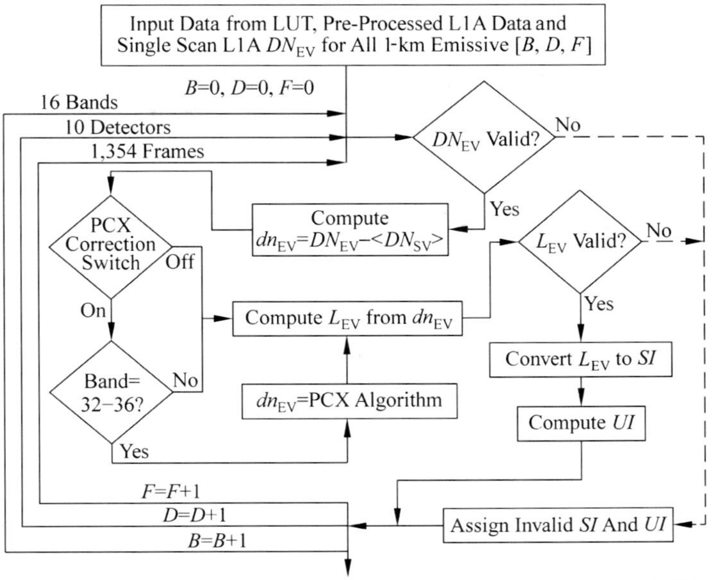

Explanation of Emissive Calibration Flow Diagram Entries	
Entry	Meaning
DN_{EV} valid?	Is the count value of the earth view digital number (DN_{EV}) in range?
Compute dn_{EV} $= DN_{EV} - < DN_{EV} >$	Subtract the space view average from DN_{EV} to form dn_{EV}
PCX Correction Switches on?	Is the switch to perform the correction of Bands 32 – 36 on?
dn_{EV} = PCX Algorithm	If the PCX correction switch is on and the Band is 32 – 36, correct it using Band 31 and lookup table values
Compute L_{EV} from dn_{EV}	Compute earth view radiance (L_{EV}) using lookup and computed calibration coefficients
L_{EV} valid?	Is L_{EV} in valid range?
Convert L_{EV} to SI	Scale L_{EV} to an integer between 0 and 32,767
Compute UI	Compute Uncertainty Index (UI)
Assign Invalid SI and UI	Fill the Scaled Integer and Uncertainty Index values with numbers indicating that the data are not valid

Figure 3.2 MODIS Level-1B emissive calibration algorithm flow diagram for one scan

however, the key calibration coefficient m_1 is not determined in the L1B from real time data sets. Instead it comes from the RSB LUTs (see Table 3.5) based on off-line analysis of the SD observations. Unlike the OBC BB data, which are collected on a scan-by-scan basis for use in TEB calibration, the SD calibration data are not available every scan.

Table 3.5 Most significant reflective solar bands LUTs

LUT Name	Description
B26_B5_Corr_Switch	Flag to turn on (1) or off (0) the Band 26 correction
B26_B5_Corr	Coefficients for the Band 26 correction
B26_B5_Frame_Offset	Frame offset to use for the Band 26 correction
E_sun_over_pi	RSR-weighted solar irradiance/pi for RSB detectors
K_FPA	Focal plane temperature correction factor
K_inst	Instrument temperature correction factor
m1	Reflectance calibration linear terms
RSB_NedL	RSB noise equivalent delta radiances
RSB_specified_uncertainty	Factor used in computing uncertainty index
RSB_UI_scaling_factor	Factor used in computing uncertainty index
RVS_RefSB	Quadratic coefficients for calculating the EV RVS for RSB
Sigma_K_inst	Uncertainty in the instrument temperature correction factor
Sigma_m1	Uncertainty in m1
Sigma_PV_Resid_Elec	Uncertainty related to electrical cross-talk
Sigma_R_Star_Lin_Resid_Ucoeff	Uncertainty related to deviations from linear behavior in R_Star
Sigma_RSB_ADC	Uncertainty in the RSB ADCs
Sigma_RVS_RSB	Uncertainty in RVS at nadir frame
Sigma_T_inst	Uncertainty in the instrument temperature
SWIR_OOB_correction_switch	Flag which turns on (1) or off (0) SWIR OOB leak correction
SWIR_OOB_sending_band	Emissive Band to be used when performing SWIR OOB correction
T_FPA_ref	Focal plane temperature reference
T_inst_ref	Instrument temperature reference
X_OOB_0	Coefficients of quadratic SWIR band correction formula
X_OOB_1	Coefficients of quadratic SWIR band correction formula
X_OOB_2	Coefficients of quadratic SWIR band correction formula

Figure 3.3 shows the flow diagram of the RSB calibration module which is executed after the TEB calibration module. Compared to the TEB calibration, an extra loop is added for the bands/detectors with sub-frames (Bands 1 – 7). For the Short-Wave Infrared (SWIR) bands (5 – 7 and 26), the algorithm includes a thermal leak correction that applies to both the Terra and Aqua MODIS data (Xiong et al., 2002b). Some of the data used in the SWIR thermal leak correction comes from the processed TEB data.

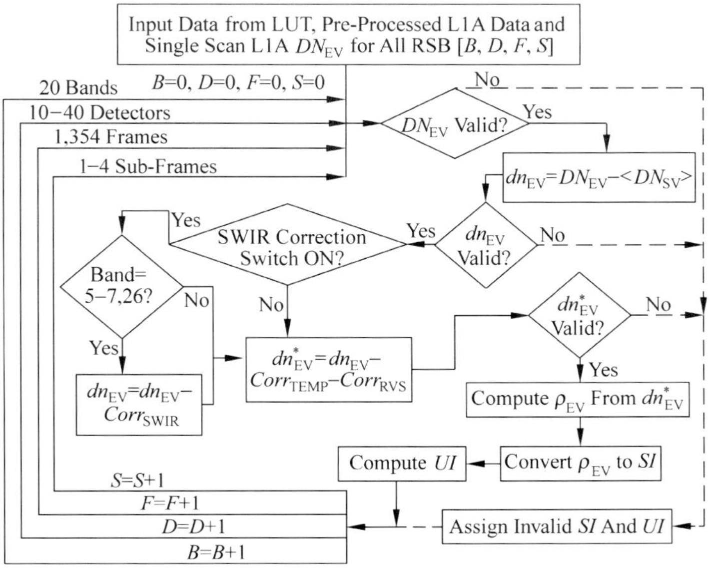

Explanation of Reflective Calibration Flow Diagram Entries	
Entry	Meaning
DN_{EV} valid?	Is the count value of the earth view digital number (DN_{EV}) in range?
Compute $dn_{EV} = DN_{EV} - < DN_{SV} >$	Subtract the space view average from DN_{EV} to form dn_{EV}
SWIR correction switches on?	Is the switch to perform the correction of the Short Wave Infrared (SWIR) Bands on?
$dn_{EV} = dn_{EV} - Corr_{SWIR}$	If the SWIR correction switch is on and the Band a SWIR Band, correct it using Band 28 (MODIS Terra) or Band 25 (MODIS Aqua) and lookup table coefficient values
$dn^*_{EV} = dn_{EV} - Corr_{TEMP} - Corr_{RVS}$	Compute and subtract temperature and RVS (response vs. Scan Angle) corrections from dn_{EV}
dn^*_{EV} valid?	Is dn^*_{EV} within valid range?
Compute ρ_{EV} from dn^*_{EV}	Compute earth view reflectance (ρ_{EV}) using lookup and computed calibration coefficients
Convert ρ_{EV} to SI	Scale ρ_{EV} to an integer between 0 and 32,767
Compute UI	Compute Uncertainty Index (UI)
Assign invalid SI and UI	Fill the Scaled Integer and Uncertainty Index values with numbers indicating that the data are not valid

Figure 3.3 MODIS Level-1B reflective calibration algorithm flow diagram for one scan

Following the radiometric calibration of each granule, a set of band dependent scale and offset terms are calculated and written into the L1B output as SDS attributes that are used to reconstruct the radiance and reflectance values from the scaled integers. In each data set, the RSB radiance values can also be derived from the reflectance factors using the global attributes "Earth-Sun distance" and Solar Irradiance weighted by the relative spectral response (RSR), "Solar Irradiance on RSB Detectors over pi" (see Table 3.6). The MODIS Level-1B Product User's Guide (MCST, 2003b) provides a detailed description of this process.

Table 3.6 Retrieval of radiance, reflectance, and uncertainty from MODIS Earth View Data Set

<table>
<tr><th>Data Sets Retrieved</th><th>Product Generation Formula</th></tr>
<tr><td>SDS Name
"EV_1KM_Emissive"
SDS Attributes
"radiance_scales",
"radiance_offsets"</td><td>Radiance Product:
Radiance = radiance_scales *
(Earth View Data-radiance_offsets)</td></tr>
<tr><td rowspan="3">SDS Names
"EV_1KM_RefSB"
"EV_250_Aggr1km_RefSB"
"EV_500_Aggr1km_RefSB"
"EV_Band26"
"EV_500_RefSB"
"EV_250_Aggr500_RefSB"
"EV_250_RefSB"
SDS Attributes
"radiance_scales",
"radiance_offsets"
"reflectance_scales",
"reflectance_offsets"
Global Attributes
"Earth-Sun Distance"
"Solar Irradiance on
RSB Detectors over pi"</td><td>Reflectance Product:
Reflectance = reflectance_scales *
(Earth View Data-reflectance_offsets)</td></tr>
<tr><td>Radiance Product:
Radiance = radiance_scales *
(Earth View Data-radiance_offsets)</td></tr>
<tr><td>Radiance Product:
Alternate formula for RSB radiance:
$\text{Radiance} = \text{Reflectance}^{*}\left(\dfrac{\text{Solar\_Irradiance}}{(\text{Earth\_Sun Distance})^2}\right)$</td></tr>
<tr><td>SDS Names
"EV_1KM_RefSB_Uncert_Indexes"
"EV_250_Aggr1km_RefSB_Uncert_Indexes "<br>"EV_500_Aggr1km_RefSB_Uncert_Indexes "<br>"EV_Band26_Uncert_Indexes "<br>"EV_500_RefSB_Uncert_Indexes "<br>"EV_250_Aggr500_RefSB_Uncert_Indexes "<br>"EV_250_RefSB_Uncert_Indexes "
SDS Attributes
"specified_uncertainty",
"scaling_factor"</td><td>Percent Uncertainty:
$\text{Percent Uncertainty} = \text{specified\_uncertainty}^{*} \exp\left(\dfrac{\text{Uncertainty\_Index}}{\text{scaling\_factor}}\right)$</td></tr>
</table>

3.4 Code Standards and Properties

The algorithms in the Level-1B code are based on the MODIS Algorithm Theoretical Basis Document (ATBD) and its supplemental documents (Guenther et al., 1998; Xiong et al., 2002a; Xiong et al., 2002b). The code also meets requirements mandated by the MODIS Software Development Standards and Guidelines Document (MCST, 1998) and by the Goddard Distributed Active Archive Center (GDAAC) where L1B production is performed. In order to ensure maximum flexibility during the lifetimes of the Terra and Aqua missions, the code was designed to be a sequential processing code organized in modules, enabling easy modification of algorithms if necessary (MCST, 2003a). All code and LUT modifications are tracked through a configuration management system. The code is written in the "C" programming language and was originally developed to run on SGI platforms. It was subsequently modified to run on Linux platforms to enhance data processing speed. Modifications are necessary in order to run successfully on other platforms.

The modifications and updates of the L1B code are based on additional information from instrument pre-launch and on-orbit calibration and characterization. Due to sensor differences between the MODIS instruments on the Terra and Aqua platforms, two separate processing versions of the L1B code, one for each sensor, are maintained. This allows insertion of platform-specific additions or modifications to the existing calibration algorithms as necessary. For example, only MODIS/Aqua Level-1B code contains a correction for premature sensor saturation in some emissive bands when the on-board calibrator blackbody's temperature is increased. The updates of the LUTs for each sensor are also made independently.

Code designers correctly anticipated that the state of the MODIS instrument would change over the life of its mission, and so the design of the Level-1B program called for a method to change the calibration coefficients used without actually changing the program itself. In consequence, almost all relevant calibration coefficients are entered into Level-1B code via the use of look-up tables (LUTs). As the telemetry and on-board calibration from the instrument are analyzed, the LUTs are revised if necessary. At the time of execution, the Level-1B program reads approximately 100 LUTs organized for the sake of convenience into three HDF files: Reflective Solar Band LUTs, Thermal Emissive Band LUTs, and Quality Assurance LUTs. While some LUTs are constant, other LUTs are implemented as time-dependent step function or piecewise linear functions. The HDF input data sets for these LUTs actually consist of many different pieces, each identified with the state of the instrument at a given time during the mission. When the Level-1B program is run, the LUT appropriate to the data time being processed is reconstructed and used.

Level-1B LUTs are also used to control code execution of various calibration processes, allowing for maximum flexibility during the mission lifetimes and between instrument platforms. For instance, an out-of-band correction to the short-wave infrared bands is controlled by an on/off switch, by which band is selected as the "correcting" band, and by what coefficients are used in the correction, all through the use of time-dependent LUTs.

Tables 3.4 and 3.5 list the principal Emissive and Reflective calibration LUTs. The MODIS LUT Information Guide for Level-1B (MCST, 2003d) provides a complete listing of each LUT. Some of the parameters in the LUTs come from the sensors' pre-launch calibration and characterization and others are derived from on-orbit performance.

3.5 Data Processing

MODIS data are initially processed at the Goddard Distributed Active Archive Center (GDAAC). Before ingest by the Level-1B program, scientific and telemetry data in digital counts, plus associated geolocation data, are organized into 5-minute segments ("granules"). These comprise the dynamic input to the Level-1B program. The LUTs and metadata configuration files make up the static input to the program.

At the GDAAC, Level-1B output for the Terra and Aqua missions is produced as the raw data (in L1A format) are received. Each data granule takes approximately 20 minutes to process on an SGI Origin 2000 or Origin 3000 machine. Because of the large volume of data, several simultaneous processing streams are maintained. In addition, data reprocessing for complete mission lifetimes is performed as understanding of the instrument deepens and algorithms are improved or modified. When data are reprocessed, the Collection number of the data is changed (values have ranged from 1 through 5 as of this writing). Detailed explanations of Level-1B software versioning, software histories, and lookup table histories may be obtained from the MODIS Characterization Support Team (currently found at http://mcstweb.mcst.ssai.biz/mcstweb/index.html). The data user may make use of the metadata items "DESCRREVISION", "PRODUCTIONDATETIME", "ALGORITHM-PACKAGEVERSION", and "PRODUCTIONHISTORY" to obtain a complete account of the Collection number, time the data were produced, the software version and lookup table versions used to produce the data, and the Level-1A software versions used to produce the input data to the calibration algorithm respectively.

Level-1B output data are used to produce MODIS Land, Atmospheres, and Oceans products, which include ocean color, sea surface temperature, cloud, snow, aerosol, precipitable water, and other products at native resolutions as well as in spatial- and/or time-binned form. The higher level science products are produced at the MODIS Data Adaptive Processing System (MODAPS) and then sent back to the GDAAC for distribution (Salomonson et al., 2002; Isaacman et al., 2003). Information regarding data distribution may be obtained from the GDAAC MODIS data Web site (http://modis.gsfc.nasa.gov/data/).

3.6 Data Product Retrieval

The calibrated MODIS Earth view data are stored as scaled integer (*SI*) scientific data sets (SDS) within the Level-1B output files. The output files also contain scale and offset terms stored as associated attributes which enable the user to convert the calibrated data from any band to TOA radiances and, in the case of the Reflective Solar Bands, to reflectance factors. The following expressions are used to convert the *SI* to scene TOA radiances or reflectance factors.

For the TEB radiance:

$$Radiance = radiance\_scales^{*}(SI - radiance\_offsets) \tag{3.3}$$

For the RSB reflectance factor:

$$Reflectance = reflectance\_scales^{*}(SI - reflectance\_offsets) \tag{3.4}$$

The RSB radiance:

$$Radiance = Reflectance \times \frac{Solar\_Irradiance}{(Earth - Sun\_Distance)^2} \tag{3.5}$$

For the thermal emissive bands, the radiance scale and offset terms are used to reconstruct the radiance values calculated within the Level-1B emissive calibration algorithm. For the reflective solar bands, the reflectance scales and offsets may be used to reconstruct the reflectances calculated within the reflective calibration algorithm, but use of the radiance scales and offsets for the reflective solar bands will result in only an approximate reconstruction of radiances. This is due to the fact that the reflective radiance scales are derived using a band-averaged value of the Solar irradiance, which actually varies from detector to detector due to the individual detectors' relative spectral response (RSR) function. An alternative method to derive reflective radiances, which yields more precise results, is to use the global attributes "Earth-Sun Distance" and "Solar Irradiance on RSB Detectors over pi" to reconstruct the radiances directly from reflectance values. This method is given by Equation 3.5 and the names of the global attributes are listed in Table 3.6 together with the product generation formula. To cover a broad range of uncertainty in percentage for the retrieved EV product, an exponential approach is used. Two SDS attributes, "specified_uncertainty" and "scaling_factor", are used to convert the Uncertainty Index (*UI*) back to the percentage uncertainty. Additional information is provided in the MODIS Level-1B Product User's Guide (MCST, 2003b).

3.7 Summary

The MODIS Level-1B processing code has worked well throughout the years to efficiently generate Earth View products which are of great use to the scientific community. A large measure of its success is due to the fact the code was deliberately designed to be easily modified when necessary and to have almost no hard-coded calibration values. In addition, human error in calculation and conversion to program input of calibration coefficients has been minimized through a series of checks and duplications at almost every step in the process. LUT updates are accomplished smoothly through cooperation between the MCST and the GDAAC, resulting in a relatively short length of time between when a MODIS instrument change is detected by analysis of on-board calibration data and when the consequent change to the calibration coefficients is implemented in Level-1B processing. As a result the code serves in the most versatile manner possible to process data from two different satellite platforms and from any phase of a satellite mission simultaneously.

References

Barnes WL, Pagano TS, Salomonson VV (1998) Pre-launch characteristics of the Moderate Resolution Imaging Spectroradiometer (MODIS) on EOS AM-1. IEEE Transactions on Geoscience and Remote Sensing 36: 1,088 – 1,100

Barnes WL, Salomonson VV (1992) MODIS: A global imaging spectroradiometer for the Earth Observing System. Critical Review of Optical Technologies for Aerospace Sensing CR 47: 1 – 23

Barnes WL, Xiong X, Salomonson VV (2003) Status of Terra MODIS and Aqua MODIS. Advances in Space Research 32: 2,099 – 2,106

Guenther B, Godden GD, Xiong X, Knight EJ, Montgomery H, Hopkins MM, Khayat MG, Hao Z (1998) Pre-launch algorithm and data format for the Level-1 calibration products for the EOS AM-1 Moderate Resolution Imaging Spectroradiometer (MODIS). IEEE Transactions on Geoscience and Remote Sensing 36: 1,142 – 1,151

Guenther B, Xiong X, Salomonson VV, Barnes WL, Young J (2002) On-orbit Performance of the Earth Observing System Moderate Resolution Imaging Spectroradiometer; first year of data. Remote Sensing of the Environment 83: 16 – 30

Isaacman A, Toller G, Guenther B, Barnes WL, Xiong X (2003) MODIS Level 1B calibration and data products. Proceedings of SPIE—Earth Observing Systems Ⅷ 5151: 552 – 562

MODIS Characterization Support Team (1998) Software Requirements Specification for the Moderate Resolution Imaging Spectroradiometer (MODIS) Level-1B Software System. PD-200-CD-001-001

MODIS Characterization Support Team (2003a) MODIS Level-1B In-Granule Calibration Code (MOD_PR02) High-Level Design, M1037; Latest release is available on-line at http://www.mcst.ssai.biz/mcstweb/L1B/product.html

MODIS Characterization Support Team (2003b) MODIS Level-1B Product User's Guide, MCM-PUB-01-U-0202- REV B; Latest release is available on-line at http://www.mcst.ssai.biz/mcstweb/L1B/product.html

MODIS Characterization Support Team (2003c) MODIS Level-1B Products Data Dictionary, MCM-02-2.3.1-PROC_L1BPDD-U-01-0107- REV B; Latest release is available on-line at http://www.mcst.ssai.biz/mcstweb/L1B /product.html

MODIS Characterization Support Team (2003d) MODIS LUT Information Guide For Level-1B; Latest release is available on-line at http://www.mcst.ssai.biz/mcstweb/L1B/product.html

Salomonson VV, Barnes WL, Maymon PW, Montgomery HE, Ostrow H (1989) MODIS: Advanced facility instrument for studies of the Earth. IEEE Transactions on Geoscience and Remote Sensing 27: 145 – 153

Salomonson VV, Barnes WL, Xiong X, Kempler S, Masuoka E (2002) An overview of the Earth Observing System MODIS instrument and associated data systems performance. Proceedings of IGARSS

Xiong X, Chiang K, Esposito J, Guenther B, Barnes WL (2003) MODIS on-orbit calibration and characterization. Metrologia 40: 89 – 92

Xiong X, Chiang K, Guenther B, Barnes WL (2002a) MODIS thermal emissive bands calibration algorithm and on-orbit performance. Proceedings of SPIE—Optical Remote Sensing of the Atmosphere and Clouds III 4891: 392 – 401

Xiong X, Sun J, Esposito J, Guenther B, Barnes WL (2002b) MODIS reflective solar bands calibration algorithm and on-orbit performance. Proceedings of SPIE—Optical Remote Sensing of the Atmosphere and Clouds III 4891: 95 – 104

4 MODIS Geolocation

Robert E. Wolfe

4.1 Introduction

The first Moderate Resolution Imaging Spectroradiometer (MODIS) (Salomonson et al., 1989) was launched in December 1999 on the polar orbiting NASA Earth Observing System (EOS) Terra satellite and the second MODIS was launched on the polar orbiting Aqua satellite in May 2002. Each MODIS acquires daily global data in 36 spectral bands—29 with 1 km, 5 with 500 m and 2 with 250 m nadir pixels. The Terra and Aqua satellites have exterior orientation (position and attitude) measurement systems designed to enable unaided navigation of MODIS data to within approximately 150 m (1σ) at nadir. A global network of ground control points has been used to improve the geolocation accuracy for MODIS/Terra to better than 45 m and for MODIS/Aqua to better than 60 m. This chapter contains an overview of the geolocation approach, a summary of nearly four years of MODIS/Terra and one and one half years of MODIS/Aqua geolocation results, an overview of ongoing work, and a discussion of the applicability of this approach to future missions. The MODIS approach allows an operational characterization of the MODIS geolocation errors and enables individual MODIS observations to be geo-located to the sub-pixel accuracies required for terrestrial global change applications (Townshend et al., 1992; Justice et al., 1998; Roy, 2000).

4.2 Background

Satellite data production systems operationally register different orbits of data by geometric correction of each orbit into a common Earth-based coordinate system. Geometric correction is necessary to remove distortions introduced by the instrument sensing geometry, the curvature of the Earth, surface relief, and perturbations in the motion of the sensor relative to the surface. Geometric correction can be considered a two-stage process: first the sensed observations are geolocated, and then they are gridded into a predefined georeferenced grid. The geometric distortions present in satellite data may be categorized into system dependent and system independent distortions. System dependent geometric distortions are introduced by the sensor. System independent distortions are introduced by the motion of the sensor, oblateness and rotation of the Earth, and surface relief.

Correction for these distortions can be performed using parametric and/or non-parametric approaches. Non-parametric approaches require the identification of distinct features that have known locations, usually termed ground control points (GCPs), to model the spatial relationship between the sensed data and an Earth based coordinate system. The spatial relationships are assumed to be representative of the geometric distortions and are used to calculate mapping functions, for example polynomial functions (Bernstein, 1983). Non-parametric approaches can correct all types of geometric distortion (Roy et al., 1997). However, non-parametric approaches are not suitable for the operational correction of satellite data because accurate GCPs are expensive to collect and may not be available over homogeneous, unstructured and cloudy scenes. In addition, the sun-target-sensor geometry, used in the generation of many of the MODIS land products (e.g. Schaaf et al., 2002; Vermote et al., 2002), must be estimated when non-parametric approaches are used (Roy and Singh, 1994). Parametric approaches require information concerning the sensing geometry (interior orientation) and the sensor attitude and position (exterior orientation) that describe the circumstances that produced the sensed image. GCPs may be used to correct errors in the sensor interior and exterior orientation knowledge (Emery et al., 1989; Moreno and Melia, 1993; Rosborough et al., 1994). Relief information is required to remove relief distortion effects that are dependent upon the sensing altitude, the terrain height, and the distance of the terrain from nadir (Schowengerdt, 1997).

4.3 Approach

The MODIS geolocation approach is a parametric approach, using accu-rate and rapid measurement of the satellite exterior orientation, augmented by GCPs to remove orientation biases and trends (Wolfe et al., 2002). The approach encompassed a number of areas all of which contributed to reaching a pre-launch accuracy goal of 50 m (1σ), including: development of a detailed preflight error analysis (Fleig et al., 1993); development of a geolocation Algorithm Theoretical Basis Document (Nishihama et al., 1997) with a detailed physical model of important geometric elements; accurate ancillary data; and detailed long-term error analysis after launch. Collaborative meetings with the instrument and satellite builders several years prior to launch enabled some design decisions to be made that made the geolocation goals easier to reach. For example, one such change was to reduce the vibrations from the Advanced Spaceborne Thermal Emission and Reflection Radiometer (ASTER) cooling system on the Terra satellite. The Terra and Aqua satellites and the MODIS instruments were built with special attention paid to the structure's thermal sensitivity and with mechanical isolation of the instruments and spacecraft components.

The interior orientation is measured precisely before launch and the exterior orientation is measured accurately on-orbit. The detailed physical model of

MODIS includes a model of the dual-sided MODIS scan mirror which proved very valuable in the error analysis (Wolfe et al., 2002). The ancillary data requirements, including accurate orbit and attitude data, the first global 1 km Digital Elevation Model (DEM) (Logan, 1999) and a global distribution of fine resolution GCPs (Bailey et al., 1997), were critical to achieving the geolocation accuracy. A detailed on-orbit error analysis enabled the initial large on-orbit geolocation errors to be quickly reduced to within the science community's accuracy goals. Results from the long-term trend analysis were then used to quantify and reduce errors caused by yearly cycles and long-term trends in the MODIS/Terra data to maintain the accuracy within the goals. Reaching these goals required the geolocation team to coordinate closely with multiple groups, in particular, the instrument and satellite builders and the flight dynamics group.

4.3.1 Instrument Geometry

The Terra and Aqua spacecrafts orbit the Earth at an altitude of 705 km in a near polar orbit with an inclination of 98.2° and a mean period of 98.9 minutes (Salomonson et al., 1989). The orbits are sun-synchronous with a 16-day repeat cycle, i.e., the orbit ground track repeats every 16 days. The dayside equatorial local crossing time is 10:30 AM (descending) for Terra and 1:30 PM (ascending) for Aqua. Periodic drag makeup and inclination adjustment maneuvers maintain the ground track and solar geometry throughout the mission lifetimes. MODIS has a 110° across-track field of view and senses the entire equator every two days with full daily global coverage above approximately 30° latitude. MODIS senses in 36 spectral bands from the visible to the thermal infrared-29 with 1 km (at nadir) pixel dimensions, five with 500 m pixels and two with 250 m pixels.

MODIS is a paddle broom (sometimes called a whiskbroom) electro-optical instrument that uses the forward motion of the satellite to provide the along-track direction of scan (Fig. 4.1). The light reflected or emitted from the Earth is reflected into the instrument telescope by a rotating two-sided scan mirror. One-half revolution of the scan mirror takes approximately 1.477 seconds and produces the across-track scanning motion. The light is then focused onto separate calibrated radiation detectors covered by narrow spectral band-pass filters. MODIS simultaneously senses in each band, 10 rows of 1-km detector pixels, 20 rows of 500-m detector pixels and 40 rows of 250-m detector pixels. Each row corresponds to a single scan line of MODIS data that is nominally composed of 1,354 1-km, 2,708 500-m, and 5,416 250-m observations. This section contains a summary of the detailed description of the MODIS instrument geometry in Nishihama et al. (1997).

The MODIS detectors are grouped on four focal planes—Long Wave Infrared (LWIR), Short/Medium Wave Infrared (SWIR/MWIR), Near Infrared (NIR) and Visible (VIS). Detectors for each band are laid out on the focal planes

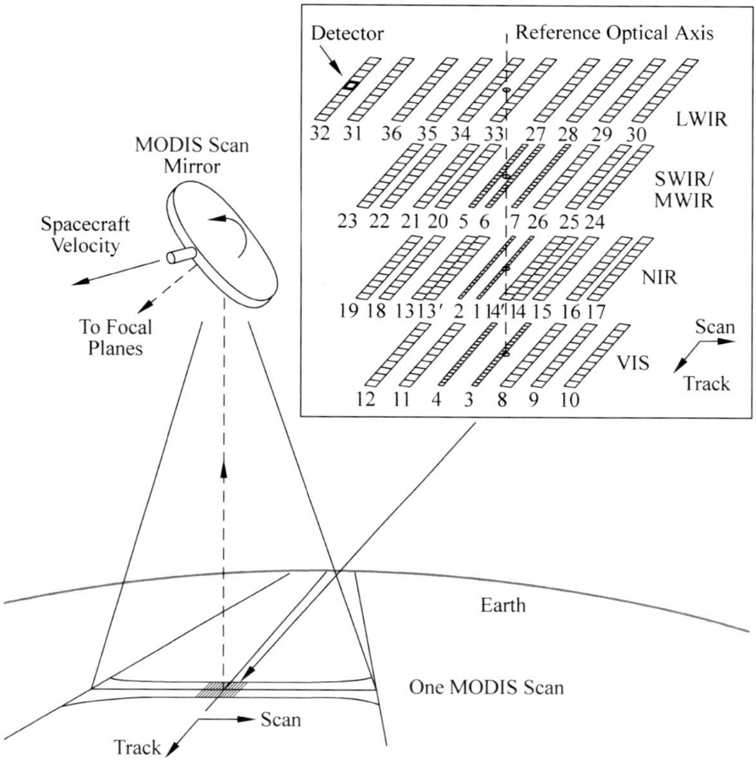

Figure 4.1 Overview of MODIS sensing geometry. A scan of MODIS data is sensed over a half revolution of the MODIS double sided scan mirror and is focused onto four focal planes containing the 1 km, 500 m and 250 m bands (36 bands total). The instantaneous sensing of the four co-registered focal planes is shown, illustrating the MODIS "paddle broom" sensing geometry

in the along-scan (cross-track) direction causing the same Earth location to be sampled at different times by different bands. Each 1-km, 500-m and 250-m observation is sampled in 333.333 μs, 166.667 μs and 83.333 μs, respectively. To allow for detector readout, the detector integration time is 10 μs less than the data-sampling rate at each of the three MODIS resolutions. To the first order, the MODIS point-spread function is triangular in the scan direction (see Fig. 4.2). The centers of the integration areas of the first observation in each scan are aligned in a "peak-to-peak" alignment. In the track direction, the point-spread function is rectangular and the observations at the different resolutions are nested, allowing four rows of 250 m observations and two rows of 500 m observations to cover the same area as one row of 1-km observations.

Each scan of the Earth's surface is elongated because of the MODIS sensing geometry and Earth curvature such that the swath width is approximately 2,340 km. The mirror angular velocity (2.127 rad/sec) and the forward velocity of the satellite

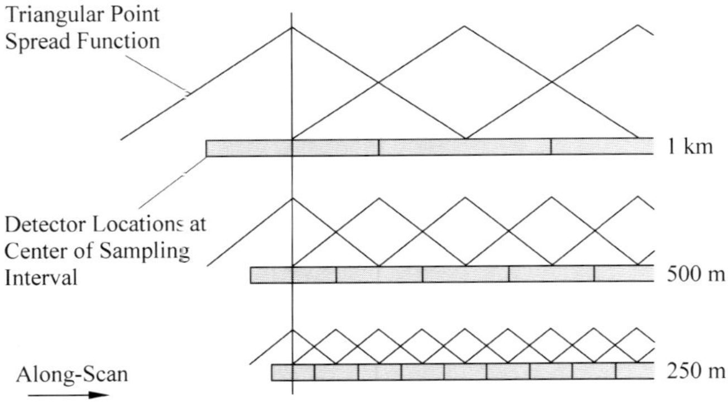

Figure 4.2 Detector along-scan triangular point spread function and the peak-to-peak alignment of the three MODIS spatial resolutions

(7.5 km/s) are configured such that at nadir, in the track direction, the leading edge of one scan abuts the trailing edge of the next scan. Adjacent scans begin to overlap away from nadir with a 10% overlap occurring at scan angles of 24° from nadir. Consequently, the detectors on the leading edge of a scan sense surface features before the detectors on the trailing edge of the following scan (see Fig. 4.3). This overlap increases to almost 50% at the scan edge. The same point on the Earth's surface may be sensed by up to three consecutive scans at the scan edge. This phenomenon is called the "bow-tie" effect and is seen in other whiskbroom wide-field-of-view sensors such as Advanced Very High Resolution Radiometer (AVHRR), though this effect is less evident for scanners with only one detector per band. At the scan edge the projection of a MODIS detector's

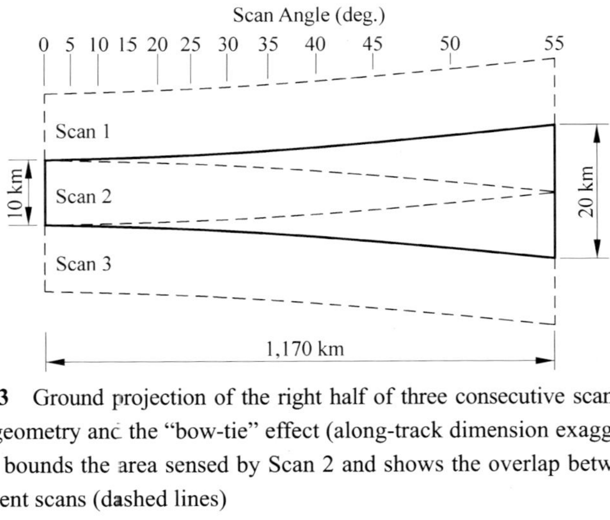

Figure 4.3 Ground projection of the right half of three consecutive scans showing the scan geometry and the "bow-tie" effect (along-track dimension exaggerated). A solid line bounds the area sensed by Scan 2 and shows the overlap between it and two adjacent scans (dashed lines)

instantaneous field of view (IFOV) onto the surface (also the ground sample distance and pixel size) is approximately 2.0 and 4.8 times larger than at nadir in the track and scan directions, respectively (see Fig. 4.4).

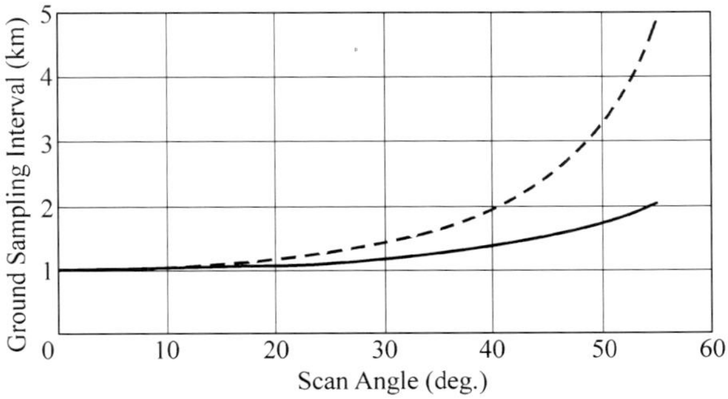

Figure 4.4 Ground sampling interval (pixel size) in the scan (dashed line) and track (solid line) directions plotted as a function of scan angle (nadir is 0°)

All MODIS products are generated in a hierarchy of levels: retrieved geophysical parameters at the same location as the MODIS instrument data (Levels-1 and -2), Earth-gridded geophysical parameters (Levels-2G and -3), and Earth-gridded model outputs (Level-4). The smallest amount of MODIS land data processed at any one time is defined at Levels-1 and -2 as a granule and corresponds to approximately 5 minutes of MODIS sensing consisting of approximately 203 scans. A granule covers an area approximately 2,340 km wide across-track by 2,030 km along-track. The MODIS geolocation product is a Level-1 product and defines for each 1km MODIS observation the geodetic latitude and longitude, terrain height, sensor zenith angle, sensor azimuth, slant range to the sensor, solar zenith angle, and solar azimuth (Nishihama et al., 1997). These geolocation data are also stored in the MODIS Level-1B (L1B) calibrated radiance products, in several MODIS Level-2 products and in the MODIS land L2G products (Wolfe et al., 1998).

The first MODIS/Terra data produced (Collection 1) were early release products of beta quality (Justice et al., 2002). Processing of MODIS/Terra Collection 3 began in June 2001 and included the first major reprocessing. Collection 3 products were of provisional quality (continually improving and partially validated). Collection 4 processing of MODIS/Terra began in November 2002, including the second major reprocessing, created validated products of high quality with well defined uncertainties. The first MODIS/Aqua data processed was also of provisional quality and designated Collection 3. This collection was completed in December 2003 and Collection 4, which included a full MODIS/Aqua reprocessing, started. A combined Collection 5 MODIS Terra and Aqua reprocessing started began in June 2005.

4.3.2 Exterior and Interior Orientation

Precise knowledge of the exterior and interior orientation of the sensor is needed for accurate geolocation using the parametric approach. This section is a summary of the sensor interior and exterior orientation, details of which, including error estimates, can be found in Wolfe et al. (2002).

The interior orientation parameters are characterized prior to launch by the instrument and spacecraft builders (Barnes et al., 1998). This is done by first measuring the instrument characteristics in ambient conditions and then in thermal vacuum with expected on-orbit temperatures. Geometric parameters measured include the focal length, focal planes' geometry and bore-sight orientation. The sensor's orientation with respect to the satellite navigation coordinate system is measured when it is mounted on the satellite. Similar characterization of the attitude sensors is performed pre-launch.

The satellite exterior orientation is described by the satellite attitude, the rotation of the satellite navigation coordinate system with respect to a celestial inertial coordinate system, and by the satellite ephemeris, the satellite's position and velocity in an earth centered inertial referenced frame.

The attitude is estimated by star trackers and inertial gyros. The data from these sensors are combined in real-time using a Kalman filter (Maybec, 1979). Soon after launch these sensors' interior orientations are characterized by a series of attitude maneuvers and by inter-comparison of various attitude sensors measurements using least-squares techniques and a-priori knowledge of the error characteristics of each sensor (Glickman et al., 2003). A ground based retrospective batch least-squares technique (Wertz, 1978) is used to help perform this calibration and to validate the real-time on-orbit navigation. This same ground based software is used to produce post-processed definitive attitude data for Terra.

Terra ephemeris is calculated using Tracking and Data Relay Satellite System (TDRSS) On-board Navigation System (TONS). Ranging from Terra to the geosynchronous TDRSS satellites is used in conjunction with earth gravity and upper-atmospheric drag models to accurately locate the Terra spacecraft in real-time (Folta et al., 1993). Because Aqua does not use TDRSS for data transmission, the on-board navigation software uses predicted ephemeris generated from ground based ranging data.

Aqua geolocation that is used for higher level science data processing is processed with retrospective definitive ephemeris because it is more accurate, even though there is some delay in producing the geolocation products. More time-sensitive users (near real-time users) must use the predicted ephemeris which accompanies the real-time sensor data. For Terra geolocation, there is no significant difference in the accuracy real-time TONS and retrospective definitive ephemeris. However, following orbit and attitude maneuvers, post-processed ground based ephemeris and attitude data is used because of inaccuracies in real-time propagation after these events.

4.3.3 Algorithm

The MODIS geolocation calculation is performed for a hypothetical ideal band known as Band 0. Band 0 is modeled as being located at the middle of the four MODIS focal planes and is used as a reference from which the positions of any band are calculated by applying appropriate offsets (Barnes et al., 1998). This section is a summary of the detailed description of the MODIS geolocation algorithm in Nishihama et al. (1997).

The geolocation data are computed for the centers of each 1 km Band 0 observation (Wolfe et al., 2002). Computing the location of the center of a single observation is performed in several steps. First, the line-of-sight $\boldsymbol{u}_{\text{foc}}$ from each detector of a band is generated in the focal plane coordinate system:

$$\boldsymbol{u}_{\text{foc}} = (x, y, f) \tag{4.1}$$

where (x, y) are the positions of the detector focal plane coordinates and f is the focal length. (Bold type is used to denote matrices throughout this chapter.) The line-of-sight is then rotated from the focal plane (foc) to the telescope (tel) coordinate system and then to the instrument (inst) coordinates system:

$$\boldsymbol{u}_{\text{img}} = \boldsymbol{T}_{\text{inst/tel}} \boldsymbol{T}_{\text{tel/foc}} \boldsymbol{u}_{\text{foc}} \tag{4.2}$$

(In this chapter the symbol $\boldsymbol{T}_{\text{sys2/sys1}}$ is used to represent a 3 by 3 transformation matrix that rotates a 3-vector from coordinate system sys1 to sys2. For instance, $\boldsymbol{T}_{\text{inst/tel}}$ rotates a vector from the telescope to the instrument coordinate system.)

One of the key elements of the MODIS geometric model is the mirror model. Because it is impossible to manufacture a two-sided mirror with perfectly parallel sides and align the mirror perfectly with the mirror rotation axis, three angles are used to characterize the mirror surfaces and to construct the normals $\hat{\boldsymbol{n}}_{\text{side}i}$ to the mirror surfaces (see Fig. 4.5).

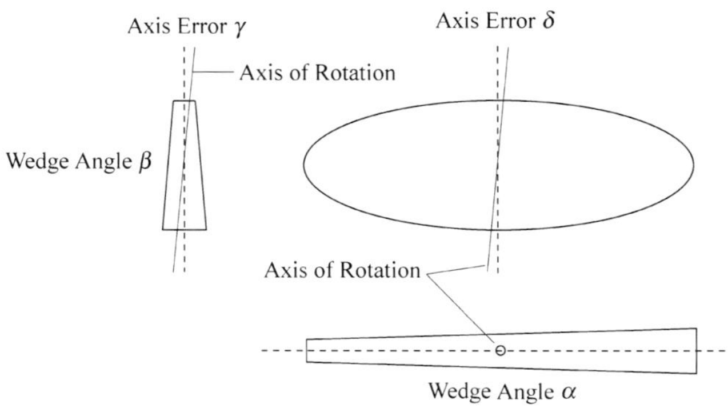

Figure 4.5 Three angles (wedge angles α and β, and axis error γ) used to characterize the two surfaces of the MODIS scan mirror. Angle δ is not used in the described error analysis because it does not have a significant geometric impact

Using the time of the observation, t, the angle of rotation of the scan mirror θ is determined and the normal to the rotating scan mirror $\hat{\boldsymbol{n}}_{\mathrm{side}i}$ for mirror side i is constructed in the scan mirror (mirr) coordinate system. The mirror normal is rotated to the instrument coordinate system:

$$\hat{\boldsymbol{n}}_{\mathrm{inst}} = \boldsymbol{T}_{\mathrm{inst/mirr}} \boldsymbol{T}(\theta)_{\mathrm{rot}} \; \hat{\boldsymbol{n}}_{\mathrm{side}i} \tag{4.3}$$

The line-of-sight vector is reflected by the scan mirror, computed by:

$$\boldsymbol{u}_{\mathrm{inst}} = \boldsymbol{u}_{\mathrm{img}} - 2\hat{\boldsymbol{n}}_{\mathrm{inst}} \left(\boldsymbol{u}_{\mathrm{img}} \cdot \hat{\boldsymbol{n}}_{\mathrm{inst}} \right) \tag{4.4}$$

The satellite position $\boldsymbol{p}_{\mathrm{eci}}$ and velocity $\boldsymbol{v}_{\mathrm{eci}}$ are defined in the Earth Centered Inertial (ECI) coordinate system, and the satellite attitude $(\xi_{\mathrm{r}}, \xi_{\mathrm{p}}, \xi_{\mathrm{y}})$ represents the rotation between the spacecraft navigation and orbital coordinate systems. At time t a composite transformation matrix is constructed to rotate from the instrument coordinate system, through the spacecraft (sc), orbital (orb), and ECI coordinate systems, to the Earth Centered Rotating (ECR) coordinate system:

$$\boldsymbol{T}_{\mathrm{ecr/inst}} = \boldsymbol{T}_{\mathrm{ecr/eci}} \; \boldsymbol{T}(\boldsymbol{p}_{\mathrm{eci}}, \boldsymbol{v}_{\mathrm{eci}})_{\mathrm{eci/orb}} \; \boldsymbol{T}(\xi_{\mathrm{r}}, \xi_{\mathrm{p}}, \xi_{\mathrm{y}})_{\mathrm{orb/sc}} \boldsymbol{T}_{\mathrm{sc/inst}} \tag{4.5}$$

The line-of-sight and the satellite position are then rotated to ECR coordinates:

$$\begin{aligned} \boldsymbol{u}_{\mathrm{ecr}} &= \boldsymbol{T}_{\mathrm{ecr/inst}} \; \boldsymbol{u}_{\mathrm{inst}} \\ \boldsymbol{p}_{\mathrm{ecr}} &= \boldsymbol{T}_{\mathrm{ecr/eci}} \, \boldsymbol{p}_{\mathrm{eci}} \end{aligned} \tag{4.6}$$

The intersection of the line-of-sight with the WGS-84 ellipsoid $\boldsymbol{x}_{\mathrm{ellip}}$ is then calculated as:

$$\boldsymbol{x}_{\mathrm{ellip}} = \boldsymbol{p}_{\mathrm{ecr}} + d\boldsymbol{u}_{\mathrm{ecr}} \tag{4.7}$$

The ellipsoid intersection is calculated by turning the problem into a unit sphere intersection problem (by independently rescaling the components of each vector by the inverse of the length of the corresponding ellipsoid axis) and is trigonometrically solved for the slant range d.

An iterative search process is used to follow the line-of-sight from the instrument to the intersection of the terrain surface represented by a Digital Elevation Model (DEM). Complex relief does not confuse this technique because the search is in a downward direction. Precompiled maximum and minimum local terrain heights are used to limit the search.

Subsequently, the slant range to the sensor, sensor zenith angle and sensor azimuth are computed and stored for each intersection. In addition, the solar zenith and azimuth are computed from the observation time and geodetic latitude and longitude using standard astronomical models (Standish et al., 1992).

4.3.4 Error Sources

A number of error sources are anticipated that include errors in the exterior and interior orientation, digital elevation model errors, and errors due to refraction and aberration (Fleig et al., 1993).

Errors in sensor attitude (due to exterior or interior orientation errors) will induce geolocation displacements that are directly proportional to the attitude error, the sensor altitude, local Earth curvature (ignoring terrain effects), and the scan angle (Nishihama et al., 1997). Along and across track sensor position measurement errors will cause displacements in the planimetric geolocation position. Altitude position measurement errors will cause along-scan displacements that increase with scan angles further from nadir.

Figure 4.6 illustrates geolocation displacement error ellipses at nadir (solid lines) and at the scan edge (dotted lines) for all the error components. Both the static and dynamic error components are illustrated. The along-scan component is 41% larger than the along-track component at nadir, and 249% larger at 5° scan angle. If the static errors are removed, the total remaining error is expected to be 47 m (1σ) at nadir and 166 m (1σ) at 55° scan angle.

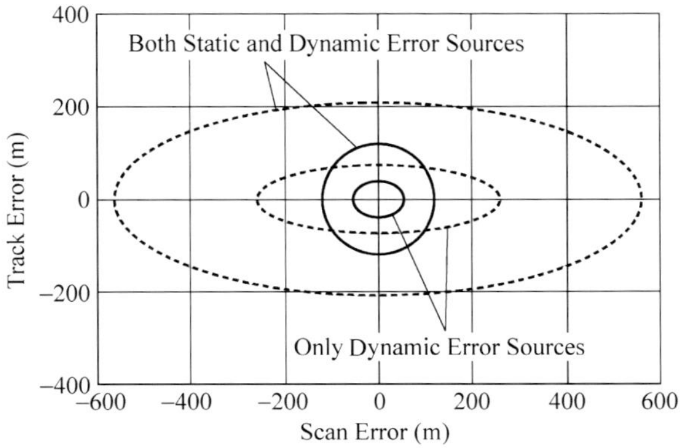

Figure 4.6 Geolocation displacement error ellipses at nadir (solid lines) and at 55° scan angles (dotted lines) for two cases: when both the static and dynamic error components are considered, and when only the dynamic components errors are considered. Errors are shown at the (1σ) level and illustrate all the known expected error components

Although the DEM is used to remove relief effects, residual geolocation errors may be introduced because of errors in the DEM, errors introduced by interpolating the line-of-sight intersection in the DEM, and high frequency relief variations occurring within each IFOV. Line-of-sight intersection interpolation errors, DEM resolution issues, and the variable MODIS IFOV dimensions combine to introduce complex unmodeled location errors.

The geolocation algorithm does not model refraction or aberration because the effects are small (Noerdlinger and Klein, 1995). Since these effects are not modeled, they will appear as small biases in the geolocation error analysis.

4.3.5 Ground Control Points

The MODIS operational geolocation error analysis and reduction methodology uses a global distribution of land Ground Control Points (GCPs) to characterize and then remove some of the error components discussed above. The MODIS geolocation group collaborated with the Terra and Landsat-7 instrument teams to develop a library of land GCPs. A global distribution of Landsat-4 and Landsat-5 precision geolocated terrain corrected Thematic Mapper (TM) scenes was obtained (see Fig. 4.7) and approximately five cloud-free GCPs were selected from each TM scene. A 24-km square image "chip", including the terrain height at each 30 m TM pixel, was extracted around the GCP location and stored in a GCP library. TM bands 3 (0.66 μm) and 4 (0.83 μm), comparable to the two MODIS 250-m bands 1 (0.645 μm) and 2 (0.859 μm), are stored. The latitude, longitude, and height of the GCPs are known to 15 m (1σ).

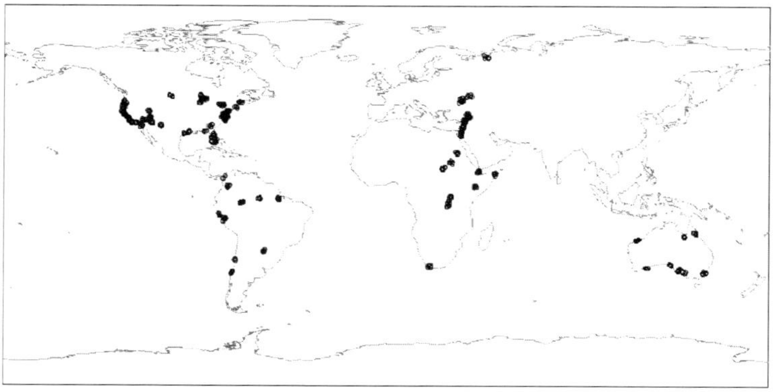

Figure 4.7 Global distribution of 420 land ground control points (GCPs) extracted from 110 Landsat TM scenes

The Landsat GCPs are located in the sensed MODIS L1B data by area-based matching (Wolfe et al., 2002). GCPs sensed by MODIS at view zenith angles greater than 45° are not used as the surface area sensed by the MODIS IFOV increases rapidly above this zenith angle (see Fig. 4.4). Residual errors between the known GCP locations and the corresponding locations in the MODIS data are used in the geolocation error and reduction analyses. In order to remove poorly matched GCPs, only those with maximum correlation coefficients greater than 0.6 are considered. In this way inaccurate, out-of-date and out-of-season GCPs, and GCPs contaminated by cloud and aerosols, are less likely to be used. GCP residuals are also used to quantify the impact of changes to the interior orientation parameters and exterior orientation biases. For this latter purpose, the

GCP residuals are normalized by dividing by the dimension of the local MODIS observation instantaneous field of view (IFOV) in the along-scan and along-track dimensions. In this way, the GCP residuals are meaningfully expressed in nadir pixel dimensions or equivalently in meters at nadir. These values can be scaled to approximate the mean residuals over the scene by multiplying by 1.38 in the track direction and 2.19 in the scan direction.

4.3.6 Geolocation Error Analysis and Reduction Methodology

A deterministic least squares (minimum variance) estimation is used to compute the sensor orientation parameters that best fit the GCP data. Models of the interior and exterior orientation parameters are described by linearized collinearity equations, building on the mathematical foundation described by Konecny (1976). These equations are used to compute the along-track, across-track and radial position, roll, pitch, and yaw attitude angles, and mirror parameters, and their rates of change (Wolfe et al., 2002). In this process, all errors are assumed to be randomly distributed with a mean of zero. Initially after launch, it is not possible to uniquely differentiate between attitude and position induced geolocation errors. Consequently, the error analysis is initially performed holding certain parameters fixed. Updates to the interior orientation parameters are performed by modifying look-up-tables in the geolocation software that define the transformation matrix elements (see Section 4.3.3). Reliable numerical solution of the linearized collinearity equations without holding certain parameters fixed may only be performed after initial updates have been made.

Corrections of any systematic exterior orientation measurement errors can only be performed after the sensor interior orientation parameters are well defined. To date, no explicit corrections of these measurement errors have been performed, though implicit corrections may have been made when the sensor interior orientation parameters were updated.

Updates of dynamic interior orientation parameters (for example, associated with day-side/night-side thermal flexing) have been performed for MODIS/Terra. These updates are currently temporally-parameterized adjustments to the transformation matrix elements, but in the future they may be matrix elements parameterized with respect to the on-board instrument engineering data stream (e.g. on-board temperature).

The temporal parameterization was found to be well represented by a linear and sinusoidal equation:

$$f(d) = u_0 + u_1 \frac{d}{P} + u_2 \sin\left(\left(\frac{u_3}{P} + \frac{d}{u_4}\right) 2\pi\right) \tag{4.8}$$

where u_i are the estimated coefficients, d is the number of days since 12:00

Universal Time Corrected (UTC) January 1, 2000 (J2000 epoch) and P is the number of days per year (365.242).

A combination of temporal and temperature based parameterization was evaluated. This parameterization was found to be well represented by a combination of linear temporal and temperature terms:

$$g(w,d) = c_0 + c_1 \frac{d}{P} + c_2 (w - c_3) \tag{4.9}$$

where c_i are the estimated coefficients, w is an on-board temperature, and d and P are defined above.

A least squares approach was used to independently estimate these coefficients for the roll and pitch biases in the spacecraft to instrument coordinate system alignment matrix $\boldsymbol{T}_{\text{sc/inst}}$ (see Eq. (4.5)).

4.4 Results

Geolocation results for the MODIS instruments on the Terra and Aqua satellites are given separately in the following two subsections.

4.4.1 MODIS/Terra Results

Analysis of MODIS/Terra geolocation was performed after launch using GCPs. This analysis revealed instrument to spacecraft alignment biases and that the on-board attitude data was better than post-processed attitude data. Significant differences were found in the mirror geometric characterization, and corrections were made to wedge angles between the two sides of the scan mirror and to the tilt of the scan mirror axis of rotation.

At the time of writing, five updates have been made to the MODIS/Terra geometric parameters (see Table 4.1). The first four updates had no time-dependence, but the fifth incorporates a slowly varying time-dependent trend in the roll and pitch axes.

The first three parameter updates were performed during the 13 months after first light (Wolfe et al., 2002) and involved three updates to the instrument to spacecraft alignment ($\boldsymbol{T}_{\text{sc/inst}}$ in Eq. (4.5)) and two updates to the scan mirror parameters (used to construct $\hat{\boldsymbol{n}}_{\text{side}i}$ used in Eq. (4.3)). After these updates the geolocation error was near to the geolocation goals. The fourth update was performed 26 months after first light and corrected an along-scan non-linearity in the residuals that was modeled by a combination of a pitch bias and an offsetting tilt (pitch) bias in the scan mirror rotation axis (modeled in the $\boldsymbol{T}_{\text{inst/mirr}}$ matrix in Eq. (4.3)).

Table 4.1 Interior orientation parameter updates (arcseconds) performed since MODIS/Terra launch. No updates were performed to the position parameters

Up-date	Start Date	Instrument to Spacecraft Alignment			Scan Mirror Coefficients			Mirror Axis Tilt
		Roll	Pitch	Yaw	α	β	δ	
—*	Dec. 1999	0	208.3	– 39.2	3.9	0	– 26.8	12.4
1	Apr. 2000	**243.1**	**– 145.5**	– 39.2	3.9	0	– 26.8	12.4
2	June 2000	243.1	– 145.5	**95.1**	3.9	**38.8**	**– 2.9**	12.4
3	Mar. 2001	**255.3**	**– 152.3**	**98.2**	**1.2**	**38.0**	**– 0.6**	12.4
4	Apr. 2002	**251.6**	**83.5**	98.2	**– 4.1**	38.0	– 0.6	**– 180.7**
5	Oct. 2002	251.6 + LTT	83.5 + LTT	98.2	– 4.1	38.0	– 0.6	– 180.7

* At-launch value
(**Bold**): Changed in each update
LTT: Long-term trend

Figure 4.8 illustrates the at-launch MODIS geolocation (a) and the geolocation after the first update to the MODIS interior orientation was performed (b). The figure shows false color Landsat-5 TM data and MODIS 250 m data for a 60-km square area of the Florida coastline. The MODIS data are shown in red and the TM data are shown in blue. The TM data were spatially degraded to 250 m using the MODIS point spread function defined for the sensing geometry of each MODIS 250-m observation. The MODIS at-launch data (see Fig.4.8(a)) clearly exhibit along-track and smaller along-scan displacements relative to the TM data. Some isolated differences are also due to differences in the cloud cover and

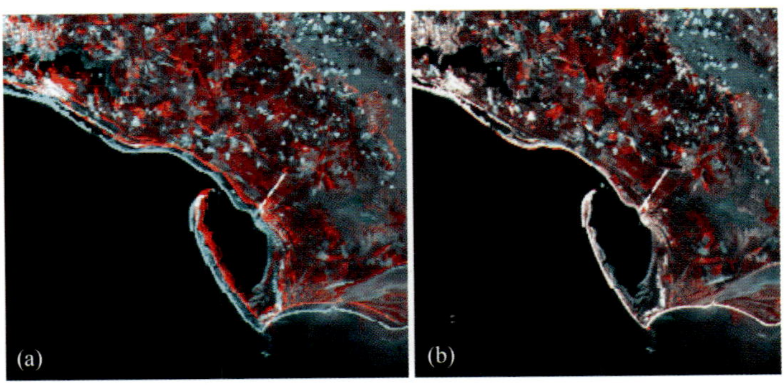

Figure 4.8 False color images illustrating at-launch MODIS/Terra geolocation (a), and the geolocation after the first update to the MODIS interior orientation was performed (b), for a 60 km square area over the Gulf of Mexico coastline, northwest Florida. MODIS Band 1 (0.645 μm) shown as red and spatially degraded TM Band 4 (0.66 μm) shown as blue, with green set to zero. The MODIS data were sensed on February 24, 2000, and the TM data were sensed on October 3, 1993

land-cover change that may have occurred between the acquisitions of these data. Figure 4.8(b) shows the impact of the first update. The images are well matched and differences at the 250-m resolution are difficult to discern visually.

The long-term time-series illustrated in Fig.4.9 reveals a significant linear trend in the track direction and a yearly cycle in both the scan and the track directions. After the fourth update, different long-term parameterizations were assessed to determine the best approach to remove the long-term trend. Parameters for the temporal and temperature based fits shown in Tables 4.2 and 4.3 were estimated based on data from November 2000 to May 2002.

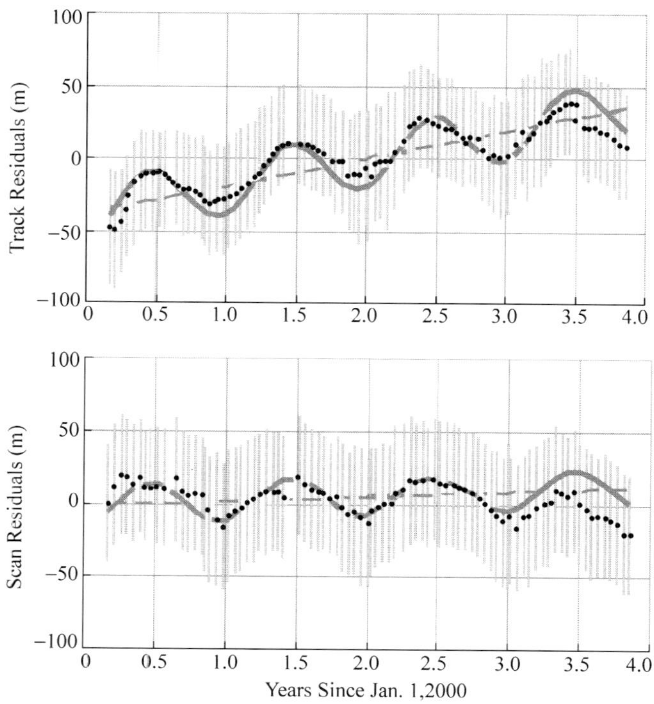

Figure 4.9 MODIS/Terra GCP residuals (solid dots) plotted as a function of time in the track (top) and scan (bottom) directions before removal of the long-term trend. These results are adjusted for scan angle and shown in nadir equivalent units. An average of 1,313 GCP residuals are plotted for each 16-day period. Error bars representing one standard deviation (1σ) about the mean are shown. Also shown are the original estimate of the linear fit (dashed grey line) and yearly sinusoidal fit (solid gray line)

Temperature analyses were performed to test if the non-linear cyclic portion of the long-term trend was related to the instrument temperature.

Of the 75 MODIS/Terra on-board passive temperature sensors, data from five were examined and the Short/Medium Wave Infrared Objective Lens Assembly Temperature sensor data was found to have highest correlation to the GCP residuals. Figure 4.10 shows the relationship between this on-board temperature and the

Table 4.2 Estimated MODIS/Terra long-term trend temporal coefficients in nadir equivalent units

	Term [units]	Track	Scan
u_0	bias [m]	– 37.7	– 1.7
u_1	linear [m/year]	19.0	3.4
u_2	amplitude [m]	– 19.9	– 13.5
u_3	phase [days]	470.0	470.0
u_4	period [days]	365.243	365.243

Table 4.3 Estimated MODIS/Terra long-term trend linear and temperature coefficients in nadir equivalent units

	Term [units]	Track	Scan
c_0	bias [m]	– 37.7	– 1.7
c_1	linear [m/year]	19.0	3.4
c_2	linear [m/degree K]	– 20.9	– 12.7
c_3	bias [degrees K]	267.63	267.54

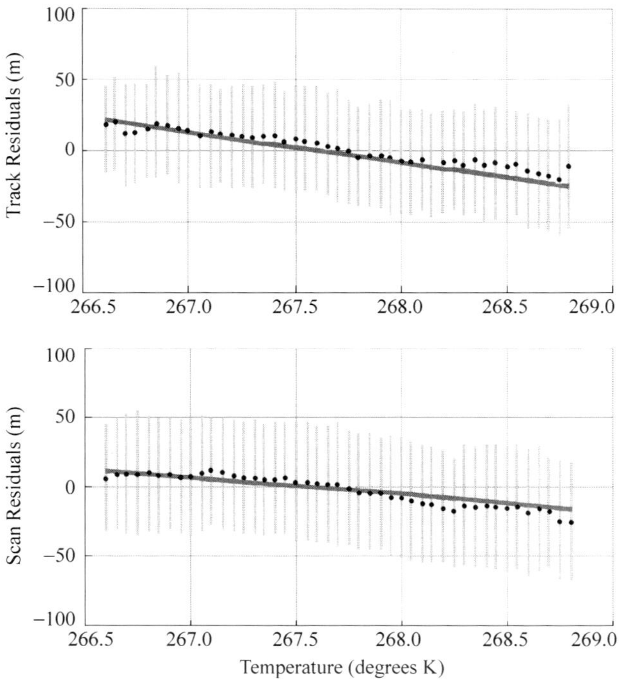

Figure 4.10 MODIS/Terra GCP residuals (solid dots) plotted as a function of on-board temperature in the track (top) and scan (bottom) directions after removal of the original linear temporal long-term trend. These results are adjusted for scan angle and shown in nadir equivalent units. Averages of 867 GCP residuals are plotted for each 0.05°K bin. Error bars representing one standard deviation (1σ) about the mean are shown. Also shown is the original estimate of the linear temperature fit (grey line)

control point residuals after the linear trend was removed and also shows the original linear temperature fit. The overall yearly variation in temperature of ± 1.25 degrees K is in-phase with the cyclic term found in the temporal analysis. This supports the idea that the yearly cycle is related to temperature.

In late 2002, the geolocation team had to choose one of four approaches to be used in the first complete MODIS/Terra reprocessing (called Collection 4): (a) fixed biases; (b) linear trend; (c) linear trend plus yearly cyclic temporal fit; and (d) linear trend plus linear temperature fit. At that time it was thought that the geolocation accuracy goal would not have been met for the entire collection if static biases were used. The three other approaches were more promising (see Table 4.4), particularly in the track direction, with better and very similar results for the cyclic temporal and the temperature approaches.

Table 4.4 Remaining GCP residual error for three approaches to removing the long-term trend in MODIS/Terra geolocation. These results are adjusted for scan angle and shown in nadir equivalent units (meters). The results are basd on over 33,000 GCP residuals acquired during 1.5 years of MODIS/Terra Collection 3

Trend Removal Approach	Track RMSE	Scan RMSE
a. None	47.3	45.5
b. Linear temporal fit	45.0	45.5
c. Linear and cyclic temporal fit*	42.9	44.5
d. Combined temporal and temperature linear fit	43.3	44.7

* Approach chosen for Collection 4 reprocessing.

The cyclic temporal approach was chosen for ease of implementation and because it was considered more robust. Use of this approach for MODIS/Terra Collection 4 resulted in an at-nadir geolocation accuracy of 37.1 m and 43.0 m Root Mean Square Error (RMSE) in the track and scan directions (see Table 4.5).

Recent analysis of Collection 4 data reassessed the different long-term trend approaches based on the MODIS/Terra data sensed from February 2000 to November 2003. The results shown in Table 4.5 reveal that the cyclic temporal approach (row c) performed about as well as the temperature approach (row d). As expected, in the track direction where the temporal trend is largest, both approaches performed significantly better than the static biases approach (row a) and linear trend approach (row b). However, in the scan direction where the temporal trend is smaller, all four approaches performed equally well. The static bias approach was not expected to perform as well as the other approaches, but it did. This can be traced to the estimate of the linear trend in the scan direction (3.4 m/year) which, based on the earlier analysis of only 1.5 years of data, did not represent the trend for the entire period, now estimated to be − 3.5 m/year. This is evident in Fig.4.9 which shows that the positive linear trend does not follow the overall negative trend in the data.

Table 4.5 Remaining GCP residual error for three approaches to removing the long-term trend in MODIS/Terra geolocation. These results are adjusted for scan angle and shown in nadir equivalent units (meters). The results are basd on over 110,000 GCP residuals acquired during 3.75 years of MODIS/Terra Collection 4

Trend Removal Approach	Track RMSE	Scan RMSE
a. None	41.6	43.2
b. Linear temporal fit	38.3	43.6
c. Linear and cyclic temporal fit	37.1*	43.0*
d. Combined temporal and temperature linear fit	37.4	42.6

* Actual results from the Collection 4 reprocessing.

Further analysis of the remaining geolocation errors reveals a latitudinal dependence in the cyclic term of the long-term trend. Since 80% of the control point match ups are in the northern hemisphere the global trend follows that of the northern hemisphere. Figure 4.11 shows the 16-day mean of the residuals for the globe, and separately for the northern and southern hemispheres. This figure shows that the cyclic trend was removed from the northern hemisphere. However,

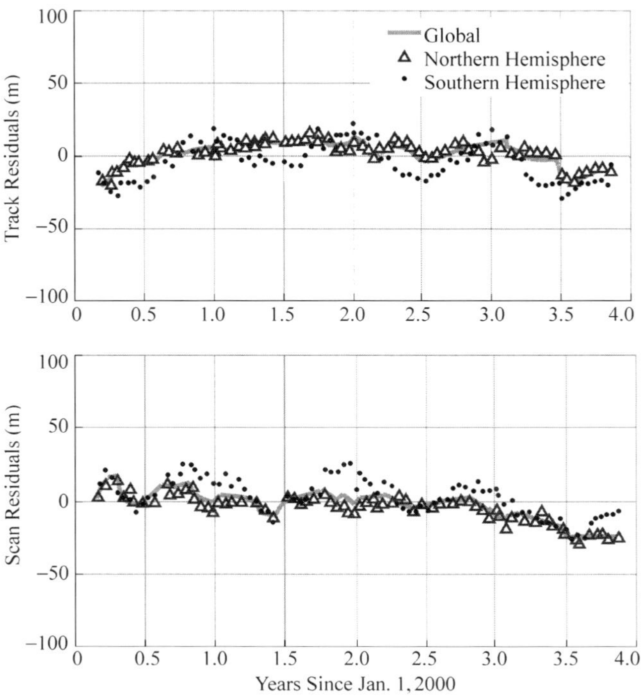

Figure 4.11 MODIS/Terra GCP residuals plotted as a function of time in the track (top) and scan (bottom) directions after removal of the long-term temporal trend. These results are for 16-day periods, adjusted for scan angle and shown in nadir equivalent units

the southern hemisphere shows a cyclic trend that can be explained by an over-correction in that hemisphere. Analysis shows that a similar hemispheric difference in the GCP residuals would also have occurred if a correction had been done using the temperature approach.

A current hypothesis is that a within-orbit thermal effect is causing both the cyclic trend and the hemispheric differences. A thermal effect combined with the yearly change in solar declination could explain the differences in the cyclic trend in the two hemispheres. An approach to correct for within-orbit thermal effects is being developed and evaluated.

The overall MODIS/Terra at-nadir geolocation error is well within the accuracy goal of 50 m (1σ). Further refinements in the estimate of the long-term trend based on the entire MODIS/Terra record will be made before the next major MODIS/Terra reprocessing, scheduled to begin at the end of 2004.

4.4.2 MODIS/Aqua Results

Initial on-orbit analysis of MODIS/Aqua revealed alignment and mirror effects similar to MODIS/Terra. The first and second updates were performed within four months after first light (see Table 4.6). However, a large within-orbit error was found and determined to have been caused by confusion of coordinate systems in the satellite flight software (Glickman et al., 2003). The third update was performed simultaneously with the flight software fix in March 2003 and was quickly followed by the fourth update. This last update included a correction to the tilt of the scan mirror rotation axis, similar to the one performed for MODIS/Terra.

Table 4.6 Interior orientation parameter updates (arcseconds) performed since MODIS/Aqua launch. No updates were performed to the position parameters

Update	Start Date	Instrument to Spacecraft Alignment			Scan Mirror Coefficients			Mirror Axis Tilt
		Roll	Pitch	Yaw	A	β	δ	
—*	May 2002	0	0	0	23.1	35.1	– 3.9	13.0
1	Aug. 2002	**424.3**	**– 26.4**	**– 50.7**	23.1	35.1	– 3.9	13.0
2	Oct. 2002	**306.7**	**0.0**	**– 38.0**	**– 5.6**	**37.1**	**– 6.4**	13.0
3	Mar. 2003	**406.8**	**79.8**	**– 76.1**	– 5.6	37.1	– 6.4	13.0
4	Apr. 2003	**407.2**	**574.9**	**– 76.3**	– 5.6	37.1	– 6.4	**– 422.2**
5**	Jan. 2004	**409.5**	**582.2**	– 76.3	– 5.6	37.1	– 6.4	– 422.2

* At-launch value

** Planned

(Bold): Changed in each update

Figure 4.12 illustrates the geolocation residuals for the 9 months of MODIS/Aqua data acquired after the flight software was fixed. For this period the at-nadir geolocation accuracy is 47.6 m and 57.1 m in the track and scan directions. The fifth update to the geolocation parameters, expected to primarily correct a small bias in the pitch axes, will be made in time for the first MODIS/Aqua reprocessing scheduled to begin in January 2004. During this reprocessing, attitude data acquired before the spacecraft flight software fix will be rotated to the correct coordinate system to remove the geolocation error caused by the original coordinate system confusion.

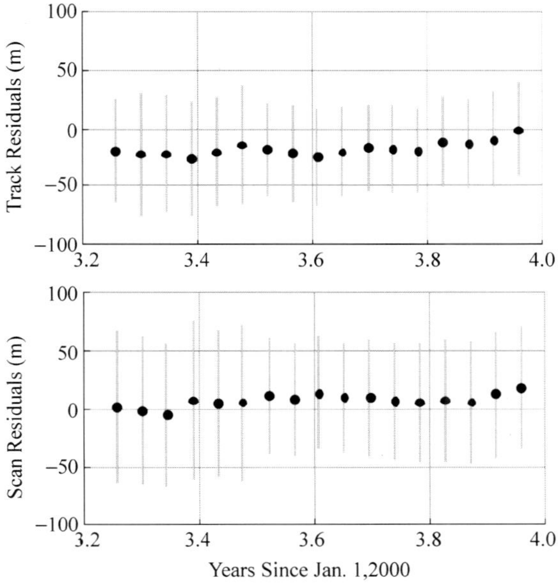

Figure 4.12 MODIS/Aqua GCP residuals (solid dots) plotted as a function of time in the track (top) and scan (bottom) directions since the Aqua spacecraft flight software fix in March 2003. These results are adjusted for scan angle and shown in nadir equivalent units. An average of 1,257 GCP residuals are plotted for each 16-day period. Error bars representing one standard deviation (1σ) about the mean are shown

Two other issues remain that have prevented the MODIS/Aqua geolocation accuracy goal of 50 m from being reached: another fix is needed for the flight software to correct a known readout timing bias in one star tracker, and a high frequency jitter caused by the Advanced Microwave Scanning Radiometer -EOS (AMSR-E), also on-board Aqua, must be modeled and removed. Removal of the start tracker bias will occur soon and is expected to primarily improve the accuracy in the track direction. It appears feasible that an algorithm can be developed to model and remove the AMSR-E induced jitter but more work is needed. Removal of this jitter in the roll axes is expected to improve the accuracy in the scan direction. Long-term trend analysis is now underway.

4.5 Conclusion and the Future

The MODIS geolocation approach has been used to quickly reduce large on-orbit geolocation errors and then to maintain the geolocation accuracy near to or within the science community's stringent goals. The accuracy goal has been met for the entire MODIS/Terra mission and for MODIS/Aqua the accuracy is now well within the specification and very close to the goal. This accurate geolocation has enabled the MODIS science team to concentrate their resources on the accurate retrieval of biophysical parameters.

Future operational moderate resolution missions such as Visible Infrared Imager Radiometer Suite (VIIRS) on the National Polar-orbiting Operational Environmental Satellite System (NPOESS) and NPOESS Preparatory Project (NPP) are using a similar geolocation approach. However, the geolocation specifications accuracy goals for VIIRS/NPOESS are not stringent enough to ensure that geolocation accuracy for the land community is achieved. Since the earth remote sensing science community is relying on these operational missions to provide high quality long-term climate data records, the accuracy required for these data is typically more stringent than is required by the operational community. There are no technological hurdles to reaching these goals for future missions but there are a number of programmatic (primarily cost) hurdles.

A better quantification of the impacts of geolocation accuracy on the retrieval of geophysical parameters is needed to justify the additional expense needed to meet the terrestrial research community's geolocation accuracy needs on future operational missions. Less accurate geolocation typically results in blurring of the data in multi-day composited products. This can cause over (or under) estimates of terrestrial geophysical quantities (Roy, 2000). Land cover and land use changes at or near the spatial resolution of the satellite are more difficult or impossible to distinguish in data sets with less accurate geolocation (Townshend et al., 1992). Also needed is a way to quantify the belief that the costs of making changes up-front in the instrument and spacecraft design are normally much smaller than those needed to fix the geolocation accuracy later (if it is even possible).

In the future we expect to use a newer Shuttle Radar Terrain Mapping (SRTM) DEM and MODIS derived land-sea mask. We plan to refresh the geolocation land control point library with the goal of a more even distribution of control points globally. We also plan to evaluate the merits of computing the mean elevation of each observation (vs. the pierce point). It may also be possible in the future to automate the long-term trend analysis using a Kalman filtering approach.

MODIS has set a new standard for geolocation of moderate resolution earth remote sensing satellite data. It is important to continue this approach in the new NASA missions and other operational missions. It should be possible to extend this approach to future finer resolution sensors as well.

Acknowledgements

Accurate MODIS geolocation required contributions from a large number of specialized groups, including: the instrument and satellite builders, flight dynamics and attitude control groups, and other instrument teams. The author acknowledges the contributions of these groups, the contributions of MODIS geolocation team members and the support provided by the MODIS science team. This work was performed under the direction of the MODIS Science Team in the Terrestrial Information Systems Branch (Code 922) of the Laboratory of Terrestrial Physics (Code 900) at NASA GSFC. The work was funded under NASA GSFC contract NAS5-32350.

References

Bailey GB, Carneggie D, Kieffer H, Storey JC, Jovanovic VM, Wolfe RE (1997) Ground Control Points for calibration and correction of EOS ASTER, MODIS, MISR and Landsat 7 ETM+ Data, SWAMP GCP Working Group final report. USGS EROS Data Center, Sioux Falls, South Dakota

Barnes WL, Pagano TS, Salomonson VV (1998) Pre-launch characteristics of Moderate Resolution Imaging Spectroradiometer (MODIS) on EOS-AM1. IEEE Trans Geosci Remote Sens 36: 1,088 – 1,100

Bernstein R (1983) Image geometry and rectification. In: Colwell RN (ed) Manual of Remote Sensing, 2nd Edition. Am Soc of Photogrammetry, Falls Church, Virginia, pp 873 – 922

Emery WJ, Brown J, Nowak ZP (1989) AVHRR Image Navigation—Summary and Review. Photogrammetric Eng Remote Sens 55(8): 1,175 – 1,183

Fleig AJ, Hubanks PA, Storey JC, Carpenter L (1993) An analysis of MODIS Earth location error, Version 2.0. NASA GSFC, Greenbelt, Maryland

Folta, DC, Elrod B, Lorenz M, Kapoor A (1993) Precise navigation for the Earth Observing System (EOS)-AM1 spacecraft using the TDRSS onboard navigation system (TONS). Adv Astronautical Sci Univelt Inc., San Diego, California 84: 103 – 123

Glickman J, Hashmall J, Natanson G, Sedlak J, Tracewell D (2003) Earth Observing System (EOS) Aqua launch and attitude support experiences. Flight Mechanics Symposium Proceedings, NASA GSFC, Greenbelt, Maryland

Justice CO, Vermote E, Townshend JRG, Defries R, Roy DP, Hall DK, Salomonson VV, Privette JL, Riggs G, Strahler A, Lucht W, Myneni RB, Knyazikhin Y, Running SW, Nemani RR, Wan Z, Huete AR, van Leeuwen W, Wolfe RE, Giglio L, Muller JP, Lewis P, Barnsley MJ (1998) The Moderate Resolution Imaging Spectroradiometer (MODIS):

Land remote sensing for global change research. IEEE Trans Geosci Remote Sens 36: 1,228 – 1,249

Justice CO, Townshend JRG, Vermote EF, Masuoka E, Wolfe RE, Saleous, N, Roy DP, Morisette JT (2002) An overview of MODIS Land data processing and product status. Remote Sens Env 83: 3 – 15

Konecny G (1976) Mathematical models and procedures for the geometric registration of remote sensing imagery. Int Arch Photogrammetry Remote Sens 21: 1 – 33

Logan TL (1999) EOS/AM-1 Digital Elevation Model (DEM) data sets: DEM and DEM Auxiliary data sets in support of the EOS/Terra platform, JPL D-013508. Jet Propulsion Laboratory, California Institute of Technology, Pasadena, California

Maybec PS (1979) Stochastic models, estimation and control, Vol 1. Academic Press, New York

Moreno JF, Melia J (1993) A method for accurate geometric correction of NOAA AVHRR HRPT data. IEEE Trans Geosci Remote Sens 31: 204 – 226

Nishihama M, Wolfe RE, Solomon D, Patt FS, Blanchette J, Fleig AJ and Masuoka E (1997) MODIS Level-1A Earth location Algorithm Theoretical Basis Document Version 3.0, SDST-092. Lab Terrestrial Phys, NASA GSFC, Greenbelt, Maryland

Noredlinger PD, Klein L (1995) Theoretical basis of the SDP Toolkit Geolocation package for the ECS project, 445-TP-002-002. Hughes Applied Information Systems, Landover, Maryland, pp 1 – 1[illegible]

Rosborough GW, Baldwin DG, Emery WJ (1994) Precise AVHRR image navigation. IEEE Trans Geosci Remote Sens 32: 644 – 657

Roy DP, Singh S (1994) The importance of instrument pointing accuracy for surface bidirectional reflectance distribution mapping. Int J Remote Sens 15: 1,091 – 1,099

Roy DP, Devereux B, Grainger B, White S (1997) Parametric geometric correction of airborne thematic mapper imagery. Int J Remote Sens 18: 1,865 – 1,887

Roy DP (2000) The impact of misregistration upon composited wide field of view satellite data and implications for change detection. IEEE Trans Geosci Remote Sens 38: 2,017 – 2,032

Salomonson VV, Barnes WL, Maymon PW, Montgomery HE, Ostrow H (1989) MODIS: Advanced facility instrument for studies of the Earth as a system. IEEE Trans Geosci Remote Sens 27: 145 – 153

Schaaf CB, Gao F, Strahler AH, Lucht W, Li X, Tsang T, Strugnell N, Zhang X, Jin Y, Muller J-P, Lewis P, Barnsley M, Hobson P, Disney M, Roberts G, Dunderdale M, d'Entremont RP, Hu B, Liang S, Privette J, Roy D (2002) First Operational BRDF, Albedo and Nadir Reflectance Products from MODIS. Remote Sens of Env 83: 135 – 148

Schowengerdt RA (1997) Remote Sensing Models and Methods for Image Processing, Second edition. Academic Press, San Diego, California, pp 100 – 109

Standish EM, Newhall XX, Williams JG, Yeomans DK (1992) Orbital ephemerides of the Sun, Moon and Planets. In: Seidelmann PK (ed) Explanatory Supplement to the Astronomical Almanac. Univ Books, Mill Valley, California, pp 279 – 374

Townshend JRG, Justice CO, Gurney C, McManus J (1992) The impact of misregistration on change detection. IEEE Trans Geosci Remote Sens 30: 1,054 – 1,060

Vermote EF, El Saleous NZ, Justice CO (2002) Operational atmospheric correction of the MODIS data in the visible to middle infrared: first results. Remote Sens of Env 83: 97–111

Wertz RW (ed) (1978) Spacecraft attitude determination and control. D. Reidel Publishing Co., Dordrecht, Holland

Wolfe RE, Nishihama M, Fleig AJ, Kuyper JA, Roy DP, Storey JC and Patt FS (2002) Achieving sub-pixel geolocation accuracy in support of MODIS land science. Remote Sens Env 83: 31 – 49

Wolfe RE, Roy DP, Vermote E (1998) MODIS land data storage, gridding and compositing methodology: Level 2 Grid. IEEE Trans Geosci Remote Sens 36: 1,324 – 1,338

5 Introduction to MODIS Cloud Products

Bryan A. Baum and Steven Platnick

5.1 Introduction

The Earth's radiative energy balance and hydrological cycle are fundamentally coupled with the distribution and properties of clouds. Therefore, the ability to remotely infer cloud properties and their variation in space and time is crucial for establishing climatologies as a reference for validation of present-day climate models and in assessing future climate change. Remote cloud observations also provide data sets useful for testing and improving cloud model physics, and for assimilation into numerical weather prediction models.

The MODerate Resolution Imaging Spectroradiometer (MODIS) imagers on the Terra and Aqua Earth Observing System (EOS) platforms provide the capability for globally retrieving these properties using passive solar reflectance and infrared techniques. In addition to providing measurements similar to those offered on a suite of historical operational weather platforms such as the Advanced Very High Resolution Radiometer (AVHRR), the High-resolution Infrared Radiation Sounder (HIRS), and the Geostationary Operational Environmental Satellite (GOES), MODIS provides additional spectral and/or spatial resolution in key atmospheric bands, along with on-board calibration, to expand the capability for global cloud property retrievals.

The core MODIS operational cloud products include cloud top pressure, thermodynamic phase, optical thickness, particle size, and water path, and are derived globally at spatial resolutions of either 1- or 5-km (referred to as Level-2 or pixel-level products). In addition, the MODIS atmosphere team (collectively providing cloud, aerosol, and clear sky products) produces a combined gridded product (referred to as Level-3) aggregated to a 1° equal-angle grid, available for daily, eight-day, and monthly time periods. The wealth of information available from these products provides critical information for climate studies as well as the continuation and improved understanding of existing satellite-based cloud climatologies derived from heritage instruments.

This chapter provides an overview of the MODIS Level-2 and -3 operational cloud products. All products described in this chapter are available from the NASA Goddard Earth Sciences Distributed Active Archive Center (GES DAAC). However, the MODIS instrument has direct broadcast capability on both the Terra and Aqua platforms. Ground stations that obtain the MODIS data as the spacecraft passes within range, can obtain free software to generate the operational cloud products in near real-time. New applications using these near real-time cloud products are being explored by a number of users.

5.2 MODIS Instrument and Calibration

MODIS is a 36-band whiskbroom scanning radiometer currently flying on the NASA Terra and Aqua platforms. Terra was launched in December, 1999, and is in a (daytime) descending orbit with an equatorial crossing of 1030 local solar time. Aqua, launched in May 2002, is in an ascending orbit (daytime) with a 1330 local crossing time. The instrument is comprised of four focal planes covering the spectral range 0.42 – 14.24 μm, with each spectral band defined by an interference filter. Spatial resolution for nadir views varies from 250 m to 1 km, depending on the spectral band. MODIS has several onboard instruments for in-orbit radiometric and spectral characterization. A solar diffuser panel is used for reflectance calibration for bands with wavelengths from 0.45 through 2.1 μm, and an accompanying diffuser stability monitor is used to assess the stability of the diffuser at wavelengths up to 1 μm. Thermal spectral bands are calibrated with an onboard blackbody.

This chapter will describe briefly the comprehensive set of remote sensing algorithms used by the MODIS atmosphere science team to infer cloud properties. The description begins with pixel-level geolocated and calibrated (radiance and/or reflectance-based) data known as Level-1B (L1B). Pixel-level cloud properties, at either 1-km or 5-km resolution, are based on the L1B data and ancillary data streams discussed later in this chapter. The pixel-level cloud product is referred to as Level-2, and is typically stored for a 5-minute granule of data. Based on the Level-2 product, a set of global gridded statistics products are developed and are labeled as Level-3.

The cloud property retrieval algorithms depend critically on accurate characterization and calibration of the L1B radiance and reflectance data from both Terra and Aqua MODIS imagers. Considerable time and attention have been given in the post-launch era to understanding and mitigating performance issues (through characterization) such as longwave infrared (IR) band cross-talk, scan mirror reflectance variation with scan angle, thermal leakage into reflective band measurements, and detector striping. Part of this effort has been to verify the radiometric integrity of MODIS measurements (Moeller et al., 2003). As performance issues have been addressed, adjustments have been incorporated into the MODIS L1B production code for use in the various processing streams (known as a Collection) of MODIS L1B and Level-2 products. Each collection includes reprocessing of historical MODIS data and forward processing of the ongoing measurements, until the beginning of the next collection. Each collection uses the best available post-launch instrument characterization, as well as science product algorithm improvements such as improved treatment of surface radiative characteristics (spectral albedo, emissivity, and skin temperature), mitigation of day-night science product biases, and improved production code efficiency.

For example, pre-launch testing of the MODIS proto-flight model (PFM) on Terra revealed a light leak from the infrared window (band 31, 11 μm) into bands 32 – 36 (wavelengths ranging from 12 to 14.3 μm). To complicate matters, the leak included both spatial and spectral components. The radiometric impact for typical radiances in these bands ranged from less than 1% in band 32 to more than 10% in band 36. These impacts are near or exceed the MODIS radiometric accuracy specification of 0.5% (band 32) and 1% (bands 33 – 36) at typical scene temperatures. A simple linear correction algorithm was developed to reduce the contamination in these bands.

As another example, the Terra MODIS scan mirror Response Versus Scan angle (RVS) characterization was not measured at the system level in pre-launch tests. Shortly after launch, it was found that an unexpectedly large asymmetry was present in the MODIS longwave infrared (LWIR) bands. In other words, there was a difference in the radiances from one side of the scan to the other, and this in turn led to systematic biases in several cloud products as a function of scan. A Terra deep space maneuver was performed in 2003; the resulting high quality RVS characterization was applied to the operational L1B processing algorithm and effectively removed the RVS asymmetry in the LWIR bands.

With careful and continual evaluation of the instrument and cloud products, production runs are of much higher quality than earlier versions. Up-to-date information on Terra and Aqua algorithm versions and reprocessing efforts are given on the MODIS atmosphere group Web site (http://modis-atmos.gsfc.nasa.gov/products_calendar.html).

5.3 Level-2 Cloud Products

The MODIS cloud products are generated on a granule basis; a granule is 5 minutes of data and typically consists of 2,030 along-track pixels. The suite of operational cloud products (Platnick et al., 2003) begins with cloud detection or masking, i.e., deciding whether or not a cloud is present. Infrared techniques are employed to estimate cloud top pressure, effective cloud amount (product of cloud fraction and cloud emittance), and cloud thermodynamic phase. In daytime data, cloud optical thickness and effective particle size are provided using solar reflectance techniques. A cirrus reflectance retrieval is provided separately. The following sections provide background on the methodology used to infer these various parameters. With the exception of the cloud mask, all cloud products are archived in a single Hierarchical Data Format (HDF) file with the product designation MOD06; the cloud mask product is in MOD35.

5.3.1 Cloud Masking

The focus of the product is to indicate a level of confidence as to whether the satellite has an unobstructed field-of-view (FOV) to a pixel's location on the surface. The MODIS cloud mask product serves as the primary ancillary input to the other cloud algorithms (Ackerman et al., 1998). In addition to the potential for obstruction in the line of sight due to clouds, heavy aerosols (e.g., smoke) and dust will also act to decrease the likelihood of finding clear-sky conditions. The product provides more information than a simple yes/no decision; there are 48 bits of output per 1-km pixel that includes information on sets of multispectral test results, the processing path, and limited ancillary information such as a land/ocean tag. The first eight bits provide a summary sufficient for most applications. Additionally, the first two bits simply offer information in four categories: confident clear, probably clear, uncertain/probably cloudy, and cloudy.

The algorithm uses a variety of multispectral tests involving combinations of up to 19 spectral bands. The use of these bands changes somewhat as calibration issues are mitigated. Different sets of tests are applied depending on the surface (land, water, snow/ice, desert, and coast) and solar illumination (day/twilight/night). In addition to the multispectral tests, a textural test is applied over ocean when initial clear-sky tests are inconclusive.

Several ancillary data sets are used in the cloud detection process. Surface snow and ice data are provided at 25 km resolution by the Near Real-Time Ice and Snow Extent (NISE) product from the National Snow and Ice Data Center. The NCEP Reynolds blended SST product (Reynolds and Smith, 1994) has been implemented to improve the product at nighttime over oceans and in areas where there are strong temperature gradients such as in the vicinity of the Gulf Stream.

5.3.2 Cloud Thermodynamic Phase

There are currently three inferences of cloud phase found in the MOD06 cloud product: (1) a bispectral IR algorithm stored as a separate Science Data Set (SDS), (2) a set of shortwave IR (SWIR) tests, and (3) a decision tree algorithm that includes cloud mask results as well as the IR and SWIR tests. The latter two phase retrievals are stored in the MODIS "Quality_Assurance_1km" output SDS and not as individual SDS's. The decision tree algorithm provides the phase used in the subsequent optical and microphysical retrieval. The current IR phase algorithm is at a 5 km spatial resolution, while the other two are at 1 km. This section will summarize the IR phase retrieval (Strabala et al., 1994; Baum et al., 2000b). Details on the SWIR and decision tree phase tests are beyond the scope of this chapter, but can be found in (Platnick et al., 2003).

The IR phase retrieval provides four categories: ice, water, mixed phase, and uncertain. A "mixed phase" cloud is thought to consist of a mixture of ice and water particles. With the IR-based method, cloud phase is inferred from the brightness temperature difference (BTD) between the 8.5 and 11 μm brightness temperatures (BTD[8.5 – 11]) as well as the 11 μm brightness temperature. The physical basis for this approach stems from the observation that the imaginary component of the index of refraction differs for ice and water at these two wavelengths. The BTD[8.5 – 11] is affected by atmospheric water vapor absorption, surface emissivity, and cloud particle size (small particles scatter more radiation than large particles). Radiative transfer simulations show that for ice clouds, the BTD[8.5 – 11] values tend to be positive in sign, whereas for low-level water clouds, the BTD[8.5 – 11] values tend to be very negative. This simple bispectral IR technique is adequate for classifying the phase as either "ice" or "water" for about 80% of the cloudy pixels on a global basis. The most problematic areas are optically thin cirrus, multilayered clouds (especially thin cirrus over lower-level water clouds), and single-layered clouds having cloud top temperatures between 233 K and 273 K (i.e., supercooled water or "mixed phase" clouds). Supercooled water or mixed-phase clouds tend to occur most frequently in the high latitude storm belts of both hemispheres.

To improve the inference of cloud phase during daytime, a decision tree algorithm is implemented during the processing step where cloud optical thickness and effective particle size is inferred (Platnick et al., 2003; King et al., 2004). The decision tree begins with the cloud mask results as well as the bispectral IR cloud phase results, and incorporates tests using reflectances obtained at a visible wavelength (e.g., 0.65 μm) and a shortwave-infrared (SWIR) wavelength (e.g., 1.64 μm or 2.15 μm). At wavelengths less than about 0.7 μm, clouds composed of either liquid or ice tend to absorb very little solar radiation. At the SWIR wavelength, the imaginary index of refraction values for both water and ice increase in comparison with those at the visible wavelength. However, the values for ice and water diverge from each other, with ice having a higher imaginary index of refraction than that of water. Even with these supplementary tests, mixed-phase clouds remain a challenge.

5.3.3 Cloud Top Pressure and Effective Cloud Amount

For the past several decades, a technique known as CO_2 slicing has been used to infer cloud-top pressure and effective cloud amount (the product of the cloud fraction and the cloud emittance) from radiances measured in spectral bands located within the broad 15-μm CO_2 absorption region. As the wavelength increases from 13.3 μm to 15 μm, the atmosphere becomes more opaque due to CO_2 absorption, thereby causing radiances obtained from these spectral bands to be sensitive to a different portion of the atmosphere. This technique has been applied to data from the High resolution Infrared Radiometer Sounder (HIRS; Wylie and Menzel, 1999) as well as the Geostationary Operational Environmental Satellite

(GOES) sounder (Menzel et al., 1992; Menzel and Purdom, 1994). The field-of-view (FOV) size for HIRS at nadir is approximately 18 km and for GOES is 10 km. MODIS provides measurements at 1-km resolution and at four wavelengths located in the broad 15-μm CO_2 band, but cloud top properties are produced at 5-km spatial resolution.

The MODIS cloud pressure is converted to cloud temperature through the use of gridded meteorological products that provide temperature profiles every 6 hours. The product used for this purpose is provided by the NCEP Global Data Assimilation System (GDAS; Derber et al., 1991). It has been noticed that differences exist between measured and calculated (i.e., model derived) clear-sky radiances. Further discussion on an effort to mitigate this issue is presented in Section 5.5.4.

There are many benefits to the CO_2 slicing algorithm, foremost of which is its unique heritage and application to data spanning a record of more than 25 years. Cloud properties are derived similarly for both daytime and nighttime data as the IR method is independent of solar illumination. This approach is very useful for the analysis of mid-level to high-level clouds, and especially semi-transparent clouds such as cirrus. One constraint to the use of the 15-μm channels is that the cloud signal (change in radiance caused by the presence of cloud) becomes comparable to instrument noise for optically thin clouds and for clouds occurring in the lowest 3 km of the atmosphere. When low clouds are present, cloud top temperature is adjusted until agreement is achieved between model-calculated and observed 11-μm radiances; cloud top pressure is then estimated by matching the temperature to the GDAS profile.

Figure 5.1 shows a true color image of an Aqua scene from 20 November 2000 at 1710 UTC, encompassing north-central South America, specifically Trinidad, Venezuela, Guyana, Suriname, French Guiana, and Brazil. This true-color image is created by superimposing three bands: 0.65 μm (Band 1 in red), 0.55 μm (band 4 in green), and 0.47 μm (band 3 in blue). Clouds are white but display a variety of textures and spatial scales. The land surfaces appear as green or brown depending on the amount of surface vegetation, while deep-ocean appears as dark blue. Shallow coastal waters are light blue or brown in appearance. Retrieved cloud-top pressures (Fig.5.2(a)) and temperatures (Fig.5.2(b)) are obtained at 5-km spatial resolution. From the range of cloud-top temperature, one gains a sense of the complexity inherent in such retrievals. Not only do the cloud properties change over short horizontal distances, but often high- and low-level clouds co-exist.

5.3.4 Cloud Optical and Microphysical Properties

Cloud optical thickness is defined as the vertical integration of extinction over the cloud physical thickness. For water clouds composed of spherical particles, effective particle size is defined as the ratio of the third moment to the second

Figure 5.1 True color image of an Aqua scene from 20 November, 2000 at 1710 UTC, encompassing north-central South America, specifically Trinidad, Venezuela, Guyana, Suriname, French Guiana, and Brazil. This true color image is created by superimposing three bands: 0.65 μm (Band 1 in red), 0.55 μm (Band 4 in green), and 0.47 μm (Band 3 in blue). Clouds are white but display a variety of textures and spatial scales. The land surfaces appear as green or brown depending on the amount of surface vegetation, while deep ocean appears as dark blue. Shallow coastal waters are light blue or brown in appearance

moment of the particle size distribution. The definition of effective particle size for ice clouds is made more difficult because ice particles tend to be non-spherical. For ice clouds, the effective particle size is defined as being proportional to the ratio of the total volume to the projected area of the ice particles for a given size and habit distribution.

The simultaneous retrieval of optical thickness and effective particle size derived from cloud reflectance measurements in solar band atmospheric windows is well known. MODIS retrievals are performed using a band that is practically non-absorbing for bulk water/ice (0.65, 0.86, or 1.2 μm) combined with three longer wavelength bands where bulk water/ice has significant absorption (1.6, 2.1, and 3.7 μm). Three separate effective sizes are provided in MOD06 corresponding to each of these absorbing bands. The 3.7-μm band includes a significant thermal emission component in addition to the solar reflectance component. A proper accounting of the thermal emission must include consideration of cloud top temperature, emissivity (a function of cloud microphysics), and above-cloud atmospheric emission and transmittance. In addition, for clouds with optical thicknesses less than about 6, transmittance of surface emission

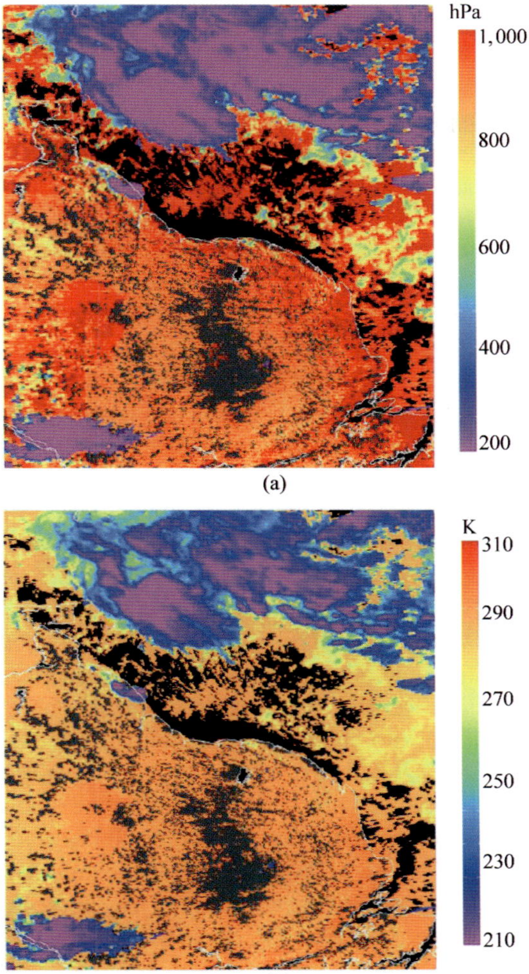

Figure 5.2 (a) Retrieved cloud-top pressure in hPa and (b) cloud-top temperature in Kelvin obtained at 5-km spatial resolution for the scene in Fig. 5.1

through the cloud can be important. Because of these additional complexities, the 3.7-μm retrievals are thought to be of higher uncertainty than those from the other two bands. However, the 3.7-μm results are included in the MOD06 product because of this band's heritage on earlier satellite imagers such as the AVHRR.

The 2.1-μm band retrieval of effective particle size for water clouds is preferred to the 1.6-μm retrieval because of the increased absorption (the imaginary index of refraction is about an order of magnitude higher for water at 2.1 than at 1.6 μm). Additionally, more than half of the 1.6-μm detectors are inoperative on the Aqua platform, rendering the use of this band problematic. The 2.1-μm band retrieval of effective particle size is the default quantity used in

Level-3 aggregations, which are discussed later in this chapter. MODIS is the first satellite imager to take measurements in each of these absorbing bands simultaneously.

The retrievals are based on library calculations of plane-parallel homogeneous clouds overlying a black surface in the absence of an atmosphere. Separate libraries have been built for water and ice clouds; currently no separate library exists for mixed phase clouds. The bulk scattering properties for the ice clouds are based on mixtures of hexagonal plates, columns, bullet rosettes, and aggregates (Baum et al., 2000a; King et al., 2004) and a set of particle size distributions developed from midlatitude cirrus field campaigns.

When clouds are not opaque, reflected sunlight for overcast pixels can be significantly affected by the reflectivity of the underlying surface. Over land, the surface albedo is highly variable, both spectrally and with surface type. Over water, surface reflectance can increase significantly due to sun glint. Ice- and snow-covered surfaces are bright at a visible wavelength but dark in SWIR bands.

Realistic estimates of the surface reflectance beneath clouds, as well as corrections for atmospheric transmittance, are now employed on a pixel basis during operational processing. Towards this end, the MODIS operational surface spectral albedo/BRDF (bidirectional reflectance distribution function) product (MOD43) provides 16-day composites of clear-sky observations at 1-km spatial resolution for both BRDF and albedo. The spectral albedo product includes both solar illumination and diffuse sky values. All further references to surface albedo refer to the diffuse sky albedo since this is the quantity most relevant to cloud property retrievals. More specifically, diffuse sky is officially referred to as the "white sky albedo" in the MOD43 product, and means the albedo with lambertian incident intensity. These properties are provided for all the relevant MODIS solar bands (except at 3.7 μm). From a 16-day composite available at the time of the last major algorithm delivery, data are composited and aggregated by land cover type to determine the extent to which ecosystem was useful as a predictor of spectral albedo. The MODIS land cover product (MOD12) is used in this effort. Dispersions in spectral diffuse albedo were generally found to be less than 20% for any given ecosystem. With the ecosystem-based albedo aggregations in three separate latitude bands, a sinusoidal fit was made between the summer/winter extremes to replicate the seasonal cycle. An upcoming delivery will include spectral albedo composited from a year of MOD43 data, and is discussed further in Section 5.5.3. For operational processing, snow and ice cover is provided by the NISE product; when snow or ice is present, spectral albedos are provided from field measurements. Further details regarding the MODIS optical and microphysical retrieval algorithm are described in Platnick et al. (2003).

Figure 5.3 shows the retrieved optical thickness (Fig. 5.3(a)) and effective particle size (Fig. 5.3(b)) for ice and water clouds for the same scene shown in Fig. 5.1. These results are obtained at 1-km spatial resolution in contrast to the cloud top pressure at 5-km resolution. Inherent in this image is the classification of cloud thermodynamic phase. Ice cloud optical thickness and effective particle radius values are shown on a logarithmic scale in colors ranging from purple to green, while water cloud values are displayed in colors ranging from yellow to red.

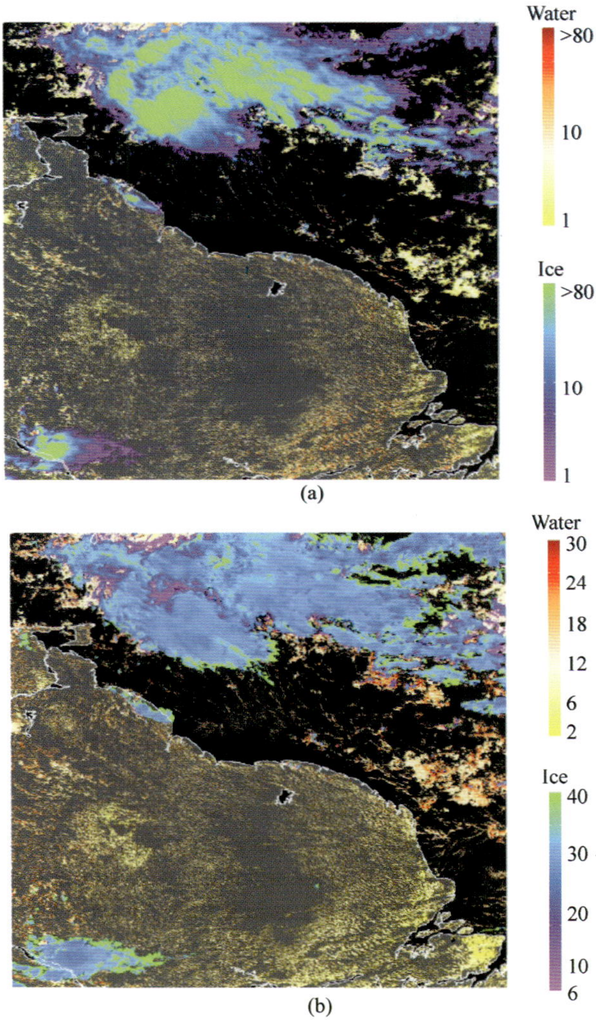

Figure 5.3 (a) Retrieved cloud optical thickness and (b) effective particle size for both ice- and water-phase clouds at 1-km spatial resolution for the scene in Fig. 5.1

Over land, note that the lower clouds are cellular, likely having sub-pixel scales that violate the plane-parallel assumption (i.e., homogeneous in the horizontal plane) used by both the optical and microphysical retrieval algorithm and the 11-μm cloud top temperature algorithm (used for low clouds). Partly cloudy pixels will fundamentally impact the cloud mask, cloud thermodynamic phase, and cloud height retrievals as well.

5.3.5 Cirrus Reflectance Algorithm

One type of ice cloud that is particularly difficult to quantify from satellite data is semitransparent cirrus due to its wide range of microphysical (e.g., particle size and shape), macrophysical (height and temperature), and optical properties (bulk scattering characteristics). The variation in these cirrus properties results in large spatial and temporal variability of optical thickness and emittance. MODIS has a unique spectral reflectance channel at 1.38-μm that is centered near a strong water vapor absorption band. The upwelling radiance is attenuated by the low-level water vapor so that there is little or no contribution by the radiation reflected by the surface and scattered by the low-level particulate atmospheric constituents (e.g., dust and aerosols). The reflectance measured at this wavelength is primarily from the presence of mid- to high-level clouds. Once the high-level cloud (cirrus) reflectance component is identified, the equivalent reflectance in the MODIS 0.65-μm band is inferred through an empirical technique. Further details may be found in Gao et al. (2002).

5.4 Global Gridded (Level-3) Products

Once the Level-2 granule-level cloud products have been produced, spatial and temporal composites are aggregated to daily, eight-day, and monthly files. Statistics are sorted onto a $1° \times 1°$ equal-angle grid (row by column) containing 180×360 individual cells. The Level-3 atmosphere product, MOD08, is derived separately for Aqua and Terra. For the daily product, every Level-2 granule that overlaps any part of the data day, defined as being from 0000 to 2400 UTC, is included in the temporal compositing process. A granule that spans either 0000 UTC or 2400 UTC may be included in two consecutive MOD08 daily products. The eight-day product is derived from the daily Level-3 products summarized over eight consecutive days. The eight-day intervals are reset at the beginning of each year similarly to the Level-3 products produced by the MODIS ocean and land discipline groups. The monthly product provides a summary of the daily products obtained over a calendar month.

While there is no separation of the cloud parameters by ascending or descending node, there is a day/night separation and a process/no process decision for a number of parameters. Cloud fraction (from the cloud mask) and cloud top properties (i.e., cloud pressure and IR phase) are processed for both day and night and are provided in the Level-3 products as daytime only, nighttime only, and combined day and night. Cloud optical and microphysical properties are summarized for daytime only since they are not derived at night. The cloud fraction derived from the cloud mask is currently provided in the SDS names beginning with "Cloud_Fraction". A daytime-only cloud fraction is derived also, from successful optical and microphysical retrievals, and aggregated by cloud thermodynamic phase. As an example, the SDS name for the liquid phase cloud fraction is "Cloud_Fraction_Water" (Platnick et al., 2003).

An example of the Level-3 monthly product is shown in Fig. 5.4 for liquid water clouds from MODIS Terra in April 2003. Cloud optical thickness is shown in the top image and effective particle radius in the bottom image. These particular aggregation statistics are weighted by the quality assurance (QA) assignments given to each retrieved pixel. Unweighted statistics are also reported in the Level-3 file. Water cloud optical thicknesses are seen to be larger in the midlatitudes and poleward; especially noticeable are the large values throughout the southern oceans. Effective radius retrievals are seen to be generally smaller over the continents. Few results are shown for Antarctica at this time of year because the retrievals are performed only for daytime observations.

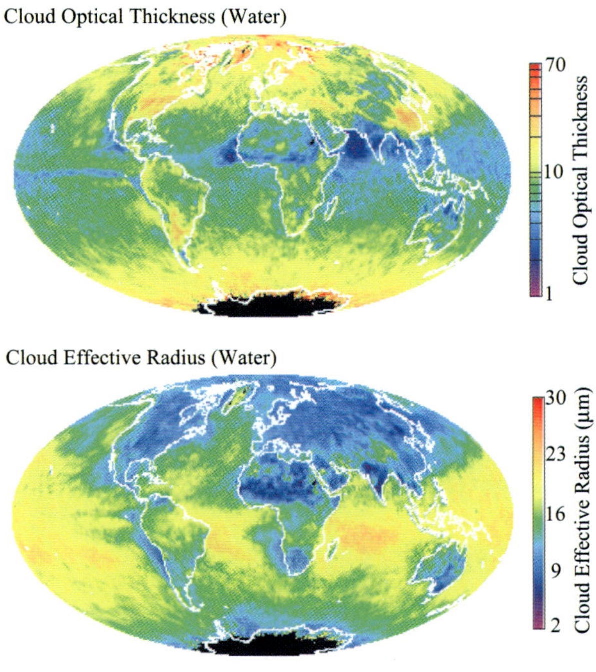

Figure 5.4 MODIS Terra Level-3 April 2003 monthly aggregations for liquid water cloud retrievals of optical thickness (top) and effective particle radius (bottom)

The Level-3 products also include histograms of cloud parameters and joint histograms derived from comparison of two parameters. Approximately 13 joint histograms are derived from the MOD06 cloud product. Note that the histogram counts are not weighted by the retrieval QA. Ten joint histograms are aggregated by cloud phase (5 for liquid water clouds and 5 for ice clouds). The joint histograms are computed from the following list: cloud optical thickness, effective particle radius, cloud-top temperature, and effective cloud amount. The other three joint histograms are aggregated by solar illumination (day, night, or combined day and night) and are built from cloud-top pressure and effective cloud amount. Examples of joint histograms for Terra and Aqua are shown in Fig. 5.5. These examples are derived from data collected in August, 2003, between 10° – 20° S, 70° – 85° W (Peru/Humboldt current regime), and show the joint histogram of liquid water cloud optical thickness and effective particle radius. The Level-3 product bin sizes are used to form the joint histogram. Further Level-3 algorithm and aggregation details, along with example images, can be found in King et al. (2003).

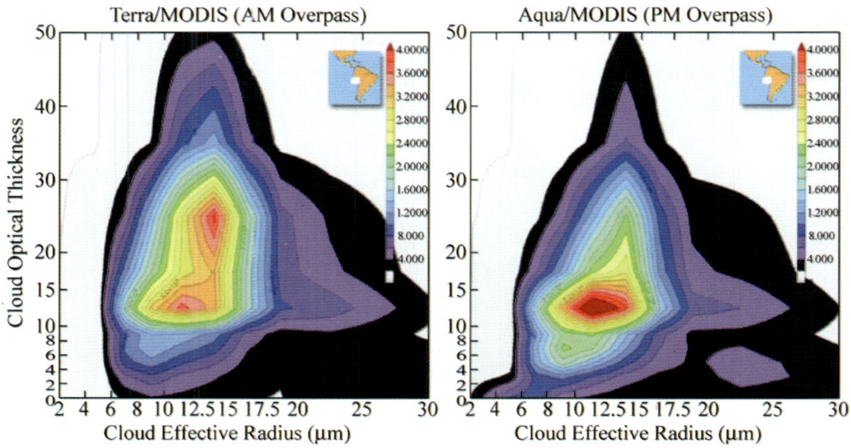

Figure 5.5 Joint histogram of liquid water cloud optical thickness and effective particle size from products derived in August 2003, between 10° – 20°S, 70° – 85°W (coastal Peru/Chile)

5.5 Future Algorithm Efforts

A number of algorithm refinement and research efforts are currently being pursued. In this section we describe a few of these activities.

5.5.1 Detection of Multilayered Clouds

Satellite-based cloud property retrievals are performed under the assumption that observations for a given field of view contain a single cloud layer. The retrieved cloud top properties will contain the least error if the uppermost cloud layer is optically thick (i.e. opaque). However, observations indicate that multilayered clouds are common, especially for the case in which semi-transparent ice cloud overlies lower-level water clouds. In this situation, the assumption of a single cloud layer results in a cloud top pressure that lies between that of the upper and lower cloud layers. Error in the cloud top pressure propagates into errors in the inferred optical and microphysical properties.

New research is being conducted to determine when cirrus overlies water clouds in daytime and nighttime conditions (Baum et al., 2003; Nasiri and Baum, 2004). While many approaches are being developed and tested, the daytime algorithm described here is based on SWIR bands at 1.6 / 2.1 and 11 μm, respectively. As noted previously in the discussion on cloud phase, ice particles absorb more radiation than water particles at the SWIR wavelengths (1.6 / 2.1 μm). Ice clouds also tend to reside at much higher altitudes than water clouds. The specific assumptions invoked for each pixel array (nominally 200×200 pixels) are that (a) at most two distinct cloud layers are present, (b) any pixels not uniquely associated with either of the two distinct cloud layers are classified as being multilayered, (c) clouds of both ice and water phase are present, (d) knowledge of the clear-sky SWIR/IR properties are known, and (e) a distance of at least 2 km in height between the layers is required to identify the separate layers.

The MOD35 cloud mask product provides information regarding clear and cloudy pixels, and cloud thermodynamic phase is assessed using the 8.5 μm and 11 μm brightness temperatures. Although either the 1.6- or 2.1-μm reflectance bands can be used in the multilayered cloud detection technique, the 2.1-μm band is used for the Aqua data because a number of the 1.6-μm band detectors on the Aqua MODIS instrument are inoperative (the band is comprised of 20 total detectors). To provide a confidence level for the assessment of whether a pixel contains more than one cloud layer, the method is applied as the pixel array, or tile, is moved gradually across the data granule, thereby testing each pixel (away from the granule borders) multiple times. The more times a pixel is flagged as containing multiple cloud layers, the higher the confidence in that assessment.

5.5.2 Improved Ice Cloud Microphysical and Optical Models

One major difference between midlatitude and tropical cirrus is that the tropical cirrus formed near centers of deep convection in mesoscale systems tend to contain more particles of larger sizes than the synoptically-generated cirrus in the midlatitudes. For most synoptically-generated midlatitude cirrus, the largest particles sizes measured are generally less than 800 μm. Larger particles tend to settle out quickly due to the relatively low updraft velocities in the cloud layer. Updraft velocities tend to be much higher in mesoscale systems associated with the generation of tropical cirrus (e.g. anvils). In situ measurements show that, in addition to high quantities of small particles, many large particles (greater than 1 mm in diameter) are present even in the uppermost regions of tropic cirrus anvils (Heymsfield et al., 2002).

Research is continuing towards the development of bulk scattering properties over a wider range of particle habit and size distributions (Yang et al.,

2001, 2003; Heymsfield et al., 2002, 2003; Nasiri et al., 2002), which will in turn expand the bulk scattering model libraries used for cloud remote sensing efforts. Scattering properties are now available for a much larger set of ice particles and include droxtals, hexagonal plates, solid columns, hollow columns, aggregates, and two- and three-dimensional bullet rosettes. Research is also underway to perform scattering calculations for more realistic large polycrystals such as those found in tropical cirrus.

5.5.3 Improved Land Spectral Albedo Maps

As discussed, the optical and microphysical retrieval algorithm requires surface albedo in visible and SWIR bands as the boundary condition for cloud radiative transfer modeling over land. Previously, a single 16-day composite of spectral albedo from the operational MOD43 product was available, requiring questionable assumptions on seasonal changes. A year's worth of MOD43 16-day spectral albedo composites have been archived in the GES DAAC since the last major optical and microphysical algorithm software delivery. With these data, a complete annual surface spectral albedo map is now available for use in the next production collection. Seasonality is built into the annual albedo maps. Cloud and seasonal snow cover, however, curtail retrievals to approximately half the global land surfaces on an annual equal-angle basis, precluding MOD43 albedo products from direct assimilation into the optical and microphysical production environment.

A temporal interpolation technique has been developed to fill missing data in the operational albedo product by imposing pixel-level and local regional ecosystem-dependent phenological behavior onto retrieved pixel temporal data in such a way as to maintain pixel-level spatial and spectral detail (Moody et al., 2005). The resulting value-added product provides spatially complete surface albedo maps and statistics for both direct and diffuse illuminations. Data are stored on one-minute and coarser resolution equal-angle grids for the first seven MODIS wavelengths (0.47 μm through 2.1 μm) and for three broadband wavelengths (0.3 – 0.7 μm, 0.3 – 5.0 μm and 0.7 – 5.0 μm).

5.5.4 Clear-Sky Radiance Maps

An effort is being made to understand and mitigate differences between calculated and measured clear-sky infrared radiances to avoid modest cloud-top pressure assignment errors. These radiance differences can lead to errors in retrieved cloud pressures (Wielicki and Coakley, 1981). The primary cause of errors in the calculated clear-sky radiance is in the assignment of surface temperature from the gridded meteorological product, especially over land. The difficulty in assigning a realistic surface temperature over land is not unexpected given the range of diurnal variation, especially in areas of sparse vegetation.

Both daily and 8-day mean differences between observed and calculated clear-sky radiances will be generated for all MODIS mid- and long-wave infrared bands beginning with the MODIS Collection 5 processing. An example of 8-day mean biases (April 1 – 8, 2003) generated for Terra bands 31 and 34 is shown in Fig. 5.6. Clear-sky pixels (1-km resolution) are determined from cloud mask output (MOD35). Calculated radiances are derived from the 101-level Pressure-Layer Fast Algorithm for Atmospheric Transmittance (PFAAST; Strow et al., 2003). Application of the PFAAST requires temperature, moisture, and ozone profiles. The GDAS gridded product provides the temperature and moisture profiles while climatologically-averaged profiles of ozone are used. Biases will be accumulated on a global 25-km equal area grid.

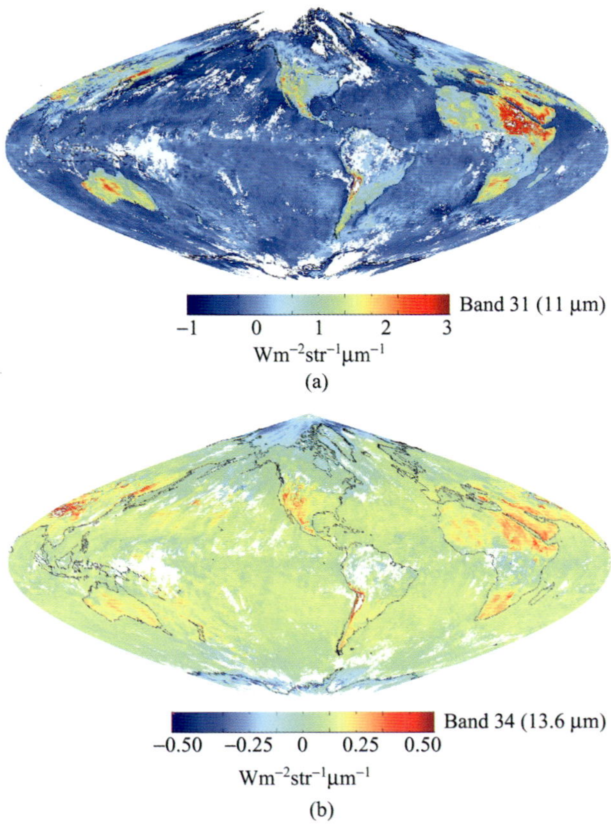

Figure 5.6 Two examples of mean daytime Terra 8-day biases are shown: (a) 11 μm (Band 31) and (b) 13.6 μm (Band 34). Notice that in the Band 31 example, large differences occur over land surfaces where diurnal heating is not adequately captured by input surface temperatures. The Band 34 example indicates that perhaps upper tropospheric temperatures or tropopause locations were not accurate in this time period (April 1 – 8, 2003). Figures depict observed minus calculated radiances

5.6 Summary

This chapter provides a brief overview of the pixel-level (Level-2) and gridded (Level-3) MODIS operational cloud products. The core operational cloud products include cloud top pressure, thermodynamic phase, optical thickness, effective particle radius, and water path. These products are derived globally at spatial resolutions of either 1 or 5 km. Alorithm descriptions and data presented in Sections 5.3 and 5.4 are applicable to the MODIS Atmosphere Team Collection 4 processing stream. The efforts described in Section 5.5 were written before software was finalized for the Collection 5 processing stream that began in April 2006 and therefore may not accurately represent the final operational algorithm status (e.g., multiplayer cloud detection algorithm), nor should the section be considered a complete summary of all Collection 5 changes and updates. Further information including documentation, algorithm details and history, and quicklook imagery can be found at the MODIS atmosphere team Web site: http://modis-atmos.gsfc.nasa.gov.

References

Ackerman SA, Strabala KI, Menzel WP, Frey, RA, Moeller CC, Gumley LE (1998) Discriminating clear sky from clouds with MODIS. J Geophys Res 103: 32,141 – 32,157

Baum BA, Kratz DP, Yang P, Ou S, Hu YX, Soulen PF, Tsay SC (2000a) Remote sensing of cloud properties using MODIS Airborne Simulator imagery during SUCCESS. I. Data and models. J Geophys Res 105: 11,767 – 11,780

Baum BA, Soulen PF, Strabala KI, King MD, Ackerman SA, Menzel WP, Yang, P (2000b) Remote sensing of cloud properties using MODIS Airborne Simulator imagery during SUCCESS. II. Cloud thermodynamic phase. J Geophys Res 105: 11,781 – 11,792

Baum BA, Frey RA, Mace GG, Harkey MK, Yang P (2003) Nighttime multilayered cloud detection using MODIS and ARM data. J Appl Meteor 42: 905 – 919

Derber JC, Parrish DF, Lord SJ (1991) The new global operational analysis system at the National Meteorological Center. Weather Forecasting 6: 538 – 547

Gao BC, Yang P, Han W, Li RR, Wiscombe WJ (2002) An algorithm using visible and 1.38 μm channels to retrieve cirrus cloud reflectances from aircraft and satellite data. IEEE Trans Geosci Remote Sens 40: 1,659 – 1,668

Heymsfield AJ, Bansemer A, Field PR, Durden SL, Stith JL, Dye JE, Hall W, Grainger CA (2002) Observations and parameterizations of particle size distributions in deep tropical cirrus and stratiform precipitating clouds: Results from in situ observations in TRMM field campaigns. J Atmos Sci 59: 3,457 – 3,491

Heymsfield AJ, Matrosov S, Baum BA (2003) Ice water path-optical depth relationships for cirrus and deep stratiform ice cloud layers. J Appl Meteor 42: 1,369 – 1,390

King MD, Menzel WP, Kaufman YJ, Tanré D, Gao BC, Platnick S, Ackerman SA, Remer LA, Pincus R, Hubanks PA (2003) Cloud, Aerosol and Water Vapor Properties from

MODIS. IEEE Trans Geosci Remote Sens 41: 442 – 458

King MD, Platnick S, Yang P, Arnold GT, Gray MA, Riédi JC, Ackerman SA, Liou KN (2004) Remote sensing of liquid water and ice cloud optical thickness, and effective radius in the arctic: Application of airborne multispectral MAS data. J Atmos Oceanic Technol 21: 857 – 875

Menzel WP, Wylie DP, Strabala KI (1992) Seasonal and diurnal changes in cirrus clouds as seen in four years of observations with the VAS. J Appl Meteor 31: 370 – 385

Menzel WP, Purdom JFW (1994) Introducing GOES-I: The first of a new generation of Geostationary Operational environmental Satellites. Bull Amer Meteor Soc Vol. 75 No 5: 757 – 781

Moeller CC, Revercomb HE, Ackerman SA, Menzel WP, Knuteson RO (2003) Evaluation of MODIS thermal IR band L1B radiances during SAFARI 2000. J Geophys Res 108:D13: 8,494

Moody EG, King MD, Platnick S, Schaaf CB, Gao F (2005) Spatially complete surface albedo data sets: Value-added products derived from Terra MODIS land products. IEEE Trans Geosci Remote Sens 43: 144 – 158

Nasiri SL, Baum BA, Heymsfield AJ, Yang P, Poellot M, Kratz DP, Hu YX (2002) Development of midlatitude cirrus models for MODIS using FIRE- Ⅰ, FIRE- Ⅱ, and ARM in-situ data. J Appl Meteor 41: 197 – 217

Nasiri SL, Baum BA (2004) Daytime multilayered cloud detection using multispectral imager data. J Atmos Oceanic Tech 21: 1,145 – 1,155

Platnick S, King MD, Ackerman, SA, Menzel WP, Baum BA, Riédi JC, Frey RA (2003) The MODIS cloud products: algorithms and examples from Terra. IEEE Trans Geosci Remote Sens 41: 459 – 473

Reynolds WR, Smith TM (1994) Improved global sea surface temperature analyses using optimum interpolation. J Climate 7: 929 – 948

Strabala KI, Ackerman SA, Menzel WP (1994) Cloud properties inferred from 8 – 12 μm data. J Appl Meteorol 2: 212 – 229

Strow L, Hannon S, Machado S, Motteler H, Tobin D (2003) An overview of the AIRS radiative transfer model. IEEE Trans Geosci Remote Sens 41: 303 – 313

Wielicki, BA, Coakley JA (1981) Cloud retrieval using infrared sounder data: Error analysis. J Appl Meteorol 20: 157 – 169

Wylie DP, Menzel WP (1999) Eight years of global high cloud statistics using HIRS. J Climate 12: 170 – 184

Yang P, Gao BC, Baum BA, Wiscombe W, Hu YX, Nasiri SL, Heymsfield A, McFarquhar G, Miloshevich L (2001) Sensitivity of cirrus bidirectional reflectance to vertical inhomogeneity of ice crystal habits and size distributions. J Geophys Res 106: 17,267 – 17,291

Yang P, Baum BA, Heymsfield AJ, Hu YX, Huang HL, Tsay SC, Ackerman S (2003) Single scattering properties of droxtals. J Quant Spectrosc Radiant Transfer 79 – 80: 1,159 – 1,169

6 MODIS Observation of Aerosol Loading from 2000 to 2004

D. Allen Chu and Lorraine Remer

6.1 Introduction

Aerosol optical depth (AOD) is a measure of aerosol loading in an atmospheric column. The satellite-based measurements can cover a large area, providing a means to study aerosol effects at a global scale. However, in the early years, such datasets are only available over the ocean, such as from the AVHRR measurements. TOMS is the first satellite sensor capable of deriving aerosol properties over both land and ocean but the measurement accuracy has been affected by its large footprint (~50 km) and strong altitude dependence. MODIS (Salomonson et al., 1989) is one of the sensors designed to measure aerosol properties onboard the EOS (Earth Observing System) Terra (launched on 18 December, 1999) and Aqua (launched on 4 May, 2002) satellites launched into Sun-synchronous orbits with equator-crossing times at around 10:30 a.m. and 1:30 p.m. respectively. The unique set of seven MODIS channels (0.47, 0.55, 0.67, 0.87, 1.24, 1.64, and 2.1 μm) with 250 – 500 m resolutions enables us not only to derive aerosol optical depth but also other parameters, for example, fine-mode fraction over land and fine-mode fraction and effective radius over ocean. It is expected to aid significantly in studying aerosol effects on Earth's energy budget and hydrological cycle.

The twice-a-day MODIS AOD measurements also provide pseudo-synoptic views of aerosol distribution and movements. The initial success of applying MODIS standard products (10 km× 10 km) to particle pollution monitoring (Chu et al., 2003; Hutchison et al., 2003; Wang et al., 2003; Al-Saadi et al., 2005) has drawn a lot of attention from air quality and public health communities. The correlations between MODIS AOD and $PM_{2.5}$ mass concentration further show that MODIS AOD can be used as an indicator for $PM_{2.5}$ mass concentration (NASA, 2003; Al-Saadi et al., 2005). Based upon MODIS AOD data, we can pinpoint the pollution source to local, interstate, international, or intercontinental origins, generated by urban/industrial pollution, forest fires, or dust storms. Because of the differences of natural variability and human-induced effects on local environments, each region tends to show different aerosol characteristics. MODIS aerosol products can be used to characterize aerosol properties over both land and ocean especially in conjunction with other sensors. For example, in-phase variation is obtained between MODIS fine-mode AOD and MOPITT CO (carbon monoxide) from biomass burning as dominated by carbonaceous aerosol, whereas out-of-phase seasonal variations are shown from anthropogenic sulfate emission (Edwards et al., 2004). The former is correlated because both are emitted from the same source. The latter is primarily attributed to the chemistry that sulfate is generated by oxidation with OH and H_2O_2, which peaks in the summer, while CO is consumed by OH and therefore has a summer minimum (Rasch et al., 2000).

6.2 Multi-Year Aerosol Datasets

MODIS-derived aerosol properties are reported in level-2 (L2) granule-based (granule: 5-minute measurements) and level-3 (L3) global daily, 8-day, and monthly gridded products. The L2 product contains retrieved aerosol properties (e.g., AOD, fine-mode fraction, Ångström Exponent, etc.) at 10 km× 10 km, and L3 contains the statistics (mean standard deviation, maximum, minimum, etc.) obtained within 1°× 1° grids. The details of MODIS aerosol products can be found at the Web site http://modis-atmos.gsfc.nasa.gov. Different file specifications (e.g., deletion and addition of parameters) are possible in different versions of the products. Three data processing cycles were implemented since the beginning of Terra MODIS data acquired on 24 February, 2000. The version (or Collection) 2 data were produced prior to November 2000 and version 3 covers from November 2000 to October 2002. In this chapter, we used the version-4 L3 1°× 1° Terra MODIS aerosol data from March 2000 to February 2004 to produce 4-year aerosol climatology, and Aqua aerosol data from July 2002 to March 2003 to compare with Terra aerosol data in the period from July 2002 to March 2003. Table 6.1 shows the dates excluded from the analysis because of no data or partial coverage measurements acquired by Terra and Aqua MODIS.

Table 6.1 Julian days excluded from the analysis because of no data or partial measurements acquired by Terra and Aqua MODIS

Year	No Data	Partial Measurements
Terra MODIS 2000	219, 220, 221, 222, 223, 224, 225, 226, 227, 228, 229, 230	116, 117, 118, 179, 231, 301
Terra MODIS 2001	167, 168, 169, 170, 171, 172, 173, 174, 175, 176, 177, 178, 179, 180, 181, 182, 183	166, 184
Terra MODIS 2002	79, 80, 81, 82, 83, 84, 85, 86	104, 105, 151
Aqua MODIS 2002	182, 183, 184, 211, 212, 213, 214, 215, 216, 217, 218	
Terra MODIS 2003	351, 352, 353, 354, 355, 356, 357, 358	32, 350
Terra MODIS 2004	—	50

6.3 MODIS Aerosol Retrieval Algorithm and Expected Accuracy

The MODIS aerosol retrieval algorithm is comprised of two different schemes to retrieve aerosol properties over land and ocean. The details of the algorithm can be found in Kaufman et al. (1997), Chu et al. (1998) (land), Tanré et al. (1997) and Tanré et al. (1999) (ocean). The enhancements and modifications of the schemes were described in Chu et al. (2003) (land), Levy et al. (2003) (ocean) and Remer et al. (2005) (land and ocean). Here, we briefly discuss the methodology of these two approaches. The aerosol ocean algorithm employs six MODIS 500 m spectral bands (0.55, 0.67, 0.86, 1.24, 1.64, and 2.1 μm) to retrieve aerosol properties including aerosol optical depth, fine-mode fraction, and effective radius etc. within 10 km × 10 km grid box (nadir) under the cloud-free and glint-free conditions (e.g., glint angle > 40°). Three procedures are executed in order to obtain "clear" pixels for aerosol retrieval. Cloud screening is first performed, including the spatial variability tests (0.55 μm reflectance measurements) (Martins et al., 2002), brightness temperature tests (6.7, 11, and 12 μm) (Ackerman et al., 1998), and 1.38 and 0.66 μm reflectance and reflectance ratio of 1.38/1.24 μm tests (Gao et al., 2002). The latter is to remove high and thin cirrus clouds. Sediment mask is applied subsequently to discard pixels due to the enhanced reflection by river sediments around river mouths (Li et al., 2003). The remaining (cloud-free, glint-free, and sediment-free) pixels of the total of 400 pixels (20×20 at 500 m resolution in 10 km×10 km box) are then sorted in ascending order to further remove 25% darkest and brightest pixels for possible residual cloud, glint, and cloud-shadow contaminations. Given the "clear" pixels within 10 km× 10 km grid box, least residual method (Tanré et al., 1997) is used to determine the fraction of fine and coarse mode aerosol models by minimizing the differences of six-paired measured and calculated spectral radiances (0.55 – 2.1 μm) of 20 combinations of 4 fine-mode and 5 coarse-mode aerosol models. For the best scenario, the normalized residual is expected to be less than 3% with optimal retrieval quality. For other cases, the retrieved aerosol properties need to take into account the processing paths permitted in less ideal conditions (e.g., cloud screening, missing spectral channels, etc.). The details of quality assessment (QA) of retrieval can be found in MODIS Atmosphere QA Plan (Chu et al., 2000) (http://modis-atmos.gsfc.nasa.gov). The major aerosol parameters retrieved over ocean include spectral aerosol optical depths (0.47 – 2.1 μm), Ångström exponents, fine-mode fraction, and effective radius. The secondary parameters such as reflected/transmitted fluxes, Cloud Condensation Nuclei(CCN), columnar mass concentration are obtained from lookup tables.

Over land, the dark target approach (Kaufman et al., 1997a) is adopted to retrieve aerosol optical depth based upon the reflectances at 0.47, 0.66, and 2.1 μm wavelengths. Aerosol optical depths are derived at 0.47 and 0.66 μm wavelengths from the mean reflectances of "clear" pixels of 10 km× 10 km areas. The use of

10 km× 10 km box is mainly to accommodate surface variability in order to better select dark pixels at the global scale. Cloud-free pixels are first selected by MODIS cloud mask (Ackerman et al., 1998), followed by the removal of water and snow/ice pixels using positive Normalized Difference Vegetation Index (NDVI) at 250 m resolution and Near real-time Ice and Snow Extent (NISE) snow mask. Surface reflectances of the selected (i.e., cloud-free, water-free, and snow/ice free) pixels at 0.47 and 0.66 μm are estimated by a fixed relationship based upon the measured reflectance at 2.1 μm ($\rho_S^{0.47\,\mu m} / \rho_S^{2.1\,\mu m} = 0.25$ and $\rho_S^{0.66\,\mu m} / \rho_S^{2.1\,\mu m} = 0.5$) (Kaufman et al., 1997b; Kaufman et al., 2002). To further eliminate the contaminations by residual clouds, water, and snow/ice, 20 – 50 percentile is used to calculate the mean surface reflectance and reflectance at top of the atmosphere. To retrieve aerosol optical depth, predetermined aerosol models obtained from the field experiments of SCAR-A (Kaufman et al., 1997a) (sulfate aerosol, $\omega_0 \sim 0.96$) and SCAR-B (Chu et al., 1998) (smoke in South America, $\omega_0 \sim 0.9$) and SAFARI 2000 (Ichoku et al., 2003) (smoke in Southern Africa, $\omega_0 \sim 0.85$) are used. The selection of urban/industrial or biomass-burning aerosol model (both dominated by fine-mode particles) is based upon geographic locations and seasons of emission sources. The determination of the fraction of fine-mode aerosol is based upon the ratio of path radiance between 0.66 and 0.47 μm—similar to Ångström exponent except that path radiance ratio is not affected by the assumption of aerosol models. The final AOD is obtained by the linear combination of AODs of fine and coarse mode aerosols. AOD, fine-mode fraction, and Ångström exponent are the primary parameters obtained from aerosol retrieval and also can be validated by AERONET sun/sky measurements. The secondary aerosol parameters include reflected/ transmitted fluxes and column mass concentration derived from lookup tables.

Despite that MODIS reports aerosol optical depths at seven wavelengths over ocean, only 0.66 μm and 0.87 μm channels are sufficiently similar for direct comparison with AERONET Sunphotometer measurements. At other wavelengths, AOD is derived by linear interpolation at the scale of log-AOD and log-wavelength. Based upon the validation using AERONET Sunphotometer and other sunphotometer/radiometers, the retrieval errors of aerosol optical depth are found within $\Delta\tau_a = \pm 0.03 \pm 0.05\,\tau_a$ (Remer et al., 2002; Levy et al., 2003; Remer et al., 2005), which can be translated to standard errors ~0.02, a half of those of AVHRR ~0.04 (Stowe et al., 1997).

The dark target method and fixed relationship of land surface reflectance has achieved satisfactory success over vegetated surface with retrieval errors within $\Delta\tau_a = \pm 0.05 \pm 0.20\,\tau_a$ at 0.47 and 0.66 μm (Chu et al., 2002; Chu et al., 2003; Remer et al., 2005) except in the coasts, mountain tops, and partial snow melting regions. The restrictions of 2.1 μm reflectance < 0.15 in version 2 and < 0.25 in version 3 have excluded aerosol retrievals over bright deserts (e.g., Sahara and Arabia) and over snow or ice. The version 4 processing algorithm extends retrieval over semi-arid region (2.1 μm reflectance $\leqslant 0.4$) but set with lowest quality ($= 0$) because of larger uncertainty $\Delta\tau_a \geqslant 0.3$ (Remer et al., 2005).

6.4 Characterization of Aerosol Optical Depth Distribution

Aerosols in the atmosphere arise from both natural and anthropogenic sources. Figure 6.1 shows the monthly mean AOD (at 0.55 μm) images obtained from L3 daily Terra MODIS aerosol products from 2000 to 2004, instead of the average

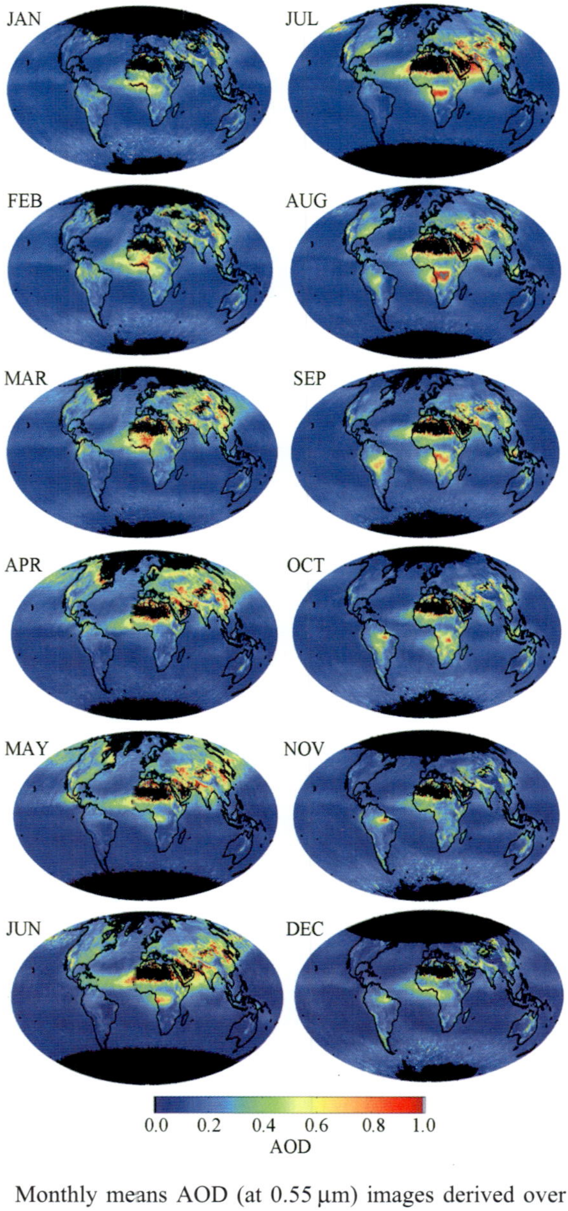

Figure 6.1 Monthly means AOD (at 0.55 μm) images derived over 4-year time span from March 2000 to February 2004 from Terra MODIS aerosol products

of monthly mean of each year. It is clear to see the springtime Asian dust outbreaks spreading over the Northern Hemisphere and the shift of African dust outflow with latitude in different seasons. Smoke and pollution aerosols are most visible in Central Africa from January to March. The dry-season biomass burning also depicts seasonal maximums in the Central America (February – May), South America (August – November) and Southern Africa (July – October). Dust outbreaks from South Asia over the Arabian Sea are severe between June and August. Summertime haze (June-August) in the Eastern US, Western Europe, and Eastern China are significant in terms of air pollution. It should be noted, however, that melting snow/ice causes caveats of high AOD values along the vegetation outline in March – April in Northern Canada and on the mountaintops in the Western United States and Central Mexico.

As pointed out by Chou et al. (2002), atmospheric circulation controls oceanic aerosol distribution and thus results in a strong latitudinal distribution. Emission, chemistry, and dry/wet deposition processes control aerosol life cycles and distribution over land before being transported to downwind regions. Figure 6.2 shows the time series of zonal average of daily MODIS AOD obtained over both land and ocean. The zonal mean AOD values are clearly peaked at mid-latitude in the Northern Hemisphere from June to August because of higher oxidation efficiency (greater illumination, higher temperature and concentration of OH and H_2O_2) for SO_2 to form sulfate aerosol in the summer than winter. Also shown but farther extended to high latitudes (> 60°N) are attributed to Asian dust outbreaks during March – May and forest fires in Southeastern Russia from May to July. Smoke generated by peat fires around Moscow, Russia in September 2002 and intense fires in the southeast of Russia in the following spring are anomalies

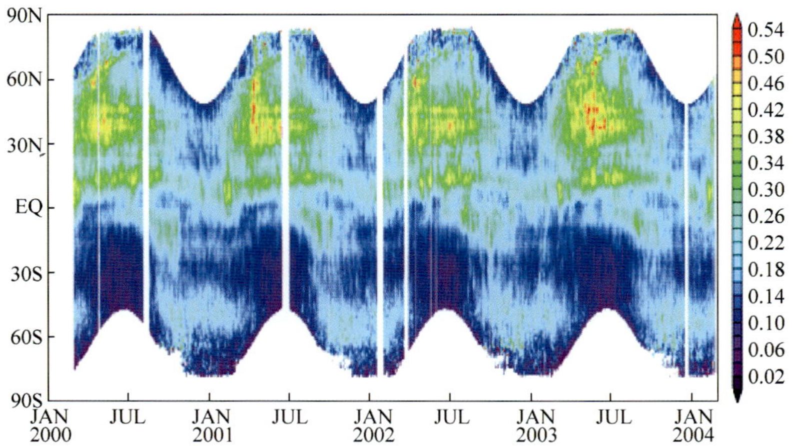

Figure 6.2 Time series of zonal average of daily Terra MODIS AOD obtained over both land and ocean

compared to previous years. Edwards et al., (2004) showed the out-of-phase distribution of MODIS AOD with MOPITT columnar CO when originated from industrial emission sources reflecting to the state of oxidation as mentioned above and in-phase distribution from biomass burning sources.

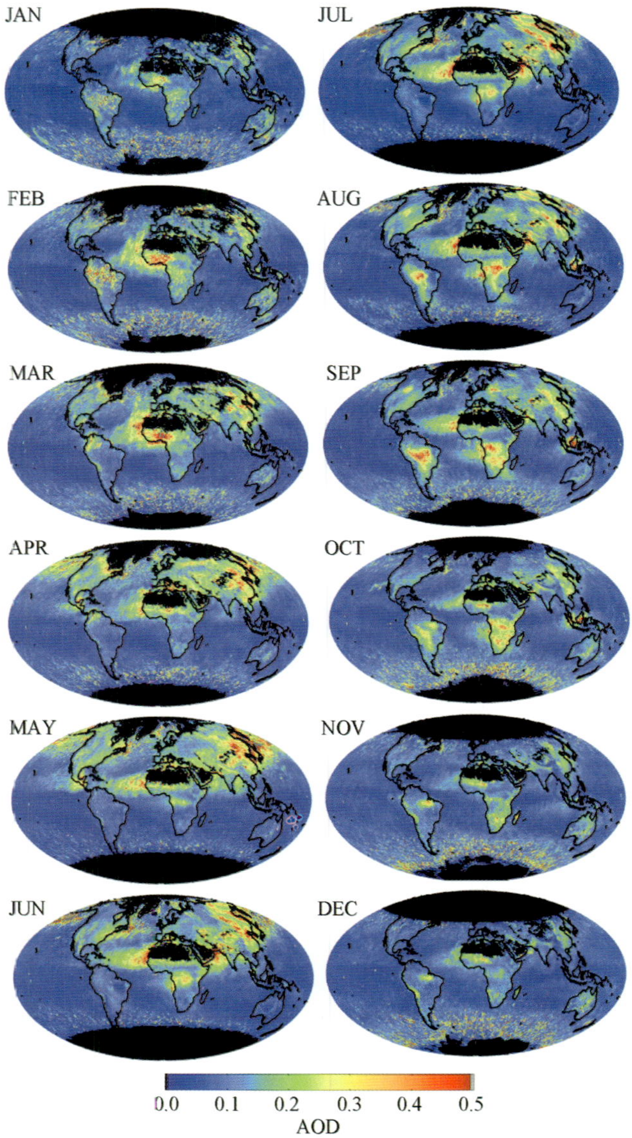

Figure 6.3 Corresponding standard deviation AOD (at 0.55 μm) images to monthly mean as shown Fig. 6.1

MOPITT columnar CO when originated from industrial emission sources reflecting to the state of oxidation as mentioned above and in-phase distribution from biomass burning sources. Just the north of the Equator (~15°N) the western African dust outbreak and biomass burning produce local maximums. However, since MODIS does not have retrievals over the Sahara where the maximum AOD is located, the zonal mean AOD in Fig. 6.2 over that latitudinal belt (15 – 30°N) is lower than reality. The minimums (AOD < 0.02) found in Southern Hemisphere from May to July are attributed to rainfall during the wet seasons in contrast to dry-season biomass burning producing distinct seasonality. The feature located in 40° – 60°S between August and October is most likely related to the outflows of biomass burning in South America and Southern Africa, whereas between November and April it is most likely attributed to cloud contamination.

Figure 6.1 depicts the monthly mean aerosol loading over the years from 2000 to 2004, and Fig. 6.2 emphasizes on the significance of latitudinal coverage and time duration of the events. As such, sporadic smoke plumes generated by forest fires in a short period of time are not seen. The AOD standard deviation (see Fig. 6.3) can rather depict clearly those events as well as events of higher variability with respect to space and time, for instance, smoke from Western US and Canada, and smoke plume outflows from South America and Southern Africa. In addition, large standard deviation found in the regions of prevalent events reflects to strong inter-annual variability. In the following sections, global, hemispheric, and regional AOD are analyzed.

6.5 Global and Hemispheric Analysis

The time series of daily global and hemispheric mean and standard deviation are derived based upon daily AOD data from March 2000 to February 2004 (see Fig. 6.4(a)). It is shown that the largest variation occurs in May – June time frame, and slightly less in the February – March period. Referring to Fig. 6.3, they can be seen as dictated by higher inter-annual variability of dust and fire emissions in the Asian/Pacific and African/Tropical Atlantic regions. Table 6.2 tabulates the global and hemispheric annual means in different yearly cycles. In general, aerosol loading is shown above the annual means from March to August and below the annual means from September to February, illustrating the annual variation pattern. The global annual means are semi-conserved (0.187 – 0.197) with standard deviation in the range between 0.01 and 0.02. The times of the occurrence of maximum AOD, however, vary from year to year causing large standard deviations, for example, in February to April 2001 as attributed to Asian dust outbreaks, and in May 2003 as attributed to smoke generated by forest fires in southeast Russia (Siberia). The minimal AOD values (~0.15) are shown consistently in December without much variation.

Table 6.2 Global, Hemispheric, and regional annual mean and standard deviation derived in a yearly cycle from March 2000 to February 2004

	Mar. 2000 – Feb. 2001	Mar. 2001 – Feb. 2002	Mar. 2002 – Feb. 2003	Mar. 2003 – Feb. 2004
Global	0.190 (0.017)	0.187 (0.020)	0.191 (0.019)	0.197 (0.022)
N. Hemisphere	0.235 (0.050)	0.231 (0.050)	0.233 (0.047)	0.247 (0.058)
S. Hemisphere	0.143 (0.030)	0.142 (0.027)	0.146 (0.027)	0.143 (0.026)
India	0.341 (0.088)	0.341 (0.100)	0.342 (0.100)	0.360 (0.113)
W. Africa	0.317 (0.095)	0.288 (0.053)	0.284 (0.058)	0.306 (0.075)
E. Asia	0.274 (0.076)	0.273 (0.080)	0.271 (0.073)	0.301 (0.106)
E. US	0.235 (0.077)	0.228 (0.091)	0.239 (0.095)	0.252 (0.090)
S. America	0.203 (0.072)	0.211 (0.077)	0.219 (0.097)	0.233 (0.096)
S. Africa	0.173 (0.070)	0.170 (0.063)	0.165 (0.050)	0.165 (0.058)
Remote Ocean	0.118 (0.020)	0.114 (0.021)	0.113 (0.024)	0.109 (0.022)

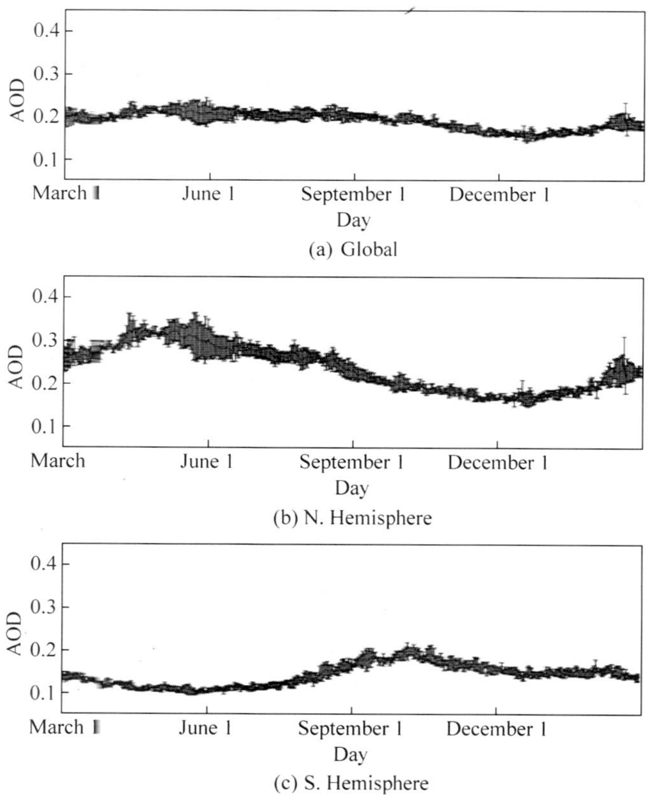

Figure 6.4 Daily mean AOD and standard deviation calculated over the 4-year time span (March 2000 – February 2004) of the (a) whole globe, (b) Northern Hemisphere, and (c) Southern Hemisphere

In the Northern Hemisphere, the magnitudes of the variation of AOD are much more distinct over the annual cycle and nearly out of phases as opposed to that in the Southern Hemisphere (see Figs. 6.4(b) and 6.4(c)). Compared to annual means ~0.23 – 0.25 (standard deviations ~0.04 – 0.06) in the Northern Hemisphere, significantly higher AOD values are found between March and August (~0.35) as opposed to September to February (~0.15). It is reversed in the Southern Hemisphere with maximums ~0.2 in September – October and minimums ~0.1 in June (c.f. annual means ~0.14 – 0.15 and standard deviations ~0.02 – 0.03). Differing from many variable aerosol sources in the Northern Hemisphere (e.g., dust storms, biomass burning, and air pollution), the biomass burning in the dry season (August – October) in South America and Southern Africa appears to dominate in the Southern Hemisphere. The smaller annual variability (0.1 – 0.2) in the Southern Hemisphere is expected because of the oceans covering most of the area and less aerosol sources (c.f. 0.15 – 0.25 in the Northern Hemisphere). In regions close to the source, the variability is believed to be larger.

6.6 Regional Analysis

Seven regions, the Eastern US (SCAR-A, 1993; TARFOX, 1996; CLAMS, 2001), South America (SCAR-B, 1995), India subcontinent (INDOEX, 1999 – 2000), South Africa (SAFARI 2000), East Asia (ACE-Asia, 2001), Western African, and an area in remote ocean, are chosen—five of them are based upon field experiments conducted in the past (as indicated inside the parentheses) and the last two are included for comparison purpose. Figure 6.5 depicts the location and coverage of each region. MODIS aerosol retrieval has greatly benefited from building aerosol models (sulfate and smoke) using the columnar sun/sky measurements acquired during these field experiments. Therefore, it is important to understand the temporal variability in each of the regions. As shown in Table 6.2, India subcontinent and surrounding ocean has the largest aerosol loading (~0.34) and the remote ocean has the smallest loading (~0.1) as expected. It is however surprising to learn that biomass burning in Southern Africa produces the second smallest loading. While Southern Africa and remote ocean reveal virtually no change (or slight decrease) in the annual means, the rest of the regions depict either an increasing trend (such as in India and South America), or rather variable variation without a trend (such as in Western Africa, East Asia, and Eastern US). Note that the MODIS inability to retrieve dust over the Saharan Desert significantly reduces aerosol loading in the region. It is also worth noting that the increase in AOD is most pronounced from 2003 to 2004 in nearly all the regions (except Remote Ocean and Southern Africa) regardless of the trend.

Daily AOD data provide much detailed information, including annual and seasonal variations as well as minimums and maximums throughout the years. Figure 6.6 shows daily AOD mean and standard deviation computed over the 4 year time span (2000 – 2004) for the selected regions.

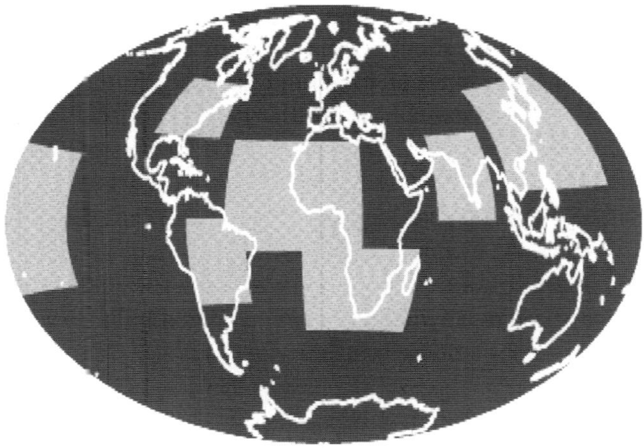

Figure 6.5 The location and coverage of each region for the field experiments being conducted in the past to study aerosol properties and the effects on cloud microphysics and radiation

Except the remote ocean, the rest of the regions show strong annual (same as seen in global and hemispheric scales) and inter-annual (as shown by the large standard deviation) variations. Over the India subcontinent and surrounding ocean, maximum AOD values of ~0.5 – 0.6 with standard deviation ~0.2 – 0.3 are found in June – July to reflect the intensity of dust outbreaks from the neighboring Saudi-Arabia Peninsula and South Asia deserts (e.g., Pakistan, Afghanistan, etc.) due to solar heating. The minimum AOD values of 0.2 – 0.3 occur in December – February with small standard deviation ~0.05 – 0.1. In Western Africa, biomass burning and pollution dominate in the region from February to March, resulting in maximum AOD values between 0.4 and 0.5 (standard deviation ~0.2), which is slightly greater than those generated by dust outbreaks (~0.3 – 0.4) during July – August. Note that dust over the Saharan Desert is excluded in the calculation. Minimums of 0.2 – 0.3 mainly occur between November and December. In Eastern US, AOD of 0.3 – 0.4 with standard deviation up to 0.15 extends from late March to the end of August, encompassing springtime biomass burning transported from Central America and local summertime haze from urban/industrial sources. Minimums of 0.1 – 0.2 (standard deviation 0.05 – 0.1) primarily occur during the winter (November – December). In East Asia, a similar pattern of daily AOD is found as that in the Eastern US but with slightly increased aerosol loading to 0.35 – 0.45. Dust outbreaks (Taklimakan and Gobi deserts), pollution (Eastern China), and biomass burning (Southeast Asia) contribute to the higher loading from March to mid-June.

The winter minimums in East Asia are comparable to those derived in the Eastern US. The sudden decrease of AOD in East Asia in the end of June should be related to the developing summer monsoon in the region.

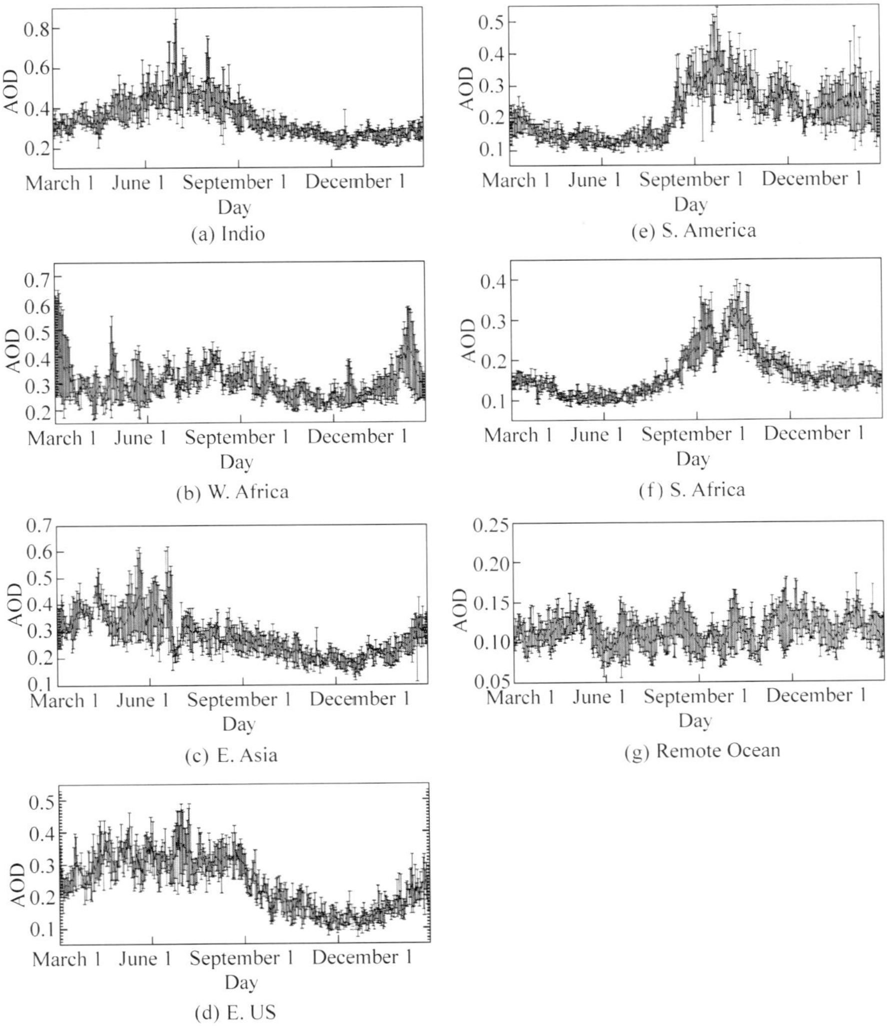

Figure 6.6 Same as in Fig. 6.4 but for the follow regions: (a) India, (b) Western Africa, (c) East Asia, (d) Eastern US, (e) South America, (f) Southern Africa, and (g) Remote Ocean

The most isolated seasonal events occur in Southern Africa as a result of dry-season biomass burning (August to November) with elevated AOD ~0.3 (standard deviation ~0.1), in contrast to the minimum AOD ~0.1 in the rainy season (April to July). The increased AOD in South America also resulted from biomass burning seems to be in-phase with that in Southern Africa except the former extends farther to December, resulting in multiple maximums in the range between 0.3 and 0.4. The elevated AOD (~0.2 – 0.3) in January – February 2004 with large standard deviations (~0.2) are anomalies caused by unknown reasons. Similar to Southern Africa, the minimums (~0.1) of South America occur during the rainy season.

6.7 Terra vs Aqua

The late morning (10:30 a.m.) and early afternoon (1:30 p.m.) satellite measurements are compared in time series to investigate possible morning-afternoon difference. The Aqua MODIS sensor began to acquire measurements in late June 2002. Up to date, the version 4 Aqua MODIS data have been generated from July 2002 until March 2003. Therefore Terra and Aqua MODIS aerosol data are only compared for this time period. We use daily AOD differences (Terra—Aqua) and root mean square difference (RMSD) as an indicator for any morning-afternoon difference. Figure 6.7 shows the time series of daily AOD means from Terra and Aqua MODIS and the corresponding difference. The daily AOD mean values of Terra and Aqua MODIS show an extremely similar pattern at the global scale, with differences less than ± 0.025 and (RMSD) ~0.0074 throughout the period. Between September and December Terra shows slightly larger AOD values than those from Aqua. Such differences are also seen in both hemispheres during the same period. The small RMSD (< 0.01) and daily AOD difference (< 0.02) over the time period compared indicate that the morning-afternoon difference is negligible at both global and hemispheric scales.

However, distinct differences are found between Terra and Aqua in India and South America associated with seasonal events (see Figs. 6.8(a), 6.8(c), and 6.8(e)). The difference in East Asia is rather weaker compared to the other two regions. The RMSD of ~0.07 (with daily AOD differences up to 0.2) in India in July – August is mainly attributed to dust outbreaks (e.g., caused by stronger instability due to solar heating in the afternoon than in the morning) and air pollution in terms of the oxidation and growth of aerosol particles. Note that it is the monsoon season that cloud cover may change significantly between morning and afternoon. The number of grid points ($1° \times 1°$) used in the calculation of mean AOD, however, does not show enough evidence to support the cause by cloud cover.

Also, it is hard to believe that the variation of cloud cover (if it exists) only causes the mean AOD differences in July – August but not other seasons. The RMSD of ~0.04 in South America is shown in August – September as a result of dry-season biomass burning and ~0.034 in East Asia due to summer urban/industrial haze in July – September. Only the latter shows much lesser intensity of the difference. The former is most likely caused by stronger fire intensity in the afternoon and the latter is due to higher oxidation state of converting gas pollutants to aerosols. In the Eastern US (RMSD ~0.03), although a similar oxidation state exists in the summer, the daily AOD mean differences are much smaller (< 0.05) compared to that in East Asia. The smaller RMSD in Western Africa (~0.02), Southern Africa (0.02), and Remote Ocean (0.01), associated with daily AOD difference < 0.05 throughout the period, suggest no apparent morning-afternoon difference in these regions. However, it may still exist in smaller regions of consideration.

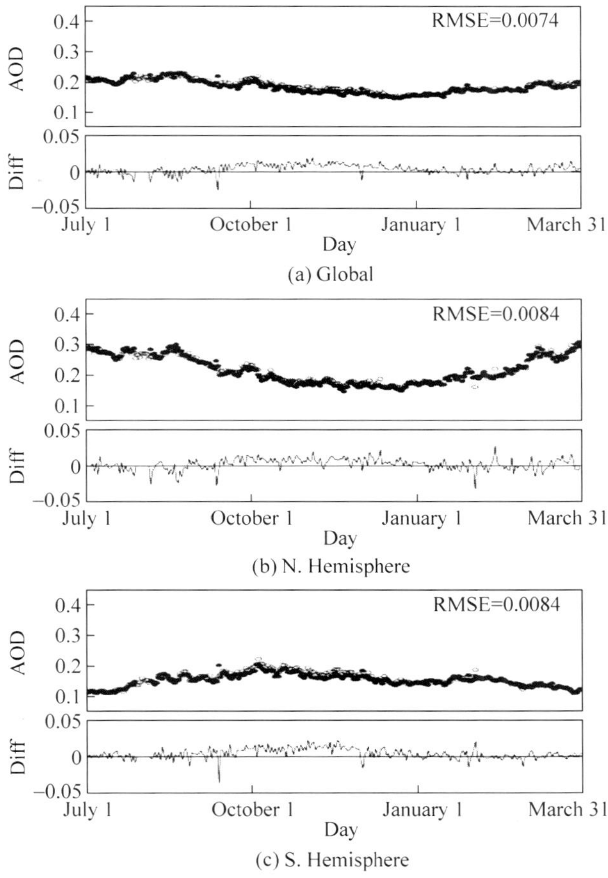

Figure 6.7 Comparison of daily mean AOD obtained from Terra and Aqua MODIS and the corresponding differences (Terra—Aqua) from July 2002 to March 2003 of the (a) whole globe, (b) Northern Hemisphere, and (c) Southern Hemisphere. Open circle denotes Terra MODIS data and closed circle denotes Aqua MODIS data

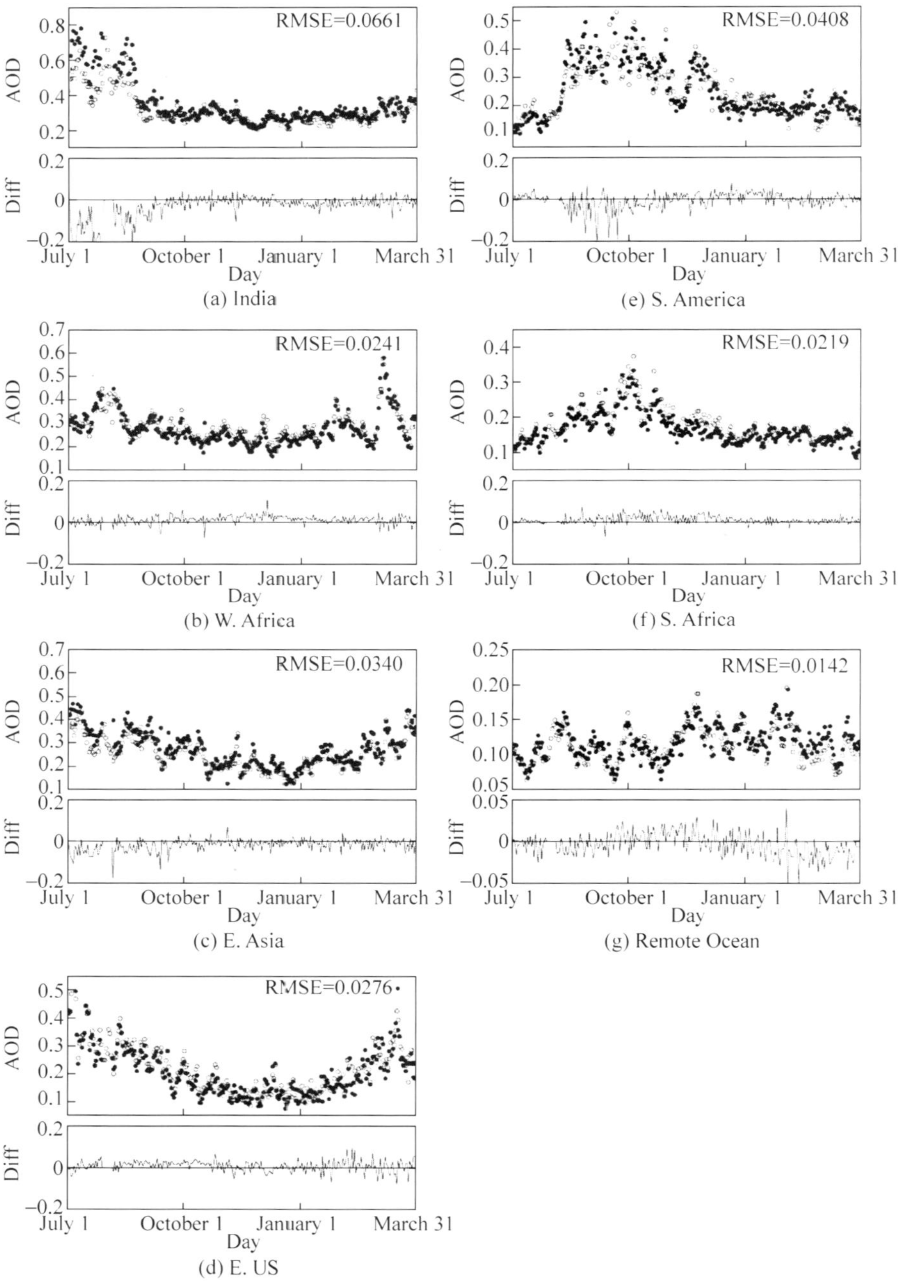

Figure 6.8 Same as in Fig. 6.7 but for the following regions: (a) India, (b) Western Africa, (c) East Asia, (d) Eastern US, (e) South America, (f) Southern Africa, and (g) Remote Ocean

6.8 Conclusions

MODIS' ability to retrieve AOD over both land and ocean represents a step forward to studying aerosol effects on the Earth system. The inter-annual, annual, and seasonal variabilities are analyzed based upon 4-year Terra MODIS AOD data. Prevalent aerosol events produce large inter-annual variability from regional to hemispheric and global scales. Sporadic fires also contribute to inter-annual variability but only at the regional scale. Semi-annual variability is not only shown at global and hemispheric scales but also in the regions of India, East Asia, Eastern US, and South America associated with much stronger magnitude. Seasonal variability is most pronounced in Western Africa and Southern Africa due primarily to biomass burning. Much more frequent variation found in the remote ocean is believed to be controlled by the movement of synoptic systems.

The variability between morning and afternoon is studied using Terra and Aqua MODIS AOD data between July 2002 and March 2003. No significant variation is found at global and hemispheric scales. In the selected regions where field experiments took place, only India, South America have shown significant AOD differences between satellite overpass times in the seasons of prevalent aerosol events. While differences are found in East Asia and Eastern US, they are much smaller compared to those in India and South America. In Western Africa, Southern Africa, and Remote Ocean, no apparent pattern is shown.

References

Ackerman SA, Strabala KI, Menzel WP et al. (1998) Discriminating clear-sky from clouds with MODIS. J Geophys Res 103(32): 141 – 158

Al-Saadi J, Szykman J, Pierce B, Kittaka C, Neil D, Chu DA, Remer L, Gumley L, Prins E, Weinstock L, Clinton MD, Wayland R, Dimmick F (2005) Improving National Air Quality Forecasts with Satellite Aerosol Observations, BAMS, 86(9), 1,249 – 1,261

Chou MD, Chan P-K, Wang M (2002) Aerosol radiative forcing derived from SeaWiFS-retrieved aerosol optical properties, J. Atmos. Sci., 59(3), 748 – 757

Chu DA, Kaufman YJ, Remer LA et al. (1998) Remote sensing of smoke from MODIS airborne simulator during SCAR-B experiment. J Geophys Res 103(31): 979 – 987

Chu DA, Strabala K, Platnick S, Mattoo S, Hucek R, Ridgeway B (2000) MODIS Atmosphere QA Plan. EOS Project Office, NASA Goddard Space Flight Center, available on http://modis-atmos.gsfc.nasa.gov

Chu DA, Kaufman YJ, Ichoku C et al., (2002) Validation of MODIS aerosol optical depth retrieval over land. Geophys Res Lett 29(12): doi 10.1029/2001GL013205

Chu DA, Kaufman YJ, Zibordi G, Chern J-D, Mao J-M, Li C, Holben HB (2003) Global Monitoring of Air Pollution over Land from EOS-Terra MODIS. J Geophys Res 108(D21):4661, doi: 10.1029/2002JD003179

Chu DA, Remer LA, Kaufman YJ, Schmid B, Redemann J, Knobelspiesse K, Chern J-D,

Livingston J, Russell P, Xiong X, Ridgway W (2005) Evaluation of aerosol properties over ocean from MODIS during ACE-Asia, J. Geophys. Res., 110, D07308, doi: 10.1029/2004JD005208

Edwards DP, Emmons LK, Hauglustaine DA, Chu DA, Gille JC, Kaufman YJ, Petron G, Yurganov LN, Giglio L, Deeter MN, Yudin V, Ziskin DC, Warner J, Lamarque J-F, Francis GL, Ho SP, Mao D, Chen J, Grechko EI, Drummond JR (2004) Observations of carbon monoxide and aerosols from the Terra satellite: northern hemisphere variability. Submitted J Geophys Res

Gao B-C, Kaufman YJ, Tanre D, Li RR (2002) Distinguishing tropospheric aerosols from thin cirrus clouds for improved aerosol retrievals using the ratio 1.38 μm and 1.24 μm channels. Geosphys Res Lett 29(18): 1890, doi: 10.1029/2002GL015475

Hutchison KD (2003) Applications of MODIS satellite data and products for monitoring air quality in the state of Texas. Atmos Environ 37(2): 403 – 412

Ichoku C, Remer LA, Kaufman YJ, Levy R, Chu DA, Tanré D, Holben BN (2003) MODIS observation of aerosols and estimation of aerosol radiative forcing over southern Africa during SAFARI 2000, J. Geophys. Res., 108(D13), 8499, doi: 10.1029/2002JD002366

Kaufman YJ, Tanre D, Remer LA, Vermote EF, Chu DA, Holben BN (1997a) Operational remote sensing of tropospheric aerosol over the land from EOS-MODIS. J Geophys Res 102(17): 051 – 061

Kaufman YJ, Wald A, Remer LA, Gao B-C, Li R-R, Flynn L (1997b) The MODIS 2.1 μm channel- Correlation with visible reflectance for use in remote sensing of aerosol. IEEE Trans Geosci Remote Sens 35(5): 1,286 – 1,298

Kaufman YJ, Gobron N, Pinty B, Widlowski J-L, Verstraete MM (2002) Relationship between surface reflectance in the visible and mid-IR used in MODIS aerosol algorithm - theory. Geophys Res Lett 29(23): 2116, doi: 10.1029/2001GL014492

Levy RC, Remer LA, Tanré D et al. (2003) Evaluation of the Moderate-Resolution Imaging Spectroradiometer (MODIS) retrievals of dust aerosol over the ocean during PRIDE. J Geophys Res 108(D19): 8594, doi: 10.1029/2002JD002460

Li R-R, Kaufman YJ, Gao B-C, Davis CO (2003) Remote sensing of suspended sediments and shallow coastal waters. IEEE Trans Geosci Remote Sens 41(3): 559 – 566

Martins JV, Tanré D, Remer L, Kaufman Y, Mattoo S, Levy R (2002) MODIS Cloud screening for remote sensing of aerosol over oceans using spatial variability. Geophys Res Lett 29(12): 10.1029/2001GL013252

NASA (2003) The Application of Satellite Data to Particle Pollution. NASA Technical Memorandum, November 2003

Rasch PJ, Barth MC, Kiehl JT, Schwartz SE, Benkovitz CM (2000) A description of the global sulfur cycle and its controlling processes in the National Center for Atmospheric Research Community Climate Model. Version 3, J Geophys Res 105: 1,367 – 1,385

Remer LA, Tanre D, Kaufman YJ, Ichoku C, Mattoo S, Levy R, Chu DA, Holben B, Dubovik O, Smirnov A, Martins JV, Li R-R, Ahmad Z (2002) Validation of MODIS aerosol retrieval over ocean. J Geophys Res 29(12): 10.1029/2001GL013204

Remer LA, Kaufman YJ, Tanré D, Mattoo S, Chu DA, Martins JV, Li R-R, Ichoku C, Levy RC Kleidman RG, Eck TF, Vermote E, Holben BN (2005) The MODIS Aerosol Algorithm,

Products and Validation, J. Atmos. Sci., 62(4), 947 – 973

Solomonson V et al. (1989) MODIS advanced facility instrument for studies of the earth as a system. IEEE Trans on Geos And Remote Sens 27(2): 145 – 153

Stowe L, lgnatov A, Singh R (1997) Development, validation and potential enhancements to the second-generation operational aerosol product at the National Environmental Satellite, Data, and Information Service, of the National, Oceanic and Atmospheric Administration, J. Geophys. Res., 102: 16,923 – 16,934

Tanré D, et al. (1999) Retrieval of aerosol optical thickness and size distribution over ocean from MODIS/EOS spectral radiances, J. Geophys. Res., 104, 2,261 – 2,278

Tanré D, Kaufman YJ, Herman M, Mattoo S (1997) Remote sensing of aerosol properties over oceans using the MODIS/EOS spectral radiances. J Geophys Res 102: 16,971 – 16,988

Wang J, Christopher SA (2003) Intercomparison between satellite-derived aerosol optical thickness and $PM_{2.5}$ mass: Implication for air quality studies: Geophys Res Lett 30(21): 2095, doi:10.1029/2003GL018174

7 MODIS Land Products and Data Processing

Robert E. Wolfe and Nazmi Saleous

7.1 Introduction

The two Moderate Resolution Imaging Spectroradiometer (MODIS) instruments on-board NASA's Earth Observing System (EOS) Terra and Aqua satellites make key measurements for understanding the Earth's terrestrial ecosystems (Salomonson et al., 1989). Global time-series of terrestrial geophysical parameters have been produced from MODIS/Terra for over 5 years and for MODIS/Aqua for over 2.5 years. These well calibrated instruments, a team of land scientists and a large data production system have allowed for the development of a new suite of high quality land product variables at spatial resolutions as fine as 250 m in support of global change research and natural resource applications. This chapter summarizes the MODIS land science team's products, describes the data processing approach and describes the process for monitoring and improving the product quality.

The original MODIS land science team was formed in 1989. The team's primary role is the development and implementation of the land geophysical algorithms. In addition, the team provided feedback on the design and pre-launch testing of the instrument and helped guide the development of the data processing system. The key challenges the science team dealt with before launch were the development of algorithms for a new instrument and providing guidance of the large and complex multi-discipline processing system. The land team with the ocean and atmosphere discipline teams drove the processing system requirements, particularly in the area of the processing loads and volumes needed to daily produce geophysical maps of the Earth at resolutions as fine as 250 m. The processing system had to handle a large number of data products, large data volumes and processing loads, and complex processing requirements. Prior to MODIS, daily global maps from heritage instruments, such as Advanced Very High Resolution Radiometer (AVHRR), were not produced at resolutions finer than 5 km.

The processing solution evolved into a combination of processing the lower level products in a NASA's EOS Distributed Active Archive Center (DAAC) and production of the higher level discipline specific products in the MODIS Science Investigator Lead Processing System (SIPS), the MODIS Adaptive Processing System (MODAPS). The DAACs also handled archive and distribution of the products to the user community.

7.2 Land Products and Characteristics

MODIS land products are divided into three overall product families: radiation budget variables, ecosystem variables and land cover characteristics (Justice et al., 1998). The radiation budget variables are useful for developing a better understanding of the surface/atmosphere energy exchange and are important in studies of the hydrological cyclic, biological productivity and climate variability. The ecosystem variables measure spatial and temporal dynamics of the Earth's terrestrial vegetation and are inputs into global productivity modeling. The land cover characteristics variables measure both anthropogenic and natural changes in the terrestrial landscape and are used to help understand the causes of the change. Table 7.1 lists the MODIS land products and gives some details on the products: the level(s), where the products are archived and distributed, spatial resolution(s) and temporal resolution(s). Figures 7.1, 7.2 and 7.3 show examples of products from each family.

Table 7.1 MODIS data products and characteristics

Product	Name	DAAC	Level(s)	Resolution(s)	
				Spatial	Temporal
Radiation balance product family					
MOD09	Surface Reflectance	EDC LP	L2G, L3	250 m, 500 m, 1 km, 0.05 deg.	1 day, 8 days
MOD11	Surface Temperature and Emissivity	EDC LP	L2, L3	1 km, 0.05 deg.	5 min., 1 day, 8 days, monthly
MOD43	BRDF/Albedo	EDC LP	L3	1 km, 0.05 deg.	16 days
MOD10	Snow Cover	NSIDC	L2, L3	500 m	5 min., 1 day, 8 days
MOD29	Sea Ice Extent	NSIDC	L2, L3	500 m	5 min., 1 day
Vegetation product family					
MOD13	Vegetation Indices	EDC LP	L3	1 km, 500 m, 250 m	16 days
MOD15	LAI and FPAR	EDC LP	L3	1 km	8 days
MOD17	GPP and NPP	EDC LP	L4	1 km	8 days, 1 year
Land cover product family					
MOD12	Land Cover and Vegetation Dynamics	EDC LP	L3	1 km, 0.05 deg.	1 year
MOD14	Thermal Anomalies and Fire	EDC LP	L2, L3	1 km	5 min.,1 day, 8 days
MOD44	VCC and VCF	EDC LP	L3	500 m, 250 m	96 day, 1 year

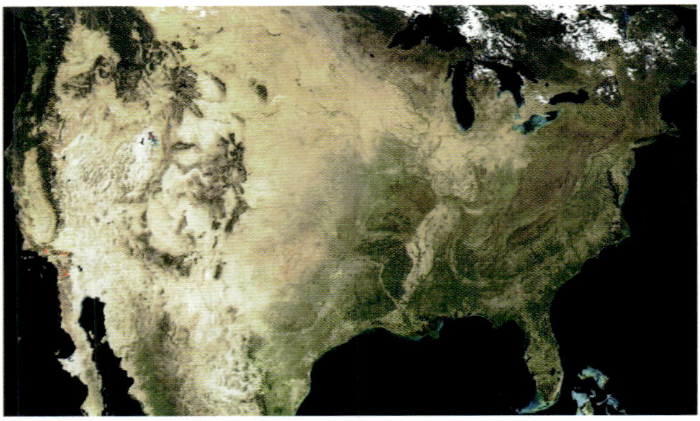

Figure 7.1 Examples from the MODIS land radiation budget and land cover families: 16-day nadir reflectance with 8-day fire and thermal anomalies. The nadir reflectance data covers the conterminous US and is for the16-day period starting October 16, 2003 (day of year 289). The fire locations (in red) are for the 8-day period starting October 24, 2003 (day of year 297) and show a large fire event in Southern California

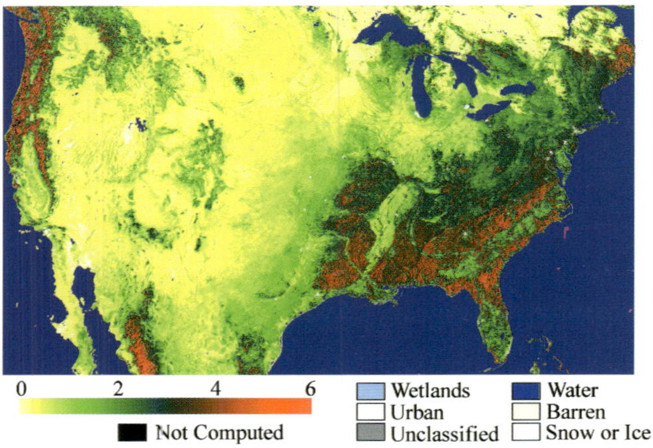

Figure 7.2 Example from the MODIS land ecosystem variables product family: leaf area index. The data is from the 8-day period starting October 24, 2003 (day of year 297) and covers the conterminous US

The radiation budget variables start with the surface reflectance product which is a building block for many of the downstream algorithms (Vermote et al., 2002). This product is the reflectance corrected for atmospheric scattering and absorption. The Bidirectional Reflectance Distribution Function (BRDF), an estimate of the surface anisotropy across the viewing and illumination geometry, is computed from 16 days of surface reflectance for each of the 7 land spectral bands (Schaaf et al., 2002). Three broadband and individual spectral band BRDF

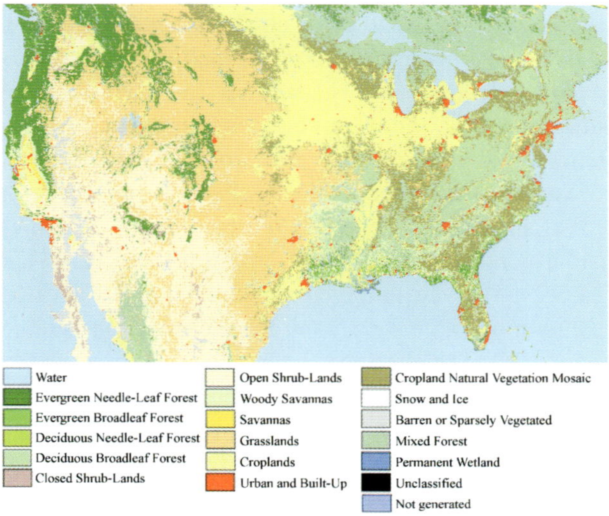

Figure 7.3 Example from the MODIS land cover characteristics product family: land cover. The data is from 2001 and covers the contiguous US

summary products are provided: bi-hemispherical (white-sky) albedo, directional hemispherical (black-sky) albedo at local solar noon, and nadir BDRF-adjusted reflectances (NBAR) at the mean overpass solar zenith angle. During the overlap between the Terra and Aqua missions, data from both MODIS instruments are combined for more coverage and a better overall retrieval accuracy.

The land surface temperature and emissivity product, a key input for studying the energy balance at the Earth's surface, is produced using split-window and statistical regression techniques (Wan et al., 2002). Both day and night data are used and provided up to four temperature measurements per day when MODIS Terra and Aqua data are combined.

Snow cover and sea-ice are the two cryospheric products (Hall et al., 2002). The snow cover product is produced at 500 m resolution, a finer resolution than previous sensors, which provides better details in mountainous regions. The improved accuracy of this product is helping scientists to better understand the inter-annual variability in global snow cover. The 1-km sea ice product includes the extent of the sea ice and a retrieval of the ice surface temperature.

The vegetation product family includes two vegetation indices, the heritage Normalized Difference Vegetation Index (NDVI) and an improved Enhanced Vegetation Index (EVI) that self-corrects for atmospheric and soil calibration factors (Huete et al., 2002). Newer more phenologically based algorithms produce the Leaf Area Index (LAI) and Fractional Photosynthetically Active Radiation (FPAR) products (Myneni et al., 2002). LAI defines an important structural property of a plant canopy which is the one-sided leaf area per unit ground area. FPAR measures the proportion of available radiation in the photosynthetically active wavelengths that a canopy absorbs. The Gross Primary Production (GPP) and Net Primary Production (NPP) use a combination of MODIS LAI/FPAR and meteorological data to estimate vegetation productivity (Running et al., 2004).

The land cover product family includes the land cover and vegetation dynamics products that build on the annual temporal signal contained in most of the other MODIS products to classify the Earth's surface into a set of discrete classes and to estimate key phenological inflection points, i.e. start of spring green-up (Friedl et al., 2002). The vegetation continuous fields (VCF) product is an estimate of the percent cover of each pixel in three basic land cover types (Hansen et al., 2002). The vegetation cover conversion (VCC) algorithm uses the two 250 m resolution bands to detect and label changes in land cover (Zhang et al., 2002). The thermal anomalies and fire product identifies the location, timing and energy of fires (Justice et al., 2002a).

MODIS land products are produced in a hierarchy of three processing levels: Level-2—geophysical parameters retrieved at the same location as the MODIS instrument data, Levels-2G and 3—earth-gridded geophysical parameters, and Level-4—modeled outputs in an earth-grid.

The smallest unit of data processed at any one time is defined at Level-2 as a granule and at Levels-2G, -3 and-4 as a tile. In addition, Levels -3 and-4 products are produced in a coarser resolution Climate Modeling Grid (CMG). A granule corresponds to 5 minutes of MODIS sensing and covers approximately 2,340 km × 2,030 km in the across track and along track directions, respectively. A tile corresponds to a 1,200 km × 1,200 km area of the grid. Level-2 products sensed over a 24 hour period are binned into an intermediate tiled data format referred to as Level-2G (Wolfe et al., 1998). The Level-2G format provides a convenient data structure for storing multiple observations over each grid cell, enabling flexibility in subsequent temporal compositing.

The tiled products are produced in global grids based on the Sinusoidal (SIN) and Lambert Azimuthal Equal Area (LAEA) map projections (Snyder, 1987). Almost all of the land tiled products are produced in the equatorial SIN grid and are stored at three different resolutions, 230 m (7.5 arcsec), 460 m (15 arcsec) and 920 m (30 arcsec). These grids are commonly referred to as the 250 m, 500 m and 1 km grids since they are close to these resolutions. The tiled sea-ice product is produced in the LAEA grid and stored at resolution of 1,002.701 km. The grid contains both Northern and Southern hemisphere centered projections. The CMG products are primarily stored in an equal-angle geographic grid with a resolution of 0.05 deg. (approximately 5.6 km at the equator). The exception is the sea-ice CMG product which is also stored in a polar LAEA grid.

7.3 Data Production

The initial design of the EOS Data Information System (EOSDIS) envisioned a tightly controlled production system among three DAACs: GSFC Earth Sciences

(GES) DAAC in Greenbelt, MD for Level-1 and Level-2 products, the National Snow and Ice Data Center (NSIDC) DAAC in Boulder, CO for cryospheric products and Land Processes (LP) DAAC at the Eros Data Center (EDC) in Sioux Falls, SD for all the other land products (Justice et al., 2002b). The MODIS science team's desire to have more flexibility along with budget constraints led to consideration of an alternative approach. The production of Level-1 products was assigned to the GES DAAC and the responsibility for producing Level-2 and higher products was given to the MODIS science team to develop a SIPS. The resulting MODAPS developed under guidance from the science team, offered more flexibility and allowed rapid updates of the product generation codes ensuring improvements in the quality of the MODIS land products. The data archive and distribution of products at all levels remained the responsibility of the three DAACs.

7.3.1 Data Flows

Figure 7.4 summarizes the Terra MODIS land data flows. The Aqua data flow is identical to Fig. 7.4 with the exception of the raw data flow from the instrument to the ground receiving station. The Terra raw instrument data are broadcast to the MODIS receiving station at White Sands, New Mexico and then sent to the EOS Data and Operations System (EDOS) before being sent to the GES DAAC. The Aqua raw data are broadcast to the Svalbard Ground Station in Norway and the Alaskan Ground Station in the US and then sent to EDOS. The MODIS land products are derived from the calibrated MODIS data (Level-1B), MODIS geolocation data, and ancillary data that include National Center for Environmental Prediction (NCEP) meteorological data and Global Modeling and Assimilation Office (GMAO) data. These inputs are delivered to MODAPS from the GES DAAC (denoted C0 and D0 in Fig. 7.5) as soon as they are acquired and processed. Under normal circumstances, over 80% of the MODIS instrument data are

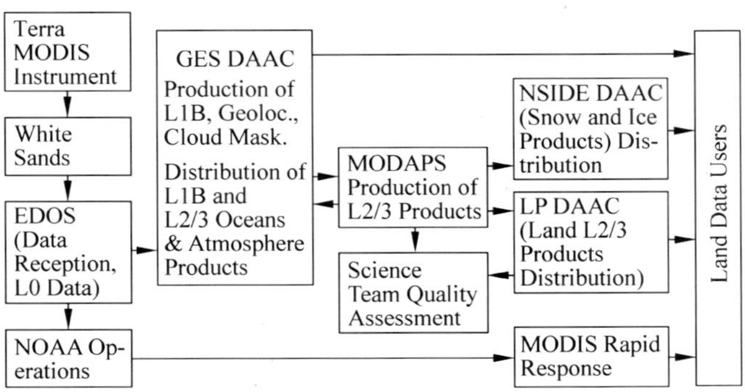

Figure 7.4 MODIS overall data flows

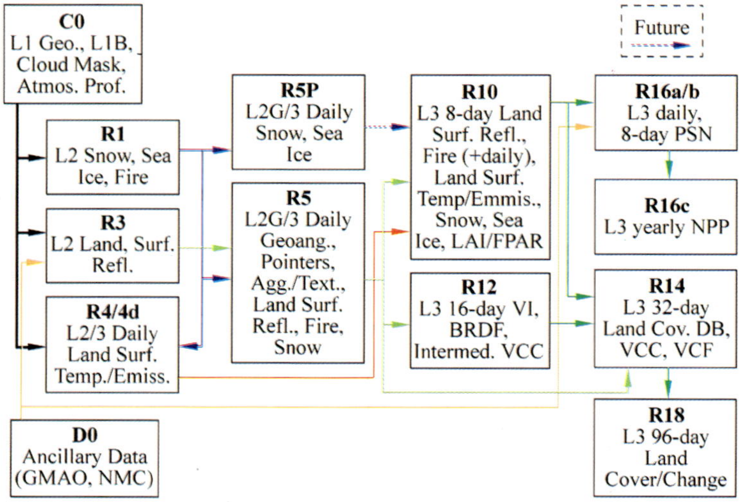

Figure 7.5 MODIS Land production sequence showing the data flow between algorithms

delivered to the GES DAAC within hours of acquisition; the remainder of the data may take days to archive into the DAAC and may require multiple iterations with the data provider. Ancillary data delivery lags behind real time by approximately 24 hours for NCEP predictions and by up to 2 weeks for certain GMAO data.

The Level-2 products are generated first in the MODAPS as they are required to feed the higher order production streams. Given the science algorithm requirements, three different recipes are used in MODAPS to create the land Level-2 products (see Fig. 7.5). The first recipe (R1) is used to create the snow, sea ice and fire products. It runs independently on each MODIS granule and so is run 288 times per day. The second recipe (R3) is used to create the Level-2 land surface reflectance product. It requires one full orbit of data as input (20 granules) and creates up to 20 output granules. This recipe runs up to 15 times per day and requires all granules from one orbit to be available before starting. The third recipe (R4) is used to create the daily land surface temperature products. This recipe, which creates both Level-2 and Level-3 daily products, divides the global into six latitudinal zones and runs one processing stream per zone.

Level-2 data created by R1 and R3 are binned into the intermediate daily Level-2G data products that in turn are composited to create many of the daily Level-3 products. These operations are performed in the R5 and R5P recipes for the equatorial SIN and polar LAEA projections, respectively. Only the MODIS land, snow and sea ice products are produced in the LAEA projection. R5 creates up to 326 tiles per day and R5P creates up to 552 tiles. The Level-2G and Level-3 daily products are further temporally composited to create: 8-day land surface reflectance, land surface temperature, snow, fire, and LAI/FPAR products (recipe R10), 16-day BRDF/Albedo, VI and intermediate land cover conversion products (recipe R12), and 32-day land cover products (recipe R14). The composited products are used to create the NPP/PSN and quarterly land cover products.

7.3.2 Algorithm Improvements

The MODIS processing priorities have evolved since the "first light" when the instrument started acquiring Earth view data. The early period of MODIS was devoted to addressing problems in the low level algorithms and ensuring proper production of higher order products. The extensive product quality assessment activities in the months that followed the "first light" led to a large number of algorithm fixes and code changes. An expedited approach to promote code changes into operational production was adopted to accommodate these frequent updates and to ensure a rapid transition to high quality products. As the algorithms matured, the processing priority shifted to the creation of consistent data sets. As a result of this change, the approach to promote algorithm updates into the operational production system evolved into a more structured procedure. Figure 7.6 shows the sequence of events used in the new approach. The impact of science code changes on downstream products and on the consistency of the data set is assessed through an extensive science test prior to accepting the change. As the products became more mature, the extent of the science test process increased. These tests now are quite extensive, involving production of two 16-day periods globally and/or an entire year of MODIS products over 33 tiles. Based on the test evaluation results, algorithm changes have to be approved by the science discipline leads and the MODIS science team leader before being used in operational production.

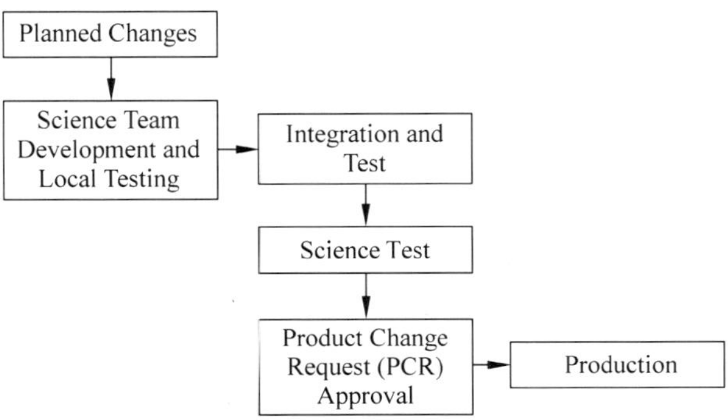

Figure 7.6 The MODIS land science algorithm update and science test process flow

MODIS/Terra land products generated in the months following the first light were of beta quality and not appropriate for scientific publication. These products were archived at the different DAACs under the Collection 1 label and were released to allow the users to gain familiarity with the new data formats and parameters (see Fig. 7.7). Within a year after the MODIS/Terra launch most of the land products matured to provisional quality. These provisional products had a quality that was sufficient for use by the general research community, but users were urged to contact the science team before using the data in publications. Although the quality of these products was not optimal, the improvement over the beta data set was substantial and so a reprocessing of the MODIS/Terra record began in July 2001. These provisional products were archived and distributed from the DAACs as Collection 3 data. Further improvements in the algorithms and an on-going validation effort lead many MODIS land products to achieve a validated stage 1 maturity in November 2002. These high quality science products had uncertainties well defined over a range of representative conditions and were ready to be used in scientific publications. The MODIS/Aqua and combined MODIS Terra and Aqua products were not released until they reached provisional quality. To bring the whole record of MODIS data to the validated stage 1 level, a reprocessing Terra and Aqua data started in December 2002 and July 2003, respectively. Products generated in this activity are archived and distributed from the DAACs as Collection 4 data. Validation activities are still on-going to bring the MODIS land products to validated stage 2 level where the product accuracy is assessed over a distributed set of validation sites. An improved data set will be created as part of a Collection 5 forward and reprocessing effort that is expected to start in the fourth quarter of 2005. New products such as Burn Scar (Roy et al., 2002b) and land cover change will be included in the Collection 5 activity.

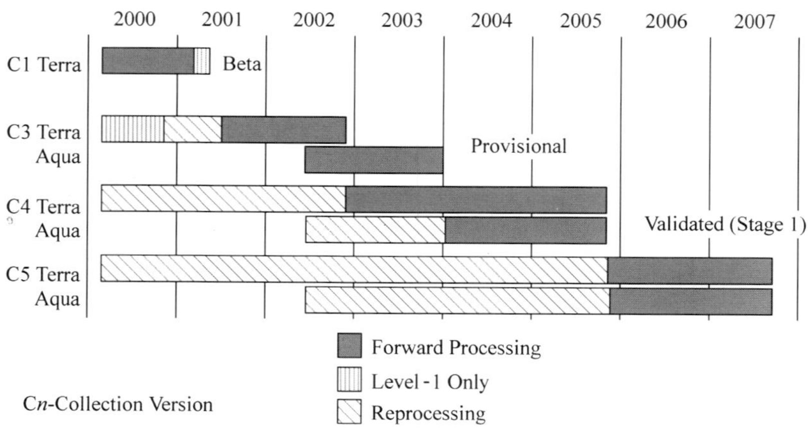

Figure 7.7 Summary of the different Collections of MODIS data sets produced or planned, the range of data included in each Collection and its production time

7.3.3 Quality Assurance Approach

Quality Assurance (QA) is a key element of producing consistent and scientifically useful datasets. The objective of the MODIS land QA activity is to identify and flag suspect or poor-quality data before their release to the public and lead to improvements in product quality. For MODIS land this is a challenge because there are a large number of data dependencies, a large product volume, numerous error sources, and different spatial and temporal resolutions (Roy et al., 2002a). Errors can occur in many areas: in the calibration and characterization of the sensor, in the algorithm itself or its implementation, and in generation of the products.

The QA approach involves activities both within a central QA team and at the science team member facilities. The central QA team developed a number of tools and techniques to help coordinate the QA activities. One tool that was particularly useful in the early stages of the mission was a web-based global browser produced for most of the products on a daily basis to allow for a quick qualitative assessment of the products. Also, key summary QA indicators for each product are stored in a database that enables visualization of the indicators temporally and spatially. The analysis tool also allowed for inter-comparison of indicators within a single product and between products. The QA team and the science team identified a small set of globally distributed key tiles that could be used for an in-depth analysis of the product quality. For each of these tiles, summary statistics are generated for each fine-resolution Level-3 product, separately for each of 7 biomes and each of 13 land cover classes. This summary data can then be plotted temporally and comparisons can easily be made between versions of the products and between MODIS on Terra and on Aqua. To help coordinate across multiple products the central QA team maintains a web page of known issues that are updated based on their analysis and analysis by individual science team members.

Ultimately, each science team member is responsible for producing high quality products. Science team members receive a flow of MODIS data for QA purposes and also order data from the DAACs. Science team members apply specialized QA approaches to this data to better understand the product quality and to test potential algorithm improvements.

7.3.4 Validation Approach

The objective of the MODIS land validation activities is to assess the land products' uncertainty against independent measurements (Morisette et al., 2002). These independent measurements include in-situ data collected over a distributed set of fixed validation test sites, and data from air-borne and other space-borne sensors. As in the QA activity, coordination of the MODIS land science team validation activities enabled resources to be pooled to allow more validation data to be collected within the available validation resources. The validation coordination effort sought to leverage existing resources such as contributing to and participation in intensive field validation campaigns and other existing validation infrastructure. In addition, working with other EOS science teams, a globally distributed set of EOS core validation sites were established to focus year-round field measurement activities. These intensively studied sites allow for tower and field measurements to be extrapolated to the land biophysical products at MODIS' resolution.

7.4 Conclusion

This chapter summarizes the MODIS Land products and describes the process by which the algorithms and the processing have evolved since the start of the EOS Terra and Aqua missions. The large science investigator-led data processing system provided an environment that initially enabled rapid improvement in the quality of the products and then later provided the stability needed to create a consistent time-series of validated products. Key to this effort was a committed science team with both the resources and the responsibility for algorithm development and refinement, as well as, product quality assurance and validation. The science team was complemented by a centralized support and data processing group that coordinated and enhanced the science team's efforts. The EOSDIS infrastructure provided at the DAACs a frame work for production, archive and distribution of the products to both the end users. Lessons learned from this successful effort are being applied to production of land products from heritage instruments such as AVHRR and to future missions such as the National Polar-orbiting Operational Environmental Satellite System (NPOESS) Preparatory Project (NPP) and NPOESS. The MODIS Land global time-series of terrestrial geophysical parameters is meeting many of the needs of the terrestrial global change research and natural resource application community.

Acknowledgements

The author acknowledges support provided by the MODIS Science Team and the MODIS Science Data Support Team. This work was performed under the direction of the MODIS Science Team in the Terrestrial Information Systems Branch (Code 614.5) of the Hydrospheric and Biospheric Sciences Laboratory (Code 614) at NASA GSFC. The work was funded under NASA GSFC contract NAS5-32350.

References

Friedl MA, McIver DK, Hodges JCF, Zhang XY, Muchoney D, Strahler AH, Woodcock CE, Gopal S, Schneider A, Cooper A, Baccini A, Gao F, Schaaf C (2002) Global land cover mapping from MODIS: algorithm and early results. Remote Sens Env 83: 287 – 302

Hall DK, Riggs G, Salomonson VV, DiGirolamo NE, Bayr KJ (2002) MODIS snow-cover products. Remote Sens Env 83: 181 – 194

Hansen MC, DeFries RS, Townshend JRG, Sohlberg R, Dimiceli C, Carrol M (2002) Towards an operational MODIS continuous field of percent tree cover algorithm: examples using AVHRR and MODIS data. Remote Sens Env 83: 303 – 319

Huete A, Didan K, Miura T, Rodriguez EP, Gao X, Ferreira LG (2002) Overview of the radiometric and biophysical performance of the MODIS vegetation indices. Remote Sens Env 83: 195 – 213

Justice CO, Vermote E, Townshend JRG, Defries R, Roy DP, Hall DK, Salomonson VV, Privette JL, Riggs G, Strahler A, Lucht W, Myneni RB, Knyazikhin Y, Running SW, Nemani RR, Wan Z, Huete AR, van Leeuwen W, Wolfe RE, Giglio L, Muller JP, Lewis P, Barnsley MJ (1998) The Moderate Resolution Imaging Spectroradiometer (MODIS): Land remote sensing for global change research. IEEE Trans Geosci Remote Sens 36: 1,228 – 1,249

Justice CO, Giglio L, Korontzi S, Owens J, Morisette JT, Roy D, Descloitres J, Alleaume S, Petticolin F, Kaufman Y (2002a) The MODIS fire products. Remote Sens Env 83: 244 – 262

Justice CO, Townshend JRG, Vermote EF, Masuoka E, Wolfe RE, Saleous, N, Roy DP, Morisette JT (2002b) An overview of MODIS Land data processing and product status. Remote Sens Env 83: 3 – 15

Morisette JT, Privette JL, Justice CO (2002) A framework for the validation of MODIS Land products, Remote Sens Env 83: 77 – 96

Myneni RB, Hoffman S, Knyazikhin Y, Privette JL, Glassy J, Tian Y, Wang Y, Song X, Zhang Y, Smith GR, Lotsch A, Friedl, M, Morisette JT, Votava P, Nemani RR, Running SW (2002) Global products of vegetation leaf area and fraction absorbed PAR from year one of MODIS data. Remote Sens Env 83: 214 – 231

Roy DP, Borak JS, Devadiga S, Wolfe RE, Descloitres J (2002a) The MODIS Land product quality assessment approach. Remote Sens Env 83: 62 – 76

Roy DP, Lewis PE, Justice CO (2002b) Burnt area mapping using mutli-temporal moderate spatial resolution data—a bi-directional reflectance model-based expectation approach. Remote Sens Env 83: 263 – 286

Running SW, Nemani RN, Heinsch FA, Zhao M, Reeves MC, Hashimoto H (2004) A Continuous Satellite-Derived Measure of Global Terrestrial Primary Production. BioScience 54(6): 547 – 560

Salomonson VV, Barnes WL, Maymon PW, Montgomery HE, Ostrow H (1989) MODIS: Advanced facility instrument for studies of the Earth as a system. IEEE Trans Geosci Remote Sens 27: 145 – 153

Schaaf CB, Gao F, Strahler AH, Lucht W, Li X, Tsang T, Strugnell NC, Zhang X, Jin Y, Muller

J-P, Lewis P, Barnsley M, Hobson P, Disney M, Roberts G, Dunderdale M, Doll C, d'Entremount R, Hu B, Liang S, Privette J, Roy D (2002) First operational BRDF, albedo nadir reflectance products from MODIS. Remote Sens Env 83: 135 – 148

Snyder JP (1987) Map projections—a working manual. U.S. geological survey professional paper 1395. Washington, DC: United States Government Printing Office.

Vermote EF, El Saleous NZ, Justice CO (2002) Atmospheric correction of MODIS data in the visible to middle infrared: first results. Remote Sens Env 83: 97 – 111

Wolfe RE, Roy DP, Vermote E (1998) MODIS land data storage, gridding and compositing methodology: Level 2 Grid. IEEE Trans Geosci Remote Sens 36: 1,324 – 1,338

Wan Z, Zhang Y, Zhang Q, Li Z (2002) Validation of the land-surface temperature products retrieved from Terra Moderate Resolution Image Spectroradiometer data. Remote Sens Env 83: 163 – 180

Zhang X, Sohlberg RA, Townshend JRG, DiMiceli C, Carroll ML, Eastman JC, Hansen MC, DeFries RS (2002) Detection of land cover changes using MODIS 250 m data. Remote Sens Env 83: 320 – 335

8 Operational Atmospheric Correction of MODIS Visible to Middle Infrared Land Surface Data in the Case of an Infinite Lambertian Target

Eric F. Vermote and Nazmi Z. Saleous

8.1 Introduction

The atmospheric correction of remote sensing data has always been a concern to the ocean color community, where the signal of interest is almost an order of magnitude smaller than the top of the atmosphere signal. The first data over ocean were corrected for gases, molecular and aerosol effect (Gordon et al., 1983). Atmospheric correction over ocean on SeaWiFS, MODIS and MERIS is now including correction for gaseous effect, an inversion of the aerosol type and amount using the near infrared bands (0.76 μm and 0.87 μm) and accounts for the coupling of the sun-glint directional reflectance with the atmosphere. Typically, over open water the accuracy of those corrections is of the order of 10% – 30% of the reflectance in the blue band (0.412 μm and 0.450 μm), which typically represents 4.10^{-3} to 1.10^{-2} absolute reflectance unit under low aerosol loading (typically optical thickness of 0.2). However, over coastal areas, the assumption that the water contribution is in 0.76 μm and 0.87 μm is no longer valid, due to the contributions of sediments. The atmospheric correction for coastal areas can only be achieved on a case by case basis and with variable accuracy.

Over land, because of the lesser impact of the atmosphere compared to ocean, and the lack of dedicated mission (AVHRR was a meteorological satellite and Thematic Mapper was mainly used in land cover studies), the use of standard atmospheric correction procedure has been slower to establish and indices and procedures to minimize atmospheric effect have been widely used. With the design and development of the Earth Observing System mission, atmospheric correction has been prototyped over land for AVHRR (Jasmes and Kallur, 1994; El Saleous et al., 2000), Thematic Mapper (Ouaidrari and Vermote, 1999) and SeaWiFS (Vermote el al., 2001). Dedicated algorithms for retrieval of aerosol over land for MODIS (Kaufman et al., 1997) and MISR (Martonchik et al., 1997) and algorithms for atmospheric correction, which take into account gases, molecular and aerosol effects, as well as surface Bidirectional Reflectance Distribution Function (BRDF) atmosphere, have been designed, documented and evaluated in the pre-launch phase (Vermote et al., 1997; Martonchik et al., 1997).

In the first phase of the mission, an initial validation and evaluation of the MODIS algorithm (Lambertian assumption) on a global basis has been performed and the accuracy established to be 5.10^{-4} or 5% relative accuracy, whichever is greater, under low aerosol optical thickness.

The chaprer describes the operational procedure for atmospheric correction over land in the case of the infinite Lambertian target. It starts first with a theoretical background section and then shows how the solution of the equation of transfer is implemented in operations, a section is devoted to the input of the atmospheric corrections, and the last section discusses the error budget and validation.

8.2 Theoretical Background

Using the formalism developed for the 5S code, the solution of the radiation transfer equation, corresponding to the problem illustrated by Fig. 8.1(a) and employing the Lambertian Uniform Target assumption for observation in spectral band *i*, assuming a standard atmospheric profile, but variable, pressure (*P*), ozone and water vapor amount (U_{O_3}, U_{H_2O}), is written as (Vermote et al., 1997):

$$\rho^i_{\mathrm{TOA}}(\theta_s,\theta_v,\phi,P,\overbrace{\tau^i_A,\omega^i_0,P^i_A}^{Aer^i},U_{H_2O},U_{O_3}) =$$

$$T^i_{g_{OG}}(m,P)T^i_{g_{O_3}}(m,U_{U_{O_3}})\left[\begin{array}{l}\rho^i_{\mathrm{atm}}(\theta_s,\theta_v,\phi,P,Aer^i,U_{H_2O}) \\ +Tr^i_{\mathrm{atm}}(\theta_s,\theta_v,P,Aer^i)\dfrac{\rho_S}{1-S^i_{\mathrm{atm}}(P,Aer^i)\rho_S}T^i_{g_{H_2O}}(m,U_{U_{H_2O}})\end{array}\right] \tag{8.1}$$

Where ρ_{TOA} = the reflectance at the top of the atmosphere;
T_g = the gaseous transmission by water vapor, $T_{g_{H_2O}}$, by ozone, $T_{g_{O_3}}$, or other gases, $T_{g_{OG}}$ (e.g. CO_2...);
ρ_{atm} = the atmosphere intrinsic reflectance;
Tr_{atm} = the total atmosphere transmission (downward and upward);
S_{atm} = the atmosphere spherical albedo; and
ρ_S = the surface reflectance to be retrieved by the atmospheric correction procedure.

The geometrical conditions are described by θ_s, the solar zenith angle, θ_v, the view zenith angle and ϕ, the difference between the solar and view azimuth angle, P is the pressure which influences the number of molecules in the atmosphere and the concentration of absorbing gases.

τ_A, ω_0 and P_A describe the aerosol properties and are spectrally dependent, τ_A is the aerosol optical thickness, ω_0 is the aerosol single scattering albedo, describing the absorption of the aerosol, ω_0 is equal to 1 for non-absorption particles and 0 for completely absorbing particles.

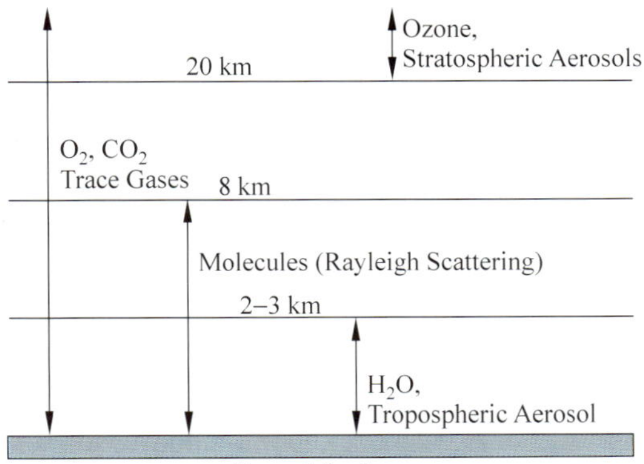

(a)

Figure 8.1(a) The atmospheric components affecting the remote sensing signal in the 0.4 – 2.5 μm range

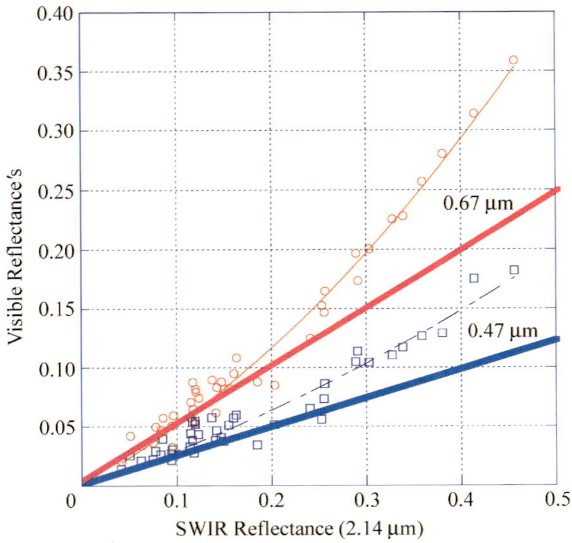

(b)

Figure 8.1(b) Empirical relationship between the visible and short wave infrared reflectance's observed over 40 sun-photometer sites a variety of land cover type and distributed globally

P_A is the aerosol phase function, U_{H_2O} is the integrated water vapor content, U_{O_3} is the integrated ozone content, m is the air-mass computed as $1/\cos(\theta_s) + 1/\cos(\theta_v)$.

The effect of the water vapor on the atmosphere intrinsic reflectance is approximated in 6S code as:

$$\rho^i_{atm}(\theta_s,\theta_v,\phi,P,Aer^i,U_{H_2O})$$
$$= \rho^i_R(\theta_s,\theta_v,\phi,P) + \left(\rho^i_{R+Aer}(\theta_s,\theta_v,\phi,P,Aer^i) - \rho^i_R(\theta_s,\theta_v,\phi,P)\right) Tg^i_{H_2O}\left(m,\frac{U_{U_{H_2O}}}{2}\right) \tag{8.2}$$

where ρ_R represents the reflectance of the atmosphere due to molecular (Rayleigh) scattering, and ρ_{R+Aer} represents the reflectance of the mixing molecule and aerosol, which is computed in 6S using the successive order of scattering method. Accounting correctly for the mixing and the so-called coupling effect (Deschamps et al., 1983) is important for achieving high accuracy in the modeling of atmospheric effect. This approximation conserves the correct computation of the coupling, and supposes that the water vapor is mixed with aerosol and that the molecular scattering is not affected by the water vapor absorption. This approximation is reasonable in most cases where observation bands are narrow and there is no strong absorption by the water vapor, as it is the case for surface remote sensing bands.

The total atmosphere transmission, T_r, is further decomposed into a downward and an upward term, which are respectively dependent on θ_s and θ_v and are computed using the same function by virtue of the reciprocity principle, that is:

$$Tr^i_{atm}(\theta_s,\theta_v,P,Aer^i) = T^i_{atm}(\theta_s,P,Aer^i)T^i_{atm}(\theta_v,P,Aer^i) \tag{8.3}$$

8.3 Operational Implementation

8.3.1 Simplification to Account for Surface Pressure

For the computer code, the functions related to atmospheric scattering and absorption, ρ_{atm}, T_{atm} and S_{atm} can be computed by interpolation from a pre-computed lookup table because they can not be simply modeled. The gaseous transmission function can be written in MODIS or VIIRS bands as simple analytical function. The molecular reflectance term can be computed very efficiently using a semi-empirical approach based on the decomposition suggested by Chandrasekhar, which is described in details in Vermote and Tanr'e (1992).

Using a subsequent approximation, we can further simplify the dependence of the key term on the pressure, by only computing $\rho_{\mathrm{R+Aer}}$ at standard pressure, P_0, enabling us to substantially reduce the dimension of the lookup tables, that is:

$$\rho^i_{\mathrm{atm}}(\theta_s,\theta_v,\phi,P,Aer^i,U_{\mathrm{H_2O}}) = \rho^i_{\mathrm{R}}(\theta_s,\theta_v,\phi,P)+\left(\rho^i_{\mathrm{R+Aer}}(\theta_s,\theta_v,\phi,P_0,Aer^i)-\rho^i_{\mathrm{R}}(\theta_s,\theta_v,\phi,P_0)\right)T^i_{g_{\mathrm{H_2O}}}\left(m,\frac{U_{U_{\mathrm{H_2O}}}}{2}\right) \tag{8.4}$$

The same approach could be applied to the transmission term, that is:

$$T^i_{\mathrm{atm}}(\theta,P,Aer^i)=T^i_{\mathrm{atm}}(\theta,P_0,Aer^i)\frac{T^i_{\mathrm{R}}(\theta,P)}{T^i_{\mathrm{R}}(\theta,P_0)} \tag{8.5}$$

where T_{R} is the atmosphere transmission function due to molecular scattering.

8.3.2 Detailed Computations

The code implements the equations detailed in (8.1) – (8.5), using a lookup table approach and analytic expression. The following section details the computation of each term in the computer code.

8.3.2.1 $T^i_{g_{\mathrm{OG}}}(m,P)$—Gaseous Transmission by Other Gases

The gaseous transmission by gases other than water or ozone in each spectral band can be written as a function of the air mass, m, and the pressure P (in atm), as:

$$T^i_{g_{\mathrm{OG}}}(m,P)=\exp\left[\begin{array}{l} m\left(a^i_0P+a^i_1\mathrm{Log}(P)\right)\\ +\mathrm{Log}(m)\left(b^i_0P+b^i_1\mathrm{Log}(P)\right)+m\mathrm{Log}(m)\left(c^i_0P+c^i_1\mathrm{Log}(P)\right)\end{array}\right] \tag{8.6}$$

8.3.2.2 $T^i_{g_{\mathrm{O_3}}}(m,U_{\mathrm{O_3}})$—Ozone Gaseous Transmission

The ozone gaseous transmission in the narrow bands (in the Chappuis band) could be simply modeled as:

$$T^i_{g_{\mathrm{O_3}}}(m,U_{\mathrm{O_3}})=e^{-ma^i_{\mathrm{O_3}}U_{\mathrm{O_3}}} \tag{8.7}$$

8.3.2.3 $T^i_{g_{\mathrm{H_2O}}}(m,U_{\mathrm{H_2O}})$—Water Vapor Gaseous Transmission

The water vapor transmission is modeled as:

$$T^i_{g_{\mathrm{H_2O}}}(m,U_{\mathrm{H_2O}})=\exp\left[\begin{array}{l} a^i_{\mathrm{H_2O}}mU_{\mathrm{H_2O}}\\ +b^i_{\mathrm{H_2O}}\mathrm{Log}(mU_{\mathrm{H_2O}})+c^i_{\mathrm{H_2O}}mU_{\mathrm{H_2O}}\mathrm{Log}(mU_{\mathrm{H_2O}})\end{array}\right] \tag{8.8}$$

8.3.2.4 $\rho_{\mathrm{R}}^{i}(\theta_{s},\theta_{v},\phi,P_{0})$—Molecular Atmospheric Reflectance at Standard Pressure

This quantity is computed by the 6S subroutine CHAND.f, described in Vermote et al. (1992), which accepts as direct input the geometrical conditions (μ_{s},μ_{v},ϕ), where μ_{s} (resp. μ_{v}) is the cosine of the solar (resp. view) zenith angle, and ϕ the relative azimuth and the molecular optical thickness in that case at standard pressure, τ_{R}, which is pre-computed (by 6S).

8.3.2.5 $\rho_{\mathrm{R}}^{i}(\theta_{s},\theta_{v},\phi,P_{0})$—Molecular Atmospheric Reflectance at Actual Pressure

The adjustment is simply done by adjusting the amount of molecules or the molecular optical thickness, according to:

$$\tau_{\mathrm{R}}(P)=P\tau_{\mathrm{R}} \tag{8.9}$$

The pressure, P, is expressed in atmospheres.

8.3.2.6 $\rho_{\mathrm{R+Aer}}^{i}(\theta_{s},\theta_{v},\phi,P_{0},Aer^{i})$—Intrinsic Reflectance at Standard Pressure

This quantity is pre-computed by 6S in a lookup table for each band and each aerosol model $(P_{\mathrm{A}},\omega_{0})$. The step in solar zenith angle is 4 deg, in view angle 4 deg corresponding to the gauss quadrature of 24 angles (with the nadir added), the step is kept constant in scattering angle (4 degree), Θ, defined as:

$$\cos(\Theta)=-\cos(\theta_{s})\cos(\theta_{v})-\cos(\phi)\sin(\theta_{s})\sin(\theta_{v}) \tag{8.10}$$

Resulting in a variable number of steps is for each θ_{s}, θ_{v} configuration. The indexing to the correct values in the lookup table is achieved through the use of the ANGLE lookup table, which keeps track of the number of azimuth angles computed for each θ_{s}, θ_{v} configuration. Though, more expensive and more complicated to interpolate within, this structure achieves a higher precision with a reduced size lookup table, for a term for which accuracy is critical to the atmospheric correction.

The step in aerosol optical depth is variable to optimize the performance of the correction with the error induced by the interpolation (i.e. finer a low optical depth).

8.3.2.7 $T_{\mathrm{atm}}^{i}(\theta,P_{0},Aer^{i})$—Atmosphere Transmission on at Standard Pressure

This quantity is pre-computed in 6S by using the successive order of scattering method and illuminating the bottom of the layer with isotropic light. The code accounts for the mixing of aerosol molecules within the atmosphere. The values are computed with a step of 4 deg in θ and for each aerosol model and each band for predefined values of τ_{A}. The interpolation for any θ and τ is relatively straightforward since this table has only 2 dimensions. The table volume is also very modest.

8.3.2.8 $T_{\mathrm{R}}^{i}(\theta, P_0)$—Molecular (Rayleigh) Transmission at Standard Pressure

The molecular transmission at standard pressure is computed using the value of molecular optical depth at standard pressure, τ_{R}. Using the two stream method, the molecular transmission could be approximated by:

$$T_{\mathrm{R}}^{i}(\theta, P_0) = \frac{\lfloor 2/3 + \cos(\theta) \rfloor + \lfloor 2/3 - \cos(\theta) \rfloor \mathrm{e}^{-\tau_{\mathrm{R}}/\cos(\theta)}}{4/3 + \tau_{\mathrm{R}}} \quad (8.11)$$

8.3.2.9 $T_{\mathrm{R}}^{i}(\theta, P)$—Molecular (Rayleigh) Transmission on at Actual Pressure

Using the same method as in Molecular Atmospheric Reflectance at Standard Pressure we simply replace, in Eq. (8.9), τ_{R} with $\tau_{\mathrm{R}}(P)$ (Eq. (8.7)).

8.3.2.10 Atmosphere Spherical Albedo at Actual Pressure

The atmospheric spherical albedo at actual pressure, $S_{\mathrm{atm}}^{i}(P, Aer^{i})$, is described as:

$$S_{\mathrm{atm}}^{i}(P, Aer^{i}) = \int_0^{\pi/2} \int_0^{\pi/2} \int_0^{2\pi} \rho_{\mathrm{atm}}^{i}(\theta, \theta', \phi, P, Aer^{i}) \sin(\theta) \cos(\theta') \mathrm{d}\theta \mathrm{d}\theta' \mathrm{d}\phi \quad (8.12)$$

By ignoring the water vapor dependence on the atmosphere intrinsic reflectance (S acting as a second order effect), we can write the same relation we have written for the atmosphere intrinsic reflectance, that is

$$S_{\mathrm{atm}}^{i}(P, Aer^{i}) = (S_{\mathrm{atm}}^{i}(P_0, Aer^{i}) - S_{\mathrm{R}}^{i}(P_0)) + S_{\mathrm{R}}^{i}(P) \quad (8.13)$$

So the $S_{\mathrm{atm}}^{i}(P_0, Aer^{i})$ is stored in a pre-calculated lookup table depending only on aerosol optical depth and model. The $S_{\mathrm{R}}^{i}(P)$ term is computed by an analytic expression based on the integral of Eq. (8.11) that is:

$$S_{\mathrm{R}}^{i}(P) = \frac{1}{4 + 3\tau_{\mathrm{R}}} [3\tau_{\mathrm{R}} - 4E_3(\tau_{\mathrm{R}}) + 6E_4(\tau_{\mathrm{R}})] \quad (8.14)$$

where E_3 and E_4 are exponential integral function (see 6S code for details; Vermote et al., 1997).

8.4 Input and Ancillary Data

The atmospheric correction approach described in Sections 8.2 and 8.3 requires key atmospheric parameters: surface pressure, ozone concentration, column water vapor and aerosol optical thickness. The surface pressure and ozone concentration are slow varying parameters both spatially and temporally. They can be estimated from the coarse resolution meteorological data. We recommend using an interpolation scheme in the temporal and spatial space to determine these parameters at the acquisition time and spatial resolution.

In general, the water vapor content and aerosols vary strongly in time and space. Where possible, they should be derived from data acquired by the same instrument for which the atmospheric correction is performed, or an instrument flying on the same platform. In the case of the MODIS surface reflectance, these parameters are derived from MODIS calibrated data.

8.4.1 Surface Pressure

The surface pressure (P) is used to compute the Rayleigh optical thickness. A primary source of surface pressure meteorological data is the National Center for Environmental Prediction (NCEP) Global Data Assimilation System (GDAS) where the surface pressure parameter is produced at a spatial resolution of 1×1 degree every 6 hours. These data are available for the period 1948 – present. The real-time data can be obtained from NCEP's FTP site (ftp.ncep.noaa.gov) and historical data can be ordered from the National Center for Atmospheric Research (NCAR) (http://dss.ucar.edu).

Other sources of a 1×1 degree surface pressure field include the National Aeronautics and Space Agency's (NASA) Global Modeling and Assimilation Office (GMAO) (http://gmao.gsfc.nasa.gov) and the European Center for Medium-Range Weather Forecast (ECMWF) where these data are available for the period 1958 – present (http://www.ecmwf.int).

The coarse spatial resolution of the meteorological data does not match the higher spatial variability of the surface pressure due to terrain elevation. To increase the spatial resolution of the surface pressure field, we use a Digital Elevation Model (DEM) to map the surface pressure at a higher resolution within each meteorological data grid cell. The GTOPO30 DEM is available globally at a resolution of 30 arc seconds (approximately 1 km) (http://edcdaac.usgs.gov/ gtopo30/ gtopo30.asp). To increase the resolution within a meteorological data grid cell where the surface pressure is P_{meteo}, we determine the set of DEM pixels that intersect the grid cell and compute for each pixel the standard pressure P^{i}_{DEM} where:

$$P^{i}_{\text{DEM}}(\text{millibar}) = 1{,}013 \bullet \frac{\text{Elevation(km)}}{8} \tag{8.15}$$

The ratio of P_{meteo} and the average pressure derived from the selected DEM pixels $(\langle P_{\text{DEM}} \rangle)$ is used to adjust the pressure at the DEM resolution. We assumed that the accuracy on the final pressure is 10 millibars.

8.4.2 Ozone

The ozone concentration is primarily obtained from the NASA's Total Ozone

Mapping Spectrometer (TOMS). These data are available daily at a spatial resolution of 1×1 degree for the period 1979 – present and can be obtained from the TOMS Website at http://toms.gsfc.nasa.gov. The nominal uncertainty of this product reported on the TOMS Web page is 3% – 4% (http://toms.gsfc.nasa.gov/eptoms/dataqual/nominal.html).

Alternatively, the National Oceanic and Atmospheric Administration's (NOAA) Total Operational Vertical Sounder (TOVS) ozone product available from NOAA's Climate Prediction Center (CPC) (http://www.cpc.ncep.noaa.gov/products/stratosphere/tovsto/) can be used.

8.4.3 Water Vapor

Where possible, the column water vapor should be derived from the instrument where atmospheric correction is performed. In the case of the MODIS instrument for example, the near-infrared bands 18 (931 – 941 nm) and 19 (915 – 965 nm) are used to retrieve the column water vapor content. The approach based on the two-band ratio is described by Gao (Gao and Kaufman, 2003). This approach determines the instantaneous water vapor content at the time of acquisition with an accuracy of 5% – 10%. Alternatively, meteorological data from NCEP GDAS can be used.

8.4.4 Aerosol Optical Thickness

The approach to aerosol optical thickness retrieval over land is based on the "dark and dense vegetation (DDV)" technique introduced by Holben et al. (1992). It is based on using an empirical relationship between the surface reflectance in the shortwave visible bands (where the aerosol effect is strong and the surface signal is low) and in the shortwave infrared bands (where the aerosol effect is negligible) to predict the surface reflectance in the visible bands. Such a relation has been used for different instruments. Ouaidrari and Vermote (1999) estimated the surface reflectance of dark target in Landsat's Band 1 (490 nm) to be 1/3 of the reflectance in band 7 (2.19 μm). El Saleous et al. (2000) estimated the surface reflectance in AVHRR's band 1 (670 nm) using the reflective component of band 3 (3.75 μm).

The original approach suggested using a linear relationship limited in scope to dark targets. Using a set of 40 AERONET sites representative of different land covers; we derived a non-linear relationship that can be applied to brighter targets. Figure 8.1(b) shows plots of the surface reflectance in blue and red bands (470 and 650 nm) as a function of shortwave infrared (SWIR) reflectance (2,130 nm) for surface with a SWIR reflectance up to 0.5. To compute the estimated surface reflectance in the blue and red bands, the relationship in Fig. 8.1(b) is applied to the reflectance in band 7 (2,130 nm) after it has been corrected for atmospheric effects.

Using the surface reflectance estimate in the blue and red bands, the estimated top of the atmosphere reflectance at the observation geometry ($\rho^i_{\text{TOA\_est}}(\theta_s, \theta_V, \phi, Aer^i)$) is computed using Equations (8.1) – (8.4) and the intrinsic atmospheric reflectance ($\rho^i_{\text{R+Aer}}(\theta_s, \theta_V, \phi, P_0, Aer^i)$ and atmospheric transmission ($T^i_{\text{atm}}(\theta, P_0, Aer^i)$, described in Sections 8.3.2.6 and 8.3.2.7. $\rho^i_{\text{TOA\_est}}(\theta_s, \theta_V, \phi, Aer^i)$ is computed for all the optical thickness values included in the 6S lookup tables. The values of the estimated TOA reflectance bracketing the observed TOA reflectance ($\rho^i_{\text{TOA\_est}}(\theta_s, \theta_V, \phi, Aer^i_1)$ and $\rho^i_{\text{TOA\_est}}(\theta_s, \theta_V, \phi, Aer^i_2)$) are identified and the aerosol optical thickness (Aer^i) is computed by linear interpolation.

8.5 Application to MODIS Data and Error Budget

The previously described algorithm has been applied to the MODIS instrument on-board the Terra (morning) and Aqua (afternoon) satellites. In that case, the retrieval of the water vapor integrated content and the aerosol optical thickness is performed using the remotely sensed data themselves using the 1-km resolution bands that enable the capture of spatial and temporal variability of those inputs. In order to examine the impact of the atmospheric effect on the MODIS land bands and estimate the accuracy of the atmospheric correction under several scenarios, we have selected three typical land covers, a forest type, a savanna and a semi arid surface. The data have been acquired by MODIS at a sun-photometer site on a day where the optical thickness was low; the correction of the level-1B data has been performed using 6S and the sun-photometer measurements (optical thickness, size distribution and refractive indices). The data have been slightly adjusted in Bands 1 (645 nm) and 3 (470 nm) to agree with the empirical relationship used for the aerosol retrieval algorithm, the error on the atmospheric correction algorithm by uncertainties in that relationship will be addressed later in this section (8.5.5).

For a variety of atmosphere and geometrical conditions (see Table 8.1(a)) the signals at the top of the atmosphere have been simulated for the three sites using the 6S radiative transfer code (Vermote et al., 1997). Figures 8.2(a) – 8.2(c) show the surface reflectance for the three sites as a function of the central wavelength of each of the seven MODIS land bands. The atmospheric impact is strong in all the bands due to the fact that all of the ranges of the aerosol models have been considered in this simulation (see Table 8.1(b)), based on the climatology established by Dubovik et al. (2002). The atmospheric effect is considerably larger at short wavelength, especially with respect to the ground surface reflectance, which is definitely to our advantage since we are using those wavelengths to retrieve aerosol properties.

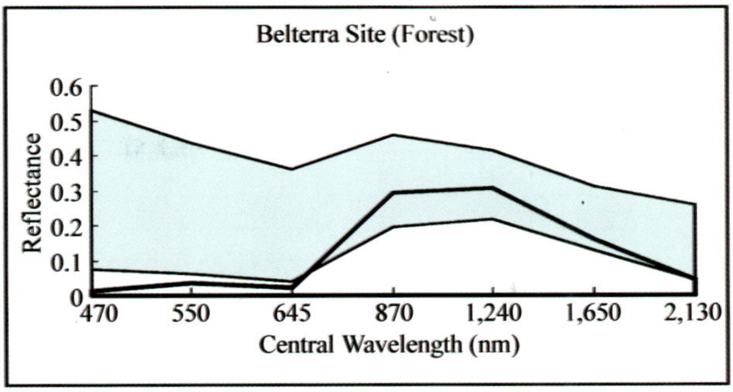

(a)

Figure 8.2(a) Surface reflectance at the Belterra site (forest) for each of the seven MODIS bands (thick line), the blue area represents the variability in the signal at the top of the atmosphere encountered by simulating the conditions described in Table 8.1

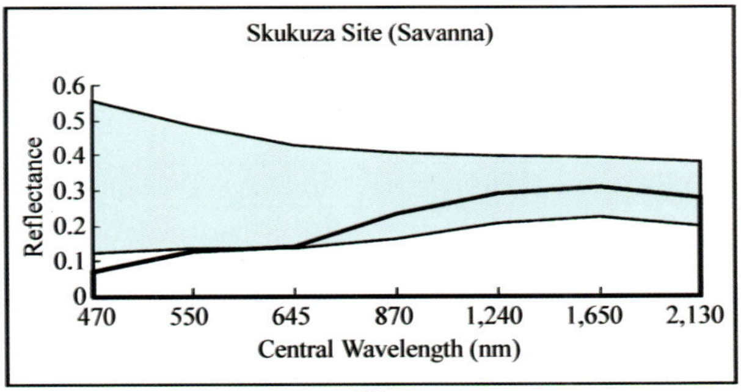

(b)

Figure 8.2(b) Surface reflectance at the Skukuza site (savanna) for each of the seven MODIS bands (thick line), the blue area represents the variability in the signal at the top of the atmosphere encountered by simulating the conditions described in Table 8.1

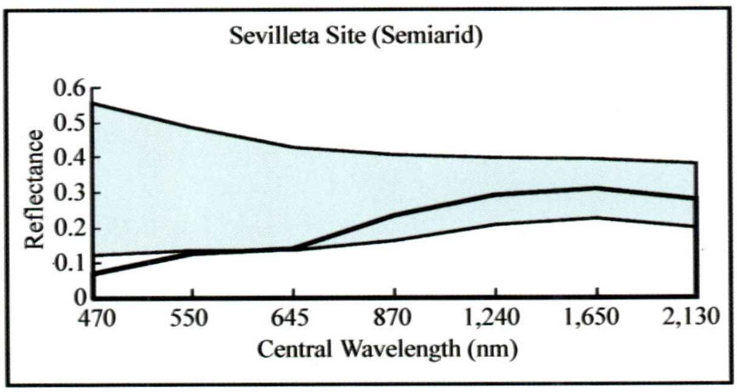

(c)

Figure 8.2(c) Surface reflectance at the Sevilleta site (semiarid) for each of the seven MODIS bands (thick line), the blue area represents the variability in the signal at the top of the atmosphere encountered by simulating the conditions described in Table 8.1

Table 8.1(a) Description of the different parameter set used to generate the top of the atmosphere reflectances and compute the uncertainties in the corrected surface reflectances

Parameter	Values			
	Solar Zenith	View Zenith	Relative Azimuth	Case Name
Geometrical Conditions	30	0	0	A
	30	30	0	B
	30	30	180	C
	30	60	0	D
	30	60	180	E
	60	0	0	F
	60	30	0	G
	60	30	180	H
	60	60	0	I
	60	60	180	J
Aerosol Optical Depth	0.05 (clear) 0.30 (average) 0.50 (high)			
Aerosol Model	urban clean, urban polluted, smoke low absorption smoke high absorption (see Table 8.1(b) for details)			
Water Vapor Content (g/cm^2)	1.0, 3.0 and 5.0 uncertainties +/− 0.2			
Ozone Content (cm · atm)	0.25, 0.3, 0.35 uncertainties +/− 0.02			
Pressure (mb)	1,013 mb, 930 mb, 845 mb uncertainties +/− 10			

Table 8.1(b) Description of the characteristics of the aerosol model used in the study (based on the climatology of Dubovik et al. (2002))

		Aerosol model			
		Urban Clean	Urban Polluted	Smoke Low Absorption	Smoke High Absorption
Refractive Index	Real	$1.41 - 0.03\,\tau_{440\,nm}$	1.47	1.47	1.51
	Imaginary	0.003	0.014	0.0093	0.021
Small Particle Mode	Volume Mean Radius (μm)	$0.12 + 0.11\,\tau_{440\,nm}$	$0.12 + 0.04\,\tau_{440\,nm}$	$0.13 + 0.04\,\tau_{440\,nm}$	$0.12 + 0.025\,\tau_{440\,nm}$
	Standard Deviation	0.38	0.43	0.40	0.40
	Volume Concentration ($\mu m^3/\mu m^2$)	$0.15\,\tau_{440\,nm}$	$0.12\,\tau_{440\,nm}$	$0.12\,\tau_{440\,nm}$	$0.12\,\tau_{440\,nm}$

Continued

		Aerosol model			
		Urban Clean	Urban Polluted	Smoke Low Absorption	Smoke High Absorption
Coarse Particle Mode	Volume Mean Radius (μm)	3.03+0.49 $\tau_{440\ nm}$	2.72+0.60 $\tau_{440\ nm}$	3.27+0.58 $\tau_{440\ nm}$	3.22+0.71 $\tau_{440\ nm}$
	Standard Deviation	0.75	0.63	0.79	0.73
	Volume Concentration ($\mu m^3/\mu m^2$)	0.01+0.04 $\tau_{440\ nm}$	0.11 $\tau_{440\ nm}$	0.05 $\tau_{440\ nm}$	0.09 $\tau_{440\ nm}$

The rest of this section presents, in detail, the impact of the sources of uncertainties, calibration, ozone and water vapor content, pressure, the relationship between the 2,130 nm and 470 nm, 645 nm bands, as well as the aerosol type, which is not inverted by the procedure which relies on a prescribed model (urban clean) and adjust the spectral dependence of the actual aerosol by using retrieval at both 470 nm and 645 nm. As a theoretical error budget, the precision is only indicative of the potential accuracy of the product and needs to be verified by independent validation (see Section 8.5.7).

8.5.1 Calibration Uncertainties

We ran a set of simulations for three different optical thicknesses (0.05:clear; 0.30:avg; 0.50:high), using the urban clean model, for an average content in water vapor and ozone at standard pressure for the 10 geometrical conditions (a through *j*), for each of the three different sites. We simulated an error of +2% and −2% in the absolute calibration across all the seven MODIS bands. The results of the simulations are summarized in Tables 8.2(a) – 8.2(c), where we report the maximum and minimum absolute error encountered as a function of aerosol optical thickness, we report the geometrical conditions at which that maximum or minimum occurred.

We also report the average error for all the geometrical conditions which will be used later when we are summing all the uncertainties. The error increases with the increase in optical thickness and the maximum error occurs where the atmospheric effects are the strongest (case *i*, sun and view at 60 deg in the backscattering directions). Generally, the overall error stays under 2% in relative for all optical thicknesses considered and does not amplified the error sources. It is true that a similar error across all wavelengths is probably favorable, but it is also representing the most likely error for MODIS which intra bands for the land bands is probably better than 2% due to the use of the solar diffuser and solar diffuser stability monitor in the calibration process. Also presented in Fig. 8.3 is the error on the retrieved optical thickness given the error on calibration.

Table 8.2(a) Error on the surface reflectance (× 10,000) due to uncertainties in the absolute calibration (±2%) for the Belterra site

Central Wavelength (nm)		470	550	645	870	1,240	1,650	2,130
Surface Reflectance × 10,000		120	375	240	2,931	3,083	1,591	480
Maximum Error × 10,000	Clear	0008j	0010i	0015i	0077i	0084i	0078i	0083i
	Avg.	0009i	0012d	0027i	0090i	0085i	0062i	0044i
	High	0012d	0013i	0047i	0112i	0106i	0089i	0071i
Minimum Error × 10,000	Clear	0003c	0005j	0005c	0059a	0059f	0031c	0014c
	Avg.	0000e	0004j	0002j	0060c	0061c	0032c	0011c
	High	0003f	0003e	0007c	0062c	0062c	0033c	0012c
Average Error × 10,000	Clear	4	7	7	62	65	44	34
	Avg.	2	8	10	67	66	39	19
	High	7	8	16	76	72	46	27

Table 8.2(b) Error on the surface reflectance (× 10,000) due to uncertainties in the absolute calibration (±2%) for the Skukuza site

Central Wavelength (nm)		470	550	645	870	1,240	1,650	2,130
Surface Reflectance × 10,000		400	636	800	2,226	2,880	2,483	1,600
Maximum Error × 10,000	Clear	0013j	0013b	0024i	0065i	0080i	0080i	0080i
	Avg.	0015i	0015i	0030i	0074i	0079i	0070i	0054i
	High	0013d	0011a	0049i	0101i	0098i	0088i	0071i
Minimum Error × 10,000	Clear	0003a	0009j	0016c	0045c	0056f	0048f	0031f
	Avg.	0006c	0004j	0016c	0046c	0058c	0049c	0032c
	High	0001e	0005j	0018c	0048c	0058c	0050c	0032c
Average Error × 10,000	Clear	9	11	17	49	61	56	45
	Avg.	8	11	19	53	63	54	37
	High	6	9	24	63	68	59	42

Table 8.2(c) Error on the surface reflectance (×10,000) due to uncertainties in the absolute calibration (±2%) for the Sevilleta site

Central Wavelength (nm)		470	550	645	870	1,240	1,650	2,130
Surface Reflectance ×10,000		700	1,246	1,400	2,324	2,929	3,085	2,800
Maximum Error ×10,000	Clear	0019j	0026i	0033i	0062i	0077i	0081i	0080i
	Avg.	0021i	0030i	0039i	0067i	0075i	0074i	0070j
	High	0017a	0040d	0046i	0086i	0090j	0086j	0081j
Minimum Error ×10,000	Clear	0014a	0021j	0027j	0047a	0058c	0061c	0055c
	Avg.	0012e	0017j	0028c	0048a	0059a	0062a	0056a
	High	0006e	0014j	0029a	0049a	0059c	0061c	0055c
Average Error ×10,000	Clear	14	24	28	49	61	65	61
	Avg.	14	25	30	53	63	64	58
	High	12	24	32	60	67	67	60

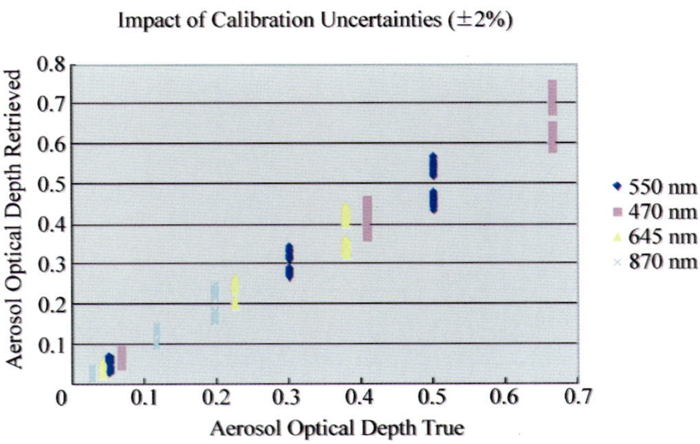

Figure 8.3 Comparison of the optical depth at 470 nm, 550 nm, 645 nm and 870 nm retrieved and input in the simulation for the 9 geometries given in Table 8.2 for an error of calibration of ±2%

8.5.2 Uncertainties on Ancillary Data Pressure

We ran a set of simulations at three different pressures, 1,013 mb, 930 mb, and 845 mb, with a variation of 10 mb for each case for an optical depth of 0.3. The three different pressures represent sites at altitude of 0 m, 700 m and 1,500 m. Tables 8.3(a) – 8.3(c) report the error in each band for the three different sites. The pressure error will influence the molecular scattering term and also the concentration of trace gases that might be affecting a specific band. However the aerosol optical thickness is also affected by the uncertainty on surface pressure (see Fig. 8.4), in such a way that eventually all the bands become affected.

Table 8.3(a) Error on the surface reflectance ($\times$ 10,000) due to uncertainties in the surface pressure ($\pm$10 mb) for the Belterra site

Central Wavelength (nm)		470	550	645	870	1,240	1,650	2,130
Surface Reflectance $\times$ 10,000		120	375	240	2,931	3,083	1,591	480
Maximum Error $\times$ 10,000	Clear	0003i	0001e	0009i	0008i	0007i	0011i	0012i
	Avg.	0003i	0001e	0008i	0008i	0007i	0011i	0011j
	High	0002i	0001e	0008i	0008i	0007i	0010i	0011i
Minimum Error $\times$ 10,000	Clear	0000c	0000a	0000c	0000a	0000a	0000c	0000c
	Avg.	0000c	0000a	0000c	0000a	0000a	0000c	0000c
	High	0000a	0000a	0000c	0000a	0000a	0000a	0000c
Average Error $\times$ 10,000	Clear	1	0	2	1	0	2	2
	Avg.	1	0	1	1	0	1	2
	High	0	0	1	1	0	1	2

Table 8.3(b) Error on the surface reflectance ($\times$ 10,000) due to uncertainties in the surface pressure ($\pm$10 mb) for the Skukuza site

Central Wavelength (nm)		470	550	645	870	1,240	1,650	2,130
Surface Reflectance $\times$ 10,000		400	636	800	2,226	2,880	2,483	1,600
Maximum Error $\times$ 10,000	Clear	0004i	0002e	0006i	0009i	0007i	0006i	0005i
	Avg.	0004i	0002e	0006i	0009i	0007i	0006i	0005j
	High	0003i	0001e	0006i	0009i	0007i	0005i	0005i
Minimum Error $\times$ 10,000	Clear	0000j	0000a	0000a	0000a	0000a	0000a	0000a
	Avg.	0000j	0000a	0000a	0000a	0000a	0000a	0000a
	High	0000c	0000a	0000a	0000a	0000a	0000a	0000a
Average Error $\times$ 10,000	Clear	1	0	1	1	0	1	1
	Avg.	1	0	1	1	0	1	1
	High	0	0	1	1	0	1	1

Table 8.3(c) Error on the surface reflectance ($\times$ 10,000) due to uncertainties in the surface pressure ($\pm$10 mb) for the Sevilleta site

Central Wavelength (nm)		470	550	645	870	1,240	1,650	2,130
Surface Reflectance $\times$ 10,000		700	1,246	1,400	2,324	2,929	3,085	2,800
Maximum Error $\times$ 10,000	Clear	0005i	0002d	0003i	0007i	0005i	0002i	0004j
	Avg.	0004i	0002d	0003i	0006i	0005i	0002i	0003f
	High	0004i	0002e	0003i	0006i	0005i	0002i	0003j
Minimum Error $\times$ 10,000	Clear	0001a	0001a	0000b	0000a	0000a	0000a	0001i
	Avg.	0000j	0000b	0000a	0000a	0000a	0000a	0000i
	High	0000a	0000a	0000a	0000a	0000a	0000a	0000i
Average Error $\times$ 10,000	Clear	1	1	1	0	0	0	2
	Avg.	1	1	1	0	0	0	1
	High	1	1	0	0	0	0	1

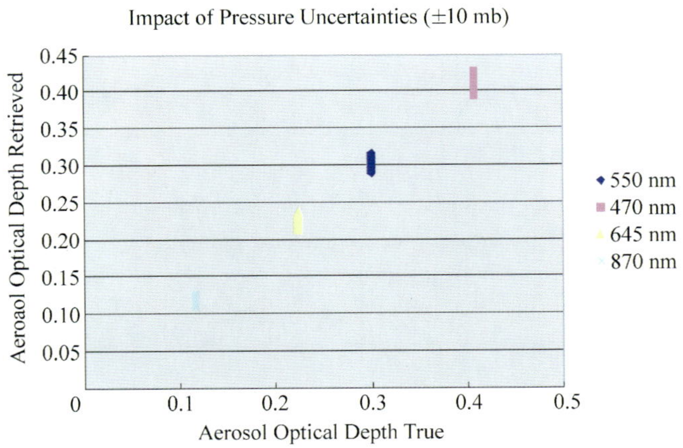

Figure 8.4 Comparison of the optical depth at 470 nm, 550 nm, 645 nm and 870 nm retrieved and input in the simulation for the 9 geometries given in Table 8.2 for an error on the surface pressure of $\pm$10 mb

8.5.3 Uncertainties on Ancillary Ozone Amount

We ran a set of simulations at three different ozone contents, 0.25 cm.atm, 0.30 cm.atm, and 0.30 cm.atm, with a variation of 0.02 cm.atm for each case for an optical depth of 0.3. Tables 8.4(a) – 8.4(c) report the error in each band for the three different sites. The uncertainties on ozone most affect the band at 550 nm, but the impact is relatively small when comparing to calibration uncertainties. However, since the band at 470 nm is also affected the aerosol optical retrieved (see Fig. 8.5) is impacted.

Table 8.4(a) Error on the surface reflectance (× 10,000) due to uncertainties in the ozone content (±0.02 cm.atm) for the Belterra site

Central Wavelength (nm)		470	550	645	870	1,240	1,650	2,130
Surface Reflectance × 10,000		120	375	240	2,931	3,083	1,591	480
Maximum Error × 10,000	Clear	0001g	0022j	0006i	0019j	0009j	0010i	0011i
	Avg.	0001g	0022j	0006i	0019j	0009j	0010i	0010i
	High	0001g	0022j	0006i	0019j	0009j	0010i	0010i
Minimum Error × 10,000	Clear	0000a	0002a	0000c	0000a	0000a	0000c	0000c
	Avg.	0000a	0002a	0000c	0000a	0000a	0000c	0000c
	High	0000a	0002a	0000c	0000a	0000a	0000c	0000c
Average Error × 10,000	Clear	0	7	1	3	1	2	3
	Avg.	0	7	1	3	1	2	3
	High	0	7	1	3	1	2	3

Table 8.4(b) Error on the surface reflectance (× 10,000) due to uncertainties in the ozone content (±0.02 cm.atm) for the Skukuza site

Central Wavelength (nm)		470	550	645	870	1,240	1,650	2,130
Surface Reflectance × 10,000		400	636	800	2,226	2,880	2,483	1,600
Maximum Error × 10,000	Clear	0002i	0021j	0005i	0024j	0012i	0011i	0011i
	Avg.	0002i	0021j	0005i	0023j	0012i	0011i	0011i
	High	0002i	0021j	0005i	0023j	0012i	0011i	0011i
Minimum Error × 10,000	Clear	0000j	0002a	0000c	0001c	0000a	0000a	0000c
	Avg.	0000j	0002a	0000c	0001c	0000a	0000a	0000c
	High	0000j	0002a	0000c	0001c	0000a	0000a	0000c
Average Error × 10,000	Clear	1	6	1	6	2	2	3
	Avg.	1	6	1	6	2	2	3
	High	1	6	1	6	2	2	3

Table 8.4(c) Error on the surface reflectance (× 10,000) due to uncertainties in the ozone content (±0.02 cm.atm) for the Sevilleta site

Central Wavelength (nm)		470	550	645	870	1,240	1,650	2,130
Surface Reflectance × 10,000		700	1,246	1,400	2,324	2,929	3,085	2,800
Maximum Error × 10,000	Clear	0004d	0024j	0011i	0030i	0024i	0018i	0014i
	Avg.	0004d	0024j	0011i	0030i	0024i	0018i	0014i
	High	0004d	0024j	0011i	0030i	0024i	0018i	0014i
Minimum Error × 10,000	Clear	0001e	0003a	0000c	0002c	0001a	0000a	0000a
	Avg.	0000j	0003a	0000c	0002c	0001a	0000a	0000a
	High	0000j	0003a	0000c	0002c	0001a	0000a	0000a
Average Error × 10,000	Clear	2	8	3	10	5	4	3
	Avg.	2	8	3	10	5	4	3
	High	2	8	3	10	5	4	3

This uncertainty affects all the bands, especially at 870 nm where the aerosol impact is important and extrapolated from 470 nm and 645 nm retrieval.

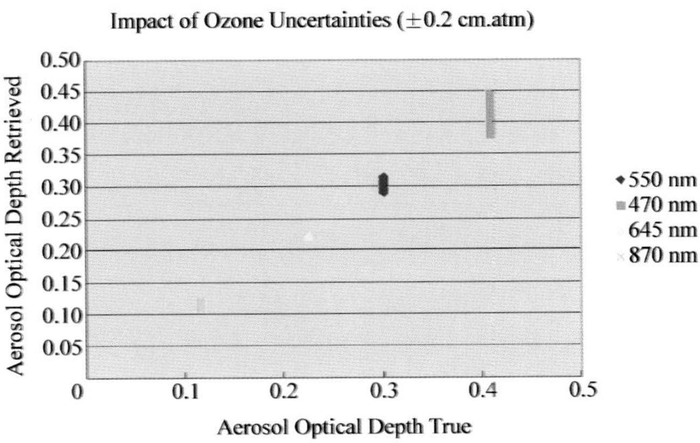

Figure 8.5 Comparison of the optical depth at 470 nm, 550 nm, 645 nm and 870 nm retrieved and input in the simulation for the 9 geometries given in Table 8.2 for an error on the ozone content of ±0.02 cm.atm

8.5.4 Uncertainties on the Water Vapor Amount

The MODIS atmospheric correction algorithm uses the values of water vapor retrieved from differential absorption technique in the near-infrared which accuracy is better than 0.2 g/cm^2. To study the impact of the possible error on the water vapor amount, we ran a set of simulations at three different water vapor contents, 1 g/cm^2, 3 g/cm^2 and 5 g/cm^2, with a variation of 0.2 g/cm^2 for each case for an optical depth of 0.3. Tables 8.5(a) – 8.5(c) report the error in each band for the three different sites.

Table 8.5(a) Error on the surface reflectance ($\times$ 10,000) due to uncertainties in the water vapor content ($\pm 0.2\ \mathrm{g/cm^2}$) for the Belterra site

Central Wavelength (nm)		470	550	645	870	1,240	1,650	2,130
Surface Reflectance $\times$ 10,000		120	375	240	2,931	3,083	1,591	480
Maximum Error $\times$ 10,000	Clear	0001d	0002i	0004i	0007i	0004i	0002i	0006i
	Avg.	0001i	0001b	0003i	0005i	0003i	0001d	0004i
	High	0001i	0001d	0003i	0005i	0003i	0001i	0003i
Minimum Error $\times$ 10,000	Clear	0000a	0000j	0001c	0004a	0002a	0000a	0003a
	Avg.	0000a	0000a	0001a	0003a	0001a	0000a	0002a
	High	0000a	0000a	0001a	0002a	0001a	0000a	0001c
Average Error $\times$ 10,000	Clear	0	1	2	5	2	0	3
	Avg.	0	0	1	3	1	0	2
	High	0	0	1	3	1	0	2

Table 8.5(b) Error on the surface reflectance ($\times$ 10,000) due to uncertainties in the water vapor content ($\pm 0.2\ \mathrm{g/cm^2}$) for the Skukuza site

Central Wavelength (nm)		470	550	645	870	1,240	1,650	2,130
Surface Reflectance $\times$ 10,000		400	636	800	2,226	2,880	2,483	1,600
Maximum Error $\times$ 10,000	Clear	0006j	0009d	0015i	0015i	0011i	0006i	0030i
	Avg.	0004j	0005b	0009d	0009i	0006i	0003i	0018i
	High	0003j	0004d	0007g	0007i	0005i	0003i	0014i
Minimum Error $\times$ 10,000	Clear	0001a	0003a	0005a	0004c	0001c	0000c	0010c
	Avg.	0001a	0002a	0003a	0003a	0001a	0000a	0006c
	High	0000a	0001a	0003a	0002a	0001a	0000a	0005a
Average Error $\times$ 10,000	Clear	2	4	6	6	2	1	13
	Avg.	1	2	4	4	1	0	8
	High	1	1	3	3	1	0	6

Table 8.5(c) Error on the surface reflectance ($\times$ 10,000) due to uncertainties in the water vapor content ($\pm 0.2\ g/cm^2$) for the Sevilleta site

Central Wavelength (nm)		470	550	645	870	1,240	1,650	2,130
Surface Reflectance $\times$ 10,000		700	1,246	1,400	2,324	2,929	3,085	2,800
Maximum Error $\times$ 10,000	Clear	0004d	0024j	0011i	0030i	0024i	0018i	0014i
	Avg.	0004d	0024j	0011i	0030i	0024i	0018i	0014i
	High	0004d	0024j	0011i	0030i	0024i	0018i	0014i
Minimum Error $\times$ 10,000	Clear	0001a	0005a	0008c	0005c	0000c	0000a	0015c
	Avg.	0001a	0004a	0006a	0004c	0000f	0000a	0010c
	High	0001a	0003a	0004c	0003c	0000c	0000a	0008c
Average Error $\times$ 10,000	Clear	3	7	11	9	4	1	21
	Avg.	2	4	7	6	2	0	13
	High	1	3	5	4	2	0	10

The band at 2,130 nm is the most affected by the error on water vapor, there is some small impact at 645 nm and 870 nm. Since 2,130 nm is affected, an error will impact the aerosol retrieval (see Fig. 8.6) and therefore all the band to a lesser extent.

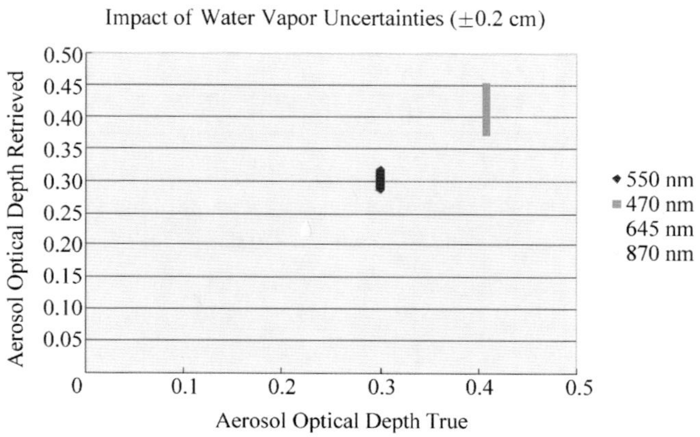

Figure 8.6 Comparison of the optical depth at 470 nm, 550 nm, 645 nm and 870 nm retrieved and input in the simulation for the 9 geometries given in Table 8.2 for an error on the water vapor content of $\pm 0.2\ g/cm^2$

8.5.5 Uncertainties on Empirical Relationship used to Determine the Surface Reflectance at 470 nm and 645 nm

The MODIS atmospheric correction algorithm uses an empirical relationship to predict the reflectance at 470 nm and 645 nm from the reflectance observed at 2,130 nm (Vermote et al., 2002), following the aerosol retrieval approach over land adopted by the atmosphere group (Kaufman et al., 1997). To account for deviation from this relationship we consider error of 0.005 in the surface estimation at 470 nm and 645 nm and run the atmospheric correction algorithm for three sites, at three different optical depths and in the nine geometries given in Table 8.1(a). The impact of the uncertainties in the empirical relationship is summarized in Tables 8.6(a) – 8.6(c). Figure 8.7 shows the impact on the retrieved optical thickness.

Table 8.6(a) Error on the surface reflectance ($\times$ 10,000) due to uncertainties in the empirical relationship between 2,130 nm and 470 nm, 645 nm for the Belterra site

Central Wavelength (nm)		470	550	645	870	1,240	1,650	2,130
Surface Reflectance $\times$ 10,000		120	375	240	2,931	3,083	1,591	480
Maximum Error $\times$ 10,000	Clear	0056j	0056j	0057a	0026j	0025f	0034b	0055a
	Avg.	0052a	0055j	0064a	0029j	0016i	0024b	0029b
	High	0053d	0063d	0066a	0029j	0018i	0027i	0027a
Minimum Error $\times$ 10,000	Clear	0050c	0044f	0049e	0003a	0000a	0000f	0024e
	Avg.	0050h	0050d	0053j	0002c	0002a	0005c	0007e
	High	0050j	0052h	0051e	0000d	0002a	0005e	0003j
Average Error $\times$ 10,000	Clear	52	49	52	10	11	17	37
	Avg.	51	52	57	9	6	13	17
	High	51	56	58	10	6	13	16

Table 8.6(b) Error on the surface reflectance ($\times$ 10,000) due to uncertainties in the empirical relationship between 2,130 nm and 470 nm, 645 nm for the Skukuza site

Central Wavelength (nm)		470	550	645	870	1,240	1,650	2,130
Surface Reflectance $\times$ 10,000		400	636	800	2,226	2,880	2,483	1,600
Maximum Error $\times$ 10,000	Clear	0056j	0057j	0060a	0039i	0037f	0038b	0066b
	Avg.	0052a	0061a	0069a	0036i	0025i	0023i	0030b
	High	0054d	0072d	0073a	0039i	0027i	0026i	0027b
Minimum Error $\times$ 10,000	Clear	0049c	0048c	0047e	0001f	0004e	0001e	0002f
	Avg.	0050f	0056h	0052j	0011f	0003e	0001e	0003c
	High	0051a	0054h	0051e	0015f	0002c	0001c	0003e
Average Error $\times$ 10,000	Clear	52	52	52	21	16	19	31
	Avg.	51	58	60	25	10	10	13
	High	51	62	62	27	10	10	14

Table 8.6(c) Error on the surface reflectance (× 10,000) due to uncertainties in the empirical relationship between 2,130 nm and 470 nm, 645 nm Sevilleta site

Central Wavelength (nm)		470	550	645	870	1,240	1,650	2,130
Surface Reflectance × 10,000		700	1,246	1,400	2,324	2,929	3,085	2,800
Maximum Error × 10,000	Clear	0056j	0056j	0063b	0056b	0088f	0172f	0183f
	Avg.	0052b	0064g	0076a	0058b	0038i	0030f	0027f
	High	0053d	0074d	0088g	0059i	0044i	0034i	0026i
Minimum Error × 10,000	Clear	0049c	0044c	0045e	0005f	0006a	0003d	0001a
	Avg.	0050f	0055h	0052j	0017f	0004e	0001a	0001e
	High	0050a	0053h	0050j	0022f	0004c	0000h	0000e
Average Error × 10,000	Clear	51	47	52	29	29	41	42
	Avg.	51	59	65	37	17	13	11
	High	51	63	68	39	18	12	10

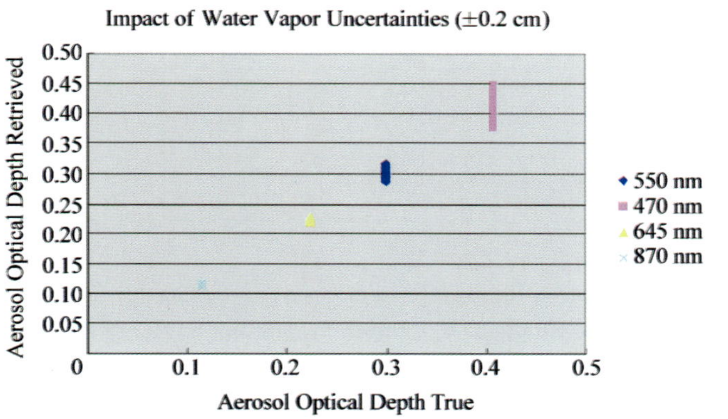

Figure 8.7 Comparison of the optical depth at 470 nm, 550 nm, 645 nm and 870 nm retrieved and input in the simulation for the 9 geometries given in Table 8.1(a) for an error of the estimation of the surface reflectance at 470 nm and 645 nm of 0.005 (empirical relationship)

8.5.6 Uncertainties on the Aerosol Model

The aerosol model is fixed to the urban clean case; it is possible to prescribe the model of aerosol as it is suggested by Kaufman et al., depending on the geographic location. However, the actual model may differ significantly from the actual aerosol. Tables 8.7(a) – 8.7(c) to 8.9(a) – 8.9(c) give an idea of the error generated by the use of the improper model. We simulated the error for three additional models, urban polluted cases, a smoke low and smoke high absorption case.

Figure 8.8 shows the associated error on the aerosol optical thickness for the smoke low absorption case, in this case the model is close to the assumed one; however, the error on the optical thickness is significant. Tracking the optical thickness is part of the validation process and enables us to estimate on a global basis the error introduced by the uncertainty on the model.

Table 8.7(a) Error on the surface reflectance (× 10,000) due to uncertainties in the aerosol model assumption (here smoke low absorption) for the Belterra site

Central Wavelength (nm)		470	550	645	870	1,240	1,650	2,130
Surface Reflectance × 10,000		120	375	240	2,931	3,083	1,591	480
Maximum Error × 10,000	Clear	0019j	0027j	0039j	0046j	0059j	0078j	0099j
	Avg.	0017i	0019j	0028i	0126j	0102j	0074j	0065j
	High	0043i	0068i	0116i	0230i	0197i	0200i	0176i
Minimum Error × 10,000	Clear	0000a	0000b	0000b	0002c	0000f	0000f	0001b
	Avg.	0000b	0003b	0000a	0050c	0033c	0013b	0000b
	High	0000b	0004a	0004f	0080a	0054b	0017b	0002e
Average Error × 10,000	Clear	2	4	5	12	10	10	13
	Avg.	4	9	7	75	52	28	16
	High	8	19	21	123	91	53	30

Table 8.7(b) Error on the surface reflectance (× 10,000) due to uncertainties in the aerosol model assumption (here smoke low absorption) for the Skukuza site

Central Wavelength (nm)		470	550	645	870	1,240	1,650	2,130
Surface Reflectance × 10,000		400	636	800	2,226	2,880	2,483	1,600
Maximum Error × 10,000	Clear	0017j	0024j	0034j	0048j	0059j	0069j	0087j
	Avg.	0018j	0008j	0019j	0089j	0092j	0076j	0065j
	High	0027i	0045i	0104i	0213i	0188i	0161i	0131i
Minimum Error × 10,000	Clear	0000a	0000b	0000f	0002c	0001f	0000f	0001c
	Avg.	0002a	0000f	0000b	0033a	0028b	0016b	0003b
	High	0002b	0000a	0001d	0042b	0042b	0023b	0001b
Average Error × 10,000	Clear	1	2	5	10	10	9	11
	Avg.	5	2	6	50	46	31	17
	High	9	10	17	86	82	57	32

Table 8.7(c) Error on the surface reflectance (×10,000) due to uncertainties in the aerosol model assumption (here smoke low absorption) for the Sevilleta site

Central Wavelength (nm)		470	550	645	870	1,240	1,650	2,130
Surface Reflectance ×10,000		700	1,246	1,400	2,324	2,929	3,085	2,800
Maximum Error ×10,000	Clear	0015j	0021j	0030j	0046j	0057j	0064j	0076j
	Avg.	0020j	0021j	0020j	0071j	0084j	0077j	0067j
	High	0020j	0038i	0053i	0127i	0119i	0094i	0060i
Minimum Error ×10,000	Clear	0000a	0001c	0001c	0002c	0002c	0002c	0001f
	Avg.	0005a	0009c	0007d	0028b	0025b	0021b	0012b
	High	0004b	0017g	0002d	0041b	0041b	0033b	0018b
Average Error ×10,000	Clear	1	4	6	10	9	10	10
	Avg.	8	13	10	41	42	36	25
	High	10	26	17	67	73	60	40

Table 8.8(a) Error on the surface reflectance (×10,000) due to uncertainties in the aerosol model assumption (here urban polluted) for the Belterra site

Central Wavelength (nm)		470	550	645	870	1,240	1,650	2,130
Surface Reflectance ×10,000		120	375	240	2,931	3,083	1,591	480
Maximum Error ×10,000	Clear	0019j	0017j	0021j	0053j	0054j	0051j	0053j
	Avg.	0025i	0044j	0081j	0255i	0175i	0090i	0086j
	High	0033j	0053d	0089j	0363i	0258i	0125d	0153j
Minimum Error ×10,000	Clear	0000a	0000b	0000a	0017c	0013c	0004b	0001a
	Avg.	0000a	0006e	0000a	0123a	0083b	0006j	0001a
	High	0000a	0016b	0001c	0197a	0137b	0015j	0001a
Average Error ×10,000	Clear	1	3	3	27	21	11	7
	Avg.	6	17	15	166	118	42	17
	High	8	29	28	272	194	73	35

Table 8.8(b) Error on the surface reflectance (× 10,000) due to uncertainties in the aerosol model assumption (here urban polluted) for the Skukuza site

Central Wavelength (nm)		470	550	645	870	1,240	1,650	2,130
Surface Reflectance × 10,000		400	636	800	2,226	2,880	2,483	1,600
Maximum Error × 10,000	Clear	0008j	0012j	0020j	0043j	0050j	0049j	0047j
	Avg.	0022i	0053j	0060j	0157i	0129i	0082f	0049j
	High	0019c	0055j	0057j	0251i	0227i	0145i	0091j
Minimum Error × 10,000	Clear	0000a	0001b	0002c	0013c	0013b	0008b	0000i
	Avg.	0004b	0000e	0000b	0069j	0067b	0027j	0003b
	High	0003b	0005h	0006b	0117b	0109b	0039j	0005b
Average Error × 10,000	Clear	1	2	4	20	19	14	8
	Avg.	8	8	15	105	102	62	25
	High	13	13	20	172	168	102	44

Table 8.8(c) Error on the surface reflectance (× 10,000) due to uncertainties in the aerosol model assumption (here urban polluted) for the Sevilleta site

Central Wavelength (nm)		470	550	645	870	1,240	1,650	2,130
Surface Reflectance × 10,000		700	1,246	1,400	2,324	2,929	3,085	2,800
Maximum Error × 10,000	Clear	0006j	0014j	0018j	0037j	0046j	0046j	0043j
	Avg.	0026i	0058g	0059f	0123f	0126f	0110f	0081j
	High	0032e	0078f	0078f	0188f	0199f	0175f	0128f
Minimum Error × 10,000	Clear	0000i	0005h	0005e	0014c	0013b	0011b	0007b
	Avg.	0000j	0012j	0020i	0043j	0068b	0048j	0007j
	High	0000j	0011j	0020j	0106j	0110b	0072j	0022j
Average Error × 10,000	Clear	1	9	10	20	20	17	13
	Avg.	15	41	41	94	97	82	52
	High	22	60	56	153	158	131	85

Table 8.9(a) Error on the surface reflectance (× 10,000) due to uncertainties in the aerosol model assumption (here smoke high absorption) for the Belterra site

Central Wavelength (nm)		470	550	645	870	1,240	1,650	2,130
Surface Reflectance × 10,000		120	375	240	2,931	3,083	1,591	480
Maximum Error × 10,000	Clear	0015j	0022j	0029j	0069j	0069j	0065j	0067j
	Avg.	0058i	0091i	0131i	0365i	0289i	0260i	0257i
	High	0043d	0143i	0240i	0495j	0345j	0247g	0222g
Minimum Error × 10,000	Clear	0000a	0002b	0000b	0020c	0015c	0006c	0001b
	Avg.	0001e	0013j	0000e	0146a	0103b	0041b	0007b
	High	0003c	0008j	0000e	0237a	0170a	0027i	0001j
Average Error × 10,000	Clear	2	5	5	33	25	15	11
	Avg.	10	29	25	211	154	82	43
	High	15	57	59	332	237	114	60

Table 8.9(b) Error on the surface reflectance (× 10,000) due to uncertainties in the aerosol model assumption (here smoke high absorption) for the Skukuza site

Central Wavelength (nm)		470	550	645	870	1,240	1,650	2,130
Surface Reflectance × 10,000		400	636	800	2,226	2,880	2,483	1,600
Maximum Error × 10,000	Clear	0013j	0017j	0027j	0057j	0064j	0061j	0060j
	Avg.	0037i	0030j	0050i	0207i	0180i	0128i	0075i
	High	0032i	0040i	0054g	0313j	0295j	0179j	0089g
Minimum Error × 10,000	Clear	0000b	0001i	0003c	0015c	0016b	0010b	0004b
	Avg.	0007b	0001e	0006b	0093b	0084b	0051b	0014b
	High	0006c	0000b	0003b	0148b	0137b	0082b	0023b
Average Error × 10,000	Clear	2	3	6	24	23	18	12
	Avg.	13	11	19	136	130	87	40
	High	18	17	26	220	212	140	63

Table 8.9(c) Error on the surface reflectance (×10,000) due to uncertainties in the aerosol model assumption (here smoke high absorption) for the Sevilleta site

Central Wavelength (nm)		470	550	645	870	1,240	1,650	2,130
Surface Reflectance ×10,000		700	1,246	1,400	2,324	2,929	3,085	2,800
Maximum Error ×10,000	Clear	0011j	0019j	0026j	0050j	0060j	0058j	0055j
	Avg.	0030i	0073g	0071f	0140f	0159j	0130j	0087f
	High	0047i	0098f	0098f	0250j	0267j	0208j	0140f
Minimum Error ×10,000	Clear	0001c	0007e	0007e	0016c	0016b	0013b	0009b
	Avg.	0014c	0014j	0006j	0095b	0085b	0070b	0045b
	High	0018b	0055j	0052j	0150b	0136b	0111b	0072b
Average Error ×10,000	Clear	2	12	13	24	24	21	16
	Avg.	20	52	48	121	122	104	72
	High	31	83	77	200	203	172	118

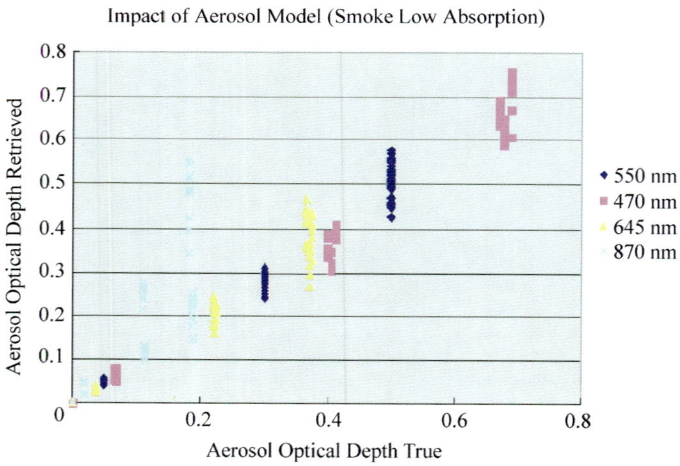

Figure 8.8 Comparison of the optical depth at 470 nm, 550 nm, 645 nm and 870 nm retrieved and input in the simulation for the 9 geometries and 3 optical depths given in Table 8.1(a) for an error on the aerosol model (actual smoke low absorption versus the urban clean used in the inversion)

Given the fact that under our assumption this error dominates any other sources, the choice of the aerosol model is critical to improve the theoretical accuracy of the current product and in particular the accuracy of the optical thickness retrieved. The dependence of the model used based on the geographic location is a first step in that direction, but one can imagine further steps involving use of aerosol and transport model such as GOCART (Chin et al., 2002) to determine the model or an attempt to invert the aerosol model using additional wavelength (i.e. 412 nm and 443 nm).

8.5.7 Overall Uncertainties

An overall uncertainty was estimated by computing the quadratic average of each average error generated by the uncertainties considered in 8.5.1 – 8.5.6 for each site. The results are presented in Table 8.10. The overall accuracy can be summarized under this term, in clear condition the average accuracy is 0.006 reflectance units or 5% relative whatever is higher, in average condition the average accuracy is 0.007 reflectance units or 7% relative whatever is higher, and in high aerosol loading conditions the average accuracy is 0.007 reflectance units or 9% relative whatever is higher. However, the minimum and maximum error observed for each category suggest a strong dependence of the error with the geometrical conditions, we are therefore planning in future version of the surface reflectance product (Collection 5 and 6) to introduce pixels and band dependent estimate of the accuracy.

Table 8.10 Overall theoretical accuracy of the atmospheric correction method considering the error source on calibration, ancillary data and aerosol inversion for 3 aerosol optical thickness (0.05: clear, 0.3: avg., 0.5: high). The uncertainties are considered independent and summed in quadratic

Central Wavelength (nm)		470	550	645	870	1,240	1,650	2,130
Belterra	Surface Reflectance × 10,000	120	375	240	2,931	3,083	1,591	480
	Clear	52	50	53	67	69	49	52
	Avg.	51	55	59	163	124	61	32
	High	52	64	65	255	189	92	46
Skukuza	Surface Reflectance × 10,000	400	636	800	2,226	2,880	2,483	1,600
	Clear	53	54	55	57	65	61	57
	Avg.	52	60	64	114	113	81	49
	High	53	64	70	174	169	116	64
Sevilleta	Surface Reflectance × 10,000	700	1,246	1,400	2,324	2,929	3,085	2,800
	Clear	53	54	61	61	70	79	78
	Avg.	55	74	79	108	109	99	78
	High	56	88	90	158	161	139	102

8.5.8 Validation of the Atmospheric Correction Algorithm

The validation of the atmospheric correction involves the validation of the atmospheric parameters used in the correction and the validation of the surface reflectance's themselves by comparison to surface reflectance estimates (derived from the use of sun-photometers data and validated against surface measurements of reflectance via high spatial resolution sensor such as ETM+. More details on the validation can be found in Vermote el al. (2002) and will not be discussed here. So far the validation has confirmed the validity of the theoretical error budget presented here but need to be extended to cover more conditions. This effort will be conducted during the validation stage 2 and 3 (Morisette et al., 2002).

8.6 Conclusions

The general approach for operational correction of the remotely sensed data in the visible to shortwave infrared spectral region assuming an infinite Lambertian target has been presented in detail. A detailed error budget has been presented in the case of the MODIS sensor. Overall, the accuracy of the atmospheric correction process has been confirmed by the validation effort which is still on-going. The error budget needs to be updated when considering non-uniform and non- Lambertian surface's as the algorithm to handle those effects becomes mature. However, the influence of those effects is probably of the second order (Vermote and Vermeulen. MODIS Atmospheric correction over land: surface reflectance, Algorithm Theoretical Basis Document. 1999, http:// modis.gsfc.nasa.gov/data/ atbd/atbd_mod08.pdf).

The accuracy is highly variable with respect to the geometrical conditions, the aerosol loading and the spectral band considered. The future version of the reflectance product will include a theoretical uncertainty estimate on a pixel, band basis.

The error budget points to the fact that improvement needs to be made in the area of the aerosol model used in the correction, especially accounting for the absorption of aerosol. This issue needs to be addressed to further reduce the uncertainties and several options are available.

References

Chin M et al. (2002) Tropospheric aerosol optical thickness from the GOCART model and comparisons with satellite and sunphotometer measurements. J Atmos Sci 59: 461 – 483

Deschamps PY, Herman M, Tanre D (1983) Modeling of the atmospheric effects and its applications to the remote sensing of ocean color. Appl Optics 22: 3,751 – 3,758

Dubovik O, Holben BN, Eck TF, Smirnov A, Kaufman YJ, King MD, Tanre D, Slutsker I (2002) Variability of absorption and optical properties of key aerosol types observed in worldwide locations. J Atm Sci 59: 590 – 608

El Saleous NZ, Vermote EF, Justice CO, Townshend JRG, Tucker CJ, Goward SN (2000) Improvements in the global biospheric record from the Advanced Very High Resolution Radiometer (AVHRR). International Journal of Remote Sensing 21(6): 1,251 – 1,277

Gao BC, Kaufman YJ (2003) Water vapor retrievals using Moderate Resolution Imaging Spectroradiometer (MODIS) near-infrared channels. Journal of Geophysical Research 108(D13), doi. 10.1029

Gordon HR, Clark DK, Brown JW, Brown OB, Evans RH, Broenkow WW (1983) Phytoplankton pigment concentrations in the Middle Atlantic Bight: comparison of ship determinations and CZCS estimates. Appl Optics 22: 20 – 36

Holben BN, Vermote EF, Kaufman YJ, Tanré D, Kalb V (1992) Aerosol Retrieval over Land from AVHRR data-Application for Atmospheric Correction. IEEE Transaction on Geoscience and Remote Sensing 30(2): 212 – 222

James ME, Kalluri SNV (1994) The Pathfinder AVHRR Land Data Set - An Improved Coarse Resolution Data Set For Terrestrial Monitoring. International Journal of Remote Sensing 15(17): 3,347 – 3,363

Kaufman YJ, Tanré D, Remer L, Vermote EF, Chu A, Holben BN (1997) Operational Remote Sensing of Tropospheric Aerosol Over the Land from EOS-MODIS. Journal of Geophysical Research 102(D14): 17,051 – 17,068

Martonchik J (1997) Determination of aerosol optical depth and land surface directional reflectances using multi-angle imagery. J Geophys Res Atmos 102: 17,015 – 17,022

Morisette JT, Privette JL, Justice CO (2002) A framework for the validation of MODIS land products. Remote Sensing of Environment 83(1-2): 77 – 96

Ouaidrari H, Vermote EF (1999) Operational Atmospheric Correction of Landsat TM data. Remote Sensing of the Environment 70: 4 – 15

Vermote EF, Tanré D (1992) Analytical Expressions for Radiative Properties of Planar Rayleigh Scattering Media Including Polarization Contribution. Journal of Quantitative Spectroscopy and Radiative Transfer 47(4): 305 – 314

Vermote EF, Tanré D, Deuzé JL, Herman M, Morcrette JJ (1997) Second simulation of the satellite signal in the solar spectrum (6S). Users Guide Version 2.0. Department of Geography, University of Maryland, Laboratoire d'Optique Atmosphérique, U.S.T.L., p 218

Vermote EF, El Saleous NZ, Justice CO, Kaufman YJ, Privette JL, Remer L, Roger JC, Tanre D (1997) Atmospheric correction of visible to middle-infrared EOS-MODIS data over land surfaces: Background, operational algorithm and validation. Journal of Geophysical Research Atmosphere 102 (D14): 17,131 – 17,141

Vermote EF, Justice CO, Descloitres J, El Saleous NZ, Ray J, Roy D, Margerin B, Gonzalez L (2001) A global monthly coarse resolution reflectance data set from SeaWiFS for use in Land, Ocean and Atmosphere applications. International Journal of Remote Sensing 22(6): 1,151 – 1,158

Vermote EF, El Saleous NZ, Justice C (2002) Atmospheric correction of the MODIS data in the visible to middle infrared: First results. Remote Sensing of Environment 83(1-2): 97 – 111

9 MODIS Snow and Sea Ice Products

Dorothy K. Hall, George A. Riggs and Vincent V. Salomonson

9.1 Introduction

The Earth's snow and sea ice cover are among the Earth's most dynamic features. Over 40% of the Earth's land surface may be covered by snow cover during the Northern Hemisphere winter, and sea ice covers 5% – 8% of the ocean surface at any given time. Both snow and sea ice covers act as an effective insulating layer between the atmosphere and land or ocean surface because the low thermal conductivity of snow and sea ice restricts exchanges of heat and chemical constituents between the atmosphere and the land or ocean.

The geographical extent of snow cover over the Northern Hemisphere varies from a maximum of $\sim 46\times10^6$ km^2 in January and February, to a minimum of $\sim 4\times10^6$ km^2 in August; between 60% – 65% of winter snow cover is found over Eurasia, and most mid-summer snow cover is in Greenland (Frei and Robinson, 1999). Seasonal snow cover in the Southern Hemisphere is generally located in relatively high mountain areas, such as in the western South America Andes. Sea ice extent varies from a maximum of roughly 15×10^6 km^2 in March, to a minimum of 8×10^6 km^2 in September in the Northern Hemisphere, and a maximum of 20×10^6 km^2 in September to a minimum of 4×10^6 km^2 in February in the Southern Hemisphere (Parkinson et al., 1987).

Numerous studies have shown the importance of accurate measurements of snow and ice extent and albedo, and snow depth and water equivalent, and sea ice concentration and thickness, as they relate to the Earth's climate and climate change (for example, see Dewey and Heim, 1981; Barry, 1983 and 1990; Ledley et al., 1999; Serreze et al., 2000). Measurements of snow and ice parameters have become increasingly sophisticated over time, and in addition, as the length of the satellite record increases, there is more potential to determine temporal trends that have climatic significance, that is, if the disparate data sets that comprise the record are well characterized and validated.

Satellites are well suited to the measurement of snow cover and sea ice because the high albedo of snow and sea ice presents a good contrast with most other natural features except clouds. Weekly snow mapping of the Northern Hemisphere using National Oceanographic and Atmospheric Administration (NOAA) National Environmental Satellite, Data, and Information Service (NESDIS) data began in 1966 and continues today in the United States, with improved spatial resolution and on a daily basis (Matson et al., 1986; Ramsay, 1998). The NOAA snow-cover record is the longest satellite snow-cover record in history, and it has been studied intensively by Frei and Robinson (1999) so that it is now a consistent record of snow-cover extent from 1966 to present. NOAA's National Operational Hydrologic Remote Sensing Center (NOHRSC) also provides a near-daily (during the snow season), 1 km resolution snow map of the United States and parts of southern Canada (Carroll, 1995). Sea ice maps derived from visible, near-infrared and infrared sensors have also been available from the NOAA National Ice Center (NIC) since 1972 (Dedrick et al., 2001), and microwave-derived maps have been available since 1975 (Carsey, 1992).

The MODIS snow and sea ice products (http://modis-snow-ice.gsfc.nasa.gov), available globally, are provided at a variety of different resolutions and projections to serve different user groups (Hall et al., 2002a and 2004a). The snow maps are available at 500 m and 0.05° resolutions on a sinusoidal projection and a latitude/longitude grid known as the climate-modeling grid (CMG). The sea ice maps are available at 1 km and 0.05° resolution, both on the Equal-Area Scaleable Earth grid (EASE-Grid) (see (Armstrong and Brodzik, 1995) for information about EASE-Grid).

The snow and sea ice products are archived and distributed through the National Snow and Ice Data Center (NSIDC), one of eight NASA Distributed Active Archive Centers (DAACs), and part of the Earth Observation System Data and Information System (EOSDIS). The products may be obtained through the EOS Data Gateway (EDG) at no charge. The EOSDIS utilizes the EOSDIS Core System (ECS) for data management across the DAACs, and the EDG, which facilitates online Web-based user access to data (Scharfen et al., 2000). Users can search and order data at NSIDC via the EDG client (http://nsidc.org/imswelcome).

Throughout the development and production of the snow and ice products, users have been consulted on the suitability of the products. Additionally, users sometimes discover problems in the data products that the developers missed, and alert the developers, making users an integral part of the ongoing QA process.

Improvements in the algorithms and data products are constantly being made based on analysis and monitoring of products. The MODIS Science Team reviews the improvements in all the products and determines when to introduce the latest, stable, functional improvements in the algorithms into the data processing streams and produce products that can best be used by the science and applications communities. Each processing of MODIS data is called a "collection" and involves using the best available algorithms for forward processing and reprocessing of MODIS data. A "collection" is generated by using the best available algorithms to process MODIS data forward from the date the "collection" processing began and to reprocess MODIS data from the start of the MODIS data record to the "collection" processing start date. In this sense the first consistent processing of MODIS data was "Collection 1" (there was no "Collection 2" due to procedural, system accounting, etc., reasons). The next major processing/reprocessing effort subsequently was called "Collection 3". At present the most up-to-date set of MODIS products is "Collection 4" which was completed in July 2004 for MODIS land products and spans the period for the Terra MODIS from February 2002 to the present and for the Aqua MODIS from June 2002 to the present. The next complete reprocessing/processing of MODIS data is known as "Collection 5" and is expected to begin in mid-2006 and continue throughout most of 2007.

There are several different data-product levels starting with Level-2 (L2). An L2 product is a geophysical product that remains in its original image swath format, with the latitude and longitude known precisely; it has not been temporally or spatially manipulated. A Level-2G (L2G) product has been gridded onto a global sinusoidal map projection as a series of 10° latitude by 10° longitude adjoining "tiles," each tile being a piece, e.g., area, of a map projection. (Unlike the original MODIS integerized sinusoidal projection, the sinusoidal projection is well-supported by off-the-shelf geographic-information system and image-processing packages.) Level-2 data products are gridded into L2G tiles by mapping the L2 pixels into cells of a tile in the map projection grid. The L2G algorithm creates a gridded product necessary for development of the Level-3 (L3) products. An L3 product is a geophysical product that has been temporally and/or spatially manipulated.

MODIS algorithms have been developed using the MODIS bands listed in Table 9.1 to map snow and sea ice and to calculate snow albedo and sea ice surface temperature. These algorithms were modified from heritage algorithms that were successfully used on earlier sensors and are described in detail in the MODIS Snow and Sea Ice User Guides (Riggs and Hall, 2006a and 2006b). In this chapter, we will describe the MODIS snow and ice products that are available for ordering in Collection 4. We will also describe validation efforts associated with the products and discuss error analysis.

Table 9.1 MODIS bands used to produce the MODIS snow and ice products

Band Number	Bandwidth (μm)	Used in Snow and/or Sea Ice Algorithms and Terra and/or Aqua
1	0.620 – 0.670	Snow & Sea Ice/Terra & Aqua
2	0.841 – 0.876	Snow & Sea Ice /Terra & Aqua
4	0.545 – 0.565	Snow & Sea Ice /Terra & Aqua
6	1.628 – 1.672	Snow & Sea Ice /Terra
7	2.105 – 2.155	Snow & Sea Ice /Aqua
31	10.780 – 11.280	Snow & Sea Ice/Terra & Aqua
32	11.770 – 12.270	Snow & Sea Ice/Terra & Aqua

9.2 Snow Products

9.2.1 Introduction

The MODIS snow-map products (see Table 9.2) provide global, daily coverage. Swath and daily products are available at 500 m resolution, while the CMG products are provided at 0.05° resolutions (~5.6 km resolution at the Equator). Fractional-snow cover or percent snow cover is already available in the CMG products, and beginning in 2006 (Collection 5), it will be provided in the 500-m

Table 9.2 MODIS standard snow data products

Long Name	Earth Science Data Type (ESDT)	Spatial Resolution
MODIS/Terra Snow Cover 5-Min L2 Swath 500 m*	MOD10_L2	500 m resolution, swath of MODIS data
MODIS/Terra Snow Cover Daily L3 Global 500 m SIN Grid (includes daily snow albedo)	MOD10A1	500 m resolution, projected, gridded tile data
MODIS/Terra Snow Cover 8-Day L3 Global 500 m SIN Grid	MOD10A2	500 m resolution, projected, gridded tile data
MODIS/Terra Snow Cover Daily L3 Global 0.05 Deg CMG	MOD10C1	0.05° resolution, lat/lon climate modeling grid
MODIS/Terra Snow Cover 8-Day L3 Global 0.05 Deg CMG	MOD10C2	0.05° resolution, lat/lon climate modeling grid
MODIS/Terra Snow Cover Monthly L3 Global 0.05 Deg CMG*	MOD10CM	0.05° resolution, lat/lon climate modeling grid
MODIS/Aqua Snow Cover 5-Min L2 Swath 500 m*	MYD10_L2	500 m resolution, swath of MODIS data
MODIS/Aqua Snow Cover Daily L3 Global 500 m SIN Grid (includes daily snow albedo)	MYD10A1	500 m resolution, projected, gridded tile data
MODIS/Aqua Snow Cover 8-Day L3 Global 500 m SIN Grid	MYD10A2	500 m resolution, projected, gridded tile data
MODIS/Aqua Snow Cover Daily L3 Global 0.05 Deg CMG	MYD10C1	0.05° resolution, lat/lon climate modeling grid
MODIS/Aqua Snow Cover 8-Day L3 Global 0.05 Deg CMG	MYD10C2	0.05° resolution, lat/lon climate modeling grid
MODIS/Aqua Snow Cover Monthly L3 Global 0.05 Deg CMG*	MYD10CM	0.05° resolution, lat/lon climate modeling grid

* An FSC enhancement at 500-m resolution will be available in mid-2006 (Collection 5).

products. Because cloud cover often precludes the acquisition of snow cover from visible and near-infrared sensors, 8-day composite products that minimize cloud obscuration complement the daily products. Quality assessment (QA) information is also included in the products.

9.2.2 MODIS Snow-Mapping Approaches

There are several heritage products that preceded the MODIS snow products. Weekly snow mapping of the Northern Hemisphere using NOAA data began in 1966 (Matson et al., 1986) and continues today in the United States, but with improved resolution and on a daily basis (Ramsay, 1998). On a more regional scale, Landsat Multispectral Scanner (MSS) (Rango and Martinec, 1979), Thematic Mapper (TM) and Enhanced Thematic Mapper Plus (ETM+) data have been used for mapping snow-covered area in drainage basins, though no community-endorsed "products" exist. The Landsat TM and ETM+ have been especially useful for measuring snow cover because of the short-wave infrared band—TM or ETM+ band 6 (1.6 μm)—which is useful for snow/cloud discrimination.

The reflectance of snow is low and the reflectance of most clouds remains high in the short-wave part of the spectrum. Various techniques, ranging from visual interpretation, multi-spectral image classification, decision trees, change detection and ratios (Kyle et al., 1978; Bunting and d'Entremont, 1982; Crane and Anderson, 1984; Dozier, 1989) have been used to map snow cover, and many of these techniques are used in the current MODIS snow-mapping algorithm along with additional spectral and threshold tests specific to the MODIS algorithm.

MODIS standard snow products include the 500-m resolution swath product, (L2G not archived for ordering), the Level-3 500-m resolution product on the sinusoidal grid, the 0.05°-resolution climate-modeling grid products (daily and 8-day composite) and in Collection 5 the monthly product. A 0.25°-resolution, non-standard product, available as a flat-binary file, is available from the data-product developers.

The automated MODIS snow-mapping algorithm (Hall et al., 1995 and 2002a) uses at-satellite reflectance in MODIS bands 4 (0.545 – 0.565 μm) and 6 (1.628 – 1.652 μm) to calculate the normalized difference snow index (NDSI) based on the heritage algorithms discussed in the previous paragraph:

$$\mathrm{NDSI} = \frac{\mathrm{band4} - \mathrm{band6}}{\mathrm{band4} + \mathrm{band6}} \tag{9.1}$$

A pixel in a non-densely-forested region will be mapped as snow if the NDSI is $\geqslant 0.4$ and reflectances in MODIS band 2 (0.841 – 0.876 μm) and MODIS band 4 (0.545 – 0.565 μm) are $\geqslant 10\%$. However, if the MODIS band 4 or band 2 reflectance is <10%, then the pixel will not be mapped as snow even if the other criteria are met. This threshold test prevents pixels containing very dark targets such as black spruce forests, or clear water bodies from being mapped as snow because very low reflectance cause the denominator in the NDSI to be quite small, and only small increases in the visible wavelengths are required to make the NDSI value high enough to classify a pixel, erroneously, as snow.

Changes that occur in the spectra of a forest stand as it becomes snow covered can be exploited to map snow cover in forests (Klein et al., 1998). The primary change in reflectance occurs in the visible wavelengths as snow has a much higher visible reflectance than soil, leaves or bark. A fundamental change that snow cover causes in the spectral response of a forest, which can be used in a global algorithm, is that the reflectance in the visible will often increase with respect to the near-infrared reflectance. This behavior is captured in the normalized-difference vegetation index (NDVI), as the presence of snow cover will tend to lower the NDVI. The NDVI (calculated using MODIS Bands 1 and 2) and NDSI are used together to improve snow mapping in dense forests. If the NDVI and NDSI values of a pixel lie in a triangular polygon region of NDVI to NDSI space then the pixel may be mapped as snow even if the NDSI is <0.4 (Klein et al., 1998).

The introduction of a "thermal mask" has improved the algorithm by eliminating much of the spurious snow cover that was found in many parts of the globe in the earlier MODIS snow maps. Some causes of the spurious snow cover are confusion with cloud cover, aerosol effects, geologic surface features and snow/sand confusion on coastlines. The thermal mask has greatly reduced errors of commission. Using MODIS infrared bands 31 and 32, a split-window technique (Key et al., 1997) is used to estimate ground temperature. If the temperature of a pixel is >283 K, then the pixel will not be mapped as snow (see also Romanov et al., (2000) wherein a thermal mask is described). In cold regions, measured snow-covered area did not change much with the introduction of the thermal mask, but misclassification in warm areas decreased dramatically.

An Aqua-specific version of the L2 snow algorithm was developed in response to MODIS instrument performance issues (Riggs and Hall, 2006a). Seventy percent of the Aqua MODIS band 6 detectors are non-functional. The Aqua-specific version of the snow-mapping algorithm uses MODIS band 7 instead of band 6 for the calculation of NDSI. Snow reflectance is similar in both bands 6 and 7 so the difference in snow reflectance in the near infrared relative to the visible is very similar and the NDSI criterion can be applied using band 7 in the ratio.

Aqua-specific settings for the snow criteria tests were determined by using the standard MODIS snow map to determine the NDSI values using band 7 data for pixels that were identified as snow. The NDSI was recalculated using band 7 for the snow pixels and the difference in NDSI values using band 6 or band 7 was determined. Based on analysis of twelve swaths, from different regions in the Northern Hemisphere, it was found that snow at-satellite reflectance is typically a few percent lower in band 7 compared to band 6 in MODIS data. The average difference in NDSI using band 7 was 0.14 lower than when using band 6. Thus, the NDSI threshold was increased from 0.40 to 0.54 (using band 7), and the NDSI vertices values of the NDSI/NDVI polygon, developed to improve snow cover mapping in forests by Klein et al. (1998), were adjusted by that difference, to produce a snow map from the Aqua MODIS that is very similar to that produced by the Terra MODIS. That approach was taken to maximize compatibility of the L2 snow maps generated with Terra or Aqua data.

Analysis of the Aqua-specific snow map algorithm along with the cloud mask reveals that in some situations along the periphery of some clouds, the cloud is not detected. In those situations the snow algorithm assumes there is no cloud but the cloud present causes erroneous snow detection by the NDSI/NDVI snow test. As of this writing it appears that the change to band 7 from band 6 has resulted in subtle changes in the ability to detect clouds and snow in such situations (Riggs and Hall, 2004). The investigation will continue in order to determine all the effects of the switch to band 7. The NDSI/NDVI test for snow was disabled for the Aqua Collection 4 reprocessing so as to decrease the occurrence of erroneous snow mapping in the Aqua L2 snow products.

9.2.3 Snow Swath Product

The snow data-product sequence begins with a 5-minute swath segment (granule) (MOD10_L2 and MYD10_L2) (see Fig. 9.1) at a spatial resolution of 500 m and covering 2,330 km (cross track) by 2,030 km (along track). Coarse resolution, 5 km, latitude and longitude data are stored as separate data layers in the product to provide geographic reference. Inputs to the swath snow-cover product are: the MODIS (Level-1B) radiance data (Guenther et al., 1998), the MODIS cloud mask (Ackerman et al., 1998; Platnick et al., 2003), and the MODIS geolocation product for the latitude and longitude, viewing geometry data and the land/water mask (Wolfe et al., 1998). Only pixels that are on land or inland water and unobstructed by clouds (according to the MODIS cloud-mask product) and in daylight are analyzed for the presence or absence of snow. The product provides a thematic map of snow, cloud (from the MODIS cloud mask) and water (from the land/water mask).

Daily Snow Fraction

Fractional snow cover (FSC) algorithms and maps have been developed utilizing Landsat, Airborne Visible/Infrared Imaging Spectrometer (AVIRIS) and MODIS data to exploit sub-pixel information. Much of this work has relied on spectral-mixture modeling (Nolin et al., 1993; Rosenthal and Dozier, 1996; Vikhamer and Solberg, 2002; Painter et al., 2003; Dozier and Painter, 2004), and neural networks (McIntire and Simpson, 2002). To date, however, these methods have not been applied for routine processing to provide global, daily snow maps.

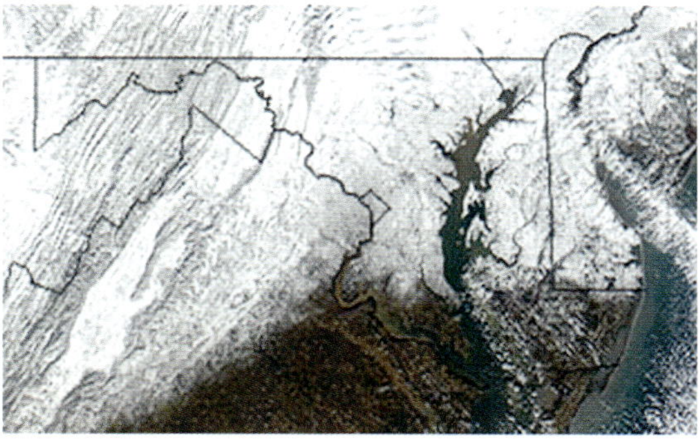

(a)

Figure 9.1(a) MODIS true color image of the eastern United States acquired on December 7, 2003, showing the first big snowstorm of the season. Courtesy MODIS Rapid Response Team

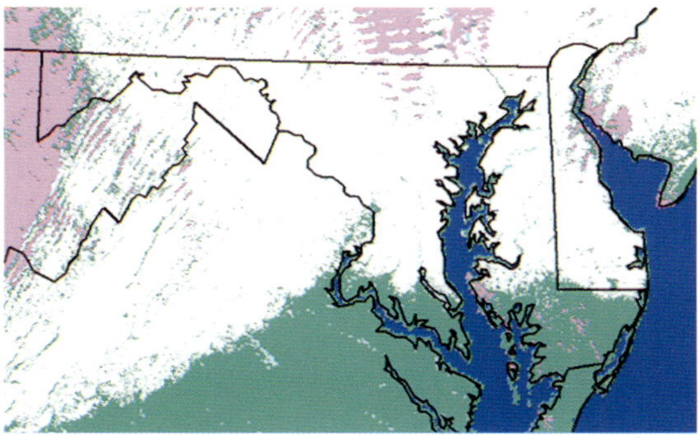

(b)

Figure 9.1(b) MODIS swath-based snow map (MOD10_L2) of the area shown in Fig. 9.1(a). Snow is white, snow-free land is green and clouds are pink. Water is shown in blue

Other work using MODIS data (Barton et al., 2001; Kaufman et al., 2002; Salomonson and Appel, 2004) has been done to develop algorithms to map sub-pixel FSC globally. Using an algorithm developed by Salomonson and Appel, FSC will be provided in the MOD10_L2 swath snow maps at 500-m resolution (see for example, Fig. 9.2). (An Aqua-specific FSC algorithm is under development necessarily using band 7.) The FSC enhancement to MOD10_L2 will be available to order in 2005. To develop the FSC algorithm, a statistical-linear relationship was developed between the NDSI from MODIS and the true sub-pixel fraction of snow cover as determined using Landsat scenes from Alaska, Canada and Russia. The fraction of snow cover within a MODIS 500-m resolution pixel can be provided with a mean absolute error of less than 0.1 over the entire range 0.0 – 1.0 in fractional snow cover (Salomonson and Appel, 2004). Déry et al.

(2005) used the algorithm to infer snow-cover areal depletion curves for the Kuparuk River Basin, and then applied the results in NASA's Catchment-Based Land-Surface Model. Overall, the algorithm appears to provide useful results over a wide range of conditions while maintaining relative computational simplicity.

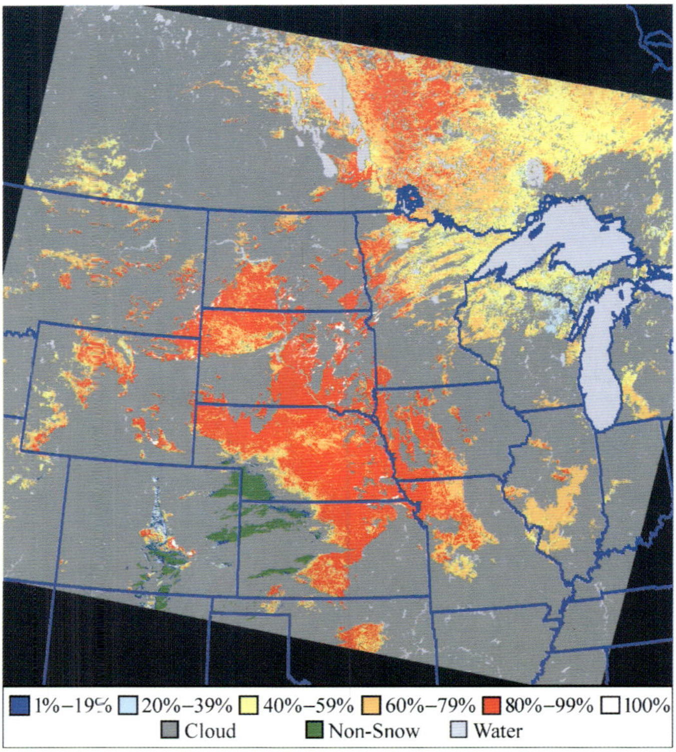

Figure 9.2 Prototype of the 500 m resolution fractional-snow cover product that will be available as part of the snow-cover swath product, MOD10_L2. This map covers parts of the mid-western and western United States and was acquired on December 14, 2000

9.2.4 Daily and 8-Day Composite Gridded Snow (Tile Products)

The second product in the sequence is created by mapping pixels from the swath product to their locations in a MODIS-specific global sinusoidal projection (Wolfe et al., 1998 and 2002) (an integerized sinusoidal projection was used for Versions 1 – 3 of the data products and a sinusoidal grid is used for Version 4). The Earth is divided into tiles of approximately 10° × 10° (latitude/longitude) and daily snow products are created for each tile. The daily L2G Terra and Aqua data products (MODL2G and MYD10 L2G, respectively) are intermediate products in which all the observations (pixels) in the snow swath product are geolocated onto the projection, resulting in a three-dimensional data array in which the observations are "stacked."

The 500-m resolution daily Terra and Aqua snow products (MOD10A1 and MYD10A1, respectively) are generated by selecting the observation with the highest score calculated from an observation's solar elevation, distance from nadir and extent of coverage in the grid cell from all the observations acquired during a day. An 8-day composite maximum snow-cover product is produced for each tile by compositing 8 days of the daily tile products. If snow were present on any day in any location on the daily tile product, it will be classified as snow covered in the 8-day composite. 8-day periods are fixed and begin from the first day of each year. The 8-day periods may go into the following year, but in each new year the 8-day period starts over again on January 1st.

Daily Snow Albedo

An algorithm has been developed to map snow albedo using MODIS data and is available in the MOD10A1 and MYD10A1 daily tile products. In deriving albedo, atmospherically-corrected MODIS surface reflectances in individual MODIS bands for snow-covered pixels located in non-forested areas are adjusted for anisotropic scattering effects using a DIScrete Ordinates Radiative Transfer model (DISORT) and snow optical properties (Klein et al., 2000; Klein and Stroeve, 2002). Currently in the algorithm, snow-covered forests are considered to be Lambertian reflectors. The adjusted spectral albedos are then combined into a broadband albedo measurement using a narrow-to-broadband conversion scheme developed specifically for snow by Shunlin Liang (written communication 2003) (Liang, 2000). Thus derived, a near-global snow albedo product at 500-m resolution (see Fig. 9.3) is available from September 15, 2003, to the present as an enhancement to the daily snow-cover product (MOD10A1). Since the product has not yet been validated, it is considered a "beta" product at this writing.

Preliminary validation activities were carried out in February 2003 in northwestern Iowa by Klein (2003). At this time, the snow cover was quite low thus providing the opportunity to evaluate the accuracy of the snow-albedo retrieval from MODIS for pixels containing a mixture of snow and other surface covers. MODIS-derived albedos were found to be within 10% of those measured in-situ.

9.2.5 Daily and 8-Day Composite Global Climate-Modeling Grid Products

The daily (MOD10C1 and MYD10C1) and 8-day composite (MOD10C2 and MYD10C2) CMG products are available to download from the National Snow and Ice Data Center (NSIDC) (see Fig. 9.4). The daily snow-cover CMG product is generated on a geographic grid, by assembling all MODIS land tiles

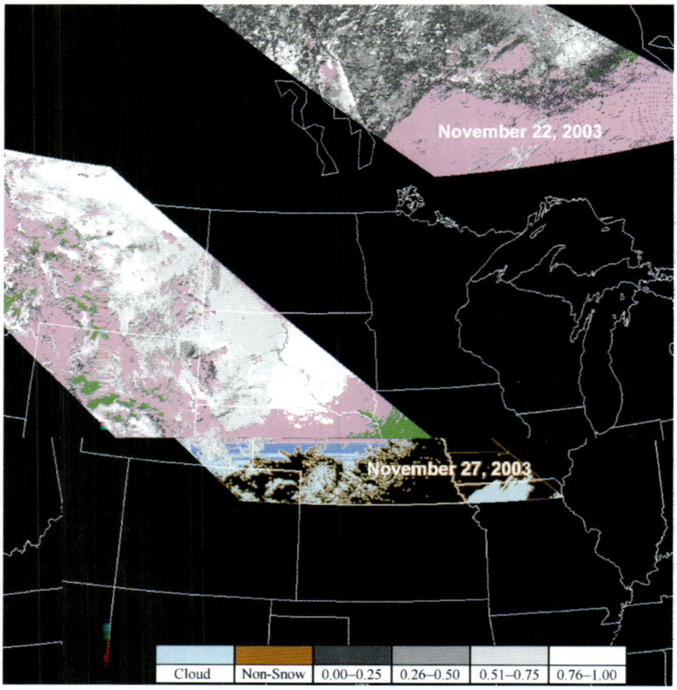

Figure 9.3 Examples of the 500-m resolution daily snow albedo product, included in MOD10A1, from two dates in November of 2003

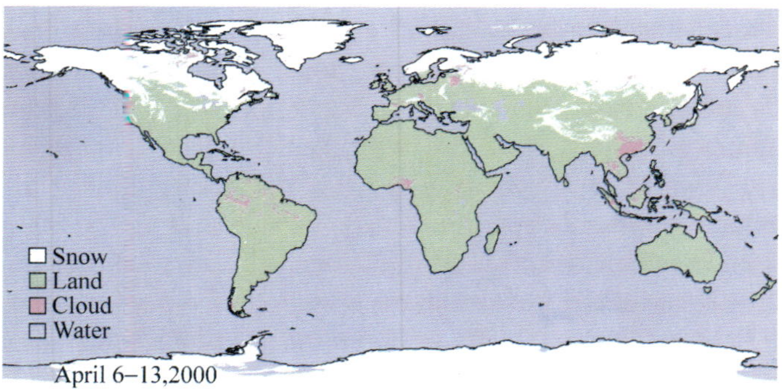

Figure 9.4 Eight-day composite (April 6 – 13, 2000) climate-modeling grid (CMG) global snow map, MOD10C2

(approximately 320) of the daily 500-m snow product and binning the 500-m cell observations into the 0.05° CMG cells. The binning technique creates an FSC array based on the number of snow obser- vations in a grid cell, a fractional cloud data array based on the number of cloud observations in each CMG grid cell, a confidence index and a QA field (Riggs and Hall, 2006). The percentages of snow and clouds in each cell are based on a count of all input observations, including non-snow and non-cloud observations. The confidence index provides a relative measure of the amount of the land surface in a cell that was viewed under clear-sky conditions. The QA field provides an indication of the quality of the 500-m data within each cell. Interpretation of FSC is problematic because the fraction of snow cover mapped in a cell is influenced not only by the fraction of snow in the cell, but by the fraction of cloud mapped.

The 8-day composite map shown in Fig. 9.4 is produced from the 8-day tile product, MOD10A2. The product is generated by merging all the MOD10A2 products (tiles) for an 8-day period and binning that 500-m data to 0.05°-resolution CMG cells to create a global CMG map of snow cover. Input values are binned into categories of snow, cloud, night, etc.. The percentages of snow, and cloud, QA and confidence index are computed, based on the binning results for each cell of the CMG, and written into the data arrays.

Errors of commission, mapping snow where none exists, in the 8-day composite global maps have been found to be very low. For example, errors of commission in Australia on three separate 8-day composite snow maps ranged from 0.02% – 0.10%.

Antarctica is mapped arbitrarily as always completely snow covered in both the daily and 8-day products. Antarctica is 99% or greater snow covered. During the summer, up to 1% of the continent may be snow-free mostly on the Antarctic Peninsula. That amount is less than the global error rate for snow mapping. Mapping Antarctica as always snow-covered was done because the MODIS cloud mask maps clouds over Antarctica most of the time, and therefore there are rarely clear views for snow mapping.

9.2.6 Monthly Snow Products

To develop the monthly Terra and Aqua snow-cover CMGs (see Fig. 9.5) (MOD10CM and MYD10CM, respectively), the mean-fractional snow is computed for each cell from the daily CMG products for the month. If a daily CMG cell has 70% or more clear (cloud free) observations, the cell will be used in the calculation for the monthly product. An average is taken of the FSC in all the cells in a month that meet this criterion.

The monthly MODIS snow products should be useful for modelers and for continuation of climatological snow-cover records (see Website for Rutgers University Climate Lab: http://climate.rutgers.edu/snowcover/) and Hall et al. (2004b) and will be available to order through NSIDC in 2006 (Collection 5).

9.2.7 Validation

The MODIS daily snow maps compare well with existing NOAA daily or near-daily snow maps such as the NOAA NESDIS and NOHRSC maps. MODIS has

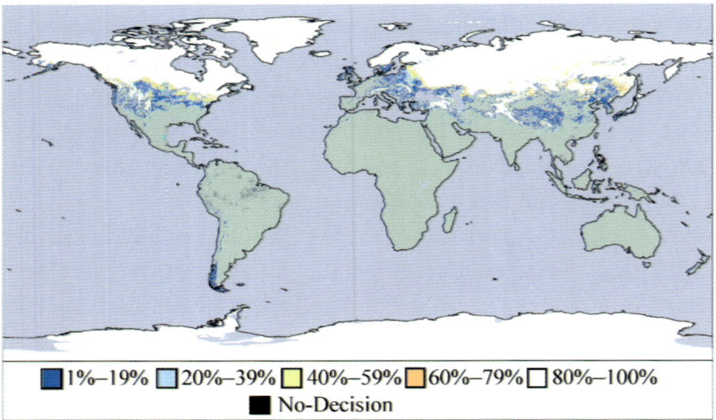

Figure 9.5 Prototype of the monthly snow map on the climate-modeling grid (CMG), MOD10CM

two major advantages over the NESDIS maps. The use of the short-wave infrared bands (MODIS bands 6 and 7) permits improved snow/cloud discrimination, and the 500-m resolution is an improvement compared to the ~5-km resolution of the NESDIS IMS snow maps. Additionally, MODIS often maps snow more frequently (i.e., more than once per day) than do the NESDIS maps thus providing information on fleeting snow cover that may have only lasted for a few hours (Hall et al., 2002b). Sometimes fleeting snow cover, and snow cover at the edges of snow-covered areas, is not mapped by NESDIS. Relative to the NOHRSC maps, the 500-m resolution of the daily MODIS maps is an improvement over the 1-km resolution of the NOHRSC maps, and MODIS maps snow globally for the entire year, while the NOHRSC maps are provided only of the United States and parts of southern Canada during the snow season. However, the important advantage of the NOAA maps, from a prediction standpoint, is that these maps utilize multiple sources of data to map snow cover including satellite and ground observations. From the standpoint of developing a climate-data record, this is more problematic than maps derived from an automated algorithm (as in the case of the MODIS products) because even if snow maps derived from automated algorithms are later determined to be incorrect, they can be reprocessed in a uniform manner.

Bitner et al. (2002) compared NESDIS and MODIS snow maps with the NOHRSC map. NOHRSC snow cover maps were compared with MODIS snow-cover maps in the Pacific Northwest and the Great Plains for 18 and 21 days, respectively, between March and June of 2001. The MODIS 500 m resolution product was degraded to match the resolution of the NOHRSC product (1 km). The agreement was reported to be 94% in the study area in the Pacific Northwest and 95% in the study area in the Great Plains. The largest disparity in snow-cover mapping between the NOHRSC and MODIS maps was at the edges of the snowpack where MODIS mapped more snow. (As mentioned previously, the MODIS maps often map more snow at the edges of snow-covered areas because several swaths may be mapped per day; if snow falls and melts within a few hours, MODIS is likely to capture it in the swath daily product.) However within the perimeter of a snowpack, the NOHRSC maps showed a more continuous snowpack when compared with the MODIS indicating that in mid-winter, the agreement between the products is good.

Klein and Barnett (2003) studied NOHRSC and MODIS snow maps and SNOpack TELemetry (SNOTEL) in-situ measurements in the Upper Rio Grande Basin for the 2000 – 2001 snow seasons, and found that the agreement between snow maps was about 86%, and the MODIS and NOHRSC snow products showed agreements of 94% and 76% with the SNOTEL measurements, respectively. A time-series comparison between MODIS retrievals and SNOTEL over the entire snow season showed an overall classification accuracy for MODIS of 88%, though the errors of commission and omission were found to be 12% and 15%, respectively, with much of the discrepancy occurring early and late in the snow season in thin snowpack conditions.

Mauer et al. (2003) compared MODIS and NOHRSC data for 46 days in the Columbia River Basin, and 32 days in the Missouri River basin during the winter and spring of 2000 – 2001. The presence or absence of snow was determined from ground observations at meteorological stations. On average, the MODIS snow maps classified fewer pixels as cloud as compared to the NOHRSC maps which permitted 15% more of the Columbia River basin area to be classified as to the presence or absence of snow relative to when NOHRSC maps were used. They found that the maps were comparable, but the MODIS maps generally provided more cloud-free data than did the NOHRSC.

MODIS and NOAA products acquired over Canada were compared with observations from almost 2,000 meteorological stations by Simic et al. (2004). Results showed similar agreement (ranging from ~80% – 100% on a monthly basis). Evergreen forests showed the lowest percentage agreement of all land-cover types studied. This is consistent with results found in Hall et al. (2001). The agreement between the MODIS snow maps and the meteorological-station data was generally found to be lower during the early part of the snow season by Simic et al., and during snow melt, particularly in forested areas (Simic et al., 2004).

Comparisons of MODIS CMG snow-cover maps with SSM/I snow-cover maps show that the MODIS maps are generally superior for mapping areal extent of snow cover especially in the early part of the snow season (Hall et al., 2002b). Earlier, other authors showed that same result using NESDIS and SSM/I snow maps (Basist et al., 1996; Armstrong and Brodzik, 1999). Hall et al. (2002b) showed differences up to 5.32 million km^2 in the amount of snow mapped using MODIS versus SSM/I, with the MODIS data being more accurate. Bussières et al. (2002) also compared MODIS-derived snow maps with SSM/I-derived snow maps over the prairie and boreal forest region in western Canada, and the taiga region in eastern Canada and found generally good correspondence, however, the SSM/I maps were not as accurate as the MODIS maps in the fall and spring. As the snow deepens during the winter, and the temperatures become consistently colder, the ability of the SSM/I to map snow improves, and the agreement between the visible and passive-microwave maps improves. This is because the passive-microwave data are not able to map wet snow cover because the penetration of the microwave signal when the snow is wet is extremely low. But of course the passive-microwave-derived maps have the potential to map snow-water equivalent and snow depth and the MODIS maps do not. It should be pointed out, however, that the spatial resolution of passive-microwave-derived images is currently only about 15 – 25 km, which makes the maps difficult to compare with the finer-resolution MODIS maps.

9.3 Sea Ice Products

9.3.1 Introduction and Algorithm Description

Daily extent and ice-surface temperature (IST) maps are produced using MODIS data at 1 km resolution (see Table 9.3). Algorithms for deriving the sea ice maps from MODIS are similar for both the Terra and the Aqua MODIS instruments, and are used to generate maps of sea ice extent as determined by reflectance and IST (Riggs et al., 1999; Riggs and Hall, 2006b). The products are produced globally at both 1-km and 0.05° (or ~ 4-km at the Equator) spatial resolution.

Table 9.3 MODIS sea ice data products

Long Name	Earth Science Data Type (ESDT)	Spatial Resolution
MODIS/Terra Sea Ice Extent 5-Min L2 Swath 1 km	MOD29	1-km resolution, swath of MODIS data
MODIS/Terra Sea Ice Extent Daily L3 Global EASE-Grid Day	MOD29P1D	1-km resolution, projected, gridded tile data
MODIS/Terra Sea Ice Extent Daily L3 Global EASE-Grid Night	MOD29P1N	1-km resolution, projected, gridded tile data
MODIS/Terra Sea Ice Extent and Ice Surface Temperature Daily L3 Global 4 km EASE-Grid Day*	MOD29E1D	4-km resolution, global, gridded
MODIS/Terra Ice Surface Temperature Daily L3 Global 4 km EASE-Grid Night*	MOD29E1N	4-km resolution, global, gridded

Continued

Long Name	Earth Science Data Type (ESDT)	Spatial Resolution
MODIS/Terra Sea Ice Extent and Ice Surface Temperature 8-Day L3 Global EASE-Grid Day*	MOD29E2D	4-km resolution, global, gridded
MODIS/Terra Ice Surface Temperature 8-Day L3 Global EASE-Grid Night*	MOD29E2N	4-km resolution, global, gridded
MODIS/Terra Sea Ice Extent and Ice Surface Temperature Monthly L3Global 4 km EASE-Grid Day*	MOD29EMD	4-km resolution, global, gridded
MODIS/Terra Ice Surface Temperature Monthly L3Global 4 km EASE-Grid Night*	MOD29EMN	4-km resolution, global, gridded
MODIS/Aqua Sea Ice Extent 5-Min L2 Swath 1 km	MYD29	1-km resolution, swath of MODIS data
MODIS/Aqua Sea Ice Extent Daily L3 Global EASE-Grid Day*	MYD29P1D	1-km resolution, projected, gridded tile data
MODIS/Aqua Sea Ice Extent Daily L3 Global EASE-Grid Night*	MYD29P1N	1-km resolution, projected, gridded tile data
MODIS/Aqua Sea Ice Extent and Ice Surface Temperature Daily L3 Global 4 km EASE-Grid Day*	MYD29E1D	4-km resolution, global, gridded
MODIS/Aqua Ice Surface Temperature Daily L3 Global 4 km EASE-Grid Night*	MYD29E1N	4-km resolution, global, gridded
MODIS/Aqua Sea Ice Extent and Ice Surface Temperature 8-Day L3 Global EASE-Grid Day*	MYD29E2D	4-km resolution, global, gridded
MODIS/Aqua Ice Surface Temperature 8-Day L3 Global EASE-Grid Night*	MYD29E2N	4-km resolution, global, gridded
MODIS/Aqua Sea Ice Monthly Global EASE-Grid*	MYD29EMD	4-km resolution, global, gridded

* Future enhancement or product.

During daylight hours, two sea ice maps were produced in Collection 4; one is derived from surface reflectances and the other from IST while at nighttime, only the IST is produced. In Collection 5, only the sea ice entent from the IST will be produced. MODIS sea ice products are generated automatically using MODIS sensor-radiance data (Guenther et al., 2002), a geolocation product (Wolfe et al., 2002), and the MODIS cloud mask product (Ackerman et al., 1998;

Riggs and Hall, 2003). In addition to sea ice extent and IST, QA, latitude and longitude, and other information are also included with the data products (Riggs and Hall, 2006b).

Sea ice extent was computed in two ways: by reflectance and by IST in Collection 4. Sea ice extent is determined by the characteristic of snow-covered sea ice to have high visible reflectance and low reflectance in the short-wave infrared. The sea ice algorithm uses the NDSI (described in Section 9.2.1) and visible reflectance criteria to map sea ice. Sea ice extent by IST is computed using the IST (see Section 9.3.2), but the user can select a temperature cutoff below which the pixel is considered ice and above which it is considered water. The MODIS cloud-mask product is relied on for masking cloudy pixels for analysis of sea ice.

Though the sea ice extent algorithms for Terra and Aqua are very similar, the major difference is that for Terra, MODIS band 6 data are used; for Aqua, MODIS band 7 data are used because of the non-functioning detectors in the Aqua MODIS band 6. Minor differences in mapping sea ice attributable to using band 7 may be observed where the ice signal is weak and confusion between sea ice and clouds may occur in some situations.

9.3.2 Calculation of Sea Ice-Surface Temperature

Measurement of Calculation of Sea Ice-Surface Temperature (IST) is possible with sensors having thermal-infrared bands such as the Temperature Humidity Infrared Radiometer (THIR), AVHRR, Landsat TM and ETM+. For the retrieval of clear-sky IST, a split-window technique is typically used when possible, where "split-window" refers to brightness-temperature differences in the 11 – 12 μm atmospheric windows. This technique allows for the correction of atmospheric effects, primarily due to water vapor (Key and Haefliger, 1992; Key et al., 1994 and 1997; Comiso, 2001) and MODIS (Hall et al., 2004a).

An identical IST algorithm is used for both the Terra and Aqua MODIS sea ice products (Hall et al., 2004a). For the retrieval of clear-sky IST, a split-window technique is used, where "split-window" refers to brightness temperature differences in the 11 – 12 μm atmospheric windows. This technique allows for the correction of atmospheric effects, primarily water vapor. First employed to determine sea-surface temperature (SST) (Prabhakara et al., 1974), the technique was subsequently used to determine IST in the Arctic with the AVHRR on NOAA polar-orbiting satellites (Key and Haefliger, 1992). For MODIS IST, the split-window technique is implemented as a simple regression model of the form:

$$T_s = a + bT_{11} + c(T_{11} - T_{12}) + d[(T_{11} - T_{12})(\sec\theta - 1)] \quad (9.2)$$

where T_s is the surface temperature, T_{11} and T_{12} are the MODIS-measured brightness temperatures in the 11 and 12 μm channels (MODIS channels 31 and 32), θ is the sensor scan angle, and a, b, c, and d are regression coefficients. The split-window temperature difference is proportional to the atmospheric water vapor amount, and the scan angle provides information on the atmospheric path length.

To determine the empirical relationship in Equation (9.2), radiosonde data from drifting ice and land-based stations were used with a radiative-transfer model, LOWTRAN (Kneizys et al., 1988), to simulate the sensor-brightness temperatures. The sensor spectral response functions were convolved with the calculated radiances and converted to brightness temperatures for each radiosonde profile. The surface temperatures used in the model calculations were then regressed against the simulated brightness temperatures to determine the coefficients in Equation (9.2). The regression coefficients in Equation (9.2) were determined separately for the Arctic and Antarctic and for three temperature ranges.

9.3.3 Swath Products

As with the snow-cover products, MODIS sea ice data products are produced as a sequence of products, beginning with a swath (scene) at a nominal pixel spatial resolution of 1 km and a nominal swath coverage of 2,330 km (cross track) by 2,030 km (along track) (about five minutes of satellite travel) (MOD29 and MYD29) (see Table 9.3). Figure 9.6 illustrates the MOD29 IST product acquired on March 12, 2003, in the northern Greenland Sea in the Arctic Ocean.

9.3.4 Daily and 8-Day Composite Gridded Sea Ice Products (Tile Products)

Automated selection of the most favorable observation from all the swaths acquired during a day generates a daily gridded sea ice product on the EASE-Grid at 1-km spatial resolution (MOD29E1D, MOD29E1N, MYD29E1D and MYD29E1N). This product includes both ice extent and IST.

9.3.5 Global-Scale Daily, 8-Day Composite and Monthly Gridded Products

A daily global sea ice product of sea ice extent and IST for the north and south polar areas in the EASE-Grid projection at 0.05°-resolution (Fig. 9.7) is available for Terra data (MOD29E1D). The Aqua product (MYD29E1D) will be available in Collection 5. Daytime and nighttime versions of the daily global sea ice product will eventually be available.

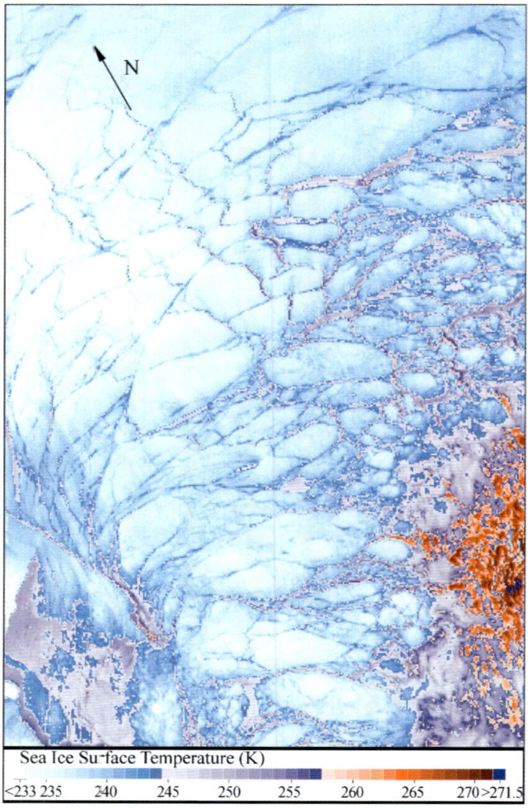

Figure 9.6 Terra MODIS ice-surface temperature (IST) product (MOD29) acquired on March 12, 2003 (18:45 UTC), in the northern Greenland Sea in the Arctic Ocean. The approximate center point of the image is 81.7°N, 1.0°E. The color scale represents IST in K (from Hall et al., 2004a)

Monthly sea ice products containing sea ice extent and IST maps (MOD29EMD, MOD29EMN, MYD29EMD and MYD29EMN) will be available in Collection 5.

9.3.6 Validation

Validation activities during the "cold period" (when meltwater is generally not present) in the Northern Hemisphere have been undertaken to assess the accuracy of the 1 km resolution MODIS IST algorithm and product (Hall et al., 2004a). Validation was also done at the Amundsen-Scott South Pole station in Antarctica using data from an automatic weather station. In the Arctic Ocean, near-surface air temperatures from the National Oceanic and Atmospheric Administration (NOAA) National Ocean Service (NOS) Center for Oceanographic Operational Products and Services (CO-OPS) Alaska tide stations, and from drifting buoys

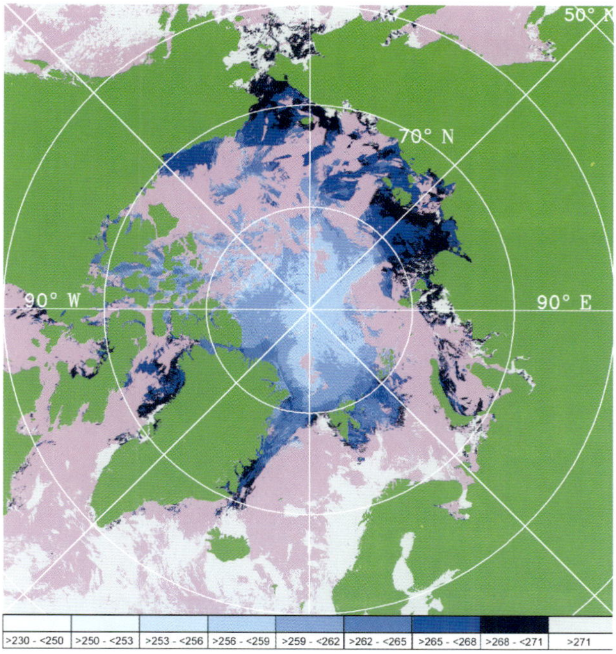

Figure 9.7 Prototype (average of May 15 – 19, 2000) of the future 8-day composite sea ice-surface temperature (IST) global EASE-Grid composite product of the north polar area MOD29C2 (prototype utilizes only five days of data). Color key shows ISTs from 230 – 271K. The color scale represents IST in °K. Clouds are shown in pink. In the compositing algorithm, each cell must be mapped as sea ice for two consecutive days otherwise it will appear as cloud. This reduces the errors of commission in the composited product (Hall et al., 2004a)

from the North Pole Environmental Observatory (NPEO) buoy program were compared with MODIS-derived ISTs. Using the standard MODIS sea ice product which utilizes the MODIS cloud mask, ISTs show a bias (mean error) of – 2.1 K and an RMS error of 3.7 K. The negative bias means that the satellite retrieval is less than the air temperature. With the bias removed, the RMS error is 3.0 K.

When additional visual cloud screening is performed to eliminate MODIS pixels thought to be contaminated by fog, results improved, with a subset of the larger data set showing a bias of – 0.9 K and an RMS error of 1.6K (see Fig. 9.8).

With the bias removed, the RMS error for the Arctic Ocean observations is 1.3 K. At the South Pole, under clear skies as determined using lidar measurements, the MODIS ISTs are also very close to near-surface air temperatures with a bias of – 1.2 K and an RMS error of 1.7 K. With the bias removed, the RMS error for the South Pole observations is 1.2 K. Thus the accuracy (RMS error) of the IST measurement is 1.2 – 1.3 K. However, it is not possible to obtain an accurate IST from MODIS in the presence of even very thin clouds or fog, making this the primary limitation of the MODIS IST product.

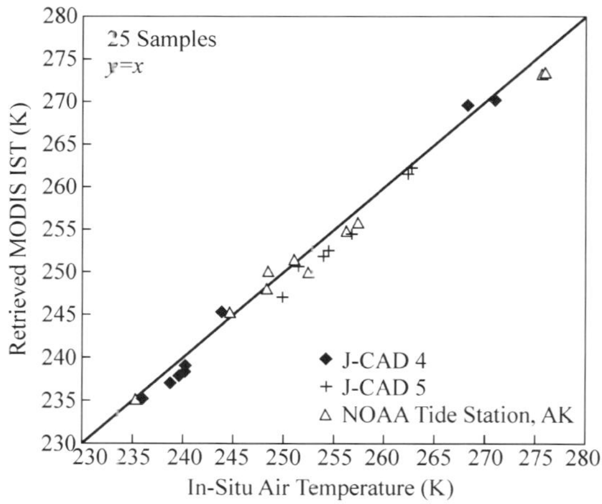

Figure 9.8 MODIS IST-retrieved skin temperatures and measured surface temperatures in three different areas of the Arctic Ocean on various dates between January 25, 2002, and May 22, 2003. In the comparison, MODIS pixels were visually screened for the likely presence of cloud or fog. The in-situ temperatures were acquired from the Japan Marine Science and Technology Center (JAMSTEC) Compact Arctic Drifter #4 and #5, or J-CAD 4 and J-CAD 5 buoys, and from the tide station at Prudhoe Bay, AK (Hall et al., 2004a)

9.4 Limitations Inherent in the Snow and Sea Ice Products

9.4.1 Land Masking in the Snow and Sea Ice Data Products

A 1-km resolution U.S. Geological Survey (USGS) global land/water mask, stored in the MODIS geolocation product, is used by all the MODIS land algorithms to determine if a pixel is land or ocean for processing (Wolfe et al., 2002). An issue with the land/water mask is that ice shelves, such as the Ross Ice Shelf of Antarctica, are classified as land and not ocean. Consequently, ice shelves have not been included in the MODIS sea ice products (Riggs and Hall, 2006b). A MODIS derived land/water mask, using several years of MODIS land cover data products has been developed by the MODIS Land Cover Product Team. That land/water mask has greater accuracy for mapping of land cover products. Integrated into that land/water mask was the Antarctic Digital Database grounding lines. That Antarctic map puts the continental land boundary at the grounding lines thus the ice sheets are properly put in the ocean and in the sea ice data products. Initial test results revealed great improvement in the quality of the sea ice products because the ice sheets were included in them. Additionally the snow products were improved because the ice sheets were removed from them. Evaluation of the MODIS land/water mask for snow mapping relative to inland water bodies is underway at the time of this writing. It is anticipated that all land products in the Collection 5 reprocessing and forward reprocessing will use the MODIS land/water mask. Following its implementation, it will be possible to monitor large icebergs such as those that have recently calved from the Ross Ice Shelf, using the sea ice products.

9.4.2 Cloud Masking

A challenging problem in snow and sea ice detection is the discrimination of snow or sea ice from clouds (Ackerman et al., 1998; Riggs and Hall, 2003). Snow has rather unique spectral characteristics compared to most other terrestrial surface features with a high reflectance in the visible and strong absorption in the infrared spectral regions. Those characteristics are key to detection of snow by use of the NDSI; however some types of clouds exhibit very similar spectral characteristics to snow which can make discrimination between snow and clouds very difficult.

The Level 2 snow and sea ice algorithms use the MODIS cloud-mask product. The MODIS cloud-mask algorithm takes a cloud-conservative approach to cloud detection. The cloud algorithm uses 14 of the 36 MODIS bands in 18 cloud spectral tests that follow different processing paths depending on surface type, geographic location and ancillary data input. Results of all cloud tests and associated processing flags are stored in the cloud-mask data product. The summary result of whether a pixel is cloudy, clear or probably clear is reported in a single summary flag of the cloud algorithm for a pixel. A complete description of the cloud mask algorithm is presented in Ackerman et al. (1998) or at the MODIS Atmospheres Website: http://www.modis-atmos.gsfc.nasa.gov/MOD35L2/index.html. The summary cloud flag is used as the cloud mask within the snow and sea ice algorithms. However, ongoing evaluation of the cloud masking performance as it relates to snow and sea ice detection has exhibited evidence that masking of clouds can be improved by selective use of individual cloud test results stored in the cloud mask product (Riggs and Hall, 2003).

Snow or sea ice detection is primarily dependent on the NDSI in the MODIS snow and sea ice algorithms. The NDSI is insensitive to most clouds except for some clouds containing ice and in some situations where the spectral signature of the cloud is similar to that of the snow. The MODIS cloud-masking algorithm contains many spectral tests; of those, we found that three can be used in the snow algorithm to minimize cloud obscuration over snow, and to mask clouds where the clouds completely obscured the surface in the L2 snow products. In MOD10_L2 products in Collection 4 there is a snow extent map using a selected group of cloud-mask tests, called the "liberal" cloud mask, in addition to the snow extent map made using the summary cloud flag. The liberal cloud mask and its resulting snow-extent map are experimental because there are some situations where clouds over a snow-covered surface are not identified as clouds by the liberal cloud mask. This results in a map that shows a bare surface when in fact the surface is snow covered. This is perhaps worse than showing cloudcover when it does not exist because it appears to a user that no snow is on the ground. Investigation continues into how to make better use of data from the cloud mask and to structure the snow algorithm to correctly identify the features in those situations. Similar lines of investigation are being pursued in the sea ice algorithm in both daytime and nighttime situations. The liberal cloud mask will not be available in Collection 5.

The accuracy of the MODIS cloud mask over snow and ice varies, with the highest accuracy generally found for daytime data and the lowest accuracy in darkness over sea ice using the thermal-infrared channels. Low illumination and the similar reflectance characteristics of clouds and sea ice at visible wavelengths cause difficulties in cloud masking using MODIS data. Additionally, atmospheric inversions can cause surface temperatures to be colder than the temperature of some clouds (McIntire and Simpson, 2002). Snow or ice/cloud discrimination problems are compounded during the polar night because the MODIS reflective bands are not useful during darkness. The absolute accuracy of the MODIS cloud mask used in the sea ice algorithm is unknown. The Aqua MODIS cloud mask algorithm was revised to use Band 7 (2.1 μm) in place of Band 6 (1.6 μm) because of the non-functional detectors on the Aqua MODIS in band 6 as described earlier. The revised Aqua cloud-mask algorithm has been in use since May 20, 2003, and has altered the masking of clouds over sea ice relative to the original cloud mask used with the Terra MODIS products.

A particular problem for mapping clouds over sea ice is the identification of thin clouds and fog. Fog often develops in sea ice covered waters when water vapor forms following the formation of leads and polynyas. While it may be possible to see the sea ice on a MODIS image, and even map the sea ice through the fog, it is not possible to retrieve an accurate IST through fog, even with the many visible, near-infrared, short-wave infrared and infrared MODIS bands.

9.5 Discussion and Conclusion

The algorithms used to develop the MODIS snow and sea ice map products were developed from heritage snow and sea ice algorithms that primarily employed Landsat TM and AVHRR data as well as aircraft data to map snow. The MODIS products compare favorably with the heritage products and represent improvement in terms of resolution, both spatial and spectral and cloud masking. Furthermore, the fully-automated nature of the MODIS algorithms makes the resultant products suitable as components of climate-data records. Through the National Snow and Ice Data Center, and the Earth Observing System Data Gateway one can order the data products and they are provided free-of-charge to the user.

At this writing, Collection 4 (or Version 4) reprocessing of the Aqua land data product suite (which includes the snow and ice products) has just been completed. Aqua MODIS reprocessing, which began in early 2004, was completed in the summer of 2004. Another reprocessing, Collection 5, for Terra and Aqua, will begin in mid 2006 and extend through 2007. These reprocessings are essential to the integrity of the long-term datasets. There are issues in any long-term dataset that need to be resolved to maximize the accuracy of the data sets, and to produce a viable and accurate segment of a climate-data record. Thus reprocessing is an ongoing process.

Collection 4 MODIS snow maps have been validated and compared quantitatively with operational snow maps as well as with field and meteorological measurements. Results show that the MODIS snow maps are providing snow extent generally within ±5% of the actual snow extent. The MODIS snow maps also provide daily snow albedo (as a beta product) and fractional-snow cover from both the 500 m (starting in Collection 5) and CMG (available currently and in Collection 5) maps. While the 500 m resolution maps are most useful for hydrological modeling, the CMG maps (daily, 8-day composite and average monthly) are useful for global-scale modeling.

The MODIS sea ice maps provide validated sea ice extent and ice-surface temperature at a resolution of 1 km during the daytime. Maps are also provided during the polar night but are not yet validated due to the extreme difficulty in reliably distinguishing clouds and snow/ice during darkness. Global-scale, 0.05° resolution sea ice extent and IST maps (daily, 8-day composite and monthly) on the EASE-Grid are under development.

Acknowledgements

The authors would like to thank Kimberly Casey/SSAI, Janet Chien/ SAIC, Nick DiGirolamo/SSAI and Hugh Powell/SAIC for programming and image-processing work on the MODIS algorithms and products.

References

Ackerman SA, Strabala KI, Menzel PWP, Frey RA, Moeller CC, Gumley LE (1998) Discriminating clear sky from clouds with MODIS. Journal of Geophysical Research 103(D24) (32): 141 – 157

Armstrong RL, Brodzik MJ (1995) An earth-gridded SSM/I data set for cryospheric studies and global change monitoring. Advances in Space Research 10: 155 – 163

Armstrong RL, Brodzik MJ (1999) A twenty year record of global snow cover fluctuations derived from passive microwave remote sensing data. 5th Conference on Polar Meteorology and Oceanography, American Meteorological Society, Dallas, TX, pp 113 – 117

Barton JS, Hall DK, Riggs GA (2001) Fractional snow cover from the MODIS snow-mapping algorithm. Proceedings of the 57th Eastern Snow Conference, 17 – 19 May 2000, Syracuse, NY

Barry R (1983) Research on snow and ice. Reviews of Geophysics and Space Physics 21: 765 – 776

Barry R (1990) Evidence of recent changes in global snow and ice cover. Geo Journal 20(2): 121 – 127

Basist A, Garrett D, Ferraro N, Grody N, Mitchell K (1996) A comparison between snow cover products derived from visible and microwave satellite observations. Journal of Applied Meteorology 35(2): 163 – 177

Bitner D, Carroll T, Cline D, Romanov P (2002) An assessment of the differences between three satellite snow cover mapping techniques. Hydrological Processes 16: 3,723 – 3,733

Bunting JT, d'Entremont RP (1982) Improved cloud detection utilizing defense meteorological satellite program near infrared measurements, Air Force Geophysics laboratory, Hanscom AFB, MA, AFGL-TR-82-0027, Environmental Research Papers No. 765, p 91

Bussières N, De Sève D, Walker A (2002) Evaluation of MODIS snow-cover products over Canadian regions, Proceedings of IGARSS'02, dates, Toronto, Canada, pp 2,302 – 2,304

Carroll TR (1995) Remote sensing of snow in the cold regions. Proceedings of the First Moderate Resolution Imaging Spectroradiometer (MODIS) Snow and Ice Workshop, 13 – 14 September, 1995, Greenbelt, MD, NASA Conf Pub 3318, pp 3 – 14

Carsey FD, ed. (1992) Microwave Remote Sensing of Sea Ice. AGU Monograph Wash., D.C., 68: 462

Comiso JC (2001) Satellite-observed variability and trend in sea-ice extent, surface temperature, albedo and clouds in the Arctic. Annals of Glaciology 33: 457 – 473

Crane RG, Anderson MR (1984) Satellite discrimination of snow/cloud surfaces. International Journal of Remote Sensing 5(1): 213 – 223

Dedrick KR, Partington K, Van Woert M, Bertoia CA, Benner D (2001) U.S. National/Naval Ice Center digital sea ice data and climatology. Canadian Journal of Remote Sensing 27(5): 457 – 475

Déry SJ, Salomonson VV, Stieglitz M, Hall DK, Appel I (2005) Hydrological Processes. 19: 2,755 – 2,774, DOI: 10.1001/hyp.5784

Dewey KF, Heim R Jr. (1981) Satellite observations of variation in Northern Hemisphere snow cover. NOAA Technical Report NESS 87, p 83

Dozier J (1989) Spectral signature of alpine snow cover from the Landsat Thematic Mapper. Remote Sensing of Environment 28: 9 – 22

Dozier J, Painter TH (2004) Multispectral and hyperspectral remote sensing of alpine snow properties, Annual Reviews of Earth and Planetary Science, 32: 465 – 494

Frei A, Robinson DA (1999) Northern hemisphere snow extent: regional variability 1972 – 1994. International Journal of Climatology 19: 1,535 – 1,560

Guenther B, Xiong X, Salomonson VV, Barnes WL, Young J (2002) On-orbit performance of the Earth Observing System Moderate Resolution Imaging Spectroradiometer; first year of data. Remote Sensing of Environment 83: 16 – 30

Hall DK, Riggs GA, Salomonson VV (1995) Development of methods for mapping global snow cover using Moderate Resolution Imaging Spectroradiometer (MODIS) data. Remote Sensing Environment 54: 127 – 140

Hall DK, Foster JL, Salomonson VV, Klein AG, Chien JYL (2001) Development of a Technique to Assess Snow-Cover Mapping Accuracy from Space. IEEE Transactions on Geoscience and Remote Sensing 39(2): 232 – 238

Hall DK , Riggs GA, Salomonson VV, DiGirolamo NE, Bayr KJ (2002a) MODIS snow-cover products. Remote Sensing of Environment 83: 181 – 194

Hall DK, Kelly REJ, Riggs GA, Chang ATC, Foster JL (2002b) Assessment of the relative accuracy of hemispheric scale snow-cover maps. Annals of Glaciology 34: 24 – 30

Hall DK, Key J, Casey KA, Riggs GA, Cavalieri DJ (2004a) Sea Ice Surface Temperature Product from the Moderate Resolution Imaging Spectroradiometer (MODIS). IEEE Transactions on Geoscience and Remote Sensing 42(5): 1,076 – 1,087

Hall DK, Foster JL, Robinson DA, Riggs GA (2004b) Merging the MODIS and RUCL monthly snow-cover records. Proceedings of IGARSS'04, 20 – 24 September 2004, Anchorage, AK

Kaufman YJ, Kleidman RG, Hall DK, Martins VJ, Barton JS (2002) Remote sensing of subpixel snow cover using 0.66 and 2.1 μm channels. Geophysical Research Letters 29(16): 1,781, doi: 10.1029.2001GLO13580

Key J, Haefliger M (1992) Arctic ice surface temperature retrieval from AVHRR thermal channels. Journal of Geophysical Research 97(D5): 5,885 – 5,893

Key J, Maslanik JA, Papakyriakou T, Serreze MC, Schweiger AJ (1994) On the validation of satellite-derived sea ice surface temperature. Arctic 47(3): 280 – 287

Key JR, Collins JB, Fowler C, Stone RS (1997) High-latitude surface temperature estimates from thermal satellite data. Remote Sensing of Environment 61: 302 – 309

Klein AG (2003) Determination of broadband albedos of partially snow-covered sites for validation of MODIS snow albedo retrievals. 60th Eastern Snow Conference, Sherbrooke, Quebec, Canada, 2003

Klein AG, Stroeve J (2002) Development and validation of a snow albedo algorithm for the MODIS instrument. Annals of Glaciology 34: 45 – 52

Klein A, Barnett AC (2003) Validation of daily MODIS snow maps of the Upper Rio Grande River Basin for the 2000 – 2001 snow year. Remote Sensing of Environment 86: 162 – 176

Klein AG, Hall DK, Riggs GA (1998) Improving snow-cover mapping in forests through the use of a canopy reflectance model. Hydrological Processes 12: 1,723 – 1,744

Klein AG, Hall DK, Nolin AW (2000) Development of a prototype snow albedo algorithm for MODIS. Proceedings of the 57th Annual Eastern Snow Conference, Syracuse, NY, May 17 – 19, 2000, pp 143 – 157

Kneizys FX, Shettle EP, Abreu LW, Chetwynd JH, Anderson GP, Gallery WO, Selby JEA, Clough SA (1988) Users Guide to LOWTRAN 7. AFGL-TR-88-0177, Environmental Research Papers, No 1010, p137

Kyle HL, Curran RJ, Barnes WL, Escoe D (1978) A cloud physics radiometer, Third Conference on Atmospheric Radiation, Davis, CA, pp 108 – 109

Ledley TS, Sundquist ET, Schwartz SE, Fellows JD, Killeen TL (1999) Climate change and greenhouse gases. EOS Transactions, American Geophysical Union 80(39): 453 – 454, 457 – 458

Liang S (2000) Narrow to broadband conversion of land surface albedo I : algorithms. Remote Sensing of Environment 76: 213 – 238

Matson M, Roeplewski CF, Varnadore MS (1986) An Atlas of Satellite-Derived Northern Hemisphere Snow Cover Frequency. National Weather Service, Wash., D.C., 75 pp

Mauer EP, Rhoads JD, Dubayah RO, Lettenmaier DP (2003) Evaluation of the snow-covered area data product from MODIS. Hydrological Processes 17: 59 – 71

McIntire TJ, Simpson JJ (2002) Arctic sea ice, cloud, water, and lead classification using neural networks and 1.6 μm data. IEEE Transactions on Geoscience and Remote Sensing 40(9): 1,956 – 1,972

Nolin A, Dozier J (1993) Estimating snow grain size using AVIRIS data. Remote Sensing of Environment 44: 231 – 238

Painter TH, Dozier J, Roberts DA, Davis RE, Green RO (2003) Retrieval of subpixel snow-covered area and grain size from imaging spectrometer data. Remote Sensing of Environment 85: 64 – 77

Parkinson CL, Comiso JC, Zwally HJ, Cavalieri DL, Gloersen P, Campbell WJ (1987) Arctic Sea Ice, 1973 – 1976: Satellite Passive-Microwave Observations. NASA SP-487, GPO, Washington, D.C

Platnick S, King MD, Ackerman SA, Menzel WP, Baum BA, Riédi JC, Frey RA (2003) The MODIS cloud products: algorithms and examples from Terra. IEEE Transactions on Geoscience and Remote Sensing 41(2): 459 – 473

Prabhakara CP, Dalu G, Kunde VG (1974) Estimation of sea surface temperature from remote sensing in the 11 – 13 μm window region. Journal of Geophysical Research 79(33): 5,039 – 5,044

Ramsay B (1998) The interactive multisensor snow and ice mapping system. Hydrological Processes 12: 1,537 – 1,546

Rango A, Martinec J (1979) Application of a snowmelt-runoff model using Landsat data. Nordic Hydrology 10: 225 – 238

Riggs GA, Hall DK, Ackerman SA (1999) Sea Ice Extent and Classification Mapping with the Moderate Resolution Imaging Spectroradiometer Airborne Simulator. Remote Sensing of Environment 68: 152 – 163

Riggs GA, Hall DK (2004) Snow mapping with the MODIS Aqua instrument. Proceedings of the 61st Eastern Snow Conference, 9 – 11 June 2004, Portland, ME

Riggs GA, Hall DK (in press b) Reduction of Cloud Obscuration in the MODIS Snow Data Product. Presented at the 59th Eastern Snow Conference, 5 – 7 June 2002, Stowe, VT, published in the Proceedings of the 60th Eastern Snow Conference

Riggs GA, DiGirolamo NE, Hall DK (2005) Comparison of MODIS daily global fractional snow cover maps at 0.05- and 0.25-degree resolutions, Proceedings of the 62nd Eastern Snow Conference, 8 – 10 June 2005, Waterloo, Canada, pp. 21 – 27

Riggs GA, Hall DK (2006a) MODIS Snow Products User Guide http://modis-snow-ice.gsfc.nasa.gov/sugkc2.html

Riggs GA, Hall DK (2006b) MODIS Sea Ice User's Guide http://modis-snow-ice.gsfc.nasa.gov/siukc.html

Robinson DA (1999) Northern Hemisphere snow cover during the satellite era. Proceedings of the 5th Conference Polar Meteorology and Oceanography, Dallas, TX, American Meteorological Society, Boston, MA, pp 255 – 260

Romanov P, Gutman G, Csiszar I (2000) Automated monitoring of snow cover over North America using multispectral satellite data. Journal of Applied Meteorology 39: 1,866 – 1,880

Rosenthal W, J. Dozier (1996) Automated mapping of montane snow cover at subpixel resolution from the Landsat Thematic Mapper. Water Resources Research 32(1): 115 – 130

Salomonson VV, Appel I (2004) Estimating the fractional snow covering using the normalized difference snow index. Remote Sensing of Environment, 89: 351 – 360

Salomonson VV, Appel I (in press) Development of the Aqua MODIS NDSI fractional snow cover algorithm and validation results, IEEE Transactions on Geoscience and Remote Sensing

Scharfen GR, Hall DK, Khalsa SJS, Wolfe JD, Marquis MC, Riggs GA, McLean B (2000) Accessing the MODIS snow and ice products at the NSIDC DAAC. Proceedings of IGARSS'00, 23 – 28 July 2000, Honolulu, HI, pp 2,059 – 2,061

Serreze MC, Walsh JE, Chapin III FS, Osterkamp T, Dyurgerov M, Romanovsky V, Oechel WC, Morison J, Zhang T, Barry RG (2000) Observational evidence of recent change in the northern high-latitude environment. Climate Change 46: 159 – 207

Simic A, Fernandes R, Brown R, Romanov P, Park W (2004) Validation of VEGETATION, MODIS, and GOES+SSM/I snow-cover products over Canada based on surface snow depth observations. Hydrological Processes, 18(6): 1,089 – 1,104, DOI: 10.102/hyp5509

Vikhamer D, Solberg R (2002) Subpixel mapping of snow cover in forests by optical remote sensing. Remote Sensing of Environment 83: 97 – 111

Wolfe RE, Roy DP, Vermote E (1998) MODIS land data storage, gridding, and compositing methodology: level-2 grid. IEEE Transactions on Geoscience and Remote Sensing 36(4): 1,324 – 1,338

Wolfe R E, Nishihama M, Fleig A, Kuyper J, Roy D, Storey J, Patt F (2002) Achieving sub-pixel geolocation accuracy in support of MODIS land science. Remote Sensing of Environment 83: 31 – 49

10 The NPOESS Preparatory Project

Robert E. Murphy

10.1 Introduction

The NPOESS Preparatory Project (NPP) is a mission designed to bridge the gap between the Earth Observing System (EOS) of NASA and the National Polar Orbiting Operational Environmental Satellite System (NPOESS), the United States' next generation of operational environmental satellites. NPP will carry four of the most important NPOESS sensors into orbit in late 2009. The goals of NPP are to provide data continuity for global change research and to provide risk reduction for the operational program as it prepares to upgrade the quality of space-borne measurements to match that of the current research measurements.

10.1.1 Origins of NPP

NPP is a unique partnership between the Integrated Program Office (IPO) and NASA. The IPO is itself a joint activity of the Department of Defense (DoD), the National Oceanic and Atmospheric Administration (NOAA), and NASA. The IPO is responsible for the development and operation of NPOESS, a continuing series of environmental satellites beginning about 2013. This new satellite series will replace the current separate civilian Polar Operational Environmental Satellite (POES) program and the military Defense Meteorological Satellite Program (DMSP) with a unified system. It incorporates advanced sensors patterned, in many cases, after those developed by NASA for its research oriented EOS.

Beginning in 1998, NASA was reassessing its approach to obtaining systematic measurements of many environmental parameters, especially those that were to be obtained from the Moderate Resolution Imaging Spectroradiometer (MODIS) on EOS Terra and Aqua. NASA participated strongly in the initial selection of the vendor to design and build the Visible Infrared Imaging Radiometer Suite (VIIRS). It became apparent that proposed VIIRS capabilities were close to those of MODIS. A NASA study conducted with the aid of the MODIS science team concluded that a number of important new capabilities, such as the capability to measure ocean fluorescence and high spatial resolution imagery in the CO_2 bands were too preliminary for flight on an operational sensor. Others, such as the need for more fine spatial resolution atmospheric and land biosphere data products, a higher precision geolocation, and a fire product were found to be consistent with operational use.

NASA funded a short study by the competing contractors to see if they could meet these additional NASA requirements. The competing contractors, the IPO and NASA concluded that the additional NASA requirements fell between the IPO's minimum requirements ("threshold") and their desired goals ("objective"), and that they were compatible with the sensor designs under consideration. The sensor requirements documents were modified and clarified to reflect this altered emphasis. The resulting VIIRS provides performance comparable to that of MODIS for those data products it makes.

The IPO plans called for a risk reduction mission prior to the main phase of the NPOESS program. The original plan was to modify the last of the NOAA operational satellites to carry prototypes of four NPOESS sensors, including VIIRS. This important step was necessitated by the considerable increase in the complexity of the sensors and their algorithms. The users communities needed to validate these new systems while they still had the current, simpler systems. However, the mass, power and data rate requirements of the new sensors were too taxing for the current NOAA spacecraft. It was more practical to build a new spacecraft suited to the new sensors.

When NASA initiated its post-2002 mission planning process and the IPO re-assessed its approach to fly a risk reduction mission prior to the main NPOESS missions, NASA and the IPO were able to define a joint mission, NPP, that met both NASA's continuity requirements for MODIS and the IPO's risk reduction requirements for VIIRS. Coupled with an assessment of the other planned capabilities of NPOESS, NASA decided to end its planned series of three copies of each of three specialized satellites (Terra, Aqua, and Aura) and planned to use NPOESS for its measurement needs beginning in 2013.

10.1.2 Program Philosophies

Successfully merging the technical aspect of the NASA and IPO programs requires an understanding of the underlying philosophies and accommodating the differences between the two. The following two paragraphs briefly summarize the key aspects of the two program philosophies. The NPP mission is responsive to both.

The operational program is based on a rigorous multi-agency requirements process (see Section 10.6.1). Once the requirements have been defined, they are placed under configuration control. The IPO is developing NPOESS under a relatively new approach, the Shared System Performance Responsibility (SSPR) contract in which major elements of the program responsibility are awarded to an industrial contractor in a competitive process. Once that award is made, the SSPR contractor has considerable authority to trade cost, schedule, and performance.

Northrop Grumman Space Technologies (NGST) and Raytheon Corporation serve as the SSPR contractor. In the case of the NPOESS contract, the SSPR contractor has direct control over sub-contracts to the sensor vendors, most of whom were originally contracted to the IPO. The sensor vendors not only developed the sensors, they also developed (or adapted) the science algorithms. The algorithms were selected for being proven. This usually means that they are not state-of-the-art. They were also selected for rapid and robust execution. Yesterday's weather is of no use to the operational community. Programmatic science advice for each science area is provided by a government appointed Operational Algorithm Team (OAT). These teams interact with NGST/Raytheon on technical issues, but their programmatic advice is to the IPO.

NASA's requirements process is considerably more informal (see Section 6.2). Whereas the IPO defined data product requirements, NASA typically defines sensor requirements that are based up community consensus science "desirements". The desirements reflect an assessment of what might be possible, and they anticipate a competitive process to select algorithm developers who will then work with the (competitively selected) sensor vendors on the implementation of the sensor design. This results in sensor requirements that are somewhat dynamic, with the science community in the loop at all times. Usually the science teams are the direct customers of the sensor vendor. This is well suited to identifying and resolving the inevitable conflicts between sensor specification, evolving scientific capabilities, and the actual needs of the scientist, and usually leads to a superior state-of-the-art sensor. It is not, however, well suited to managing to cost and schedule. The algorithms, coming from the research community, are by definition, state-of-the-art. Speed of execution is not a driving consideration. Technical and programmatic direction for the sensors is provided by the competitively selected science teams. The typical NASA program uses a NASA center to fulfill the system level management roll that is equivalent in many ways to that of the SSPR contractor.

10.2 Sensor Payload—Providing Continuity and Evolution

NASA and the IPO jointly selected the NPP payload based on NASA's research priorities and the IPO's need for risk reduction. Most of the selected NPP sensors are considerably more complex than their operational predecessors, and somewhat evolutionary from the EOS sensors.

10.2.1 VIIRS and Terra MODIS Continuity

The most critical sensor for continuity with EOS Terra and the original motivating force for the NPP mission is the Visible Infrared Imaging Radiometer Suite (VIIRS). It will continue the measurement series initiated by the Moderate Resolution

Imaging Spectroradiometer. MODIS on Terra is especially critical for cloud-free measurements of the land surface at a time scale appropriate to detect seasonal variations in the biosphere. A morning equator crossing time was chosen for the NPP orbit in order to extend the Terra time series. (The number of cloud-free or nearly cloud-free days at the time of the morning equator-crossing orbit is nearly double what it is at the time of the afternoon equator-crossing orbit, allowing a cloud-free composite to be acquired in less than ten days.)

VIIRS makes measurements in the visible, short-wave and mid-wave infrared and the far infrared to produce 21 geophysical data products including land and ocean biospheric measurements, cloud cover, aerosols and sea surface temperature (see Table 10.1). It has a swath width of 3,000 km, permitting complete global coverage in one day at a spatial resolution of <1 km. A more complete description of VIIRS is given in Chapter 11. The spectral coverage of VIIRS is slightly less than that of MODIS, while its spatial resolution is superior. The radiometric quality of VIIRS is slightly reduced from MODIS, but consistent with the measurement needs. VIIRS obtains complete global coverage in one day, whereas MODIS requires two days for complete coverage.

Table 10.1 VIIRS data products

Atmosphere	Land
Aerosol Optical Thickness	Active Fires
Aerosol Particle Size	Albedo
Cloud Base Height	Land Surface Type
Cloud Cover	Land Surface Temperature
Cloud Effective Part Size	Vegetation Index
Cloud Optical Thickness	
Cloud Top Height	Oceans
Cloud Top Pressure	Ocean Color
Cloud Top Temp	Sea Surface Temperature
Suspended Matter (Dust)	
Cryosphere	**Imagery**
Ice Surface Temp	Imagery
Sea Ice Characterization	
Snow Cover/Depth	

10.2.2 VIIRS and Continuity of Operational Imagers

VIIRS represents a major new development. With 22 spectral bands, up to 370 m resolution at nadir and 12-bit quantization, VIIRS is far more complex and powerful than the Advanced Very High Resolution Radiometer (AVHRR) on POES, which has a 6-band radiometer with a 1 km nadir resolution and 10-bit quantization. It is an even greater advance over the Operational Linescan System (OLS) on the DMSP, which has only two bands, 2.5 km nadir resolution and 6 or 8-bit quantization. Early flight of VIIRS is needed to test out the sensor and algorithms and to provide the operational user community an opportunity to adapt to the use of this new sensor while they still have access to the current generation of sensors. With 22 channels and 12-bit quantization, the users may have to handle more than ten times the data volume they experienced with AVHRR and almost 100 times more than OLS. VIIRS has bands that closely match all of the AVHRR and OLS bands for those users who do not wish to deal with the larger amount of data and the more complex algorithms.

10.2.3 Aqua and Aura Continuity

The remaining payload provides continuity with aspects of NASA's Aqua and Aura missions. With Aqua's launch on May 4, 2002 and Aura's launch on July 15, 2004, there is less concern over the risk of gaps in coverage than there was for the terrestrial biosphere measurements from EOS MODIS. The possibility of overlapping measurements by EOS and NPP sensors offers a unique opportunity for cross calibration of the sensors.

10.2.4 CrIS and AIRS Continuity

Infrared sounders provide high spectral resolution data to be used to determine atmospheric temperature and moisture profiles for the dual purpose of initializing numerical weather forecast using general circulation models and for monitoring climate variability and trends (see Table 10.2). The temperature data are also used in the identification of correlation between climate variation and atmospheric processes, the validation of global climate models and the detection of long-term global temperature trends. The Aqua satellite carries the Atmospheric Infrared Spectroradiometer (AIRS) to fulfill this function while the Cross-Track Infrared Sounder (CrIS) is used on NPP and NPOESS. CrIS is a Michelson interferometer with 1,142 channels in the thermal infrared region. It has a spatial resolution at nadir of 14 km and a swath width of 2,300 km. Its spectral, spatial, radiometric, and temporal characteristics are virtually identical with those of AIRS.

10.2.5 CrIS and Continuity of Operational Sounders

Previous operational systems used the 20-band High Resolution Infrared Sounder (HIRS). The much higher spectral resolution of CrIS (and AIRS) allows much more sophisticated algorithms and improved accuracy. CrIS is expected to achieve an accuracy of about 1°K per kilometer through much of the atmosphere; an accuracy comparable to that of radiosondes but with much greater coverage in space and time. The increase in data volume compared to HIRS is more than a factor of 10. A more complete description of CrIS is given in Chapter 14.

In addition to measuring temperature and moisture profiles, CrIS measures total column ozone independently of the ultraviolet-based methods of OMPS. This allows it to obtain data during the nighttime when ultraviolet methods won't work. Because of its high spectral resolution, CrIS can also be used to measure a wide range of other trace gases such as methane.

10.2.6 ATMS and Continuity of Operational Sounders

The Advanced Technology Microwave Sounding Unit (ATMS) is the companion to CrIS; it provides a "first guess" of temperature and moisture profiles needed for the CrIS algorithm and it provides profiles under cloud conditions where the CrIS algorithms will not work. Unlike the other NPP sensors, it has capabilities that are essentially identical to those on the current operational satellites. The 22 channels, which cover the range from 23 GHz to 183 GHz, are nearly identical to those on the Advanced Microwave Sounding Unit (AMSU). ATMS uses MMIC electronics and other technical improvements to provide AMSU-like capabilities in a sensor with 50% of the mass and power of AMSU. The footprint of ATMS is typically 50 km at nadir and the swath width is 2,300 km. At 30 kbps, the data rate is essentially the same as AMSU. A more complete description of ATMS is given in Chapter 13.

ATMS and CrIS have very different scan characteristics. CrIS actually consists of a 3 by 3 array of detectors, each with a 14 km field of view. One complete scan takes 8 seconds. During that time, ATMS makes 3 scans with its nominal 50 km field of view. Only one 50 km field of view is seen by ATMS at exactly the same time as the 9 fields of view. Other fields of view must be re-sampled from slightly different times to match, and the two sensors must be synchronized every 8 seconds to make sure that the re-sampling can be done in a consistent and efficient manner.

10.2.7 OMPS and Continuity of Research Sounders

The Ozone Mapping and Profiling Suite (OMPS) uses two sets of optics to perform the functions of three different heritage sensors. One set of optics collects photons from the nadir direction to measure the total column ozone abundance in a manner similar to that of the Total Ozone Monitoring Sensor (TOMS) and the ozone profile in a manner similar to the Solar Backscatter Ultraviolet (SBUV) sensor. A second set of optics collects photons from the limb to provide a more detailed set of ozone profiles in a manner similar to the shuttle experiment SOLSE/LORE. With the limb sensor, ozone profiles will be measured to a much finer vertical resolution (5 – 7 km) than was possible with the heritage operational sensors (see Table 10.2). Although there is a long series of research quality measurements, the OMPS design is radically different from its predecessors in terms of the fore optics and the use of CCD arrays in place of photomultiplier detectors. A more complete description of OMPS is given in Chapter 15.

Table 10.2 CrIS, ATMS & OMPS data products

EDR	Sensor(s)
Atmosphere Vertical Moisture Profile	CrIS/ATMS
Atmosphere Vertical Temperature Profile	CrIS/ATMS
Atmosphere Vertical Pressure Profile	CrIS/ATMS
Ozone Total Column & Vertical Profile	OMPS

10.3 Spacecraft and Launch Vehicle

With the large number of civilian, military, and commercial users of space, it is now possible to buy a spacecraft from a catalog. NPP's "bus" is a BCP 2000, supplied by the Ball Aerospace Corporation (Ball Aerospace, 2004). It is designed for a five-year mission lifetime, but will carry consumables sufficient to last 7.5 years. The standard bus has been upgraded in several ways. Rather than the usual single star-tracker, it carries two in order to be assured of meeting the demanding pointing requirements. (VIIRS pixels must be geolocated to ± 200 m). An on-orbit GPS receiver supports the needed precision orbit requirements.

Another upgrade is the use of the IEEE 1394 (Firewire) protocol for data communication between the spacecraft and the VIIRS and CrIS sensors. While the Firewire protocol is now standard with personal computers, this will be its first use in the harsh environment of low Earth orbit. Specially designed, radiation-hardened components are required. The older Military Standard 1553 protocol is used for OMPS and ATMS.

During the initial on-orbit checkout of NPP, the entire system can be inverted, allowing the sensors to view cold, dark space instead of the Earth. This is needed in order to assure that the thermal emission from warm scan optics was properly characterized during the pre-launch testing. Without such verification, some critical climate goals, such as determining if there is a gradual increase in global sea surface temperature, cannot be met. It is unlikely that this can be done with the NPOESS, so it is essential that the accuracy of the ground testing be verified. This maneuver would be conducted during spacecraft night by driving the spacecraft backwards at a constant rate in pitch.

In addition to the "deep space" maneuver, the spacecraft will also support a roll of up to –25° with a five minute dwell time in order to use the Moon as a calibration source. By making this type of roll, the Moon can be viewed over a limited range of phases and librations, yielding a more accurate calibration (Godden and McKay, 1998). Another key capability is to yaw the spacecraft up to ±25° in order to rapidly map out the variations in the solar diffuser bidirectional reflectance distribution function to correct for longer term variations that will be observed as the solar beta angle changes.

All data will be stored on a solid-state recorder for transmission to a ground receiving station in Svalbard, Norway. The recorder can store 3.5 orbits of data. Transmission is at 300 Mbps at a frequency of 8.2125 GHz (X-band), using an "Earth-shaped" omni antenna.

Continuous direct broadcast of all sensor and associated spacecraft data is provided at 7.812 GHz (X-band) using an "Earth-shaped" omni antenna, QPSK, lossless compression, and convolutional coding. Ground stations already receiving the direct broadcast data from Terra or Aqua will be able to capture the data with minimal changes to their hardware and software. There is no low rate data stream similar to the L-band HRPT or S-band APT data streams of the POES satellites.

Software for reading the direct broadcast signal and producing SDR's and a subset of the EDR's will be available through the NASA Direct Readout project (NASA, 2005).

A single axis articulated solar array with 204 triple junction solar cells provides 2.4 kW of power (worst case) at the end of the mission. The solar array charges an 85 ampere-hour Nickel Hydride battery that feeds power to the spacecraft and sensors over five power buses. The system operates at 28 V.

A hydrazine propulsion system is used to correct for launch insertion errors, to maintain ground track (to ±20 km), and to perform a controlled de-orbit at the end of the mission. The equator crossing time will be maintained at 10:30 AM local time. Periodic adjustments to the orbital inclination will be made to keep the crossing time to within ±10 minutes of 10:30 AM.

Launch will be from the Western Test Range (Vandenberg Air Force Base) using a Delta II rocket.

10.4 Orbit

The NPOESS will operate in sun synchronous orbits at 833 km ± 17 km. In this altitude range, wide field of view sensors such as VIIRS can achieve full global coverage in one day. Small changes to the mean altitude result in dramatically different ground tracks. For example, at a mean altitude of 820 km the satellite ground track will be an exact repeat every 5 days. At 836 km, the ground track repeats every 20 days. Both of these extremes have the potential for biases in the observations. With a 5-day repeat there will be locations on the Earth that will never be viewed at nadir angles less than 30 degrees and therefore never seen at the highest spatial resolution. With a 19-day repeat there will be extended periods when a given location on the surface will only be viewed in the back-scatter direction while other locations are only viewed in the forward scatter direction. Persistent cloud cover can cause these directional biases to persist for more than a month when attempting to achieve cloud free coverage of the surface.

NPP operates at the lower range of the NPOESS orbits with a mean altitude of 824 km. This specific altitude was chosen to provide an exact 16-day ground track repeat in a "skipping orbit" similar to those of the Landsat and EOS satellites. This approach minimizes directional biases in the observed radiances and averages the results over a bi-weekly time frame. Many climate records are compiled as monthly averages, and this will minimize the aliasing of directional variations into the monthly averages. Measurements of the land biosphere need to be compiled over shorter intervals to capture seasonal phenomena such as the spring green-up or crop emergence. The 16-day repeat time supports that usage. (as of the publication date, consideration us being given to moving NPP to an afternoon orbit with similar properties but with a crossing time of 13:30.)

10.5 Ground Segment

10.5.1 Data Downlink

Like the EOS Aqua and Aura satellites, NPP will use a high speed X-band transmitter to send stored mission data to a high latitude receiving station once per orbit. Operating at a rate of 300 Mbps, all mission data can be downloaded in a single pass for 20% of the passes. NPP will utilize a new receiving system with a 13-meter antenna located on the island of Svalbard, Norway. A new fiber optic cable system will relay the data to the operations center at the NOAA facility in Suitland, MD within minutes.

10.5.2 IDPS

All operational products will be produced in the NPOESS Interface Data Processing Segment (IDPS). During the NPP mission, there will be two instantiations of the IDPS at the operational satellite "weather centrals". One will be at the NOAA Environmental Satellite Data, and Information Service (NESDIS) in Suitland, MD and the other at Air Force Weather Agency (AFWA) in Omaha, NE. They will run standard products using algorithms developed by the NPOESS system contractor, Northrop Grumman Space Technologies (NGST) and their sub-contractors. Later in the program, in preparation for the fully operational NPOESS, there will be two additional instantiations of the IDPS at Fleet Numerical Modeling Operations Center (FNMOC) in Monterey, CA and Naval Oceanographic (NAVO) in Bay St. Louis, MS.

Data latency is a key requirement for operational products. Since the data products are used in weather forecasts, they must be made quickly and they must be robust. During NPP, 95% of the EDR's are to be produced within 120 minutes from the time the sensors acquire them. The majority of the latency is caused by the time between data transmission to the single high latitude ground station. The IDPS is designed with a highly parallel processing architecture. In order to meet the latency requirements, there may be trades between EDR accuracy and speed. Quality flags and intermediate products will be kept to a minimum. As an operational system, there will be no reprocessing.

The "weather centrals" are responsible for provision of data products, including value added products, to their respective user communities. Copies of the raw data as well as the standard products from the NESDIS IDPS will be submitted to a long-term archive that is being developed as a part of NOAA's Comprehensive Large Array-data Stewardship System (CLASS) (NOAA, 2005). CLASS will be the primary source of NPP data for all research users.

10.5.3 SDS

NASA will have a separate Science Data Segment (SDS) that will be used to assess the suitability of the operational algorithms and their associated EDR's for use in research. The SDS will be capable of producing any of the EDR's, and comparing them with ground truth data products from alternative algorithms and data products produced with alternative ancillary data. It will also support the development and implementation of alternative, more detailed sensor calibrations and level-1 algorithms. The SDS is a test and development environment that will provide the means for improvements to the sensor calibration and upgrades to the EDR algorithms that are compatible with the operational timelines. The SDS is operated by the Project Science Group (see Section 10.7). It will be used to test algorithm improvements suggested by the research community as discussed in the next paragraph.

External to the NPP, NASA is planning to develop a set if discipline-based processing centers that will develop multi-sensor CDR's. Whereas the focus of the IDPS and SDS products will be on the NPP sensors, these Climate Research Analysis Systems (CARS) will combine data from a number of separate missions and sensors. The SYMBIOS project is an example of such an activity, combining ocean color measurements from OCTS, SeaWiFS, and other sensors. These CARS will be involved in the evaluation of the NPP EDR's in terms of their suitability as CDR's, the development of new algorithms and fused data products, and the ultimate production of a to-be-determined set of CDR's. The CARS will provide the Project Science Group and its SDS their inputs on calibration issues and suggested algorithm improvements.

10.6 Measurement Requirements

10.6.1 IORD

The operational requirements for NPP are defined by the Integrated Operational Requirements Document (IORD). These requirements were developed beginning in the 1990's and last updated in 2002 (JARG, 2001). The requirements definition process was led by the IPO with major rolls played by NOAA, DoD, and NASA, and it involved all government agencies that require operational satellite data.

The IORD defines 55 Environmental Data Record (EDR's) that must be produced by NPOESS, 25 of which will be produced by NPP. For each EDR (e.g. sea surface temperature, cloud top height, atmospheric temperature profile) the IORD specifies up to 9 attributes such as the required accuracy and precision, the horizontal cell size, mapping accuracy, and measurement range. Tables 10.1 and 10.2 list the EDR's than will be produced by NPP.

The IORD was initially developed for operational users, but it is recognized that the measurements can also support long-term climate studies and other research activities. Most EDR's also have long-term stability requirements and some have specific higher accuracy requirements for climate products. The planned capabilities of EOS are clearly reflected in the requirements.

10.6.2 NASA Science Requirement

NASA science planning has evolved over the course of the development of NPP. In 1998, NASA conducted a planning workshop on post-2002 missions. That workshop, known as the Easton Workshop after the venue in Easton, MD, identified the need for systematic measurements of key variables and endorsed the concept of transitioning these measurements to NPOESS through the bridging mission that is now known as NPP (Kennel et al., 1998). The measurement requirements were not spelled out in detail, but were implicitly defined to be of the quality expected from the EOS Terra, Aqua, and Aura missions.

Since 1998, NASA has developed a research plan in 6 broad categories based, in part, on the achieved capabilities of EOS Terra and Aqua (NASA, 2004). NPP is a critical part of the plan for 5 of the 6 focus areas; Atmospheric Composition, Climate Variability, Ecosystems and Carbon, Global Energy and Water Cycles, and Weather. Each of these focus areas requires continued systematic measurements from EOS-like sensors and to accuracies that are comparable to or better than those obtained from EOS. VIIRS supports the needs of four of the focus areas, while the CrIS/ATMS combination is critical to the Global Water and Energy and the Weather focus and OMPS is key to the Atmospheric Composition focus.

Ultimately NASA's measurement requirements are those defined in the EOS program plan (EOS, 1995). Because this document is not generally available, the relevant entries are extracted in Tables 10.4 and 10.5. The EOS program relied on a sensor specification that is best captured in terms of the Level-1B data products (see Table 10.3).

Table 10.3 Orbital parameters

Orbital Altitude	824 km
Orbital Inclination	98.74°
Equator Crossing Time (Ascending node)	10:30 ±10 minutes
Orbital Period	101.498 minutes
Repeat Cycle (Days)	16
Skip Factor	2

EOS Level-2 products are geophysical products in their native spatial resolution, e.g. cloud top temperature at 1 km at nadir. In addition, research users require data products that have been set on a standard grid, usually at a lower spatial resolution and often averaged or composited over a time period of weeks to months. These level-3 products are critical to the meeting of global change researchers as they are more closely coupled to the temporal and spatial scales used in global models. Neither NPOESS nor NPP will produce data products that are equivalent to EOS Level-3 products. They must be produced in the CARS or in other non-mission facilities.

The expected product accuracies are summarized in Table 10.4. In general, the requirements are close to those defined in the IORD.

Table 10.4 EOS sensor Level-1 accuracy requirements

Sensor	Units	Absolute Accuracy	Relative Accuracy	Horizontal Resolution
MODIS	$W/m^2/sr/\mu m$	5%(1s)	$\mathrm{RMS}<\mathrm{NE}\Delta\mathrm{L}$	0.25, 0.5, 1 km
AIRS	$W/m^2/sr/\mu m$	0.2 K NEDT	0.2 K NEDT	15 km
AMSU	K	0.2 K NEDT	0.2 K NEDT	40 km
MHS	K	0.2 K NEDT	0.2 K NEDT	15 km
OMI				

Table 10.5 EOS Level-2+ product requirements

Product Name	Sensor	Accuracy	Horizontal Res
Aerosol Product	MODIS		50 km (Land), 5 km (Ocean)
Optical Thickness (Ocean)		< 0.05	5 km
Optical Thickness (Land)		0.05	50 km
Cloud Product	MODIS		5, 50 km
Cloud Cover (Fraction)		Better than 0.1	5 km
Cloud Top Height		0.5 km-cu, 1 km-ci	5 km
Cloud Top Pressure		50 mb	5 km
Cloud Top Temperature		2 – 5 K	5 km
Cloud Optical Thickness		10%	1 km
Cloud Effective Particle Size		10%/(1 – 3 km)	1 km
Land Surface Reflectance	MODIS	0.03	1 km
Snow Cover	MODIS	10% (1)	1 km
Land Surface Temp./Emissivity	MODIS		1 and 5 km
Land Cover Type	MODIS	10% (2)	1 km
Chlorophyll A Concentration	MODIS	30% – 60% (Case Ⅰ) 50% (Case Ⅱ)	1 km
Water-leaving Radiance	MODIS	0.0015 (3)	1 km
PAR (Ocean)	MODIS	5% – 10%	1 km
SST	MODIS	0.3 – 0.5 K	1 km
Cloud Mask	MODIS		250 m, 1 km
Moisture Profile	AIRS/AMSU/HSB		50 km
Temperature Profile	AIRS/AMSU/HSB		50 km
Ozone Total Column			
Ozone Vertical Profile			

10.6.3 Stratification

A key aspect of the interpretation of these requirements is the need to meet them over properly stratified conditions. In general, the IORD is silent on the details of stratification. The required accuracies cannot be obtained under *all* conditions, but must be met under *most* conditions. For example, the atmospheric temperature sounding requirement can be summarized by calling for an accuracy of 1°K per km from the surface to a pressure of 300 mb when there are no clouds present. It is not possible to obtain that accuracy when there is an unusually strong temperature inversion, but it must be achievable most of the time under the conditions that often prevail in the summer in the mid-Atlantic region of the U.S. Further, the accuracy should be achievable for a global ensemble of typical conditions over all seasons.

10.6.4 CDR's and EDR's

Following the nomenclature of the National Academy of Sciences, NASA is interested in the production of Climate Data Records (CDR's). The term CDR is used to designate a time series data (level-2 or level-3) product that is of sufficient accuracy for climate (and other global change) research. This is in contrast to the operational products, which are not intrinsically a time series. For weather forecast, the best currently available product is needed.

Time series will initially involve data from a single sensor, with due attention being paid to slow changes in the sensor characteristics, for example, the slow degradation of a solar reflectance calibration target or increased noise in the detectors. Allowance must also be made for changes in the sensor configuration such as the shift to redundant electronics, changes to detector biases, or focal plane operating temperatures, all of which can cause changes in the coefficients of the calibration equation (see Vol. 2 Chap. 5, Xiong, 2005). Ultimately data products from multiple copies of a single sensor type, or even from widely different sensors, such as sea surface temperature infrared and microwave sensors will be combined.

The NPP mission will not directly produce CDR's but it will concentrate on understanding the time dependent behavior of the NPP sensors and their data products. This will be done by analysis of pre-launch characterization data and on-board calibration systems, comparison to ground truth data, and data from other space-borne sensors over periods of weeks to years. The operational component will produce EDR's to the quality defined in the IORD. During NPP, NASA will participate in the validation of the EDR's to determine their suitability for climate studies. Improvements to the understanding of sensor characterization and calibration as well as suggested improvements to the operational algorithms that are compatible with the operational timelines will be provided to the IDPS by the NPP Project Science Office, making use of the capabilities of the SDS to demonstrate the accuracy and stability of these suggested changes. The production of CDR's will be accomplished by the operational system using appropriate calibration trending and EDR algorithms, by Climate Analysis Research Systems (CARS) developed by NASA outside the NPP mission or by other research entities.

Some of the key activities needed to enable the production of CDR's are given in Table 10.6.

Table 10.6 Activities needed to make CDR's

Activities	Sensor Vendors	NGST	IPO	NASA NPP	Other
Pre-Launch Sensor Characterization Measurements	××				
Pre-Launch Sensor Calibration Measurements	××				
Development of SDR & EDR Algorithms	×	××			
Review of SDR & EDR Algorithms			××	×	
Pre-Launch Sensor Performance Verification	××	×			
Incorporation of Characterization Data into EDR Algorithms		××			
Pre-Launch Verification of EDR Performance		××			
Analysis of Characterization Data for Climate Purposes				××	
Post-Launch Performance Verification:					
Daily Products		×	××		
Long Term (weeks-months)				××	
Post-Launch Adjustments to SDR & EDR Algorithms	×	××		×	
Post-Launch Adjustments to Sensor Operating Parameters	×	××			
Improvements to SDR & EDR Algorithms for Climate Needs				××	
Monitor On-Orbit Sensor Performance		××			
Maintain On-Orbit Calibration		××			
Assure Consistency of SDR & EDR Algorithms Over Time				××	
Forward Production of CDR's					××
Reprocessing of CDR's at 1 – 2 Year Intervals					××
Long Term (Multi-Mission) Data Set Production					××
	××	Leadership Roll			
	×	Contributing Roll			

A CDR can be produced with the same algorithms as are used for EDR's if the following conditions are met:

(1) The sensor behavior (including calibration, noise, scattered light, etc.) is sufficiently stable and understood;

(2) The EDR is of sufficient accuracy and precision, including appropriate stratification over geographic, diurnal, seasonal, and weather domains to meet research needs.

The operational EDR can be used as a CDR if the following additional conditions are met:

(1) Drifts in sensor calibration and other behaviors can be predicted in advance to sufficient accuracy;

(2) Dynamic ancillary data (e.g. ozone amounts, initial temperature and pressure fields) are consistently available at the time of processing.

10.7 Science Guidance

The unique dual role of NPP as a source of continuity of data for research and as a means of risk reduction for the operational program calls for an unusual approach for provision of science direction. NASA and the IPO have developed separate advisory teams with overlapping memberships. The IPO uses an Operational Algorithm Team (OAT) for each set of measurements to provide advice on the algorithms, sensor characterization, and sensor calibration.

The IPO also funds studies to directly support the sensor and algorithm development and to advise the IPO under its Internal Government Studies (IGS) program.

NASA has formed a pre-launch science team to advise the NASA and the NPP Project Scientist on the suitability of the operational data for climate studies. Specifically, they are testing the hypothesis that the EDR's are suitable as CDR's. NASA supports an internal team, the Project Science Group, which conducts specific technical studies in the area of sensor characterization and calibration and acts as an interface between the NASA NPP Science Team and contractor project activities.

10.8 Summary

NPP is a joint mission between the IPO and NASA that provides risk reduction for the next generation of operational environmental satellites and continuity of observations for a number of important parameters needed for the study of global change. The data will be generally of the same quality as the similar EOS data products. The 824 km polar orbit will provide EOS-like ground track repeats, facilitating the extension of weekly to monthly composite data products. The raw data for all sensors will be available through direct broadcast at X-band at no cost, and both raw data and higher level products will be available from government archives. NASA will test the suitability of the operational products for long-term global change studies and systems are available that could permit upgrade to the operational algorithms in the future.

References

Ball Aerospace & Technology Corp (2004) The spacecraft is described on the Ball Aerospace Web site; www.ballaerospace.com/npp

Execution Phase Project Plan for Earth Observing System (EOS) (1995) 170-01-01 Revision A, May

Godden GD, McKay CA (1998) Precise relative calibration of Earth-orbiting remote sensors, using the Moon. Metrologia 35: 649 – 655

Joint Agency Requirements Group (JARG) (2001) Integrated Operational Requirements Document (IORD-II). 10 December, 2001 (Online—Referenced February 22, 2005) http://npoesslib.ipo.noaa.gov/IPOarchive/MAN/IORDII_011402.pdf

Kennel C, Frieman E, Moore B and Shaffer L (1998) Report of the Workshop on NASA Earth Science Enterprise Post-2002 Missions. NASA Headquarters, Washington, D.C., November 12, 1998. (Referred to as Easton Workshop Report and frequently as Kennel Report. Available online at http://www.earth.nasa.gov/visions/Easton/index.html)

NASA Science Mission Directorate (2004) Science Strategy http://science.hq.nasa.gov/strategy/

NASA Goddard Space Flight Center (2005) Direct Readout Laboratory http://directreadout.gsfc.nasa.gov/index.cfm?section = technology

NOAA Satellite and Information Service (2005) Comprehensive Large Array-data Stewardship System (CLASS) http://www.saa.noaa.gov/nsaa/products/welcome

Xiong X, Barnes W (2005) MODIS Characterization. Earth Science Satellite Remote Sensing Vol 2, Chap 4. Springer (submitted)

11 The Visible Infrared Imaging Radiometer Suite

R. E. Murphy, Phillip Ardanuy, Frank J. Deluccia, J. E. Clement and Carl F. Schueler

11.1 Introduction

The Visible Infrared Imaging Radiometer Suite (VIIRS) is used to obtain measurements of the Earth's oceans, land surface and atmosphere to make a wide range of Environmental Data Records (EDR's). These standard products, which have been defined in the Integrated Operational Requirements Document (IORD) (JARG, 2005), are listed in Table 11.1. VIIRS is designed to provide global coverage at least once per day, with moderate (better than 1 km) spatial resolution. This combination of spatial and temporal scales has been chosen to provide needed input to operational weather and environmental models while sampling the natural variability of biological processes on the land surface and in the oceans. It has very high radiometric and geometric fidelity, enabling its use in the acquisition of long-term data records suitable for the study of climate and climate change as well as being a powerful tool for studies designed to increase our understanding of specific geophysical processes such as the interaction of the biological and physical mechanisms in ocean plankton blooms.

VIIRS will fly on the National Polar-Orbiting Operational Environmental Satellite System (NPOESS) in two polar sun-synchronous orbit planes with equator crossing times of 1730 and 1330. It will also fly on the NPOESS Preparatory Project (NPP) with a 1330 Ascending Node. NPP will launch in late 2009; the NPOESS satellites are scheduled to be launched beginning about 2013.

While the sensor is described as a "suite", it consists of a single sensor; the original planning allowed for meeting the measurement requirements with either a single sensor or a small suite of sensors. VIIRS draws heavily on the experience gained in building and operating the Moderate Resolution Imaging Spectroradiometer (MODIS) that also combined diverse functions into a single sensor. It also draws on the experience with various sensors with specific tasks such as the Sea-viewing Wide Field-of-view Sensor (SeaWiFS) for measuring ocean color, the Along Track Scanning Radiometer (ATSR) for measuring sea surface temperature, and the Operational Linescan System (OLS) for terminator imaging.

Table 11.1 List of data products from VIIRS

Product Grouping	EDR/Other Product	Comments
Imagery	Imagery*	Manually interpreted
Surface Temperatures	Sea Surface Temperature* Land Surface Temperature Ice surface Temperature	
Clouds	Cloud Base Height Cloud Cover/Layers Cloud Particle Size Cloud Optical Thickness Cloud Top Height Cloud Top Pressure Cloud Top Temp Cloud Mask	Supports CMIS Intermediate Product
Aerosols	Aerosol Optical Thickness Aerosol Particle Size	Separate alg. for land/ocean Separate alg. for land/ocean
Ocean Biosphere	Ocean Color/Chlorophyll	
Land Biosphere	Vegetation Index Vegetation/Type Albedo	
Snow & Sea Ice	Snow Cover Sea Ice age/motion	Supports CMIS Works w/ CMIS
Fire	Fire Detection	Not an EDR
Other	Soil Moisture Net Heat Flux Suspended Matter Geolocation	Supports CMIS Supports CMIS Intermediate Product

* Key performance parameter.

11.1.1 Spectral Band Compliment

VIIRS uses 4 Focal Plane Assemblies (FPA's) to hold the 21 spectral bands plus a "Day-Night Band" (DNB), a high sensitivity, large dynamic range, visible panchromatic band. The DNB and the 16 "M" bands have a moderate resolution of 740 m at nadir while the 5 "I" bands (for imagery) have a nadir resolution of 370 m. The visible and near infrared (VISNIR) FPA has 9 spectral bands with central wavelengths between 412 nm and 865 nm. The DNB FPA shares the same optical "focal plane" as the VISNIR FPA. The shortwave and midwave infrared

(SWMWIR) FPA has 8 bands with central wavelengths between 1.24 μm and 4.05 μm. The long-wave infrared (LWIR) FPA has just four bands, with central wavelengths between 8.55 μm and 12 μm. Comparisons to predecessors are shown in Fig. 11.1. The principle characteristics and these 22 spectral bands, including their principle uses, are shown in Table 11.2.

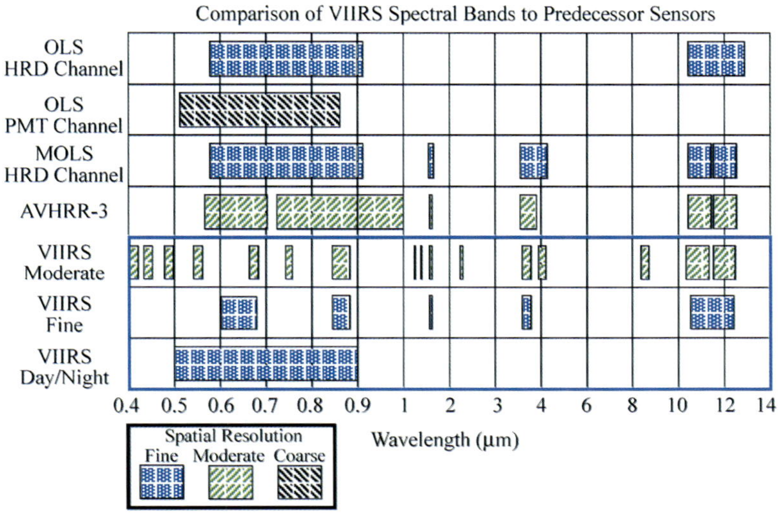

Figure 11.1 VIIRS and other imagers compared

Table 11.2 VIIRS Spectral Bands and Their Principal Uses

ID	Band Center (nm/μ)	Band Width (nm/μ)	Imagery	Surf. Temp.	Clouds	Aerosols	Ocean Color	Land Bio.	Snow/ Sea Ice	Fire	Other
DNB	700	400	M		M						
M-1	412	20				M	M	M	M		M
M-2	445	18	M			M	M	M	M		M
M-3	488	20				M	M	M	M		M
M-4	555	20				M	M	M	M		M
I-1	645	50	I		I			I	I		
M-5	672	20			M	M	M		M		M
M-6	751	15					M				M
M-7	865	39			M	M	M		M		M
I-2	865	39	I		I		I	I	I		

Continued

ID	Band Center (nm/μ)	Band Width (nm/μ)	Ima-gery	Surf. Temp.	Clouds	Aero-sols	Ocean Color	Land Bio.	Snow/ Sea Ice	Fire	Other
M-8	1.240	0.020			M			M		M	
M-9	1.378	0.015	M		M		M				
M-10	1.610	0.060			M	M	M		M	M	
I-3	1.610	0.060	I		I			I	I		
M-11	2.250	0.050	M		M	M	M	M	M	M	
M-12	3.700	0.180		M	M	M		M			
I-4	3.740	0.380	I		I					I	
M-13	4.050	0.155	M	M	M					M	
M-14	8.55	0.30	M		M						
M-15	10.76	1.00	M	M	M			M		M	
I-5	11.45	1.90	I		I					I	
M-16	12.01	0.95	M	M	M			M			

DNB = Day-Night Band
M = Moderate Resolution Band (740 m)
I = Imaging Band (340 m)

11.2 Design Philosophy

Before describing the sensor design in detail, we begin with an overview of some of the factors that drove the design of VIIRS. These are also discussed in further detail in Ardanuy et al. (2001).

11.2.1 Spatial/Temporal Design Drivers

VIIRS is responsible for two Key Performance Parameters (KPP), the Sea Surface Temperature (SST) EDR's and the Imagery EDR. These EDR's are so critical that a replacement satellite will be launched if the VIIRS performance degrades in such a way as to prevent their production at the required fidelity. The design of VIIRS was strongly driven by the particular requirements of these two EDR's and, to a lesser extent, by the other EDR's.

SST is a critical component of weather and climate models and a key indicator of long-term climate stability and climate change. It is derived from measurements in two spectral bands in the thermal IR region. Nighttime measurements of SST may utilize two additional bands in the mid-wave infrared. SST varies significantly from day to day and must be measured at least once in the daytime and once at night. The intrinsic spatial scale of variability for SST is on the order of a few kilometers, especially around the edges of ocean currents. This requires individual measurements to be at a slightly smaller scale, driving the spatial resolution to the sub-kilometer level even though the final standard data product will be reported at a horizontal cell size of 2.5 km. SST will be derived to an accuracy of better than 0.5° K.

The imagery EDR uses a large number of spectral bands in all spectral regions to produce images which can then be interpreted by skilled operators and/or automated computer algorithms to derive assessments of cloud cover and type, as well as ice edge location and concentration. Measurements are acquired in the daytime using both visible and infrared bands, while nighttime measurements rely entirely on the thermal infrared bands. The DNB provides near constant contrast imagery across the terminator and into the nighttime region at relatively coarse radiometric quality. It is capable of providing usable images under illumination by the quarter Moon. The DNB also provides a capability for the production of nighttime cloud masks, though there is no standard data product defined.

Some operational users of the VIIRS data have a need for nearly constant spatial resolution over the full field of regard. The distance from the spacecraft to a nadir location is 833 km, but to the edge of scan, five degrees below the limb, it is over 1,800 km. As a result, the down-track length of a pixel doubles and, due to the curvature of the Earth, the cross-track length triples. VIIRS is designed to meet the spatial resolution requirements at the edge of scan, and aggregate pixels up to the required spatial resolution as the field of view approaches the nadir. As a result, no pixel is larger than 1.3 km within the central 1,700 km of the scan.

A set of EDR's address the measurement of land surface properties, including albedo, land cover, vegetation index, snow cover and land surface temperature. Biologically driven variations in the surface ecosystem depend on complex interactions among seasonal forcing functions, supply of nutrients, weather events that range from short- and long-term drought to the passage of frontal systems, and episodic events such as hurricanes or human-driven land cover change. Cloud free imagery of any region can usually be acquired within 5 to 10 days, just capturing such phenomena as the emergence of crops in the spring and the senescence of vegetation in the fall. Human-induced changes such as crop harvesting and deforestation, as well as some natural phenomena such as fires and floods can cause changes that are only partly sampled by the VIIRS sensor. Land data are needed at the scale of a few kilometers for some biospheric studies and at 10 km to 20 km for radiation budget purposes. But the high spatial variability of the land surface and the non-linear nature of the algorithms dictate obtaining data at the sub-kilometer level, even down to a few hundred meters.

Ocean biological productivity is assessed through the measurement of ocean color. Seven spectral bands in the visible region are used for this purpose. The scale of ocean biospheric variability is well captured with 1-km measurements. Ocean plankton blooms have more rapid time scales than do land ecosystems, but the frequency of cloud cover makes it impractical to obtain the desired time resolution with optical sensors. The ocean biosphere has the additional complication that its ecosystems are moving and there is a complex interaction between ocean dynamics and biological processes. Further, the "sun glint" obliterates much of the needed data. A combination of two VIIRS sensors in different orbits with different viewing geometries reduces the impact of the sun glint and permits a good sampling of the natural variability.

The ocean and land surface properties are used for a wide range of operational purposes including fisheries, agricultural and land use assessments, as well as the scientific study of the global carbon cycle.

Another set of EDR's address the measurement of cloud and aerosol properties. The cloud EDR's rely primarily on infrared bands to measure cloud cover, cloud top temperature and pressure, effective particle size and cloud base height. Reflected bands provide cloud optical properties during the day, as well as cirrus cloud detection. Aerosol properties include the aerosol optical thickness and particle size, as well as suspended matter (e.g. dust). Aerosol measurements rely primarily on visible and near infrared bands, along with the 2.25 μm band over the land. Both clouds and aerosols are critical components of the study of long-term climate change. These data are used for both operational (e.g. aircraft routing) and science (e.g. climatic studies) purposes. The timescale of natural variability, especially for clouds is quite rapid, and cannot be fully encompassed by a single VIIRS. By placing VIIRS on three satellites with different equator crossing times the diurnal variability is well sampled. This coverage is suited to weather models when coupled with data assimilation. Cloud and aerosol properties are needed at the 10 km level, but the spatial heterogeneity of clouds and the non-linearity of the cloud property algorithms again drive the scale of measurements to the sub-kilometer level.

Many of the measurements needed for the study of aerosols are also used for a critical intermediate product, the atmospheric correction. Land and ocean surface properties are determined from the surface reflectance or surface temperature, but VIIRS measures the reflectance or temperature of the combination of the surface and atmosphere. For example, more than 90% of the reflectance measured in the critical ocean color bands comes from the atmosphere, and is, in effect, a bias that must be removed during the retrieval.

Additionally, a cloud mask is needed to determine whether or not a given pixel actually contains information about the surface. All algorithms work on the basis of spectral signatures, and the presence of even a small amount of bright cloud over the typically dark land or ocean surface can severely bias the algorithm's performance. (The same is generally not true for small amounts of surface reflectance in an otherwise cloud-covered pixel.) While clouds exhibit a continuum of spatial scales, the occasions where the sky is 90% cloud free generally include only small amounts of fair weather cumulus. Driven by the height of the boundary level, they have characteristic scales of a few hundred meters. Thus it is necessary to have a few bands optimized for cloud detection and surface property determination at the highest spatial resolution possible.

VIIRS also measures a number of cryospheric properties, including snow cover, fresh water ice concentration, sea ice edge and edge motion, and ice surface temperature. A combination of visible and infrared bands is used to obtain these measurements. The cryosphere measurements are used by operational users for navigation and for weather and climate modeling. The spatial requirements are similar to that of the land EDR's.

Finally, VIIRS works in conjunction with the Conical Microwave Imager/Scanner (CMIS) to provide measurements of net heat flux and soil moisture. The VIIRS role in these EDR's is supportive rather than primary.

11.2.2 Spectral/Radiometric Design Drivers

Land surface properties are primarily derived from the spectral information in the visible and near infrared region. The spectral features are diffuse and complex. The major difference amongst various soils is in their overall color and brightness. The spectral information in vegetation is overwhelmingly the signature of chlorophyll, which is measured with two well-placed spectral bands with moderately sharp bandpasses. The dynamic range has to allow for the typically low reflectances of vegetation and the higher reflectance of some bare soils and, of course, the very high reflectance of snow and ice. Since the same bands can be used for some cloud measurements, the dynamic range has to include even the brightest of clouds. The level of quantization (12 bits) is driven by the need to encompass this wide dynamic range while still providing adequate radiometric resolution for vegetated surfaces (reflectances are typically less than 3%) near the terminator.

For ocean color the situation is more complex as there are a number of pigments besides chlorophyll that are measured. The spectral features are diffuse and overlapping, requiring well placed moderate bandwidth filters. Water is extremely dark, and as much as 90% of the energy reaching the sensor comes from light scattered in the atmosphere. Ocean color sensors often allow the spectral bands to saturate over clouds and land in order to preserve the required radiometric resolution over ocean surfaces. In the case of VIIRS, it was decided to use the same spectral bands for both ocean and land to minimize data rate and to minimize sensor electronic and optical complexity. The bands that serve this dual purpose have dual gains, yielding the required radiometric resolution over the full dynamic range. These bands automatically switch from the high gain state to the low gain state when the radiance illuminating a detector exceeds a given value intermediate between the low and high ends of the full dynamic range. Dual gain switching occurs independently from detector to detector and sample to sample, so that the correct gain state is used regardless of the spatial or temporal variability of the scene radiance.

Many cloud and aerosol properties require information in the visible and short- to mid-wave infrared bands. The dynamic range required for land and ocean is not quite sufficient for clouds, as clouds exhibit a strong backscatter, resulting in reflectances above 100% of a diffuse reflectance. The spectral features used by the algorithms are again somewhat diffuse, requiring only moderate spectral resolution. A powerful way to obtain information on the presence of cirrus clouds requires a very well tailored spectral band at 1.38 microns. This band is deliberately placed in the midst of a strong water-vapor absorption region such that no radiance is sensed from a column unless high cloud is present to reflect the light before it is absorbed.

Other cloud properties make use of the thermal infrared region to measure cloud top temperatures. These same bands are used for determination of land, ice and sea surface temperatures. Inclusion of both cloud top temperatures and surface temperatures places a strong demand on performance over a wide dynamic range, from 180 K to 340 K. Due to practical considerations, the calibration is best suited to the mid- to high-temperature end of the range.

The ensemble of these spatial, temporal, spectral and radiometric design constraints leads to the basic concept of the VIIRS sensor architecture, with some bands offering a nadir spatial resolution of 370 meters and others with 740 meters. Global daily coverage is obtained by having a large field of regard, scanning nearly from limb to limb. A scanning system rather than a step-staring system using large area arrays was chosen because of the large number of spectral bands required. With a large field of regard and with certain other sensor properties that have been chosen to meet the overall requirements, scattered light from bright areas into dark ones and from hot areas into cold ones becomes an issue. This scattered light issue then leads to the concept of using a well-baffled rotating telescope rather than a rotating scan mirror to gather the Earth scene and calibration data.

The spectral and radiometric properties of the 22 spectral bands are given in Table 11.3.

The basic building blocks of VIIRS (see Fig. 11.2) are:

(1) Rotating Telescope Assembly (RTA) that scans the Earth, on-board calibration targets, and deep space.

(2) Half-Angle Mirror (HAM) that de-rotates the image and directs photons into the aft optics.

(3) Aft Optics Imager, consisting of:

—A Four Mirror Anastigmatic (FMA) to image the target scenes onto the focal planes.

—Two dichroic beam splitters to direct the photon stream onto 3 focal planes (one shared by the DNB and the Visible and Near IR FPA's).

Table 11.3 Spectral and radiometric properties of VIIRS spectral bands

Band ID	Band Center	Band Width	SNR/ NEDT	L_{typ}	L_{min}	L_{max}	SNR/ NEDT	L_{typ}	L_{min}	L_{max}
			Single or High Gain				Low Gain			
M-1	412 nm	20	352	44.90	30	135	316	155	135	615
M-2	445 nm	18	380	40	26	127	409	146	127	687
M-3	488 nm	20	416	32	22	107	414	123	107	702
M-4	555 nm	20	362	21	12	78	315	90	78	667
I-1	640 nm	80	119	22	5	718				718
M-5	672 nm	20	242	10	8.60	59	360	68	59	651
M-6	746 nm	15	199	9.60	5.30	41				41
M-7	865 nm	39	215	6.40	3.40	29	340	33.4	29	349
I-2	865 nm	39	150	25	10.30	349				349
M-8	1.240 μm	0.020	74	5.40	3.50	164.9				95
M-9	1.378 μm	0.015	83	6	0.60	77.1				41
M-10	1.610 μm	0.060	342	7.30	1.20	71.2				72.5
I-3	1.610 μm	0.060	6	7.30	10.30	72.5				56.7
M-11	2.250 μm	0.050	167	0.12	0.12	31.8				31.8
M-12	3.70 μm	0.180	0.396 K	270 K	230 K	353 K				
I-4	3.74 μm	0.380	2.500 K	270 K	210 K	353 K				
M-13	4.05 μm	0.160	0.107 K	300 K	230 K	343 K	423 K	380 K	343 K	634 K
M-14	8.55 μm	0.300	0.091 K	270 K	190 K	336 K				
M-15	10.80 μm	1.000	0.070 K	300 K	190 K	343 K				
M-16	12.00 μm	1.000	0.072 K	300 K	190 K	340 K				
I-5	11.50 μm	1.900	1.500 K	210 K	190 K	340 K				
DNB	700 nm	400	6	6.7E-05	3.0E-05	2.0E+02				

Data are from VIIRS Sensor Specification (Jan, 2002)

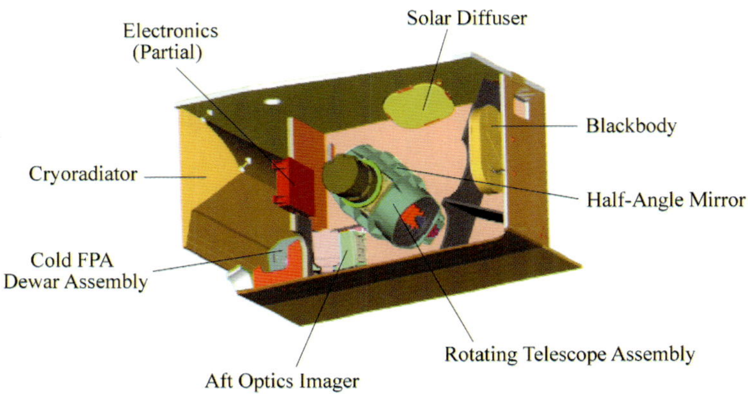

Figure 11.2 Basic building blocks of VIIRS. (The Main Electronics Module (MEM) and the Solar Diffuser Stability Monitor (SDSM) are not shown)

(4) Focal planes and associated support:
—2 ambient FPA's (the DNB and the Visible and Near IR FPA's);
—A dewar;
—2 cooled FPA's (Mid-Wave and Long-Wave IR),
(5) On-board calibrators consisting of:
—A heated On-Board Blackbody Calibrator (OBC) for IR calibration;
—A Solar Diffuser (SD);
—A Solar Diffuser Stability Monitor (SDSM).
(6) Cryoradiator;
(7) Associated electronics consisting of:
—Power Supplies;
—Signal processing electronics;
—Data handling electronics.
(8) Mechanical systems including an optical bench main frame and doors.

11.3 Follow the Photons

The key optical components are show in Fig. 11.3 to allow the reader to follow the path of photons as they enter the rotating telescope and pass through the various components. The sequence of views seen by the RTA is shown in Fig. 11.4 and additional details of the components are shown in Figs. 11.5 – 11.8.

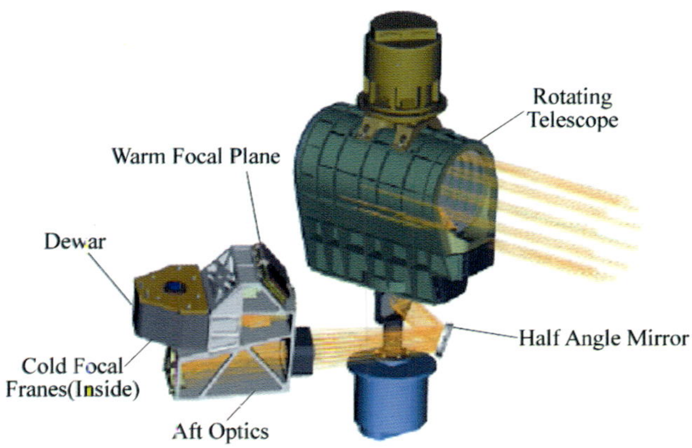

Figure 11.3 The VIIRS optical path

11.3.1 Rotating Telescope Assembly

VIIRS uses a rotating telescope to image the Earth nearly from limb to limb, deep space, and two calibration sources. The limits of the field of regard are at scan angles of ±56.06°, which corresponds to a swath width of 3,040 km. This field of regard nearly covers the Earth from limb to limb, since the horizon is at ±62.19°. After scanning the Earth, the RTA scans the blackbody calibrator, which provides a high radiance reference for the thermal infrared bands, and then the solar diffuser, which provides a high radiance reference for the visible and near infrared bands. Finally, just before scanning back onto the Earth limb, it scans deep space to obtain a low radiance reference for all bands. See Fig. 11.4. About once per month the Moon will be visible through this "space view port", allowing the use of the Moon as an alternate calibration source for the reflective bands. Each complete scan takes 1.7864 seconds; the Earth scan portion takes 0.55 seconds.

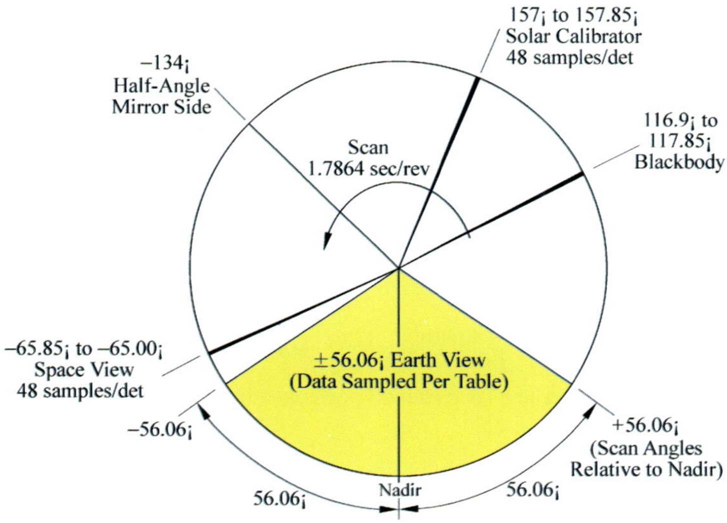

Figure 11.4 Sequence of views obtained by the RTA

The RTA is an f/6.2 a-focal, off-axis Three-Mirror Anastigmat (TMA) with an effective aperture of approximately 8 inches. In addition to three powered elements, the RTA contains a fold mirror to constrain the overall dimensions. The primary mirror is a truncated circle (or D-shape) corresponding to the shape of the aperture stop, which is also a truncated circle designed to keep a clear path for the converging optical bundle while minimizing the *Z* dimension of the sensor. (The *Z* dimension of the sensor is the distance between the side of the sensor that mounts to the underside of the spacecraft and the side that faces nadir.) The RTA is extensively baffled and has an intermediate image plane (IIP) for additional baffling. Stay light is a concern for imaging across the terminator and over the dark ocean in the presence of both in-field and out-of-field bright clouds. The D-shaped aperture stop of the system is located between the HAM and the Aft Optics in order to limit the field of view of the infrared detectors to the low-emissivity mirrors in the fore-optics. With this optical design the infrared detectors never view any of the warm, emissive structures in fore-optics at any point in the scan, thereby providing a stable infrared background due to internal instrument emission. The stability of the instrument background is essential to allow VIIRS to meet its stringent calibration requirements. Figure 11.5 shows the fore-optics and location of the system aperture stop. This and other optical schematics are simplified and do not show out-of-plane aspects of the ray traces.

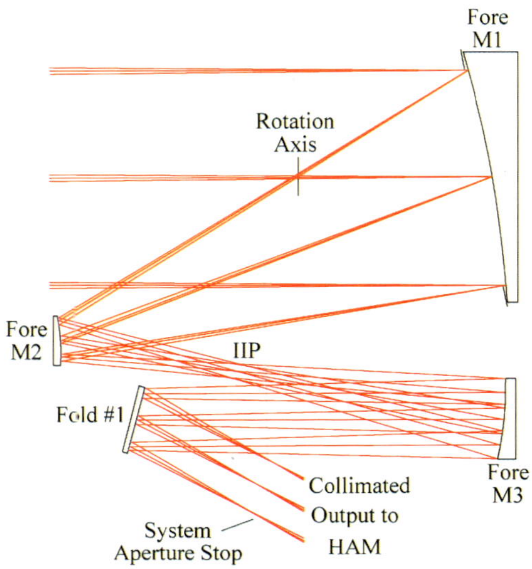

Figure 11.5 The RTA features 3 powered mirrors and a fold mirror

11.3.2 Half Angle Mirror

The half-angle mirror (HAM) rotates at half the speed of the RTA, effectively de-rotating the image and directing the photons to the fixed aft-optics. See Fig. 11.6. Since it rotates at half the speed of the RTA, it is necessary to use one side of the mirror for one scan and the other for the next. The small differences in the optical properties of the two sides of the HAM must be accounted for in the Level-1 algorithm that produces calibrated radiances. Experience has shown that, even with extremely careful control of the manufacturing and sensor characterization processes, there is still the possibility of some striping. The temperature of the HAM must be known to within about ±1 K to calibrate the thermal infrared bands.

All mirror surfaces in the first flight model are aluminum with an overcoating of Denton FSS99-500 silver. Subsequent units may use an ion assisted deposition coating. The FSS99-500 coating has better polarization properties than do current ion assisted deposition coatings, but there are concerns over how it may deteriorate over the years of pre-launch storage that the later units must accommodate. Because of the wide range of angles of incidence at the HAM as the RTA scans the Earth and calibration targets, a low polarization sensitivity coating is required.

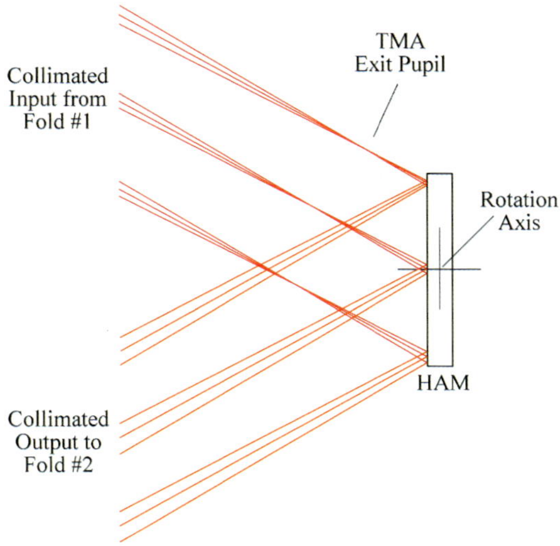

Figure 11.6 The HAM receives collimated light from the fold mirror, which corotates with the RTA, and directs it to the fixed aft optics via another fold mirror

The maximum polarization at a 45° scan angle is required to be less than 3% in the ocean color bands and is expected to be considerably less than that except for the 412 nm band.

11.3.3 Aft Optics

The aft optics imager is a Four Mirror Anastigmat (FMA) that images the incident scene on to the 4 FPA's. Like the TMA, it is an off-axis aspheric design. It has another intermediate image plane (IIP) for additional baffling of stray light (see Fig. 11.7).

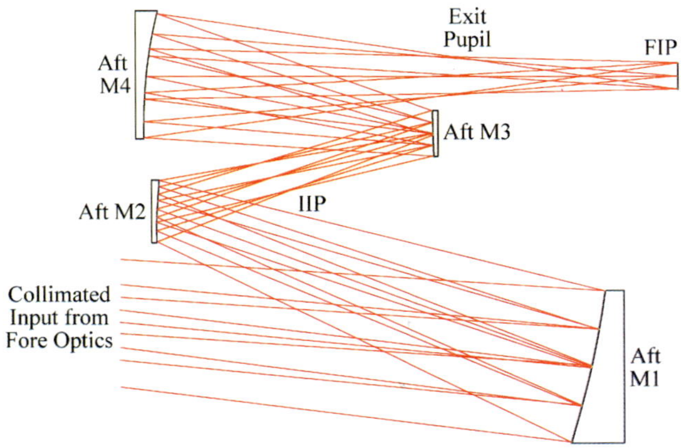

Figure 11.7 Aft optics imager

The remainder of the aft optics consists of two dichroic beam splitters, blocking filters and the focal plane arrays (see Fig. 11.8). The converging beam from the aft optics imager strikes a dichroic that reflects the VISNIR photons to the un-cooled FPA that contains the 9 VISNIR detector arrays and the slightly cooled FPA with the CCD for the near constant contrast imagery over the terminator. Photons passing through the first dichroic strike a second dichroic that reflects the short and midwave infrared photons up to a fold mirror and then into a dewar containing the cooled focal plane operating at 80 K. Longer wavelength photons passing through the second dichroic and into the same dewar to the second focal plane operating at 80 K.

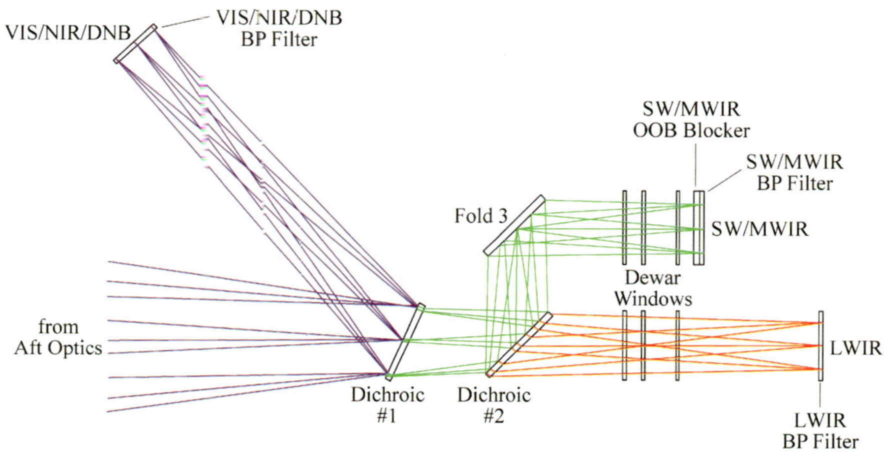

Figure 11.8 Beam splitters and focal plane assemblies

11.3.4 Focal Planes and Dewar

Each rotation of the VIIRS RTA sweeps out a path on the Earth that is just under 12 km down-track (at nadir) by 3,000 km across track. For the M (moderate resolution) bands there are 16 detectors for each band while the I (imaging resolution) bands have 32 each. One band, M-16 at 12 μm, uses two sets of detectors for Time Delay Integration (TDI) in order to achieve the required signal to noise ratio. All together VIIRS has 432 individual detectors plus the CCD array for the DNB.

The detectors are effectively rectangular in shape, with an instantaneous field of view (IFOV) of approximately 0.890 mr (740 m at nadir) downtrack by 0.445 mr (318 m at nadir) across-track for the M bands. The I bands are approximately 0.445 mr (370 m) by 0.114 mr (95 m). At nadir three adjacent IFOV's are aggregated on-board to yield a single value corresponding to a square pixel (740 m for the M bands and 370 m for the I bands). As the scanner moves off-nadir the along-track geometric IFOV (GIFOV—the linear footprint in meters on the ground) grows due to the increased distance and, in scan (cross-track), due primarily to the view-angle, and to a lesser extent at the very edge of scan, also due to Earth curvature. At 530 km from nadir the electronics shifts to a two-pixel aggregation and restores the transmitted pixel to an approximate square. From about 850 to the edge of scan at 1,500 km there is no aggregation (see Fig. 11.9). Only the single gain bands are aggregated on board in this manner. Dual gain bands are aggregated in ground processing, since the two gain states must be separately calibrated prior to aggregation. The CCD pixels are approximately 0.890 mr by 0.890 mr, corresponding to 740 m by 740 m at nadir.

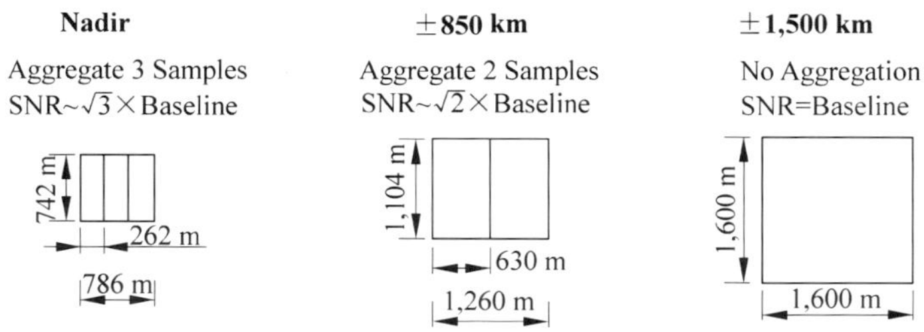

Figure 11.9 Pixel aggregation scheme for the moderate resolution bands

The nine bands in the VISNIR FPA are silicon p-type intrinsic n-type (PIN) diode arrays. The SW/MWIR and LWIR bands are all photovoltaic Mercury Cadmium Telluride (HgCdTe) detectors. In order to reduce the size of the IR detectors and thereby reduce dark current and 1/f noise, the detectors have an integrated microlens array that converts the imaging bundle to approximately f/1. All detectors except the DNB CCD are hybridized to Read Out Integrated Circuits (ROIC) that contain their initial signal processing circuitry. Bandpass filters are mounted directly above the detectors or the microlens arrays for all but the DNB.

The DNB uses a monolithic CCD array. It has two identical high gain stages and on-board processing electronics to detect and correct for anomalous readings when detectors are struck by cosmic rays. The high gain stages use 250 detectors in TDI for each pixel, while the intermediate gain stage uses 3; the low gain stage has no TDI and uses a 45X neutral gain filter. The data are spatially aggregated on-board and transmitted with a gain-state bit.

The VISNIR FPA is uncooled and operates at an instrument ambient temperature of approximately 300 K. The DNB CCD is cooled by the cryoradiator and maintained at a temperature of 251 K ± 2 K by a heater. This heater can also raise the temperature of the DNB to 300 K in order to repair damage due to space background radiation.

Six of the VISNIR bands are dual gain, allowing them to be used for dark ocean measurements or brighter land and cloud measurements depending on what is imaged. As stated above, these bands cannot be aggregated on-board, as the gain state is not known a priori.

Since all algorithms rely on the comparison of two or more spectral bands, it is essential that each spectral band view the same location. The general requirement for band-to-band registration is that they match 64% of the area or better. For the most critical bands used for SST, the requirement is better than 80% of the area. Figure 11.10 shows the layout of the VISNIR FPA and the adjacent DNB FPA. The imaging bands are placed near the center of the focal plane to minimize optical distortions and to facilitate registration with imaging bands on the other two focal planes. M-3 and M-5 are also near the center because of the criticality of co-registration of these bands for some data products. Similarly, the 10.8 and 12.1 μm bands (M-15 and M-16) used for SST are located near the center of the LWIR FPA.

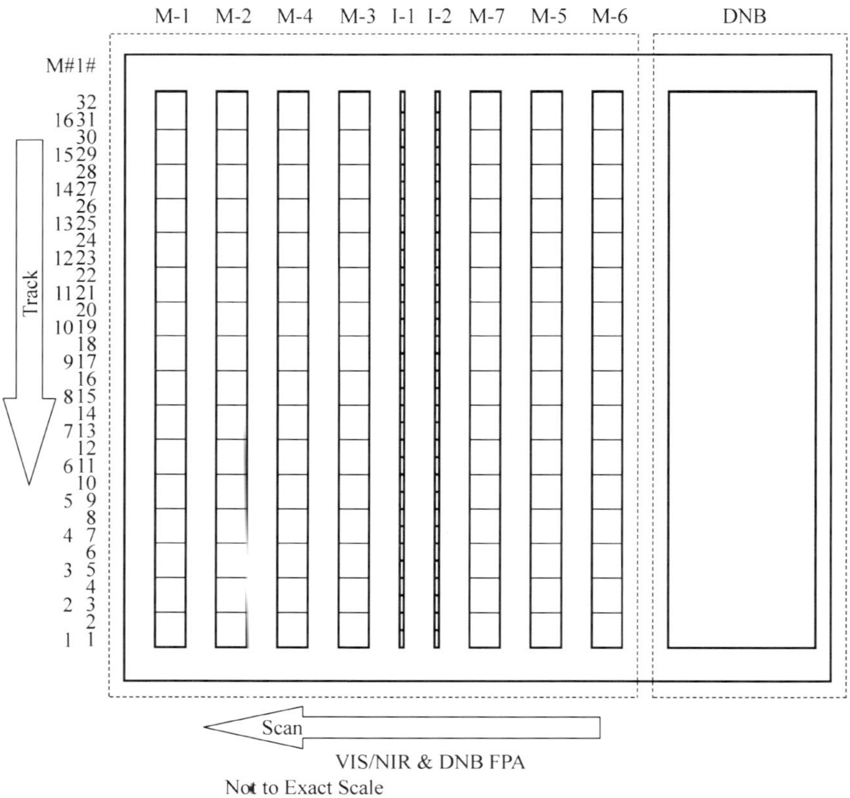

Figure 11.10 VisNIR focal plane layout

Note that as VIIRS scans the Earth, a given location will be seen first by band M-1, then M-2, M-4, M-3, I-1, I-2, M-7, M-5, M-6 and the DNB. At the moment that M-1 detectors are imaging one area, M-6 is imaging an area about 6 km away. The SWMWIR focal plane is larger, spanning about 12 km in the scan direction.

The gain, noise and spectral bandpass characteristics of each detector will differ in small ways, and there will usually be a few detectors that have characteristics significantly "out of family" in their behavior. As a result, even when imaging a completely uniform surface, there will be some striping observed. Some, but not all, of this striping is removed by the calibration performed in ground processing for all bands. Residual striping in imagery products can be further reduced by specialized destriping algorithms, but with a cost to radiometric accuracy.

The RTA rotates at a constant speed such that the spacecraft's down track motion equals the distance imaged by the 16 M-band detectors or the 32 I-band detectors at nadir during one scan period. On a given pass, detector 16 of an M band will provide a ground pixel followed by one from detector 15, and so on. At the beginning of next scan, the pixel adjacent to the one from detector 1 from the previous scan will be made by detector 16, but with the opposite side of the HAM in the optical train. This introduces a second level of striping, once per scan, in addition to the detector-to-detector striping.

The detector numbering system is natural for the flight software, but is inverted for image processing. While the operational software allows for this, those who choose to write their own software must keep this in mind.

The 16 detectors in the M-bands (and the 32 in the I-bands) match the down-track motion of the spacecraft at nadir, but off nadir the pixel sizes grow due to the increased range to the ground. At the edge of scan, the nominal 12 km down-track distance covered by the detector array has grown to nearly 24 km. This is known as the "bow-tie phenomenon" since the projected scan is shaped like a bow tie, narrow and the center and wider at the edges. This means that more than one detector will image a given spot on the Earth. While this redundant information is useful for reducing detector noise, about half of that data is trimmed from the data stream before it is transmitted to the ground in order to constrain the data rate. For the M-bands, data from detectors 1 and 16 are dropped at a scan angle of 31.59°. From 44.68° to the end of scan at 56.06° data is dropped from detectors 1, 2, 15 and 16.

11.3.5 On-Board Calibrators

VIIRS has very demanding requirements for radiometric accuracy. In general, the relative accuracy of the Level-1 radiances in the VISNIR region is to be ±2% for scenes of "typical radiance". Calibration is accomplished by viewing a Solar Diffuser (SD). The SD is a full aperture calibrator that sits in the near field of the RTA. Its Space-grade Spectralon™ coated surface is carefully characterized in the laboratory prior to launch. Characterization includes both the absolute reflectance and the bidirectional reflectance distribution function (BRDF). A screen with a transmission of 13% is permanently placed directly in front of the SD to reduce the intensity to levels that correspond to the expected intensity of Earth scenes. It consists of about 5,000 small (<1 mm) slotted apertures. Signal to noise for the VISNIR bands during calibration are in excess of 1000:1 for all bands except I-3, which is approximately 500:1.

There is no door protecting the SD. The nadir doors and the use of a nitrogen purge during ground testing will limit contamination of the SD. In flight the solar diffuser screen will provide some protection for the SD.

Over time the reflectivity of the SD is expected to change, so there is a Solar Diffuser Stability Monitor (SDSM) to monitor the spectral reflectance in eight spectral bands between 0.4 and 0.86 μ. The SDSM measures the ratio of the reflectance from the SD to that of a Spectralon™ coated integrating sphere that is directly illuminated by the Sun. Changes in the spectral reflectivity as small as 0.3% can be tracked.

The SD and SDSM are measured once per orbit, near the South Pole as the satellite approaches the terminator. When VIIRS flies in the terminator orbit, the SDSM cannot be illuminated, so VIIRS calibration stability in the VISNIR is monitored by cross-calibration with the mid-AM and mid-PM sensors. Moreover, the terminator VIIRS is less crucial for radiometric measurements than either the mid-AM or mid-PM sensors.

In the thermal infrared, the system is to meet an absolute accuracy requirement of 0.4 K in the LWIR. During each rotation the RTA scans a full aperture V-grove blackbody calibrator that is maintained at a constant temperature of 290 K±0.200 K. It can be heated up to 315 K to verify linearity of the IR detectors. The effective temperature of the blackbody can be determined from six embedded thermistors to within a calibration uncertainty of 0.05 K over its operating temperature range of instrument ambient to 315 K. The blackbody temperature is traceable to NIST with an initial accuracy of ±0.006 K. The drift over a 17-year lifetime (including pre-launch storage and on-orbit operations) is predicted to be ±0.011 K. The V-grove design is similar to that used on MODIS. The internal groove surfaces have a specular black coating and geometry such that more than 90% of the light entering from the direction of the RTA undergoes 4 internal reflections before exiting the structure. The entire incident light from the direction of the RTA undergoes at least 3 internal reflections. The multiple internal reflections from the black surfaces ensure that the structure has a reflectance of less than 0.004 and therefore an emissivity greater than 0.996 in the direction in which the blackbody is viewed by the RTA. The blackbody is viewed on every rotation of the RTA. The average of 48 scans is used to provide the reference radiance for the IR bands. Like the SD, the on-board blackbody provides a full aperture calibrator.

There are no on-board calibrators for the DNB, which has relatively lax calibration requirements. Pre-launch calibration combined with cross comparisons to the VISNIR bands is sufficient. Signal offset data can be collected when it is viewing dark space and gain data can be determined for its least sensitive stage when it is viewing the solar diffuse.

The Moon can be used as an additional calibration source. Approximately 7 times per year the Moon can be viewed directly by VIIRS during the deep space look at the end of the Earth scan. If the spacecraft were to be rolled less than 5°, the Moon could be viewed once per month, providing a completely stable external calibration source. Since the Moon's brightness varies strongly as a function of phase and libration angle, it is necessary to apply utilize an extensive database of ground-based observations and a model to obtain the reflectance at the time of the VIIRS observation.

11.4 Opto-Mechanical Systems

VIIRS consists of two modules that are independently mounted to the spacecraft, the VIIRS Opto-Mechanical Module (OMM), which contains the optics, focal planes, on-board calibrators and the cryoradiator, and the VIIRS Electronics Module (EM). The OMM is a trapezoidal solid that fits together with the electronics module within a rectangular envelope of 134 cm (X or velocity) by 141 cm (Y or anti-solar) by 85 cm (Z or nadir). The total mass is less than 275 kg.

11.4.1 Structures

There are four structural bulkheads, made from honeycombed aluminum, that comprise the folded optical bench. The telescope and HAM bulkheads are rectangular plates on the $+X$ (down-track) and $-X$ (up-track) sides of the Opto-Mechanical Module. They are joined by the aft-optics bulkhead, forming a strong bent optical bench. A zenith bulkhead reinforces this strong bench, but has no optical elements attached to it. Additional panels, made from lighter weight honeycombed aluminum house the doors and baffles.

VIIRS is attached to the spacecraft by four kinematic mounts. Four mounts are used rather than the typical three in order to reduce certain vibration modes during launch.

VIIRS has two deployable door assemblies, the nadir aperture door that consists of two separate panels, and the cryoradiator door. Both are onetime deployments. Figure 11.11 shows VIIRS once the doors have been deployed. When deployed, the nadir door acts as a baffle against extraneous light from the sun and the cryoradiator door acts to shade the radiator from heat load from the Earth.

11.4.2 Cryoradiator

The cryoradiator is a simple flat panel design with 3-stages. On orbit, the cold stage can reach a minimum temperature of less than 76 K, which allows the SWMWIR

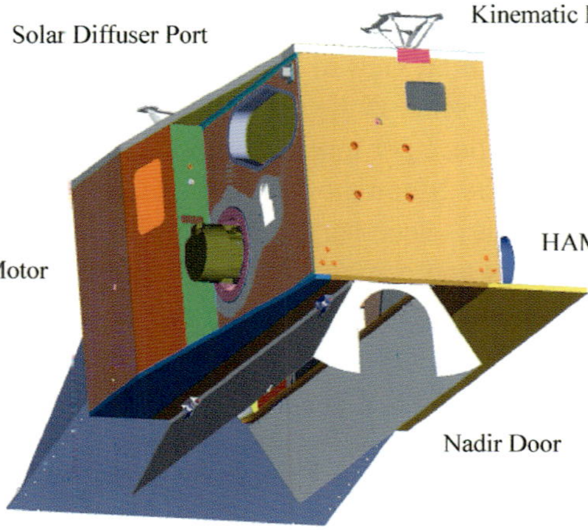

Figure 11.11 VIIRS with doors deployed

and LWIR detectors to be operated at a temperature of 80 K. The DNB is connected to the warm stage of the cooler. Once on orbit, the cryoradiator will be heated to a temperature of 270 K to drive off contamination from the spacecraft's hydrazine thrusters and residual vapors from entrapped water and other volatiles used in fabricating nonmetallic parts of the sensor. This process can be repeated occasionally, if necessary, to maintain necessary cooler performance.

11.4.3 Thermal Control and Stray Light

Although VIIRS is planned for the three specific orbits specified in the introduction, it has been designed so that any build of the sensor can be flown in any of these orbits (or in other less favorable orbits) without modification to assure operational continuity. Passive thermal control of VIIRS is maintained with gold covered Kapton multi-layered insulation (or MLI) on the sunlit portions of VIIRS, which include the $-Y$ side (solar), $+X$ side (ram), $-X$ side (anti-ram), and $-Z$ side (zenith). The $+Z$ (nadir) side of the OMM is covered with a silvered Mylar and the $+Z$ panel of the EM with white paint (YB71). The aft optics assembly, including the cold dewar, is thermally isolated from the main frame. Operational heaters are used to maintain the temperature of the scan motors above a minimum operating temperature of -5℃.

Stray light rejection is accomplished by numerous means. Given the wide range of illumination geometries encountered in the three nominal orbits over the course of a year, no single design can eliminate all scattered light. The VIIRS sensor includes many design features to eliminate or significantly reduce its sensitivity to stray light including sources originating from direct solar or Earth illumination. The nadir aperture doors on VIIRS are angled to eliminate the scatter path from the Sun to the scan cavity for all orbits. The RTA includes an integral baffle coated with a very low reflectance black paint that rejects scattered light using geometric blockages. The optical system includes two intermediate image planes, one in the RTA and one in the FMA, where aperture stops can be employed to limit the transmission of off-axis stray light. Finally, the mirror surfaces are polished to provide better than a 25-angstrom surface roughness such that all illumination except direct sunlight on the primary will result in acceptable levels of scatter within the sensor. However, even with all of these design considerations taken into account, there will still be brief times near the vernal equinox when sunlight can impinge directly on the primary mirror as the satellite passes near the South Pole. This effect, which is expected to amount to less than 0.5% of the operational time, is observed only in the terminator orbit.

The all-reflecting optical system uses diamond turned optics with nickel coatings. This relatively low cost optics is then polished to achieve the required surface to within 25 angstroms. The fore-optics assembly is made of aluminum. The aft-optics are made from aluminum, titanium, tungsten, and, in the dewar, beryllium.

11.5 Electronics

Most of VIIRS electronics are contained in the EM that is independently mounted to the spacecraft adjacent to the OMM. The EM includes the power supplies, the signal processing electronics (including the Analog Signal Processors (ASP) and the digital pre-processors), the Single Board Computer (SBC) and IEEE 1394 interface electronics, the command and telemetry electronics, and the scan servo controllers and mechanisms chain. The system is fully redundant with options to cross-strap between the primary and redundant electronics in blocks (power supply A or B, signal processing chain A or B, and servo mechanisms A or B). Extensive use is made of Field Programmable Gate Arrays (FPGA) throughout.

11.5.1 Signal Processing and Transmission

The data flow through the electronics module is shown in Fig. 11.12.

The Analog Signal Processor (ASP) receives the analog output of the detectors and applies corrections for gain and offset for each of the 432 individual detectors and then converts the analog signal to a 12-bit digital signal. The ASP can adjust the detector offsets in real time to keep the analog signals from the detectors within the desired input range of the Analog-to-Digital Converter (ADC). This is done using an on-board lookup table that provides offset data to the ASP for each detector based on the response of the detector when viewing the on-board blackbody. This function, called "DC restore", ensures that the digital output for all detectors is approximately 200 counts during the deep space view. The gains

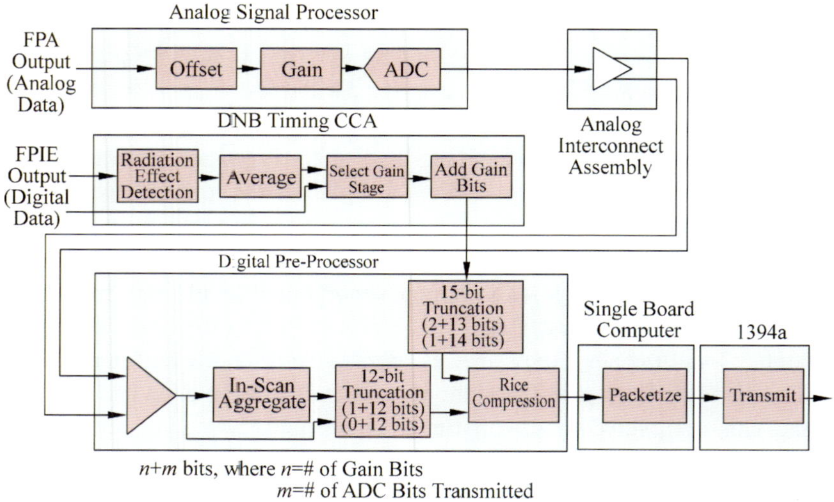

Figure 11.12 Data flow

for each FPA are set by selection of resistors. The gain is chosen to maximize the use of the ADC input range, while providing adequate margin to avoid saturation for detectors that are more responsive than the average detector in an FPA.

For the DNB, there is a separate unit, the Focal Plane Interface Electronics (FPIE) that resides in the main OMM and performs the functions of the analog signal processor and then digitizes the signals prior to transmitting directly to the digital pre-processor in the EM. The FPIE also conditions the clock and bias signals that are applied to the DNB FPA and passes the power from the EM to the DNB. Additional circuitry in the EM compares the corresponding signals form the two independent high gain stages to detect erroneous signals due to cosmic ray hits and then either averages the output of the two stages if no anomalies are detected or transmits only the lowest amplitude signal from either stages if a significant difference is detected. Finally, the appropriate gain stage is selected and coded into the output digital signal that then contains 15 bits. These 15 bits consist of either 12 or 13 bits of detector data plus one or two bits that indicate the stage that the data is from.

The digital pre-processor performs the in-scan aggregation, combining 3 raw pixels into one transmitted pixel at nadir, two into one at 530 km from nadir. No aggregation is preformed from 850 km to the edge of scan. The six bands with dual gains are transmitted without aggregation. Science data is then compressed using a spectral differencing method (where the signals from the detectors in one band are subtracted from the signals from detectors sensing the same point on the Earth in one or more other bands) followed by a lossless compression method based on the USES implementation of the Rice algorithm developed by NASA's Goddard Space Flight Center (GSFC).

All data is output on an IEEE 1394 (Firewire) data bus. VIIRS and several other NPOESS sensors will be the first space-borne sensors to use this high-speed data protocol. The development of space-qualified chips to execute this popular protocol was in itself a significant achievement.

Within the digital pre-processor there are a number of programmable adjustments to the timing that are needed to assure proper geolocation and band-to-band registration. In effect, the adjustments change the time that a given set of detectors is read out, effectively shifting the Earth location of the pixels. Two adjustments are available to correct for misalignment between the encoders and the actual pointing of the RTA and the HAM. Other programmable adjustments allow the three focal planes to be brought into alignment, fine tuning the basic mechanical alignment. Since they are accomplished through timing adjustments, these corrections are available only in the scan direction.

The Single Board Computer (SBC) is a radiation-hardened version of the 100 MIPS Motorola 603e microprocessor at its core. It carries 64 MB of RAM and 4 MB of EEPROM. Its functions include overall control of the VIIRS Sensor, collection of sensor telemetry, execution of commands that have just been received from the spacecraft/ground control or that were previously stored in a time-tagged command table, and packaging and routing compressed science data from the digital pre-processor to the IEEE 1394 electronics.

Some of the key sensor parameters that are transmitted to Earth include the bias voltages applied to the detectors, the readings from the 6 thermistors on the blackbody calibrator, the temperatures of the blackbody calibrator, the HAM and various sensor panels and subassemblies, and the cold FPA, and the rate data from the scan controllers. Note that the RTA temperature is not directly measured, but is inferred from a pair of nearby thermistors on the blackbody shield. The HAM temperature is also not directly measured, but is inferred from a thermistor inserted within the shaft that rotates the HAM, and is radiatively coupled to the rotating shaft. The HAM and RTA temperatures are indirectly measured because of the difficulty of instrumenting moving parts with thermistors in a system that must have very high reliability over a long mission life (seven years).

11.5.2 Power Supplies and Control Systems

Separate power supply modules power the digital, analog, and motor electronics and operational heaters, and outgas heaters. A total of 9 different voltages are required. The power supplies for the digital and analog electronics must be very well conditioned. When in normal operation mode the sensor is expected to dissipate approximately 170 Watts, however, up to 200 Watts is required when the on-board blackbody is being heated to 315 K. The sensor requires approximately 320 Watts of power when the outgas heaters are being operated.

Cold focal plane temperatures can be controlled to three set points: 76 K, 78 K, and 80 K. Accuracy is ±0.05 K. Current plans are to operate the system at the highest temperature as this simplifies the pre-launch calibration and predicted performance if the detectors at 80 K meet all requirements.

11.5.3 Operational Modes

VIIRS is designed to operate autonomously for up to 60 days, though more frequent command uploads are planned. Its primary modes of operation are given in Table 11.4 together with some of the key properties of each mode. Once operations have begun, VIIRS is expected to remain almost exclusively in the operational day or operational night. In the operational day mode, data are collected from all bands. In the operational night mode, only the data from the DNB and bands I-4, I-5, M-7, M-8, M-10, and M-12 through M-16 are transmitted. The operational day mode is used whenever any portion of a scan line views the Earth with solar illumination at a solar zenith angle of 95° or less. In the diagnostic mode, un-aggregated data can be transmitted.

Table 11.4 VIRS Operational mode characteristics

	RTA Power	Telescope View	Science Data	House-keeping Data	Compression	Doors
Operational Day	On	Earth/Cal	Yes	Yes	On	Open
Operational Night	On	Earth/Cal	Yes*	Yes	On	Open
Diagnostic	Var	Var	Var	Yes	On or Bypass	Open
Outgassing	Off	Anti-Nadir	No	Yes	N/A	Open**
Safe Hold	Off	Anti-Nadir	No	Yes	N/A	Open
Survival	Off	N/A	No	No	N/A	Var
Activation	Off	N/A	No	Yes	N/A	Open

* Bands I-4, I-5, and M-12 through M-16 only

** Initial outgassing will have nadir door closed

Acknowledgements

The authors were all involved in the evolutionary design of the VIIRS instrument, either as the primary developers within the Raytheon organization or as part of the government team that represented the user community interests.

The authors wish to acknowledge the contributions of many individuals who made major creative input to the basic design of the VIIRS sensor.

Within the Raytheon team, Rod Durham has served as the Program Manager from the project's inception through critical design review. The late Gerry Godden as well as Richard Julian and Jim Young played critical roles in many aspects of the design of VIIRS. Pete Kealy and Shawn Miller played equally critical roles in the development of the VIIRS algorithms and the tradeoff between sensor design and algorithms. Early in the program Marty Greenfield and John Pollicastro led a study to accommodate additional NASA requirements, which in turn led to the joint IPO-NASA NPP mission.

Others from the Raytheon team who played major roles include Robert Aguilera, Vernon Alfred, Ron Estes, Craig Kent, Jay Neuman, Cliff Nichols, Dan Pelham, Don Pies, Joe Walker, Mike Whalen, and Jeanette Woo.

From the government team, members of the IPO staff, members of the VIIRS Operational Algorithm Team (VOAT) and other technical advisors have contributed to the design of VIIRS. Mike Crison who was the first IPO Instrument Manager and later the Payloads Manager played an essential role in defining the nature of the VIIRS sensor as did Carol Welsch. Major contributions were made by technical advisors Neal Baker and Hilmer Swenson. Other contributors were Bill Barnes, Wayne Esaias, David Glackin, Bruce Guenther, Mike Haas, Alex Lyapustin, Paul Menzel, Peter Minnette, Jeff Privette, Tanya Scalione, Art Schwalb, and Ken Speidel.

From the Northrop Grumman team, which serves as the shared system performance responsibility contractor to the government, technical director Lane Darnton played a major role during the detailed design and early implementation phase. Other key contributors in this phase include Ed Hess, Jim Klein, and Steve Mills.

References

Ardanuy PE, Schueler CF, Miller SW, Kealy PS, Cota SA, Haas JM, Welsch C (2001) NPOESS VIIRS Design Process. Society of Photo-Optical Instrumentation Engineers Proceedings 4483:3 August 2001

Joint Agency Requirements Group (2005) Integrated Operational Requirements Document (IORD- II). 10 December, 2001 [Online-Referenced February 22, 2005] http://npoesslib.ipo.noaa.gov/IPOarchive/MAN/IORDII_011402.pdf

12 Conically Scanned Microwave Imager Sounder

Fuzhong Weng

12.1 Introduction

During the past 20 years there has been a dramatic increase in the use of satellite based-microwave products by the worldwide community of meteorological and oceanographic organizations. This greater emphasis is primarily attributed to the launch of the Special Sensor Microwave/Imager (SSM/I) in 1987 on board the first of a series of Defense Meteorological Satellite Program (DMSP) satellites. SSM/I contains six channels in window regions (19, 37, 85 GHz) with dual polarization, and a seventh channel centered on the 22.23 GHz water vapor line with vertical polarization. Products such as rainfall, snow-cover, cloud liquid water, water vapor, sea surface winds and sea ice concentration are produced each day on a global basis using all of the SSM/I channel measurements. SSM/I was developed by the U.S. Navy and first launched on the F-8 DMSP satellite. As a result of a shared processing agreement, all of the SSM/I products are generated by the U.S. Navy and distributed to both the U.S. Air Force and NOAA. The three agencies utilize the products in various ways to improve the analysis and forecast of weather systems.

The experience learned from the SSM/I accelerated the developments of more operational products from other microwave sensors. Using the Advanced Microwave Sounding Unit (AMSU) on-board NOAA-15, -16, and -17 satellites, NOAA generates operational products including cloud liquid, water vapor, rain rate, snow cover and sea ice concentration. This is despite the fact that the AMSU was primarily designed for temperature and water vapor sounding. The new products such as cloud ice water and ice particle size are also derived due to the unique AMSU millimeter wavelength measurements.

The success of uses of AMSU in deriving cloud and precipitation products is remarkable for various applications. These products, combined with the DMSP SSM/I derived, offer more global observations in time and space and reduce the uncertainty for operational weather and climate analyses. In the satellite data assimilation system, microwave cloud and precipitation products have been directly utilized for a quality control of sounding channel data to be assimilated into the numerical prediction models. The algorithms developed for the AMSU are being tested for applications of future microwave sensors.

This paper provides an overview for the Conically-scanned Microwave Imager

Sounder (CMIS) on board future National Polar-orbiting Operational Environmental Satellite System (NPOESS). The NPOESS is a satellite system used to monitor global environmental conditions, and collect and disseminate data related to: weather, atmosphere, oceans, land and near-space environment. It carries 10 major satellite sensors, and provides the retrievals for 55 environmental data records (EDR). CMIS is used to image the Earth's surface through clouds, and it alone will measure more than 20 EDRs. The EDRs are especially important for ocean surface wind speed and direction, and for soil moisture measurements—key performance parameter for the Department of Defense (DoD) and useful for civilian agricultural and flood warning applications. Its sounding channels will provide information on atmospheric profiles of temperature and water vapor and clouds that can be directly assimilated into numerical weather prediction models.

12.2 Instrument Overview

The first CMIS sensor will be flying on-board NPOESS C-1 in 2009. The C-1 satellite will fly in an ascending node across the equator from south to north at 7:30 pm. The CMIS system consists of a space-borne instrument, ground-based test and support equipment, and the associated ground-based science algorithms. It is tasked to support 20 of the NPOESS-assigned EDRs (see Table 12.1); for each EDR it must satisfy the requirements of multiple parameters, as applicable, such as horizontal cell size (HCS), horizontal reporting interval, refresh rate, vertical cell size, vertical reporting interval, measurement range, measurement uncertainty, measurement accuracy, measurement precision, and swath width. Some of these requirements are summarized in Table 12.2.

Table 12.1 CMIS environmental data records (EDR) and proxy data sensors

Priority	CMIS EDR	Key EDR	TYPE	Proxy Data Sensor
1A	Atmospheric Vertical Moisture Profile	Key	Atmosphere	AMSU, SSMIS
1A	Soil Moisture	Key	Land	WindSat
1A	Global Sea Surface Winds (Speed)	Key	Ocean	WindSat
2A	Global Sea Surface Winds (Direction)	Key	Ocean	WindSat
2A	Sea Surface Temperature	Key	Ocean	WindSat
2A	Precipitable Water Vapor		Atmosphere	SSMIS
2A	Atmospheric Vertical Temperature Profile	Key	Atmosphere	AMSU, SSMIS
2A	Cloud Liquid Water		Cloud	SSMIS

Continued

Priority	CMIS EDR	Key EDR	TYPE	Proxy Data Sensor
2A	Cloud Ice Water Path		Cloud	AMSU, SSMIS
2A	Total Water Content		Atmosphere	SSMIS
2A	Precipitation (Type, Rate)		Atmosphere	SSMIS
2B	Land Surface Temperature		Land	SSMIS
2B	Ice Surface Temperature		Ocean	SSMIS
2B	Sea Ice Characterization		Ocean	SSMIS
3B	Snow Cover/Depth		Land	SSMIS
3B	Global Sea Surface Wind Stress		Ocean	WindSat
3B	Pressure (Surface/Profile)		Atmosphere	SSMIS
3B	Cloud Base Height		Cloud	SSMIS
3B	Surface Type		Land	SSMIS
3B	Imagery	Key	Imagery	WindSat, SSMIS

Table 12.2 CMIS EDR requirements

	HCS (km)	Performance (U) uncertainty or (A) accuracy		Measurement Range
Atmospheric Vertical Moisture Profile	15	**Clear (U)** Surface to 600 mb 600 mb to 300 mb 300 mb to 100 mb **Cloudy (U)** Surface to 600 mb 600 mb to 300 mb 300 mb to 100 mb	 18% or 0.2 g/kg 22% or 0.1 g/kg 22% or 0.04 g/kg 20% or 0.2 g/kg 40% or 0.1 g/kg 40% or 0.04 g/kg	0 – 30 g/kg
Soil Moisture	40	8.5% (U)		0 – 100%
Sea Surface Winds Speed	20	1 m/s (A)		0 – 25 m/s
Sea Surface Winds Direction	56×35	6 deg (A)		0 – 360 deg
Sea Surface Temperature	50	0.2 K (A)		271 K – 313 K
Precipitable Water	25	Ice free ocean -1 mm (A) Land/ice - Greater of 5% or 2 mm (A)		0 – 75 mm

Continued

<table>
<tr><th></th><th colspan="2">HCS (km)</th><th colspan="2">Performance
(U) uncertainty or (A) accuracy</th><th>Measurement Range</th></tr>
<tr><td rowspan="2">Atmospheric Vertical Temperature Profile</td><td>Surface to 20 mb</td><td>40</td><td>**Clear (U)**
Surface to 700 mb
700 mb to 300 mb
300 mb to 30 mb
30 mb to 6 mb
6 mb to 1 mb
1 mb to 0.3 mb
0.3 mb to 0.01 mb</td><td>
1.6 K/1 km layers
1.4 K/1 km layers
1.3 K/3 km layers
1.5 K/5 km layers
2.4 K/5 km layers
3.5 K/5 km layers
6.5 K/5 km layers</td><td rowspan="2">162 K – 335 K</td></tr>
<tr><td>20 to 0.01 mb</td><td>200</td><td>**Cloudy (U)**
Surface to 700 mb
700 mb to 300 mb
300 mb to 30 mb
30 mb to 6 mb
6 mb to 1 mb
1 mb to 0.3 mb
0.3 mb to 0.01 mb</td><td>
2 K/1 km layers
1.4 K/1 km layers
1.3 K/3 km layers
1.5 K/5 km layers
2.4 K/5 km layers
3.5 K/5 km layers
6.5 K/5 km layers</td></tr>
<tr><td>Cloud Liquid Water</td><td colspan="2">20</td><td colspan="2">**No precipitation (U)**
Ice free ocean-0.08 kg/m^2
Land/ice-0.21 kg/m^2
Precipitation (U)
Ice free ocean-0.23 kg/m^2
Land/ice-0.45 kg/m^2</td><td>0 – 5 kg/m^2</td></tr>
<tr><td>Cloud Ice Water Path</td><td colspan="2">50</td><td colspan="2">25% (A)</td><td>0.3 – 2.6 kg/m^2</td></tr>
<tr><td>Total Water Content</td><td colspan="2">20</td><td colspan="2">**Point Measurement: (U)**
Non-precipitating-1.2 kg/m^2
Precipitating-Greater of 2.2 kg/m^2 of 15%
Global Average (U): 0.25 kg/m^2</td><td>0 – 60 kg/m^2</td></tr>
<tr><td>Precipitation (Type, Rate)</td><td colspan="2">15</td><td colspan="2">Ice free ocean-Greater of 1 mm/hr or 10% (A)
Land/ice-Greater of 1 mm/hr or 50% (A)</td><td>Rate:
0 – 50 mm/hr
Type: Rain and Ice</td></tr>
<tr><td>Land Surface Temperature</td><td colspan="2">50</td><td colspan="2">0.5 K (A)</td><td>213 K – 343 K</td></tr>
<tr><td>Ice Surface Temperature</td><td colspan="2">25</td><td colspan="2">3 K (U)</td><td>213 K – 275 K</td></tr>
<tr><td>Sea Ice Age and Ice Concentration</td><td colspan="2">20</td><td colspan="2">Probability of Correct Typing (Ice Age)-80%</td><td>Age: Ice Age Classes first year, multi-year 1/10 to 10/10</td></tr>
</table>

Continued

	HCS (km)	Performance (U) uncertainty or (A) accuracy	Measurement Range
Snow Cover/ Depth	20	20% (snow/no snow) (U)	Snow Depth Ranges: >0 cm (Any Snow Thickness) Measurement range 0 – 100%
Surface Wind Stress	20	0.015 N/m^2 for wind speed $\leqslant$ 5 m/s (U) $\alpha + \beta W + \gamma W^2$ for 5 m/s< wind speed (W) < 25 m/s (U) $\alpha = -6.58 \times 10^{-4}$ $\beta = 2.28 \times 10^{-3}$ $\gamma = 1.83 \times 10^{-4}$	0 – 2.2 N/m^2
Pressure Profile	25	0 – 10 km: 3% (A) 10 – 30 km: 5% (A)	10 – 1,050 mb
Cloud Base Height	25	2 km (U)	0 – 15 km
Fresh Water Ice	20	Ice Edge boundary: 6 km (U) Ice Concentration: 0.08 (U)	0/10 to 10/10 concentration
Vegetation / Surface Type	20	Correct Typing Probability: 75%	8 CMIS aggregated types
Imagery	Consistent with related EDRs (15, 20, 25, 40, 50, 36×25, 84×53)	Derived	Dynamic range of all measurement channels

CMIS is a microwave radiometer consisting of 77 primary channels spanning the spectrum from 6.6 to 183 GHz. Optimal beam pattern for each channel is ensured by use of 15 different, separate feedhorns, with at least one feedhorn for each frequency band. CMIS has "window" channels, at 6.6, 10.7, 18.7, 23.8 (both a window and a sounding channel), 36.5, 89 and 166 GHz, and sounding channels at 50.3 – 60.4 GHz (for O_2), 23.8 and around 183 GHz (for water vapor). Figure 12.1 shows the ensemble of CMIS channels superimposed over the frequency spectrum of the atmospheric gases and the spectral coverage (indicated by the arrows) of a few precursor space-borne sensors.

The low-frequency channels (6.6 to 36.5 GHz) use direct RF amplification and detection; the ensemble is sometimes indicated as Direct Detection Assembly (DDA). The intermediate-frequency channels (50.3 to 89 GHz) use direct RF amplification, with down-conversion (with fundamental frequency oscillators), and base-band detection.

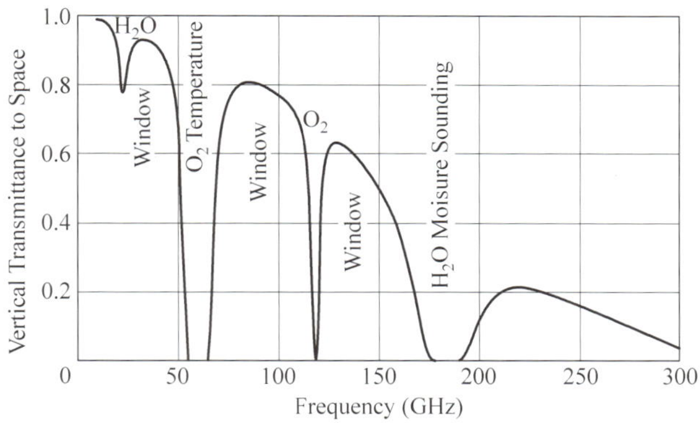

Figure 12.1 The micro- and millimeter wave spectrum of the atmospheric transmittance with the CMIS channels identified

The family of channels between 50 GHz and 60 GHz is identified as the 50/60 A. The high-frequency channels (166 GHz and 183 GHz) use a front-end mixer (with sub-harmonic oscillators) and base-band detection; the ensemble, inclusive of the 89 GHz channels, is identified as High Frequency Assembly (HFA).

CMIS employs a dual-reflector system: the channels from 6.6 GHz to 89 GHz use an off-axis reflector, 2.2 meter in diameter, to produce spatial resolution between 64 km and 15 km. A similar but smaller antenna is used at the higher frequencies. Figure 12.2 shows the CMIS instrument as deployed on the NPOESS spacecraft. All elements shown in the figure, except for the cold sky (or cold space) reflector, the warm load, the momentum wheels and the instrument control processor (ICP) assembly rotate at the spin period of 1.8987 sec.

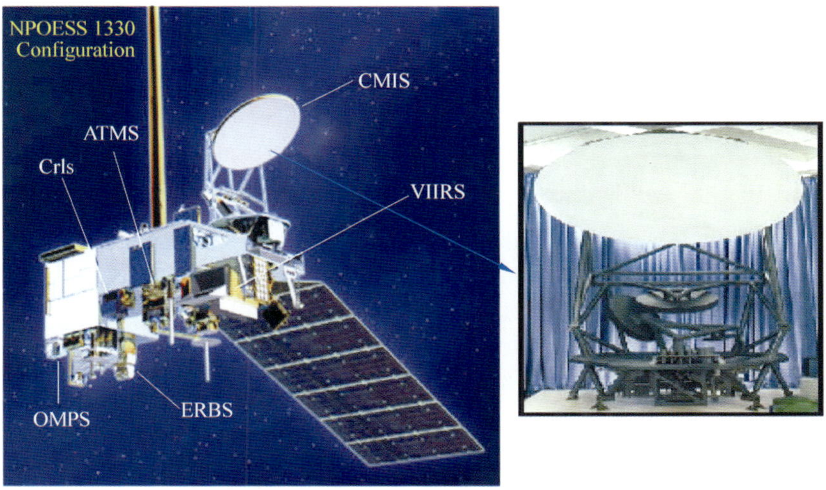

Figure 12.2 Schematic view of the CMIS antenna design and configuration on NPOESS spacecraft

The instrument consists of four separate subsystems: antenna, receiver, scan control and data handling, and thermal. The antenna subsystem includes the reflector parabolic antennas, the warm load calibrator and the feedhorns. These components collect electromagnetic radiation from the targets of interest (cold space, warm load, or the scene) and convert it from a plane wave freely propagating in space to one propagating inside a transmission line. The receiver subsystem amplifies and/or down-converts this electromagnetic wave, multiplexes it into several channels of data, rectifies it, and transforms the analog signal into a digitized stream. Physically, the receiver subsystem consists of four components: three-receiver assemblies and one receiver power supply unit. The receiver assemblies are part of the spinning section of CMIS. Each receiver assembly contains "slices", and each slice (really an integrated, almost self-contained module) contains the circuits required to detect one or more channels; circuits may consist of low noise amplifiers (LNAs), downconverters, filters, detectors, video amplifiers, multiplexers, and analog-to-digital converters (ADCs). The power supply unit contains the pulse-width-modulators, transformers, and rectifiers required to generate the DC voltages required for the receiver assemblies from the spacecraft bus power. This subsystem has both electrical and mechanical interfaces for the antenna subsystem, the ICP, and the structure. The scan control and data handling subsystem includes the electronics and mechanical components that determine the rotation speed of the instrument (and ultimately its pointing) as well as the recording, averaging, and processing of the data before they are transmitted to the ground stations. The thermal subsystem provides temperature control to the CMIS sensor. Its performance has been verified by analysis. It provides thermal isolation from the spacecraft. Heat is radiated to space to maintain thermal isolation to the spacecraft. It also interfaces with spacecraft power to provide survival heating to the CMIS, if necessary. The mounting and supporting struts are not included in this discussion, as they more properly belong to a description of the mechanical structure of the instrument.

The CMIS scan rate is 31.6 rpm. The spin axis of the CMIS is aligned parallel to the spacecraft nadir axis. The scanning of the antenna beams defines a cone, with vertex at the spacecraft and base on the ground (see Fig. 12.3). Observations are taken at the intersection of this cone and the Earth's surface, along an almost semi-circular arc which is centered on the ground track of the spacecraft. This arc encompasses about 1/3 of the rotation period of the CMIS instrument. Successive arcs are spaced by about 12.5 km on the ground (the exact value of the separation depends on the orbit altitude).

The scan geometry for the CMIS instrument is illustrated in Fig. 12.4. As the spacecraft progresses in its orbital motion, the sensor's beam sweeps a series of parallel arcs, of swath 1,700 km. The arc in the direction of the rear of the spacecraft is shown only for reference, to show the relative pointing of the low and high frequency beams. The high frequency beams observe at the front of the spacecraft, but are offset by 1/2 rotations from the low frequency ones. The angular extent of the observation arc is substantially larger than what is shown here.

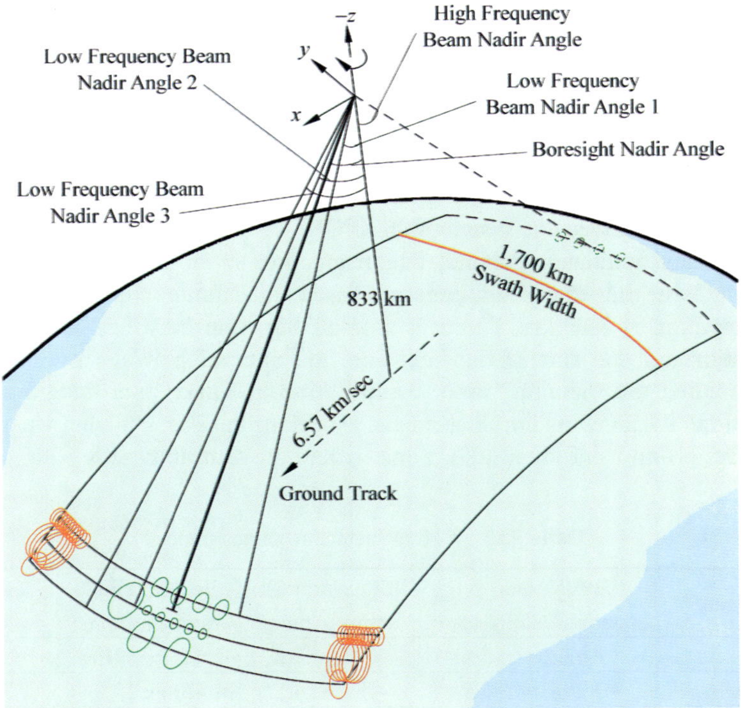

Figure 12.3 CMIS scan geometry for low and high frequencies and scan angle distribution and sub-satellite ground tracks (Boeing/AER, 2001)

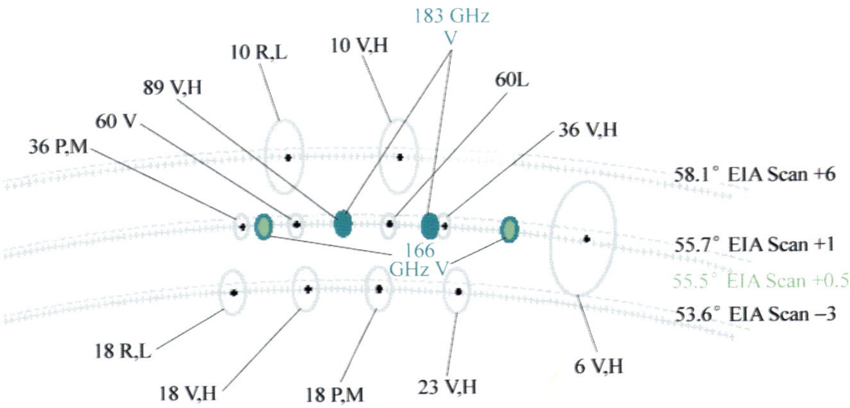

Figure 12.4 Distribution of the instrument's footprints on the ground (Boeing/AER, 2001)

Calibration data are collected from a warm source and a deep-space reflector (cold source). Both sources are viewed during the non-earth-viewing portion of the rotation cycle. The number of data points collected for each calibration source depends on the frequency of the observation (which in turn dictates the size of the feedhorn and the footprint on the ground). The 6.6 GHz channel will collect just 4 samples per scan, the 10.7 GHz one will collect 8 contiguous samples, and all other channels will collect 16 samples/scan. The finite integration time for each sensor causes the instantaneous field of view (IFOV) to be smeared across an effective field of view (EFOV) (see Fig. 12.4 for definitions). The size of the EFOV depends on both the integration time and the size and shape of the IFOV.

Table 12.3 gives the beamwidth, IFOV, EFOV, and earth incident angle (EIA) for each frequency channel. The beams line up on the ground as shown in Fig. 12.4. Note that these numbers are based on calculations that assumed an orbit elevation of 833 km. There is a small lag time between the end of one observation and the start of the next one; the interval between two successive starts is called the "sample" time. Sample time is longer than integration time. Sample time is 1.2659 ms for all channels, except for the 10 GHz one (where sample time is 2.5318 ms) and the 6.6 GHz one (where the sample time is 5.0636 ms).

Table 12.3 CMIS channel beam parameters

Channel	HPBW (deg) (1) (across × along scan)	IFOV (km) (2) across × along	EFOV (km) (2) across × along	EIA (deg) (1)
6V, 6H	1.632 × 1.659	67.7 × 37.8	67.7 × 39.3	55.75
10V, 10H	0.986 × 0.972	45.5 × 24.2	45.5 × 24.8	58.13
10R, 10L	0.98 × 0.97	45.5 × 24.2	45.5 × 24.8	58.13
18V, 18H	0.595 × 0.646	23.4 × 15.3	23.5 × 15.5	53.64
18P, 18M	0.598 × 0.614	23.4 × 15.3	23.5 × 15.5	53.64
18R, 18L	0.637 × 0.66	23.4 × 15.3	23.5 × 15.5	53.64
23V, 23H	0.569 × 0.619	23.4 × 15.3	23.5 × 15.5	55.64
36V, 36H	0.411 × 0.419	16.6 × 9.9	16.7 × 10.3	55.75
36P, 36M	0.411 × 0.419	16.6 × 9.9	16.7 × 10.3	55.75
60V Complex	0.36 × 0.315	15.0 × 7.7	14.9 × 8.2	55.75
60L	0.371 × 0.319	15.0 × 7.7	14.9 × 8.2	55.75
89V, 89H	0.377 × 0.314	15.6 × 7.8	15.3 × 8.1	55.75
166V	0.359 × 0.349	14.6 × 8.4	14.6 × 8.8	55.49
183V Complex	0.355 × 0.371	16.1 × 8.7	16.1 × 9.0	55.49

The feed antennas of the lower frequency channels (6.9 GHz to 89 GHz) are arranged near the focus of the main reflector, along three parallel rows, so that their footprints fall on one of three arcs described by the main reflector. The higher frequency channels (166 GHz and 183 GHz) fall on the single, secondary reflector arc (see Figs. 12.3 and 12.4).

The footprints of the CMIS instrument are shown as projected on the ground. The CMIS channels are described in Table 12.4. The notation for the polarization is as follows: V—linear vertical, H—linear horizontal, LC—left-hand circular, RC—right-hand circular, + 45—linear rotated 45 degrees from the vertical, and – 45—linear rotated – 45 degrees from the vertical. The 60L FFT channel is divided up into 40 closely-spaced channels whose bandwidths are listed. The 183 GHz channels are made up of pairs of frequency bands with the center frequencies as specified and the bandwidth of each member of the pair listed separately in the table.

Table 12.4 CMIS baseline channel specifications

Channel Name	Center Frequency (GHz)	Polarization	NEDT (K)	Bandwidth (MHz)
6V	6.625	V	0.40	350
6H	6.625	H	0.40	350
10V	10.65	V	0.94	100
10H	10.65	H	0.94	100
10R	10.65	RC	0.98	100
10L	10.65	LC	0.98	100
18V	18.70	V	1.18	200
18H	18.70	H	1.18	200
18P	18.70	+ 45	1.18	200
18M	18.70	– 45	1.18	200
18R	18.70	RC	1.32	200
18L	18.70	LC	1.32	200
23V	23.80	V	1.00	400
23H	23.80	H	1.00	400
36V	36.50	V	0.63	1,000
36H	36.50	H	0.63	1,000
36P	36.50	+ 45	0.63	1,000
36M	36.50	– 45	0.63	1,000
60VA	50.30	V	2.70	134
60VB	52.24	V	0.92	1,280
60VC	53.57	V	1.05	960
60VD	54.38	V	1.51	440
60VE	54.905	V	1.69	350
60VF	55.49	V	1.71	340
60VG	56.66	V	1.82	300
60VJ	59.38	V	1.88	280
60VK	59.94	V	1.51	440
60LL	60.3712	LC	3.35	57.6

Continued

Channel Name	Center Frequency (GHz)	Polarization	NEDT (K)	Bandwidth (MHz)
60LM	60.408	LC	6.34	16.0
60LU	60.4202	LC	8.74	8.4
60LV	60.5088	LC	3.80	44.8
60L FFT 1 – 40	60.43476	LC		25.0
60L FFT 1, 40			20.67	1.50
60L FFT 2, 39			22.65	1.25
60L FFT 3, 4, 37, 38			25.32	1.00
60L FFT 5, 36			29.23	0.75
60L FFT 6 – 8, 33 – 35			35.80	0.50
60L FFT 9 – 32			50.63	0.25
89V	89.0	V	0.76	4,000
89H	89.0	H	0.76	4,000
166V	166.0 ± 0.7875	V	2.37	1,425 each
183VA	183.31 ± 0.7125	V	2.62	1,275 each
183VB	183.31 ± 3.1	V	1.61	3,500 each
183VC	183.31 ± 7.7	V	1.43	4,500 each

12.3 CMIS Risk Reduction Studies with Heritage Sensors, and Proxy Data

To evaluate the performance CMIS EDR processing algorithms during the pre-launch period, we have identified three major microwave sensors to be used. These are SSMIS on DMSP, WindSat on Coriolis, and AMSU on NOAA-15 to 17. The CMIS EDRs that the data from these sensors can help evaluate are given in Table 12.1 (the right most columns). Important features of the sensors are given in Table 12.5. DMSP SSMIS was successfully launched in October 21, 2003, and the others were launched recently. The channels are listed in Table 12.6 where commonalities are identified.

Table 12.5 Available heritage sensors for CMIS

Sensor	CMIS	SSMIS	AMSU	WindSat
Spacecraft	NPOESS	DMSP	NOAA-16	Coriolis
Agency	NPOESS	NRL/USAF	NOAA	NRL
Launch Date	–	Oct. 2003	Sept. 2001	Jan. 2003
Altitude (km)	833	833	850	830
EIA (deg)	53.6 – 58.1	53.1	– 57 to 57	49.9 – 55.3
Swath (km)	1,700	1,700	2,200	1,025/350 (Fore/Aft)

Table 12.6 Proxy data for CMIS channel frequencies and other polar orbiting microwave radiometers

CMIS (53°EIA – 58°EIA)	SSMIS (53°EIA)	AMSU (x-scan)	WindSat (50°EIA – 55°EIA)	Purpose
6.625/350/H,V			6.8/125/H,V	Window
10.65/100/H,V,R,L			10.7/300/H, V, + 45, – 45, R,L	Window
18.7/200/H,V, + 45, – 45,R,L			18.7/750/H, V, + 45, – 45, R,L	Window
	19.35/355,357/H,V	23.8/251		Window
	22.235/401/V	31.4/161		Window
23.8/400/H,V			23.8/500/H, V	Sounding (H_2O)
36.5/1,000/H,V, + 45, – 45	37.0/1,545 – 1,615/H,V		37.0/2,000/H, V, + 45, – 45, R,L	Window
50.300/134/V	50.3/380/H	50.300/161		Sounding (O_2)
52.240/1,280/V				Sounding (O_2)
	52.8/389/H	52.800/380		Sounding (O_2)
53.570/960/V	53.596/380/H	53.600/168		Sounding (O_2)
54.380/440/V	54.4/383/H	54.400/380		Sounding (O_2)
54.905/350/V		54.940/380		Sounding (O_2)
55.490/340/V	55.5/391/H	55.500/310		Sounding (O_2)
56.660/300/V				Sounding (O_2)
	57.29/330/R	fo=57.290/310		Sounding (O_2)
59.380/280/V	59.4/239/R	fo ± 0.217/76		Sounding (O_2)
59.940/440/V		fo ± 0.322 ± 0.048/ 34		Sounding (O_2)

Continued

CMIS (53°EIA – 58°EIA)	SSMIS (53°EIA)	AMSU (x-scan)	WindSat (50°EIA – 55°EIA)	Purpose
60.371200/57.6/L		fo ± 0.322 ± 0.022/15		Sounding (O_2)
60.4080/16.0/L		fo ± 0.322 ± 0.010/8		Sounding (O_2)
60.4202/8.4/L		fo ± 0.322 ± 0.004/3		Sounding (O_2)
60.5088/44.8/L				Sounding (O_2)
60.434776/25.0,FFT 40 Channels/L				Sounding (O_2)
	60.792668 ± 0.357892 ± /1.3 – 26.5/R (5 Channels)			Sounding (O_2)
	63.283248 ± 0.285271/ 1.35/R			Sounding (O_2)
89.0/4,000/H,V		89.0/3,000		Window
	91.655/1,411 – 1,418/ H,V			Window
	150.0/1,642/H	150/4,000		Sounding (H_2O)
166.0 ±0.7875/1,425/V				Sounding (H_2O)
183.31±0.7125/1,275/V	183.31 ± 1.0/513/H	183.31 ± 1.0/1,000		Sounding (H_2O)
183.31±3.1/3,500/V	183.31 ± 3.0/1,019/H	183.31 ± 3.0/2,000		Sounding (H_2O)
183.31±7.7/4,500/V	183.31 ± 6.6/1,526/H	183.31 ± 7.0/4,000		Sounding (H_2O)

Key			
XXX/YYY/V,H, + 45, – 45, L,R	Center Freq. (GHz)/Bandwidth (MHz)/ Polarization		Window Channel Match
V,H, + 45, – 45, L,R	Vertical, Horizontal, + 45°, – 45°, Left Circular, Right Circular Polarization		Sounding Channel Match
EIA	Earth Incidence Angle		

12.3.1 DMSP-SSMIS

SSMIS on DMSP is a conical imager/scanner. It includes all the channels on SSM/T-1, the temperature sounder, SSM/T-2, the moisture sounder, and SSM/I, the imager. Additionally, it has 6 channels in the O_2 band near 60 GHz for stratospheric temperature soundings.

In the past, SSM/I measurements have been exploited to derive atmospheric and surface products such as cloud liquid water (Greenwald et al., 1993; Liu and Curry, 1993; Weng and Grody, 1994; Weng et al., 1997; Wentz, 1997 Weng et al., 2000), precipitable water (Alishouse et al., 1990), precipitation and snow (Grody, 1991) and sea ice concentration (Ferraro et al., 1996). Recently SSM/I data over land were also used to derive land surface temperature (Weng and Grody, 1998) and surface emissivity (Jones and Vonder Haar, 1997; Prigent et al., 1997; Weng et al., 2001).

The DMSP SSMIS Calibration/Validation (Cal/Val) program (Poe, 2001) and instrument performance will provide risk reduction for the NPOESS CMIS program and corresponding Cal/Val. The CMIS and SSMIS have similar measurement channels, which makes the SSMIS a good candidate for inter-comparison with CMIS during the Cal/Val for NPOESS. The EDRs retrieved by the SSMIS are similar to many NPOESS EDRs, however, the actual performance requirements are different in some cases, but inter-comparisons should still prove valuable. The first SSMIS was successfully launched in October 2003 on board the DMSP F16 satellite and is now undergoing a fairly extensive Cal/Val program.

12.3.2 Coriolis/WindSat

WindSat is a polarimetric radiometer and has channels at 6.8, 10.7, 18.7, 23.8 GHz, and 37.0 GHz. The 10.7, 18.7, and 37.0 channels are fully polarimetric and the 6.8 GHz and 23.8 GHz channels have only V and H. All these channels are very similar to the lower frequency window channels on the CMIS. The WindSat sensor will provide risk reduction for the CMIS program in several ways. First, the development of the WindSat polarimetric radiometer requires the development of new sensor pre-launch characterization and calibration techniques. These techniques and the results will be reviewed and where appropriate for CMIS will be incorporated in the NPOESS program. Secondly, on-orbit verification and validation of microwave polarimetry for ocean surface wind measurements will provide valuable lessons learned that may be incorporated not only for the improvement in the CMIS design and testing, but in the techniques and methodologies used for post-launch Cal/Val of the CMIS. WindSat was launched in January 2003 on the Coriolis spacecraft.

To test the accuracy of wind products from WindSat, Liu and Weng (2003) used the WindSat measurements simulated through a radiative transfer model.

This model produces simulations under global atmospheric and surface conditions. In their retrieval algorithm, the upwelling and downwelling radiative components are treated as one variable since the two components are highly correlated. As a result, under rain-free conditions, the full polarimetric measurements obtained at a single frequency can form a closure set of radiative transfer equations so that both surface and atmospheric column parameters can be simultaneously derived (Liu and Weng, 2003).

The first guess to the solutions in the retrieval is normally required because the retrieval is non-linear. In doing so, the third and fourth Stokes components and their ratio are utilized to estimate the wind direction. It was found that this ratio alone can determine the wind direction with an accuracy of 30 degrees (Liu and Weng, 2003). Without the sensor noises being added to the simulations, the algorithm can produce the wind direction with an rms error of 6.5 degree and the wind speed with an error of 0.3 ms^{-1}, respectively. If the realistic random noises are taken into account with 0.1 K for vertical and horizontal polarizations and 0.15 K for the third and fourth Stokes components, the errors increase to about 10 degrees and 0.6 ms^{-1} for the wind direction and speed, respectively. This wind speed error is much smaller than that from the statistical algorithms (1.4 ms^{-1}) applied to the same data set. Thus, the physical retrieval significantly improves the uses of the microwave polarimetric data for remote sensing of ocean wind (Liu and Weng, 2003).

Uses of microwave measurements at lower frequency window channels for soil moisture retrievals have not been very successful due to a radio frequency interference (RFI) occurring in these channels. The RFI sources are primarily associated with local transmitters from mobile and active sensors. For example, RFI observed from the EOS Aqua AMSR-E which has lower frequencies similar to WindSat and CMIS, the magnitudes of the RFI could be as large as 50 – 60 K. Without carefully removing the RFI from microwave measurements, the soil moisture product would be significantly degraded (Li et al., 2002).

12.3.3 NOAA-15, 16 and 17 AMSU

The Advanced Microwave Sounding Unit (AMSU) are flown onboard NOAA polar-orbiting environmental satellites. The instrument has been operational since 1998. The AMSU contains two modules: A and B. The A module (AMSU-A) has 15 channels and is mainly designed to provide information on atmospheric temperature profiles while the B module (AMSU-B) allows for profiling the moisture structure. The AMSU-A has an instantaneous field-of-view of 3.3° and scans ±48.7° from nadir with 15 different viewing angles at each side. The AMSU-A measures thermal radiation at microwave frequencies ranging from 23.8 GHz to 89.0 GHz. Atmospheric temperature profiles are primarily based on the measurements obtained at channels near 60 GHz, which is an oxygen absorption line. In particular, the AMSU-A sounding channels (3 – 14) respond to the thermal radiation at various altitudes because of their weighting function distributions, whereas channels 1 and 2 are primarily designed for obtaining the information on surface properties. Since the satellite provides a nominal spatial resolution of 48 km at its nadir, the temperature perturbations from synoptic to mesoscale can be reasonably depicted. In addition, several AMSU imaging channels at frequencies of 31.4, 89 GHz and 150 GHz are utilized to determine cloud liquid and ice water contents because they directly respond to the emission from liquid droplets and the scattering from ice particles (Weng and Grody, 2000; Zhao and Weng, 2002; Weng et al., 2003).

Atmospheric temperatures at any pressure level are also derived as a linear combination of brightness temperatures measured at various sounding channels, i.e.,

$$T(p) = C_0(p,\theta_s) + \sum_{i=1}^{N} C_i(p,\theta_s) T_b(v_i,\theta_s) \tag{12.1}$$

where p is the pressure; θ_s is the scanning angle; v_i is the frequency at channel i; and T_b is the AMSU brightness temperature. The coefficients, C_0 and C_i, are determined by regression with collocated rawinsonde temperature soundings matched and the AMSU-A brightness temperatures. Since the AMSU is a cross-track scanning instrument, the temperature retrieval at each pressure level is derived separately for each scanning angle. More details for the algorithm are discussed in a separate article (Zhu et al., 2002). For the CMIS applications, temperature retrieval is further simplified by fixing the observing angle in Equation (12.1). However, an advanced algorithm remains to be developed to take into account clouds and precipitation in the retrievals.

Microwave remote sensing offers some advantages of measuring cloud ice water compared to visible and infrared techniques. Visible methods require a number of gross assumptions about ice particles shape, size distribution, and cloud spatial homogeneity to convert from radiance to optical depths to *IWP*. In addition, thermal infrared techniques require accurate knowledge of cloud temperature. Microwave radiation, to the contrary, interacts with ice particles, primarily through scattering. The emission and cloud temperature are relatively unimportant. Since ice clouds are above the absorbing part of the atmosphere, they simply modulate the upwelling microwave radiation from below. While the effects of particle shape and size distribution are also important for microwave remote sensing of clouds since they determine the relation between optical depth and *IWP*, they are more amenable to calculation because the particle sizes are comparable to, and smaller than the wavelength. Weng and Grody (2000) proposed an algorithm to derive both *IWP* and D_e using dual millimeter-wavelength measurements. They found that for a given particle bulk volume density, the brightness temperature at millimeter wave frequencies can be uniquely related to

IWP and D_e through a two-stream radiative transfer solution. However, the retrievals of cloud ice water path suffer a great amount of uncertainty due to an unknown bulk volume density of ice particles (Weng and Grody, 2000). Thus, it would be of a major impact on the ice cloud remote sensing to utilize other sensors such as visible and infrared to identify a possible range of the bulk volume density.

An operational *IWP* is generated from the Advanced Microwave Sounding Unit (AMSU) measurements (Zhao and Weng, 2002). In the algorithm, both *IWP* and D_e are related to the ice particle scattering parameters, which are determined from the AMSU 89 GHz and 150 GHz measurements. The ratio of the scattering parameters measured at two frequencies provides a direct estimate of D_e. The *IWP* is then derived from the scattering parameter at 150 GHz with the derived D_e and the constant bulk volume density. A screening procedure is developed to discriminate the scattering signatures between atmospheric clouds and surface materials (Zhao and Weng, 2002).

12.4 Discussions

The CMIS EDR algorithms will be delivered from the instrument vendor as part of the acquisition contract. The entire retrieval algorithms are of unprecedented complexity as compared to any of previously developed operational algorithms (Boeing/AER, 2001). The big challenges that follow after NPOESS launch and in assessing the CMIS algorithms will include a combination of these following factors:

(1) The CMIS algorithms have been so far evaluated using mostly simulated radiances. Since the same forward model is used for both simulations and retrievals, the impacts of instruments and forward model biases on the retrievals have been excluded as none-relevant issues. A viable approach is yet to be developed to show how the biases of the forward model to observations affect the retrievals under different environmental circumstances.

(2) In the CMIS algorithm core module (Boeing/AER, 2001), variational retrieval has accommodated many CMIS channels to produce a group of EDRs. However, an optimal channel selection for a particular EDR has not been fully assessed. Our experience indicated that some EDRs can be best derived with fewer channels closely connected to the physical quantity being retrieved. If the retrieval for an EDR with a set of channels approaches linear, it is better to utilize this property because the solution is unique. Some retrieval problems at microwave frequencies such as cloud liquid water and water vapor path can also be linearized (Weng et al., 1997). However, two linear retrievals mixed together through the variational method may become a non-linear problem which not only complicates retrievals but also degrades accuracy significantly.

(3) Understanding the background error covariance matrix of the first guess and the error covariance matrix of the forward model are important. These matrices

must be developed within the CMIS configuration in terms of channel allocation and scanning geometry using collocated satellite and in-situ measurements after the forward model and instrument biases are removed.

(4) Despite our improved microwave technology on sensing the Earth environmental parameters, there remains a lack of understanding in radiative transfer theory associated with various media. Some CMIS EDR algorithm (e.g. soil moisture) is still based on a roughness theory that was mostly applicable at the frequencies lower than 6 GHz.

(5) Validation of the CMIS EDRs is always an issue. Some EDRs such as atmospheric temperature and surface wind can be better assessed since there are global ground network established for measuring these parameters. There remains difficulty in evaluating those EDRs related to clouds and surface properties. Some error models should be explored for analyzing the sensitivity of retrievals to the uncertainties associated with various input parameters.

Acknowledgements

Author would like to express sincere thanks to Dr. Stephen Mango for his support to our CMIS project through NOAA/IPO internal government study funds.

References

Alishouse JC, Snider JB, Westwater ER, Swift CT, Ruf CS, Snyder SA, Vongsathorn J, Ferraro RR (1990) Determination of cloud liquid water content using the SSM/I. IEEE Trans Geosci Remote Sen 28: 817 – 822

Boeing/AER (2001) CMIS algorithm theoretical base document. Vol. 1 – 17, NOAA Integrated Program Office, Silver Spring, MD

Ferraro RR, Weng F, Grody N, Basist A (1996) An eight year (1987 – 1994) time series of rainfall, clouds, water vapor, snow and sea ice derived from SSM/I measurements. Bull Amer Meteor Soc 77: 891 – 905

Greenwald TJ, Stephens GL, Vonder Haar TH (1993) A physical retrieval of cloud liquid water over the global oceans using Special Sensor Microwave/Imager (SSM/I) observations. J Geophys Res 98: 18,471 – 18,488

Grody NC (1991) Classification of snow cover and precipitation using the special sensor microwave imager. J Geophys Res 96: 7,423 – 7,435

Jones AS, Vonder Haar TH (1997) Retrieval of microwave surface emittance over land using coincident microwave and infrared satellite measurements, J Geophys Res 102: 13,609 – 13,626

Li L, Im E, Chang PS, St. Germain KM, Njoku E (2002) Radio-Frequency Interference (RFI) identification and suppression of the 7-GHz AMSR-E channels for land sensing. JPL report D-24839

Liu G, Curry J (1993) Determination of characteristic feature of cloud liquid water from satellite microwave measurements. J Geophys Res 98: 5,069 – 5,092

Liu Q, Weng F (2003) Retrieval of sea surface wind vector from simulated satellite microwave polarimetric measurements. Radio Sci 38: 8,078 – 8,088

Poe G (2001) DMSP Calibration/Validation plan for the Special Sensor Microwave Imager Sounder (SSMIS). NRL, Monterey, CA.

Prigent C, Rossow WB, Matthews E (1997) Microwave land emissivities estimated from SSM/I observations. J Geophys Res 102: 21,867 – 21,890

Weng F, Ferraro RR, Grody NC (1994) Global monthly precipitation estimation using DMSP F-10 and F-11 Special Sensor Microwave Imager (SSM/I). J Geophys Res 99: 14,493 – 14,502

Weng F, Grody NC (1994) Retrieval of cloud liquid water over oceans using special sensor microwave imager (SSM/I). J Geophys Res 99: 25,535 – 25,551

Weng F, Grody NC, Ferraro RR, Basist A, Forsyth D (1997) Cloud liquid water climatology derived from the Special Sensor Microwave Imager. J Climate 10: 1,086 – 1,098

Weng F, Grody NC (1998) Physical retrieval of land surface temperature using the special sensor microwave imager. J Geophys Res 103: 8,839 – 8,848

Weng F, Grody NC (2000) Retrieval of ice cloud parameters using a microwave imaging radiometer. J Atmos Sci 57: 1,069 – 1,081

Weng F, Zhao L, Ferraro R, Poe G, Li X, Grody N (2003) Advanced Microwave Sounding Unit Cloud and Precipitation Algorithms. Radio Sci 38: 8,086 – 8,096

Weng F, Yan B, Grody NC (2001) a microwave land emissivity model. J Geophys Res 106: 20,115 – 20,123

Wentz FJ (1997) A well calibrated ocean algorithm for special sensor microwave imager. J Geophys Res 102: 8,703 – 8,718

Zhao L, Weng F (2002) Retrieval of ice cloud parameters using the Advanced Microwave Sounding Unit (AMSU). J Appl Meteorol 41: 384 – 395

Zhu T, Zhang D, Weng F (2002) Impacts of Advanced Microwave Sounding Unit measurements on hurricane prediction and forecast. Mon Wea Rev 130: 2,416 – 2,432

13 Advanced Technology Microwave Sounder

Mitchell D. Goldberg and Fuzhong Weng

13.1 Introduction

The next generation of microwave sounders—the Advanced Technology Microwave Sounder (ATMS)—will be flown on the National Polar-orbiting Operational Environmental Satellite System (NPOESS) starting with the NPOESS Preparatory Program (NPP) in 2006. The ATMS has nearly the same functional requirements as those from the current operational microwave sensor suite of the Advanced Microwave Sounding Unit (AMSU)-A and -B flown on the NOAA Polar-orbiting Operational Environmental Satellites (POES). However the instrument design of ATMS is very different. The significant differences are the much smaller volume, weight and power consumption of ATMS and the extended off-nadir view angle which results in no gaps between successive orbits.

The assimilation of AMSU observations in numerical weather prediction models has resulted in significant improvements in forecast skills. To sustain these improvements, the ATMS instrument will need to meet its performance specifications and a carefully planned pre-launch instrument calibration and characterization will be needed. The post-launch calibration and validation (Cal/Val) is equally important part. In this chapter, we will highlight the differences between ATMS and AMSU instruments and present some current applications of AMSU measurements.

13.2 Instrument Overview

The ATMS, which provides 22 channels for atmospheric temperature and moisture soundings, is a combination of AMSU-A and AMSU-B channels with three additional channels at 51.76 GHz, 183.31∀ 4.5 GHz, and 183.31∀1.8 GHz, respectively. In addition, the 89 GHz channel that appears in AMSU-A and AMSU-B is replaced by the 88.2 GHz channel and the 150 GHz channel is changed to 165.5 GHz, respectively.

The ATMS uses two antenna systems of parabolic reflectors, one of which provides Channels 1 – 15 in the range of 23.8 GHz to 57.3 GHz. The 13 channels in the range of 50.3 GHz to 57.3 GHz oxygen band are used for retrieving atmospheric temperature profiles from the Earth's surface to about 50 kilometers (or

1 mb). The two window channels at the lowest frequencies at 23.8 GHz and 31.4 GHz, are used to identify precipitation and correct the effect of surface emissivity, atmospheric liquid water, and water vapor on temperature soundings. These window channels are also used to derive rain rate, sea ice concentration, and snow cover (Weng et al., 2003). The second antenna system furnishes the high frequency channels 16 – 22 that are used for humidity soundings.

Table 13.1 lists the requirement of main ATMS channel characteristics, including channel frequency, bandwidths, frequency stability, calibration accuracy, nonlinearity, radiometric temperature sensitivity (or NEΔT), and 3-dB band width for individual channels. More detailed information on the ATMS radiometers is reported in (Boeing /AER, 2001). Figure 13.1 shows a schematic view of the ATMS scan geometry. The ATMS antenna systems is required to have a nominal field of view (FOV) ranging from 1.1° to 5.2° at the half-power points and to scan across the Earth to cover an angle range of ∀52.73E (beam centers) from the nadir direction. The antenna reflectors rotate one complete revolution every 8/3 seconds, during which 96 Earth FOVs and 4 views of the cold space and the blackbody target (each separated by 1.11°), respectively, will be sampled. Beam positions 1 and 96 are the outermost scan positions of the Earth views, while beam positions 48 and 49 (at ∀0.56° from nadir) straddle the nadir. Views of cold space and a blackbody target with four samples at the end of scan provide onboard calibration once each scan (8/3 sec). The calibration algorithm for ATMS is based on the heritage AMSU methodology (Mo, 1996).

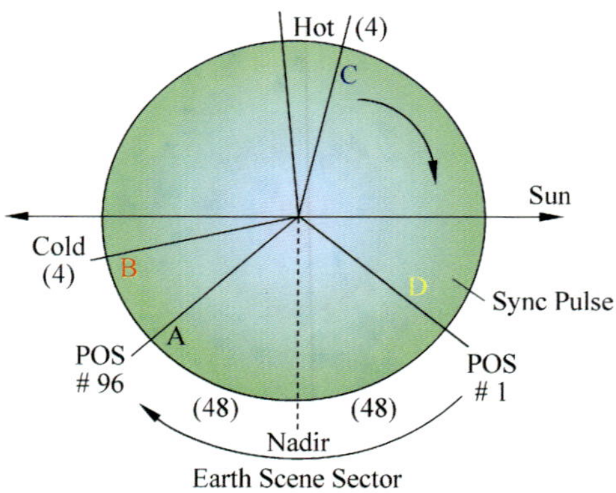

Figure 13.1 Schematic view of the ATMS scan geometry

As reported in Shiue (2001), one of the design goals of the ATMS is to reduce the required "spacecraft resource," e.g., the instrument weight, power, and volume, so as to enable the ATMS to fit into future smaller operational satellites. The combined AMSU-A and AMSU-B weight is 160 kg, and they consume 200 watts. The two AMSUs needed four reflectors driven by four separate scan motors. They are housed in four separate housing units, spread-out on the earth-facing side of the host satellites; together they occupy about 1 m^3 in volume.

Table 13.1 ATMS channel characteristics and calibration requirements. The shaded channels are the new ones that are neither in AMSU-A nor AMSU-B

Ch #	Channel Freq (MHz)	Band Width (MHz)	Freq. Stability (MHz)	Calibration Accuracry (K)	Nonlinearity Max (K)	NEΔT (K)	3-dB BW* (deg)	Romax
1	23,800	270	10	0.83	0.1	0.5	5.2	AMSU-A
2	31,400	180	10	0.83	0.1	0.4	5.2	AMSU-A
3	50,300	180	10	0.47	0.075	0.7	2.2	AMSU-A
4	51,700	400	5	0.47	0.075	0.5	2.2	
5	52,800	400	5	0.47	0.075	0.5	2.2	AMSU-A
6	53596×115	170	5	0.47	0.075	0.5	2.2	AMSU-A
7	54,400	410	5	0.47	0.075	0.5	2.2	AMSU-A
8	54,940	410	10	0.47	0.075	0.5	2.2	AMSU-A
9	55,500	350	10	0.47	0.075	0.5	2.2	AMSU-A
10	f_0=57,290.344	350	0.5	0.47	0.075	0.75	2.2	AMSU-A
11	f_0×217	78	0.5	0.47	0.075	1.0	2.2	AMSU-A
12	f_0×322.2×48	34	1.2	0.47	0.075	1.0	2.2	AMSU-A
13	f_0×322.2×22	14	1.6	0.47	0.075	1.5	2.2	AMSU-A
14	f_0×322.2×10	8	0.5	0.47	0.075	2.2	2.2	AMSU-A
15	f_0×322.2×4.5	3	0.5	0.47	0.075	3.4	2.2	AMSU-A
16	38,200	2,000	210	0.95	0.1	0.3	2.2	
17	105,500	3,000	210	0.95	0.1	0.4	1.1	
18	183,310×7,000	2,000	30	0.95	0.1	0.8	1.1	AMSU-B
19	183,310×4,500	2,000	30	0.95	0.1	0.8	1.1	
20	183,310×3,000	1,000	30	0.95	0.1	0.8	1.1	AMSU-B
21	183,310×1,800	1,000	30	0.95	0.1	0.8	1.1	
22	183,310×1,000	500	30	0.95	0.1	0.9	1.1	AMSU-B

By comparison, the ATMS, weights only 78 kg, takes 98 watts, and occupies one single box of about 0.2 m^3. Its dimension is approximately 70 cm (velocity direction) by 60 cm (nadir direction) by 40 cm. Such a drastic reduction in volume, weight and power by the ATMS is realized through the use of compact front-end "optics" design, and by using the Microwave Monolithic Integrated Circuit (MMIC) technology. The recent advances in solid state broad-band Low Noise Amplifiers (LNA) provide low noise figures, which help to compensate the higher radio frequency losses due to the ATMS' compact design and multi-band multiplexers. As a result, the ATMS can maintain reasonable NEΔT values even with its densely packaged design. Another key feature in ATMS is that its temperature sounding channels all have 2.2-degree beam-width, and they are sampled at 1.1-degree intervals in both down-track and cross-track directions (see Table 13.2). This is the so-called Nyquist sampling rate, and it is the first time that such a feature is used in any operational microwave sounders. Both the smaller 2.2-degree (static-IFOV) beam-width and the Nyquist sampling rate will contribute to a better spatial resolution feature for the temperature sounding channels. These new features are expected to enhance the ATMS's capability in differentiating sharp boundaries, such as the weather fronts.

Table 13.2 ATMS and AMSU scan and beam width characteristics

ITEMS	ATMS	AMSU	
		AMSU-A	AMSU-B
Number of Earth FOVs Per Scan	96	30	90
Blackbody and Space Samples Per Scan	4	2	4
Scan Movement	Continuous	Steps	Continuous
Scan Period (sec)	8/3	8	8/3
3-dB Beam Width	5.2°(Ch.1 – 2) 2.2°(Ch.3 – 16) 1.1°(Ch. 17 – 22)	3.33°	1.11°
Separation Hetween Adjacent FOVs	1.11°	3.33°	1.11°
Earth-Viewing Angular Width*	± 52.73°	± 48.33°	± 48.33°

* From nadir.

13.3 ATMS Studies with a Heritage Sensor: AMSU

The main objective of the ATMS and AMSU instruments is to provide atmospheric temperature and moisture profiles in all-sky conditions. The ATMS and AMSU instruments are also needed to accompany infrared sounders in order to mitigate the impact of clouds on infrared measurements. On NOAA polar orbiters, AMSU observations are collocated with measurements from the High resolution InfraRed Sounder (HIRS). The AMSU observations are also collocated with the Atmospheric Infrared Sounder (AIRS) on EOS Aqua and Infrared Atmospheric Sounding Interferometer (IASI) on EUMETSAT METOP, respectively. NPOESS ATMS will be collocated with the Cross-track Infrared Sounder (CrIS). These infrared sounders use the microwave sounders to detect and account for infrared cloud contamination and also to provide profiles in complete overcast conditions. With the exception of heavy precipitation, the microwave temperature sounding channels between 52.8 GHz and 58 GHz are largely not attenuated by clouds. The channel temperature weighting functions for AMSU-A at near nadir and largest scan angle positions are given in Fig. 13.2. The following subsections will discuss climate and weather applications of AMSU-only observations.

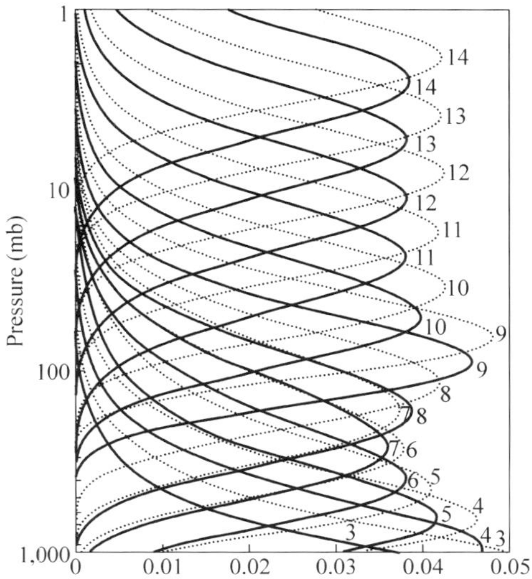

Figure 13.2 AMSU temperature weighting functions for nadir and largest scan angle

13.3.1 AMSU-A Temperature Profiles for Climate

Because microwave temperature sounding channels are not sensitive to clouds and that their radiances are linear with temperature, microwave observations are ideal for studying climate variability and long-term trends in temperature. The linear relationship also allows the interpretation of the AMSU-A observations as deep layer mean temperatures where each layer is defined by the channel weighting function. An AMSU-A temperature retrieval algorithm and processing system was specifically developed for climate applications (Goldberg, 1999 and 2001). The underlying difference between a system for climate and one for weather applications is to avoid the use of time varying a priori information for the background state. In other words, for climate applications the emphasis is on reducing systematic bias errors, whereas for weather applications the emphasis is to reduce the standard deviation with respect to the truth. For climate applications we used a fixed set of regression coefficients applied to limb adjusted (i.e. off-nadir observations adjusted to nadir) AMSU-A brightness temperatures to derive the profiles in order to ensure that climate changes are due to the observations and not due to coefficient updates or systematic errors in the background state. The limb adjustment methodology is described in Goldberg et al. (2001). The regression coefficients to derive the atmospheric temperature profiles are generated from a collocated ensemble of radiosonde temperature profiles and AMSU-A limb adjusted brightness temperatures. Channels 4 through 14 are used to derive the coefficients. For high altitudes where there are no radiosonde reports, the coefficients are derived from radiances simulated from an ensemble of rocketsondes.

The channels used in the regression vary as a function of atmospheric height. For example, when generating coefficients to derive temperatures at pressures lower than 100 mb, channels 4 through 7 are not used in order to avoid potential contamination from the surface, precipitation and high terrain.

Temperature retrievals from AMSU-A have been generated and validated routinely since July 1998. To validate the temperature retrievals, AMSU-A derived temperature retrievals are collocated with radiosonde profiles that were within a distance and time window of 50 km and 3 hours, respectively. Above 10 mb there is a lack of radiosonde data needed to assess the retrieval accuracy. Instead temperature retrievals from the Halogen Occultation Experiment (HALOE) are used as the "truth". HALOE is a solar limb sounder and derives a pressure profile from CO_2 transmission measurement at 2.8 micrometers. A temperature profile is then computed using the hydrostatic relationship. HALOE retrievals have relatively poor horizontal resolution, but very good vertical resolution (about 2 km). Root mean square (rms) errors of the differences between the retrieval and radiosonde and of differences between the retrieval and HALOE derived temperatures for March 1999 to April 2000 are shown in Figs. 13.3(a) and 13.3(b), respectively. The retrieval bias errors are not shown, but are generally well within 0.5 K. Also shown in these figures are the sample size, the Pearson correlation coefficient between the retrieval and radiosonde, and the ensemble's standard deviations for each layer. The results clearly show that the rms retrieval errors within these layers, which have a thickness of about 3 km, are generally well within 2 K. In Fig. 13.3(a) the rms errors are within 1.5 K, with the exception of the lowest layers which is due to uncertainties in surface properties and cloud liquid water. It is interesting to note that in Fig. 13.3(b), the retrieval errors begin to increase rapidly above 1 mb due to the lack of information from AMSU-A (Channel 14 peaks at about 2 mb).

The AMSU-A climate retrieval products consist of limb adjusted brightness temperatures, atmospheric temperatures, cloud liquid water and total precipitable water. The latter two are based on algorithms derived by Grody et al. (1999) and are produced only for ocean.

The products are averaged at 1 degree latitude/longitude resolution. There are three types of grids—daily, pentad and monthly. Each type is generated for ascending and descending orbits.

13.3.2 AMSU-A Weather Application

To provide the most accurate information for weather applications, including extreme events such as hurricanes, a microwave integrated retrieval system (MIRS) at NESDIS was developed to derive the profiles of atmospheric parameters such as temperature, water vapor, and cloud hydrometeors over land and oceans by using the measurements from AMSU window and sounding channels. Included

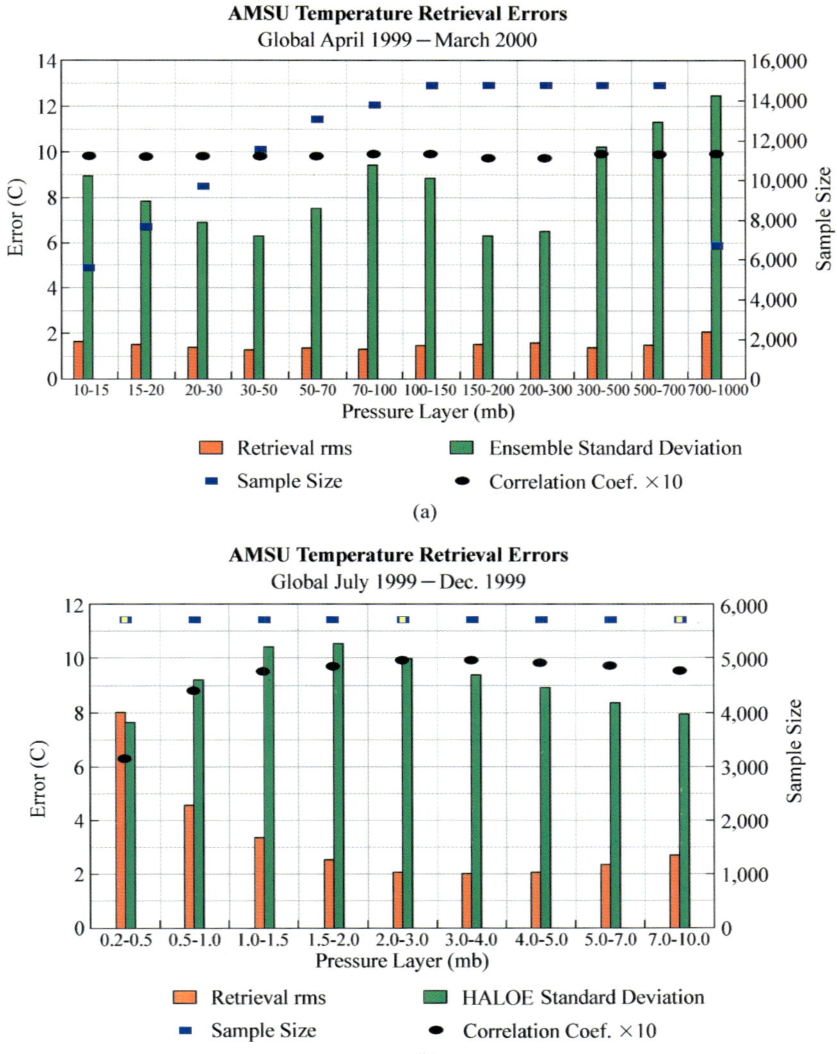

Figure 13.3 AMSU temperature retrieval errors using radiosonde (a) and HALOE collocations (b) for "truth"

in MIRS is an advanced radiative transfer model which includes atmospheric and surface scattering and polarization. With the microwave surface emissivity model (Weng et al., 2001); water vapor and cloud water can be retrieved over land. With the scattering radiative transfer model, the algorithm is now generally applicable under all weather conditions. A combined use of both microwave window and sounding channels makes it possible to simultaneously derive the cloud water profiles in addition to temperature and water vapor profiles. This integrated approach will lead to more robust advanced microwave products from current and future satellite microwave instruments having both imaging and sounding channels.

The MIRS is composed of several key components such as advanced radiative transfer models for computing radiances and gradients of radiances (or jacobians), and a scheme of minimizing the cost function that weighs the relative contribution of background (a priori) information and satellite observations (see Fig. 1 in Liu and Weng, 2004). For the retrieval of geophysical parameters, background, first guess, and observations play an important role. Backgrounds and observations define a minimization problem of the cost function. The background contains a priori information while the first guess provides a start point for solving the minimization problem. The background and the first guess can be the same or different. The first guess can be obtained from several methods: numerical weather prediction model outputs, climatological data, and regression algorithms. In this study, the initial profiles of temperature, water vapor, and non-precipitating cloud liquid water are estimated from the regression algorithm using the AMSU-A measurements. In the MIRS, the cloud liquid and ice water profiles associated with precipitation are derived through the additional variational procedure by adding AMSU-B 89 GHz and 150 GHz and using the profiles of temperature, water vapor, and cloud liquid water derived from the first variational procedure. This approach is physically sound because these two AMSU-B window channels are insensitive to the atmospheric temperature and water vapor profiles. The final step refines the temperature and water vapor profiles by using all AMSU-B 183 GHz water vapor sounding channels. The 183 GHz water vapor channel measurements can improve the accuracy of water vapor profiles throughout the troposphere. Note however that the second variational procedure is only required when the scattering signatures at 89 and 150 GHz exceed the threshold as discussed in Liu and Weng (2004).

An important process in MIRS is to correct the various biases from the AMSU instrument, within the forward model and the first guess. It was found that three AMSU instruments on board NOAA-15, -16 and -17 satellites all display some level of asymmetric radiances across the scan lines (Weng et al., 2003). The causes for this asymmetry have not been fully understood. Nevertheless, an empirical algorithm was developed and utilized in this study (Weng et al., 2003). In addition to a correction of the instrumental bias, the biases of the first guess and forward model to the truth are also corrected. When using forecast model outputs (e.g. 6 – hour forecast) as a first guess, we collocate in-situ temperature and water vapor profiles observed from radiosondes with forecast model outputs, and then calculate the global mean difference between the first guess and in-situ data. The biases in the forward radiative transfer model are derived from the mean difference between measured brightness temperatures and simulated brightness temperatures that are calculated from radiosonde data.

The effects of cloud and precipitation scattering on the temperature retrievals are readily demonstrated by comparing the results from a scattering radiative transfer model to those from the emission-based model. In Fig. 13.4(a), the magnitude of the cold temperature anomaly at 850 hPa derived from the scattering model is more agreeable with other results (Hawkins and Rubsam, 1968) than that from the emission-based in Fig. 13.4(b). Because of the scattering of liquid phase hydrometeors, the microwave brightness temperatures at sounding channels are strongly depressed. If this scattering effect would not be properly taken into account in the retrieval process, the physical temperatures at 850 hPa would have been forced to the colder values which become highly non-physical (see Fig. 13.4(b)). Using the scattering model, the retrieved structure of cooling at 850 hPa is smooth and the anomaly in a magnitude of $3°\text{K} - 5°\text{K}$ is more realistic according to Hawkins and Rubsam (1968).

The quality of the temperature distribution at 200 hPa has been also improved significantly using the scattering model. Note that the temperature at the center is about 10 degrees warmer than that of its environment (see Fig. 13.4(c)) and is favorably comparable to other early observations (Hawkins and Rubsam, 1968). It is also interesting to see that the retrieved temperature from the emission-based

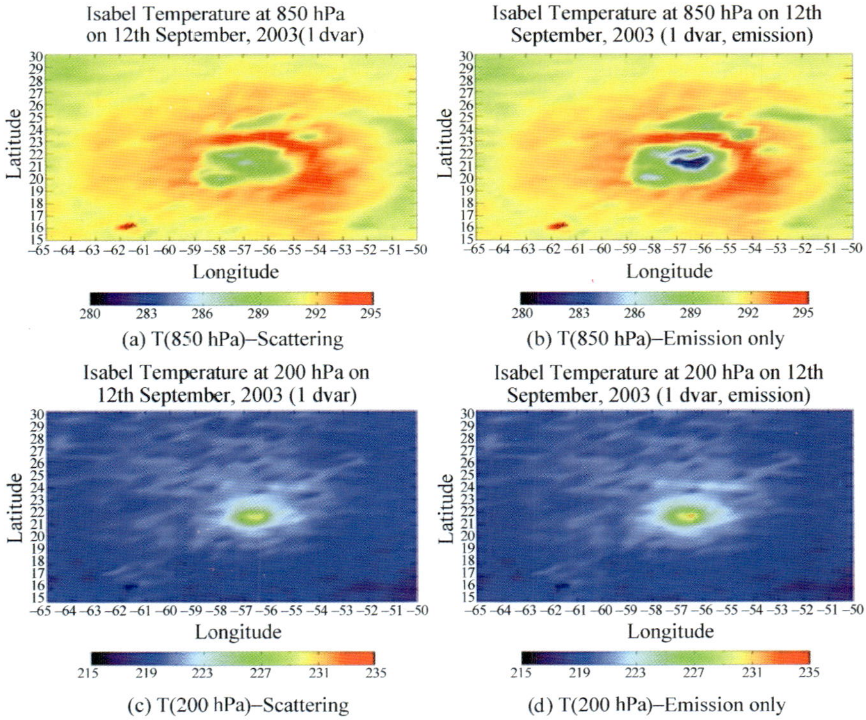

Figure 13.4 Retrieved atmospheric temperature at 850 hPa through (a) scattering radiative transfer model, and (b) emission radiative transfer model; and the retrieved atmospheric temperature at 200 hPa through (c) scattering radiative transfer model and (d) emission radiative transfer model, for Hurricane Isabel on 12th September, 2003

model (see Fig. 13.4(d)) is generally warmer than that from the scattering model (see Fig. 13.4(c)). This is because the ice cloud whose scattering is neglected in the retrievals would behave more like an absorber, attenuate warm radiation from the lower troposphere and re-emit with their own colder temperatures. The algorithm can only enforce warmer temperatures in the upper troposphere because the AMSU observed brightness temperatures are generally higher than those predicted by the emission-based model.

13.4 Discussions

The ATMS will continue the current weather and climate applications from AMSU. Comparing the channel information between AMSU and ATMS sensors, we anticipate that retrievals of water vapor from ATMS will be more accurate due to two additional water vapor channels at 183.31∀4.5 GHz and 183.31∀1.8 GHz. The improved water vapor retrieval together with the temperature channel at 51.76 will also improve lower tropospheric temperature accuracy. The shift of 150 GHz from AMSU to 165.5 GHz for ATMS has some consequence on the precipitation detection because this new ATMS channel tends to be more affected by water vapor absorption. Meanwhile, the over-sampling data from the ATMS might be of great applications in weather and climate studies due to its high spatial resolutions at wider scan swath.

Acknowledgements

The authors would like to express sincere thanks to Drs. Jim Shiue of NASA and Tsan Mo of NOAA/NESDIS for providing information on the ATMS and AMSU instruments. Some of the study results presented in this chapter were funded by the Integrated Program Office and the Joint Center for Satellite Data Assimilation.

References

Boeing/AER (2001) CMIS algorithm theoretical base document. Vol 1 – 17, NOAA Integrated Program Office, Silver Spring, MD

Goldberg MD (1999) Generation of retrieval products from AMSU-A: methodology and validation. The Proceedings on the 10th International TOVS Study Conference, pp 215 – 229

Goldberg MD (2001) Climate quality datasets from AMSU-A: methodology and validation. The Proceedings on the 11th International TOVS Study Conference

Goldberg MD, Crosby DS, Zhou L (2001) The Limb Adjustment for AMSU-A Observations: Methodology and Validation. J Appl Meteo 40: 70 – 83

Grody NC, Weng F, Ferraro R (1999) Applications of AMSU for obtaining water vapor, cloud liquid water, precipitation, snow cover and sea ice concentration. The Proceedings on the 10th International TOVS Study Conference, pp 230–240

Hawkins HF, Rubsam DT (1968) Hurricane Hilda, 1964. II. Structure and budgets of the hurricane on October 1, 1964. Mon Wea Rev 96: 617–636

Liu Q, Weng F (2004) One-dimensional variation retrieval algorithm for temperature, water vapor and cloud. IEEE Trans Geosci & Remote Sensing, (Accepted)

Mo T (1996) Pre-launch Calibration of the Advanced Microwave Sounding Unit-A for NOAA-K. IEEE Trans Microwave Theory and Tech 44: 1,460–1,469

Shiue JC (2001) Advance Technology Microwave Sounder, a new atmospheric temperature and humidity sounder for operational polar-orbiting weather satellites. Proceedings of SPIE 4540: 159–165

Weng F, Yan B, Grody NC (2001) A microwave land emissivity model. J Geophys Res 106(20): 115–123

Weng F, Zhao L, Ferraro R, Poe G, Li X, Grody N (2003) Advanced Microwave Sounding Unit Cloud and Precipitation Algorithms. Radio Sci 38(8): 86–96

14 Introduction to AIRS and CrIS

Joel Susskind

14.1 Introduction and Overview

AIRS/AMSU/HSB is a state of the art advanced infra-red microwave sounding system that was launched on the EOS Aqua platform in a 1:30 AM/PM sun synchronous orbit on May 5, 2002. An overview of the AIRS instrument and the objectives for AIRS/AMSU/HSB is a given in Pagano et al. (2003). The sounding goals of AIRS are to produce 1 km tropospheric layer mean temperatures with an rms error of 1 K, and 1 km tropospheric layer precipitable water with an rms error of 20%, in cases with up to 80% effective cloud cover. Aside from being part of a climate mission, one of the objectives of AIRS is to provide sounding information of sufficient accuracy such that when assimilated into a general circulation model, significant improvement in forecast skill would arise. The pre-launch algorithm to produce level-2 products (geophysical parameters) using AIRS/AMSU/HSB data, and expected results based on simulation studies, are given in Susskind et al. (2003). The results of that simulation indicate that the sounding goals of AIRS/AMSU/HSB should be achievable.

CrIS, like AIRS, is an advanced IR sounder that will fly on the NPP mission scheduled for launch in 2006, as well as on NPOESS, the future low earth orbiting operational satellite series. CrIS will be accompanied by ATMS, a microwave sounder with characteristics similar to those of AMSU-A/HSB. CrIS/ATMS is expected to produce products of similar accuracy as can be obtained from AIRS/AMSU/HSB.

Both AIRS and CrIS are high spectral resolution ($\nu / \Delta\nu \approx 1{,}000$) IR sounders with many channels covering the spectral range 650 cm^{-1} – 2,700 cm^{-1}. Most of these channels are opaque, due to atmospheric absorption. Opaque channels are referred to as sounding channels, while channels in which the atmosphere has little absorption are called window channels. Observations in the sounding channels are sensitive to atmospheric temperature and constituent profiles and can be used to determine these parameters. The spatial resolution of AIRS and CrIS is on the order of 15 km at nadir and is coarse compared to that of imagers such as MODIS and VIIRS, which is on the order of 1 km. Accurate soundings can be done in the presence of broken clouds with a coupled IR/microwave sounding system, and while very high spatial resolution is desirable, it is not critical for the production of accurate soundings. On the other hand, higher spatial resolution is required to the extent that sounding products at higher spatial resolution are desired. Both AIRS and CrIS contain a 3×3 array of observations within a single AMSU A like footprint (called field of regard—FOR), and a single sounding will be produced within a given FOR, with a spatial resolution of 50 km at nadir (growing to roughly 80 km by 120 km at the largest satellite zenith angle).

High spectral resolution observations in the infra-red are desirable for a number of reasons. Both AIRS and CrIS contiguously cover the spectral interval 650 cm^{-1} to 2,650 cm^{-1}, with some planned spectral gaps. AIRS has 2,378 channels with a resolving power $\nu / \Delta\nu \approx 1,200$ where $\Delta\nu$ is the spectral half-width of a channel at frequency ν, and CrIS has 1,286 channels with $\nu / \Delta\nu \approx 650$. Both AIRS and CrIS are sampled with a frequency spacing of roughly $\Delta\nu / 2$. Having a large number of channels with high spectral resolution allows for isolation of absorption features primarily due to a single gaseous absorption line, or in between absorption lines. This allows for selectivity of channels to be used when analyzing the data. The best channels are primarily sensitive to absorption by a single species. For temperature sounding, it is optimal to use channels in which absorption is primarily due to CO_2 and N_2O, whose concentrations are very close to constant in space and time. Absorption features due primarily to H_2O, O_3, CH_4, and CO are used to determine the concentration profiles of these gases, which vary significantly in space and time. Depending on the purpose, the best channels for sounding purposes are usually either between lines or on line centers.

The fact that adequate information about the surface and atmosphere can be obtained from analysis of only a subset of channels demonstrates that there is much more independent spectral information about the earth than there are degrees of freedom in the geophysical space of surface and atmospheric parameters. As a consequence of this, the effective noise of the spectrum is actually lower than that of the individual channels as observations in channels with redundant geophysical information can be used together so as to reduce the effective observational noise.

The AIRS science team is currently deriving a number of geophysical parameters from analysis of AIRS/AMSU sounding data. The primary products are atmospheric temperature profiles from the surface to 0.1 mb, water vapor profile for the surface to 100 mb, and ozone profile from the surface to 0.1 mb; land/ocean surface skin temperature and IR and microwave spectral emissivity; AIRS clear column radiances, which are the radiances that would have been seen if the atmosphere had no clouds; cloud top pressure and effective fractional cloud cover for up to two cloud types; and Outgoing Longwave Radiation (OLR) and clear sky OLR (the longwave radiation which would have gone to space if no clouds were present (Mehta and Susskind, 1999)). Atmospheric profiles of CO and CH_4, as well as total integrated atmospheric CO_2 burden are being derived in a research mode.

Figure 14.1 shows a sample AIRS spectrum. Channels used for different purposes are indicated as stars, with different colors depicting the geophysical parameter derived from that channel. AIRS has a number of mini spectral gaps, which are portrayed as gaps in the spectra shown. There is also a large spectral gap between 1,614 cm^{-1} and 2,181 cm^{-1}. Very little additional information about the Earth's surface and atmosphere is contained by channels in these spectral gaps. The meaning of the features shown in the AIRS spectrum will be discussed in detail later in this Chapter.

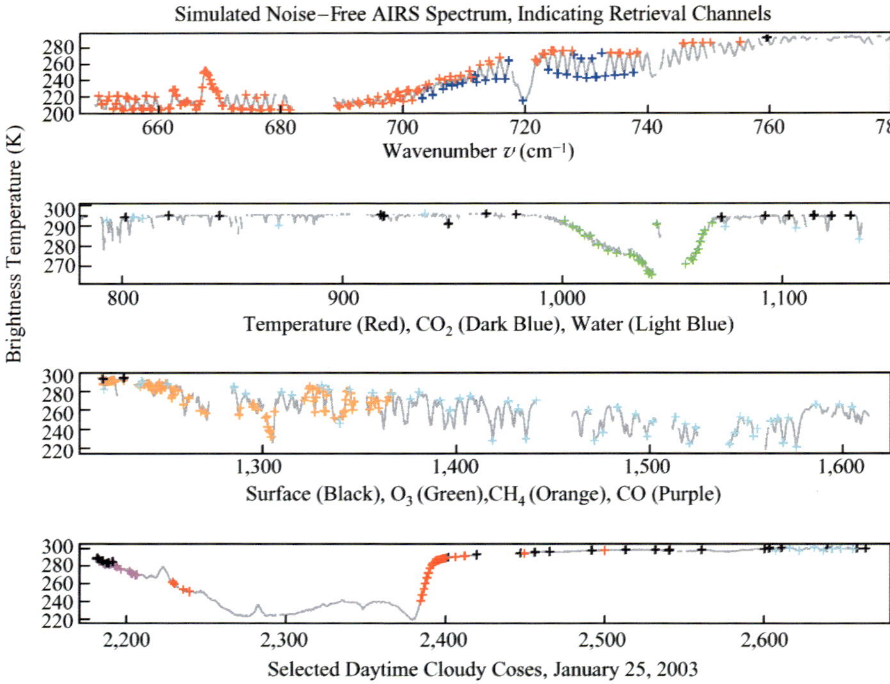

Figure 14.1 Simulated AIRS noise free spectrum. Gaps in the spectrum indicate planned spectral gaps for AIRS or dead detectors. The colored symbols show channels used for different purposes in the physical retrieval of geophysical parameters

Both AIRS and CrIS are accompanied by a microwave sounding instrument. IR and microwave observations are very complementary and the strengths of a combined system are greater than the sum of the strengths of the individual components. As shown in the next section, high resolution IR observations produce higher vertical resolution, and hence better accuracy, of mid-lower tropospheric temperature profile than the microwave observations. In addition, high spectral resolution IR observations provide the best information about surface skin temperature and constituent profiles. A major complication with regard to IR observations is that most IR channels are affected by scattering due to clouds in the FOR. Therefore, cloud effects must be well accounted for in order to take full advantage of the inherent capabilities of high spectral resolution IR observations.

Microwave observations are not appreciably affected by scattering due to (non-precipitating) clouds. Combined microwave and IR observations can be used to accurately account for the effects of clouds on the IR observations and produce accurate soundings in up to 80% partial cloud cover in the FOR (Susskind et al., 2003). Microwave soundings can also be used to produce temperature-moisture profiles under up to overcast conditions with non-precipitating clouds. The inherent accuracy of these is poorer than what would have been obtained if the high spectral resolution IR observations could have been used effectively however.

The following sections describe the basic physics giving rise to the AIRS channel radiances, including how different channels are sensitive to different surface and surface quantities; the general accuracy of geophysical retrievals determined by the AIRS science team both from simulated and real AIRS observations; results of experiments demonstrating how assimilation of temperature soundings obtained from AIRS/AMSU observations improve forecast skill; and the relationship between observations of AIRS and CrIS. The details of how surface and atmospheric soundings are obtained from analysis AIRS or CrIS data, in conjunction with either AMSU or ATMS observations, are beyond the scope of this chapter, and are given in Susskind et al. (2003).

14.2 The Radiative Transfer Equation

The monochromatic radiance going to space at frequency ν, in the IR and microwave spectral regions, under clear sky conditions is given in Equation (14.1).

$$R_{\nu,\mathrm{clr}} = \varepsilon_\nu B_\nu(T_\mathrm{S})\tau_\nu(p_\mathrm{S}) + \int B_\nu\left[\mathrm{T(p)}\right]\left(\frac{\mathrm{d}\tau_\nu}{\mathrm{d}\ell np}\right)\mathrm{d}\ell np + \rho_\nu H_\nu \tau_\nu'(p_\mathrm{S}) + (1-\varepsilon_\nu)R_\nu \downarrow \tau_\nu(p_\mathrm{S}) \tag{14.1}$$

The first term in Equation (14.1) represents radiation emitted by the surface transmitted through the atmosphere in the direction of the satellite, and the second term represents radiation emitted by the atmosphere transmitted to the satellite. These are the two main contributions to the observed radiances. The third and fourth terms represent downwelling radiation, reflected by the surface and transmitted to the satellite, coming either from the sun (third term) or the atmosphere (fourth term). The third term is significant only during the day and primarily at frequencies greater than 2,400 cm^{-1}. The fourth term is generally very small and will not be discussed here. This small term, as written, is appropriate for a specular surface, and can be modified for other types of surfaces (such as Lambertian).

The most important terms in Equation (14.1) are the Planck Black Body function $B_\nu(T)$, where ν is the frequency and T is the temperature; $\tau_\nu(p)$, the atmospheric transmittance at frequency ν from pressure p to the satellite; and $\frac{d\tau_\nu}{d\ell np}$, the derivative of $\tau(p)$, also referred to as the weighting function $W(p)$. Two surface parameters found in Equation (14.1) are the spectral surface emissivity, ε_ν, and the surface spectral bidirectional reflectance of solar radiation, ρ_ν. In addition, the surface skin temperature T_S is a dominant factor affecting the first term.

The atmospheric transmittance $\tau_\nu(p)$ is given by

$$\tau_\nu(p) = e^{-\int_0^p \sum_\ell k_{\nu\ell}(p)c_\ell(p)dp} \tag{14.2}$$

where ℓ refers to different absorbing species, $k_{\nu\ell}(p)$ is the absorption coefficient for species ℓ at frequency ν, and $c_\ell(p)$ is the fraction of air (mixing ratio) of species ℓ at pressure p.

The radiance at frequency ν reaching the satellite hence depends on the surface parameters $\varepsilon_\nu, \rho_\nu$, and T_S; the atmospheric temperature profile $T(p)$; and the constituent profiles through their effects on $\tau(p)$ and $\frac{d\tau}{d\ell np}$. $\tau(p)$ also depends on the temperature profile as $k_{\nu\ell}$ is temperature dependent. $\tau_\nu(p)$ varies from 1, at the satellite, to $\tau_\nu(p_S)$ at the surface, where p_S is the surface pressure. At frequencies in which the absorption coefficients $k_{\nu\ell}$ are very small, $\tau_\nu(p_S)$ is close to 1.0. These frequencies are referred to as windows, and the first (and third) terms of Equation (14.1) dominate the observed radiances. At frequencies in which values of $k_{\nu\ell}$ are large, $\tau(p_S)$ can be extremely small. At these frequencies, radiation emitted or reflected by the surface is not transmitted back to satellite, and only the second term contributes to the observed radiances.

In the vicinity of opaque channels, the spectral shape of the radiance spectrum R_ν depends on the weighting function $W_\nu(p)$. $B_\nu\left[T(p)\right]$ in Equation (14.2) is weighted in the integral by $\frac{d\tau_\nu}{d\ell np}$. The integral of this weighting, $\frac{d\tau}{d\ell np} d\ell np$, is equal to the integral of $d\tau$ which is $1-\tau(p_S)$. Hence term 1 contributes $\tau(p_S)$ to the total radiance and term 2 contributes $1-\tau(p_S)$ to the total radiance. For opaque channels, term 2 is the sole contributor. Moreover, the contribution to the integral is significant only over the atmospheric pressure range in which $\tau(p)$ is changing. High in the atmosphere (how high depends on k_ν),

$\tau_\nu(p)$ is close to 1.0 and $\mathrm{d}\tau/\mathrm{d}\ell np$ is close to zero. Near the surface (how near depends on k), $\tau_\nu(p)$ is close to 0.0 and $\mathrm{d}\tau/\mathrm{d}\ell np$ is close to 0. The pressure range in which $\tau_\nu(p)$ changes from near 1 to near 0 is the range which contributes to the radiance, and hence the range in which $T(p)$ and $c_\ell(p)$ contribute to the radiance in that channel.

A simplified expression for $W_\nu(p)$ is obtained for the case in which only one gas, ℓ, is absorbing, and c_ℓ and $k_{\nu\ell}$ are constant in pressure. Under these conditions,

$$\tau_\nu(p) = \mathrm{e}^{-k_\nu cp} \quad \text{and}$$
$$W_\nu(p) = k_\nu cp\mathrm{e}^{-k_\nu cp} = x\mathrm{e}^{-x} \tag{14.3}$$

The maximum value of $W_\nu(p)$ is then given by 0.37 when x=1. The pressure $p_{\nu\max}$ at which the weighting function is maximum is given by $p_{\nu\max} = 1/k_\nu c$.

Thus, for a given c, $p_{\nu\max}$ is lower (higher in the atmosphere) for large values of k_ν and higher for lower values of k_ν. The radiance will depend on the Planck function of temperature averaged over the vicinity of $p_{\nu\max}$.

A weighting function of the form $x\mathrm{e}^{-x}$ is quite broad in terms of ℓnp. Therefore, the radiance for such a channel corresponds to the temperature distribution over a wide range of the atmosphere. For the purpose of inferring temperature profile from the observed radiance spectrum, it is better to use observations in which the appropriate weighting functions are narrow (Kaplan et al., 1977). The weighting function will be narrower (that is τ drops from 1 to 0 faster) if k increases with p, as occurs at frequencies in the wing of a Lorentz broadened line, or k increases with T and T increases with p. Observations in such frequencies are particularly good for temperature sounding purposes. Observations on channel line centers are particularly poor to use because at these frequencies, k decreases with increasing p and the weighting functions are broader than $x\mathrm{e}^{-x}$. The weighting function also will be sharper if c increases with p (as it does for tropospheric water vapor). Observations at frequencies for which water vapor is the main absorber are not optimal for temperature sounding purposes because while the weighting functions are sharp, one does not know their peaks as they depend on c, which is highly variable for tropospheric water vapor in space and time.

Instruments do not measure monochromatic radiances however. Instruments are characterized by a set of channels, i, with characteristic channel spectral response functions, $f_i(\nu)$. The channel radiance R_i measured by channel i is given by

$$R_i = \int R_\nu f_i(\nu)\mathrm{d}\nu \Big/ \int f_i(\nu)\mathrm{d}\nu \tag{14.4}$$

The spectral resolution $\Delta\nu$ alluded to earlier for AIRS and CrIS refers to the full width at half-maximum (FWHM) of $f_i(\nu)$. If $\Delta\nu_i$ is narrow, as it is for AIRS and CrIS channels, then, to a very good approximation,

$$R_i \approx \varepsilon_i B_i(T_S)\tau_i(p_S) + \int B_i[T(p)]W_i(p)\mathrm{d}\ell np + \rho_i H_i \tau_i'(p_S) \\ + (1-\varepsilon_i)R_i \downarrow \tau_i p_S \quad (14.5)$$

where every term with subscript i is the channel average of the analogous term with subscript ν, as done in Eq. (14.4). To first order, the spectral response function for AIRS channel i is a Gaussian function centered at ν_i with FWHM = $\nu/1,200$.

Figure 14.1 showed the observations for the AIRS channels in terms of brightness temperatures. The brightness temperature Θ_i for channel i, with a radiance R_i, is defined as

$$\Theta_i = B_i^{-1}(R_i) \quad (14.6)$$

that is, Θ_i is the temperature at frequency ν_i for which the Planck function would be R_i. There is a unique monotonic relationship between R_i and Θ_i.

Under partial cloud cover, the radiance is given by

$$R_i = \left(1 - \sum_j \alpha_j\right) R_{i,\mathrm{clr}} + \sum_j \alpha_j R_{i,\mathrm{cld}j} \quad (14.7)$$

where α_j is the fraction of the sky, as seen from the satellite, covered by cloud type j and $R_{i,\mathrm{cld}j}$ is the channel i radiance going to the satellite which would be seen if the whole field of view were covered by cloud type j. For cloud clearing purposes, it is not necessary to know or be able to calculate $R_{i,\mathrm{cld}j}$. In constructing the simulated radiance shown in Fig. 14.1, clouds were treated as black bodies with a temperature equal to the atmospheric temperature at the cloud top pressure. This scene has only 10% cloud cover and is mostly clear.

Window regions exist between roughly 760 cm^{-1} and 1,000 cm^{-1}, 1,080 cm^{-1} and 1,250 cm^{-1}, and 2,440 cm^{-1} and 2,670 cm^{-1}. In the first two spectral regions, the brightness temperatures between absorption lines are approximately 294 K. Brightness temperatures at 2,600 cm^{-1} are closer to 300 K because of the contribution of solar radiation reflected by the surface to the observed radiances. On the weak line centers in these windows, brightness temperatures are colder, because the satellite sees less of the warmer surface (weighted by $\tau_i(p_s)$), and more of the colder air temperature $\overline{T}(p)$ above the surface (weighted by $(1-\tau_i(p_s))$, where $\overline{T}(p_i)$ is an effective atmospheric temperature averaged over the channel weighting functions. Thus the brightness temperature is the weighted average of the surface skin temperature and the effective atmospheric temperature. As $\tau_i(p_s)$ goes to zero, the brightness temperature becomes equal to the effective atmospheric temperature, roughly approximated by the temperature at the peak of the channel weighting function.

Figure 14.2 shows sample weighting functions for selected AIRS-like channels and for the temperature sounding channels of AMSU A. The sharpest temperature weighting functions for AIRS are in the lower troposphere and are in channels in the vicinity of 2,390 cm^{-1}. These channels have sharp weighting functions primarily because the absorption coefficient of the absorbing gas, CO_2, increases rapidly with increasing temperature at these frequencies (Kaplan et al., 1977). These channels provide the most important information about lower tropospheric temperature and are a subset of the red temperature sounding channels indicated in Fig. 14.1. Most of the other channels used for temperature sounding are between lines (locally warmer than line centers) in the spectral region 750 cm^{-1} – 700 cm^{-1}, and locally cooler than the line centers between 650 cm^{-1} and 700 cm^{-1}. In this latter area of the spectrum, emission is coming primarily from the stratosphere, in which temperature increases with increasing height. Hence more weakly absorbing channels (between lines) see lower in the atmosphere, which is cooler, than do channels on adjacent line centers. The local maximum in brightness temperature at 667 cm^{-1} (called the CO_2 Q branch) is a

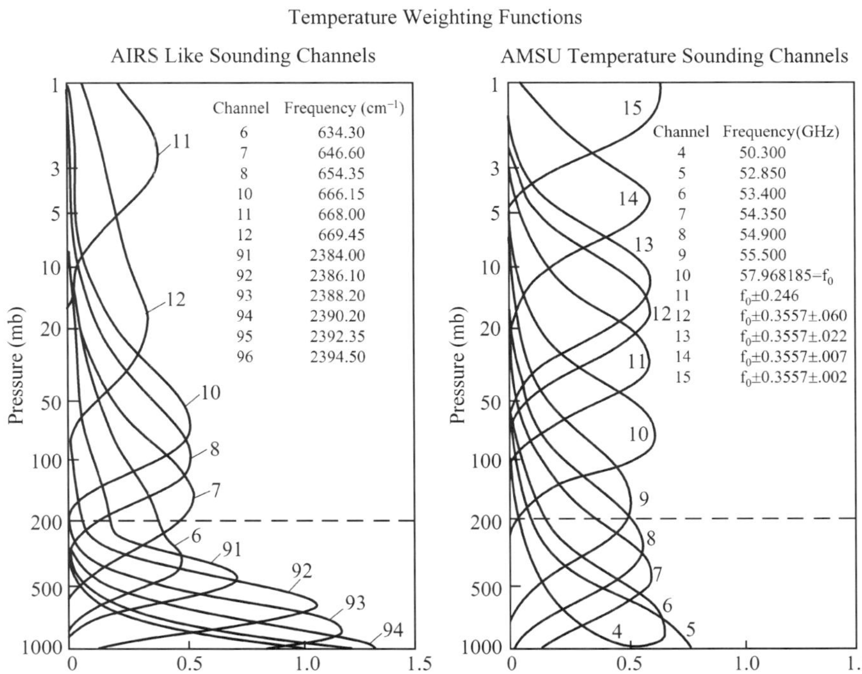

Figure 14.2 Temperature weighting functions for sample AIRS like channels and AMSU A channels

very opaque spectral region in which emission comes from as high as 1 mb. Channels 10 – 12 in Fig. 14.2 give examples of weighting functions in the CO_2 Q branch. Upper tropospheric temperatures (300 mb and lower pressures) can be determined about as well from AMSU A channels 8 – 15 as they can be determined from AIRS channel observations, up to channel noise considerations. CO_2 absorption features are also used to determine total CO_2 column amount. These are shown in dark blue in Fig. 14.2, and generally lie on the line centers of CO_2 absorption lines between 710 cm^{-1} and 740 cm^{-1}.

The spectral feature in the vicinity of 1,000 cm^{-1} – 1,070 cm^{-1} is due primarily to O_3 absorption, and a number of channels, shown in green, are use to determine the O_3 profile. Water vapor absorption is prominent in the spectral region 1,300 cm^{-1} – 1,600 cm^{-1}. Water vapor is determined using the light blue channels, looking both between lines and on line centers. Water vapor lines are also used in window regions, both on weak water vapor lines as well as between lines. The strong absorption feature at 1,306 cm^{-1} is due to methane, as are a number of nearby weaker absorption features. Channels used to determine the methane profile are shown in orange. Absorption by CO occurs near 2,200 cm^{-1}, and a number of CO sounding channels are shown in purple.

14.3 Results using AIRS/AMSU Data

The results shown below represent the state of products delivered by the AIRS Science Team algorithm as of January 30, 2004. Improvements to the algorithm continue to be made. The AIRS Science Team retrieval algorithm is basically identical to the pre-launch algorithm described in Susskind et al., (2003). The major difference is the addition of terms to account for systematic errors in computed channel radiances resulting from an imperfect parameterization of the physical processes affecting the radiances, as well as addition of a term in the channel noise covariance matrix expressing residual errors in computed channel radiances after the systematic errors are accounted for. The need for these terms was alluded to in (Susskind et al., 2003), in which the simulation study did not account for errors in the parameterization of the radiative transfer physics.

The key steps of the AIRS science team algorithm are listed below:

(1) Start with an initial state consistent with the AMSU A and HSB radiances (Rosenkranz, 2000); (2) Derive IR clear column radiances R_i^0 valid for the 3×3 AIRS Fields of View (FOVs) within an AMSU A Field of Regard (FOR) consistent with the observed radiances and the initial state; (3) Obtain an AIRS regression guess (Goldberg et al., 2003) consistent with R_i^0 using 1,504 AIRS channels; (4) Derive R_i^1 consistent with the AIRS radiances and the regression guess; (5) Derive all surface and atmospheric parameters using R_i^1 for 415 AIRS channels and all AMSU and HSB radiances; (6) Derive cloud parameters and

OLR consistent with the solution and observed R_i; (7) Apply quality control, which rejects a solution if the retrieved cloud fraction is greater than 80% or other tests fail. In the event that a retrieval is rejected, cloud parameters are determined using the initial microwave state and observed AIRS radiances.

Figure 14.3 shows the number of cases for each retrieved effective fractional cloud, in 0.5% bins, for the whole day September 6, 2002. The effective fractional cloud cover is given by the product of the fraction of the field of view covered by clouds and the cloud emissivity at 11 μm. The average global effective cloudiness was determined to be 41.20%. Also shown is the percent of accepted retrievals as a function of retrieved effective cloud cover. Roughly 90% of the cases with retrieved effective cloud cover 5% were accepted, falling to 35% at 40% effective cloud cover, and to 18% at 80% effective cloud cover. All cases with retrieved effective cloud cover greater than 80% are rejected. The average effective fractional cloudiness for all accepted cases was 24.61%.

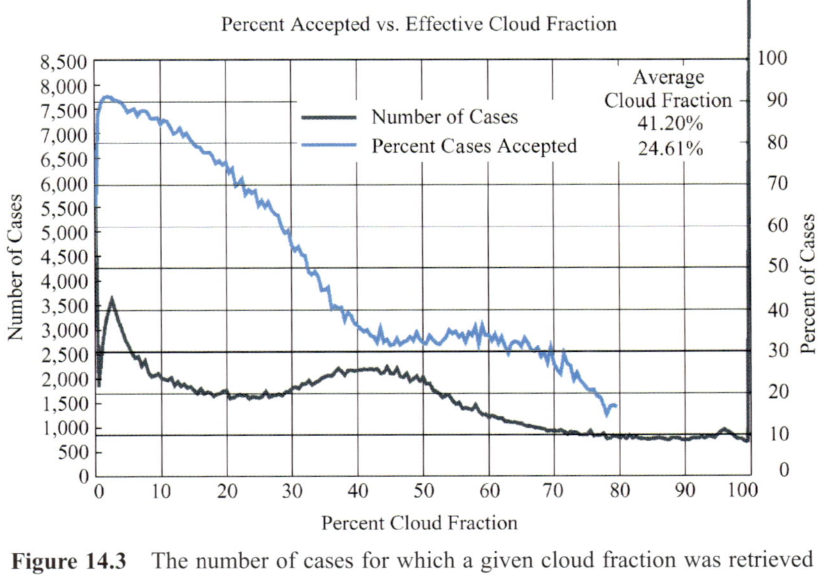

Figure 14.3 The number of cases for which a given cloud fraction was retrieved (black) and the percentage of those cases in which a successful retrieval was performed (blue). The average effective cloud fraction for all cases and accepted cases is also indicated

Figure 14.4 shows the RMS difference between retrieved 1 km layer mean temperatures and the collocated ECMWF forecast for all accepted cases as a function of retrieved effective cloud fraction. Results are shown for each of the lowest 8 km of the atmosphere. Agreement degrades with increasing cloud cover, but only very slowly except in the lowest 1 km of the atmosphere. RMS temperature differences from ECMWF at all levels are somewhat larger than the 1 K goal for retrieval accuracy. Part of this difference can be attributed to the fact that the ECMWF forecast is not perfect. It is also possible that the accuracy of the ECMWF forecast may be somewhat poorer with increasing cloud cover. The increase in RMS temperature differences at 0% cloudiness is somewhat misleading because a large percentage of clear cases occurred over Antarctica on this day. Retrievals over Antarctica are particularly difficult because of high variable terrain.

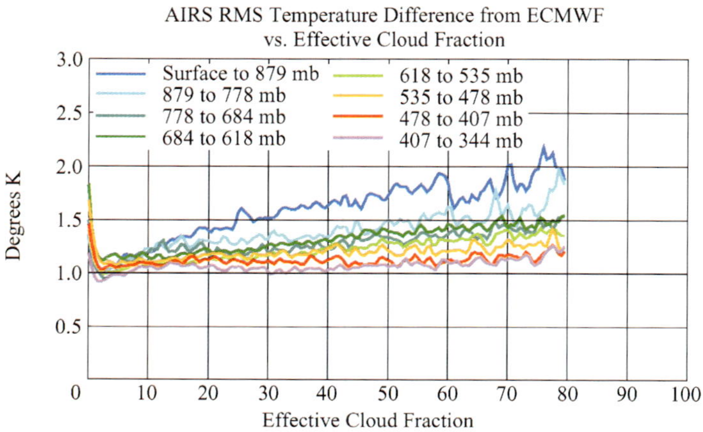

Figure 14.4 The RMS differences from ECMWF in 1 km layer mean temperatures for accepted cases as a function of retrieved effective cloud fraction

Figures 14.5(a) and 14.5(b) show RMS differences of temperature and moisture profiles from the "truth" with both simulated and real data. The gray and black curves reflect all accepted cases, and the pink and red curves are cases identified as clear, for simulated and observed radiances respectively.

For temperature, 1 km layer mean differences from the truth are shown, and for water vapor, % differences in total integrated water vapor in 1 km layers are shown. In simulation, the "truth" is known perfectly, while with real data, the 3-hour ECMWF forecast is taken as a proxy for "truth". For real data, as in simulation, temperature retrievals under cloudy conditions (roughly 47% of all cases are accepted) degrade by only a few tenths of a degree compared to cases identified as clear (roughly 3% of the cases are identified as clear), while water vapor retrievals do not degrade at all. Differences from "truth" are poorer with real data than in simulation. Two major causes of degradation are: (1) perfect physics was assumed in simulation; and (2) the "truth" has errors in real data. The degradation of soundings in the presence of "real clouds", as compared to soundings in clear cases, appears to be similar to that implied by simulation. In both cases, water vapor retrievals are of comparable accuracy in clear and partially cloudy conditions. This implies that inherent errors in retrieval accuracy,

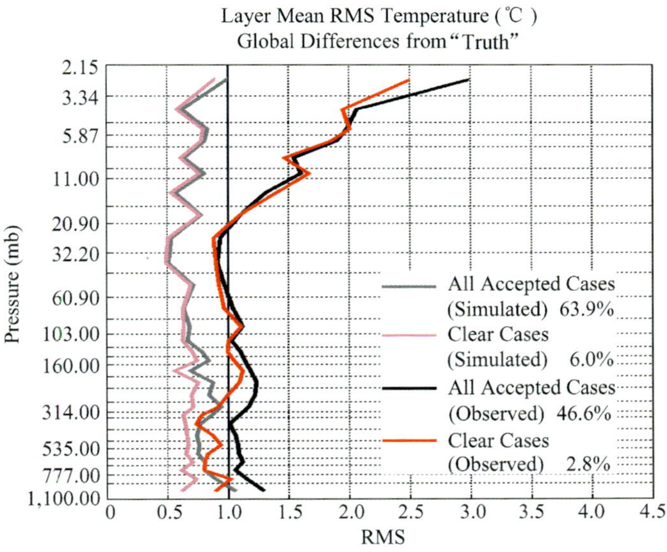

(a)

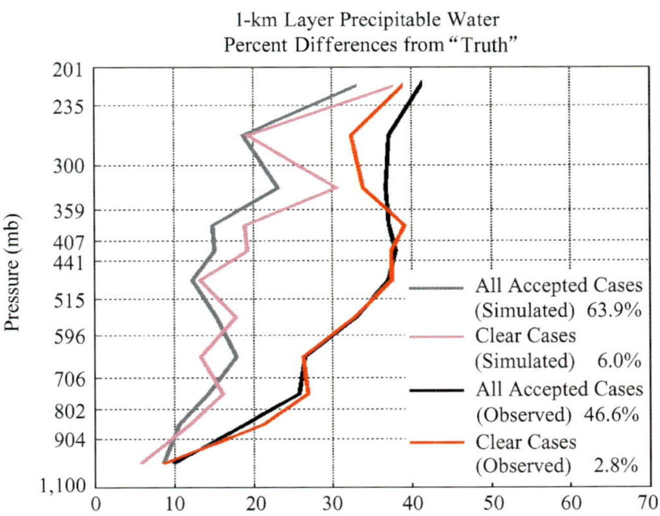

(b)

Figure 14.5 (a) RMS layer mean temperature differences from "truth" for all accepted cases and cases identified as essentially clear in both simulation and real data. (b) RMS percent differences from ECMWF in 1 km layer precipitable water for the same cases shown in Fig. 14.5(a)

due to limited vertical resolution (water vapor has more vertical structure than temperature), are larger than residual errors due to imperfect cloud clearing. In addition, cloudier cases tend to be moister than dry cases, and are "easier" in terms of percentage error.

Figure 14.6(a) shows RMS layer mean temperature differences between accepted retrievals, the ECMWF forecast, and collocated radiosonde reports (± 1 hour, ± 100 km) for September 6, 2002. The number of cases included in each

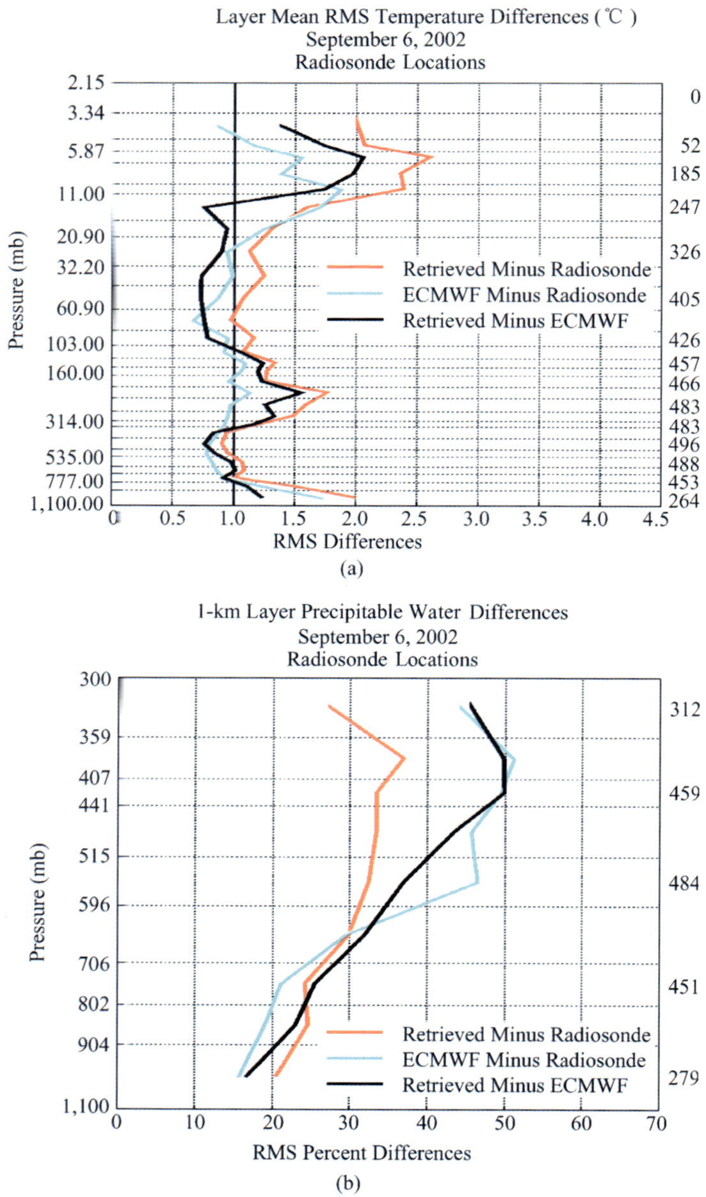

Figure 14.6 (a) RMS layer mean temperature differences with observed data in radiosonde locations. (b) RMS layer precipitable water percent differences in radiosonde locations

of the layers is indicated at the right of the figure. It is interesting to note that the RMS temperature differences between the retrievals and ECMWF are generally smaller in the vicinity of radiosondes than they were globally (see Fig. 14.5(a)). This is because the ECMWF forecast is more accurate in the vicinity of radiosondes than it is globally.

The 3-hour ECMWF forecast agrees with radiosondes to 1 K between roughly 750 mb and 20 mb. Spatial and temporal sampling differences between ECMWF, retrievals, and radiosondes contribute to some extent to the increased differences between both ECMWF and retrievals as compared to radiosondes beneath 750 mb, as spatial and temporal variability of the atmosphere is greatest near the surface. Retrieval accuracy near radiosondes is somewhat poorer than that of the forecast at all levels, especially in the vicinity of 200 mb. This is most likely due to limitations in the current methodology used to account for systematic errors in the radiative transfer used in the calculations and accounting for residual physics errors in the channel noise covariance matrix. Improvement is expected in this area with further research.

Figure 14.6(b) shows analogous results for percent differences in 1 km layer mean precipitable water, for which the sounding goal for AIRS is 20%. With regard to water vapor, it is clear that AIRS retrievals are significantly more accurate than the ECMWF forecast above 700 mb. AIRS differences from radiosondes are greater than the 20% goal. Spatial and temporal sampling differences between AIRS and radiosondes may contribute significantly to the apparent water vapor "errors" as water vapor changes rapidly in space and time.

Figure 14.7(a) shows the retrieved effective cloud top pressure and effective cloud fraction for ascending orbits on January 25, 2003. The results are presented in terms of cloud fraction in 5 groups, 0% – 20%, 20% – 40%, etc. with darker colors indicating greater cloud cover. These groups are shown in each of 7 colors, indicative of cloud top pressure. The reds and purples indicate the highest clouds, and the yellows and oranges indicute the lowest clouds. Cloud fields are retrieved for all cases in which valid AIRS/AMSU observations exist. Gray means no data was observed. Figure 14.7(b) shows the retrieved 500 mb temperature field. Gray indicates regions where either no valid observations existed or the retrieval was rejected, generally in regions of cloud cover 80% – 100%. Retrieved temperature profile fields are quite coherent, and show no apparent artifacts due to clouds in the field of view. Figures 14.7(c) and 14.7(d) show retrieved values of total precipitable water vapor above the surface and above 300 mb. Note the high values of upper tropospheric water vapor to the east of extensive cloud bands attributed to cold fronts.

Figure 14.8(a) shows the retrieved 700 mb temperature field for ascending orbits on January 25, 2003. Figure 14.8(b) shows the collocated ECMWF 3-hour forecast 700 mb temperature field. These fields appear very similar to each other. Their difference is shown in Fig. 14.8(c), in which white shows agreement to ± 0.5 K, each shade of red shows AIRS warmer than ECMWF in intervals of 1 K (0.5 – 1.5, 1.5 – 2.5, etc.), and each shade of blue shows AIRS colder then

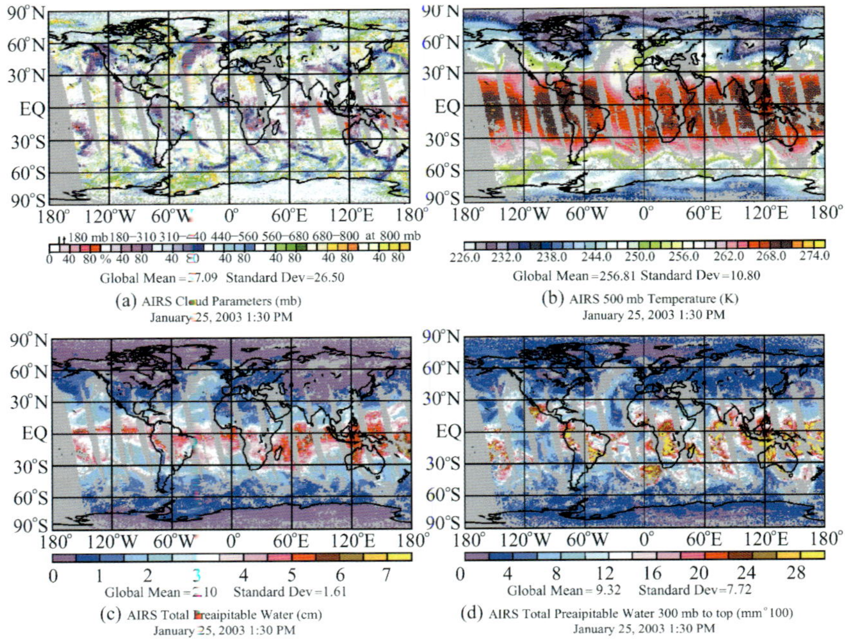

Figure 14.7 Sample retrieved geophysical parameters for ascending orbits on January 25, 2003. Gray indicates missing data due to orbit gaps (clouds) as well as rejected retrievals (other fields)

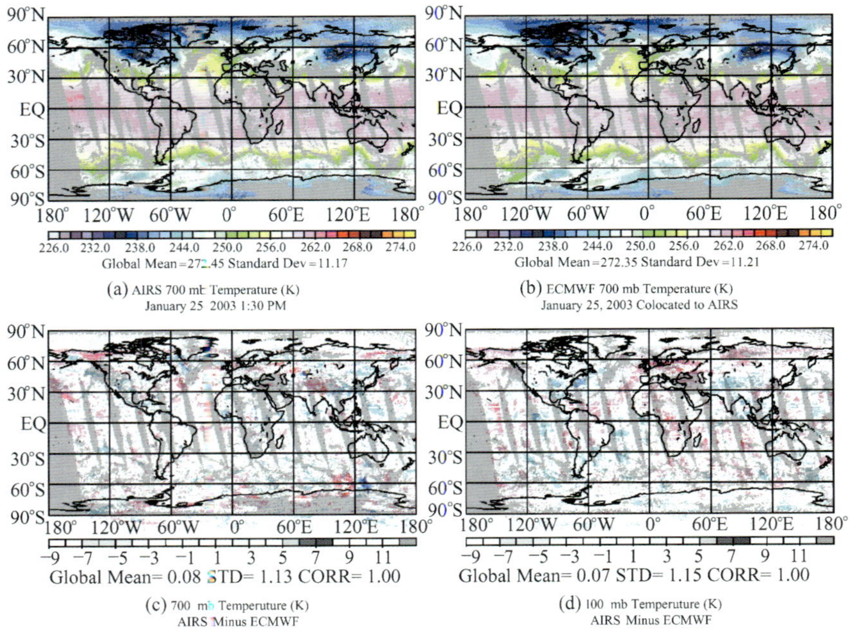

Figure 14.8 Comparison of AIRS retrieved temperatures and those predicted from the ECMWF three – hour forecast

ECMWF in intervals of 1 K. The area weighted global mean difference of the two fields is 0.08 K, and the area weighted standard deviation is 1.13 K. Most areas are white or the first shade of red or blue. The largest differences between the two fields occur in the vicinity of 35°S – 55°S, 100°E – 140°E, and show up as a dipole, with AIRS warmer to the west of 120°E and colder to the east. Figures 14.8(a) and 14.8(b) show this to be an area of a cold air mass, extending from the polar region to the mid-latitudes. This cold air mass is coherent in both the retrieved and forecasted fields, but is centered further east in the retrieved field compared to the forecast field, corresponding to a phase error in the 3-hour ECMWF forecast. This is precisely the type of information that satellite data can provide (if accurate enough) to help improve forecast skill. Figure 14.8(d) shows the difference between the retrieved and forecast 100 mb temperature fields. At the 100 mb level, a corresponding warm front (not shown) exists in both the retrieved and forecast fields in the area discussed above, with an analogous phase error to that found at 700 mb. Consequently, the retrieved 100 mb field is cooler than ECMWF to the west and warmer to the east in the region discussed above. This out of phase relationship of patterns of differences from ECMWF at 700 mb and 100 mb is indicative of phase errors in the ECMWF forecast, as there is no reason for retrieval errors to be out of phase with each other at 700 mb and 100 mb. This out of phase relationship in spatial patterns of differences between retrieved and forecast temperatures at 700 mb and 100 mb is found in numerous places in Figs. 14.8(c) and 14.8(d) and indicates many areas where the satellite data should improve the ECMWF forecast.

14.4 Forecast Impact Experiments

The forecast impact experiments shown in this section were conducted by Dr. Robert Atlas and his research group at the GSFC Laboratory for Atmospheres. The data assimilation system used in the experiments is FVSSI, which represents a combination of the NASA Finite Volume General Circulation Model (FVGCM) with the NCEP operational Spectral Statistical Interpolation (SSI) global analysis scheme implemented at lower than the operational horizontal resolution—T62. The basics of the finite-volume dynamical core formulation are given in DAO's Algorithm Theoretical Basis Document (see http://polar.gsfc.nasa.gov/sci_research/atbd.php), and the FVGCM has been shown to produce very accurate weather forecasts when run at high resolution (Lin et al., 2004). The AIRS temperature profiles produced by SRT were presented to the SSI analysis as rawinsonde profiles with observational error specified at 1° K at all vertical levels.

Results are presented for three sets of experiments in which data was assimilated for the period January 1 – January 31, 2003. Five-day forecasts were run every two days beginning January 6, 2003 and forecasts every 12 hours were verified against the NCEP analysis, which was taken as "truth". In the first experiment, called "control", all the data used operationally by NCEP was assimilated, but no AIRS data was assimilated. The operational data included all conventional data, TOVS and ATOVS radiances for NOAA-14, 15, and 16, cloud tracked winds, SSM/I total precipitable water and surface wind speed over ocean, QuikScat surface wind speed and direction, and SBUV ozone profiles. In the second and third experiments, called "clear AIRS" and "all AIRS", temperature profiles retrieved from AIRS soundings were assimilated in addition to the data included in the "control" experiment. "Clear ocean" included all accepted temperature retrievals derived from AIRS over ocean and sea ice in cases where the retrieved cloud fraction derived from AIRS was less than or equal to 2%, while the "all ocean" experiment assimilated accepted AIRS temperature soundings over ocean and sea ice for all retrieved cloud fractions.

Figure 14.9 shows anomaly correlation coefficients of forecast sea level pressure verified against the NCEP analysis for both Northern Hemisphere extra-tropics and Southern Hemisphere extra-tropics for both the "control" and "all AIRS" experiments. An anomaly coefficient of 0.6 or greater indicates a skillful forecast. In the Northern Hemisphere, addition of all AIRS soundings

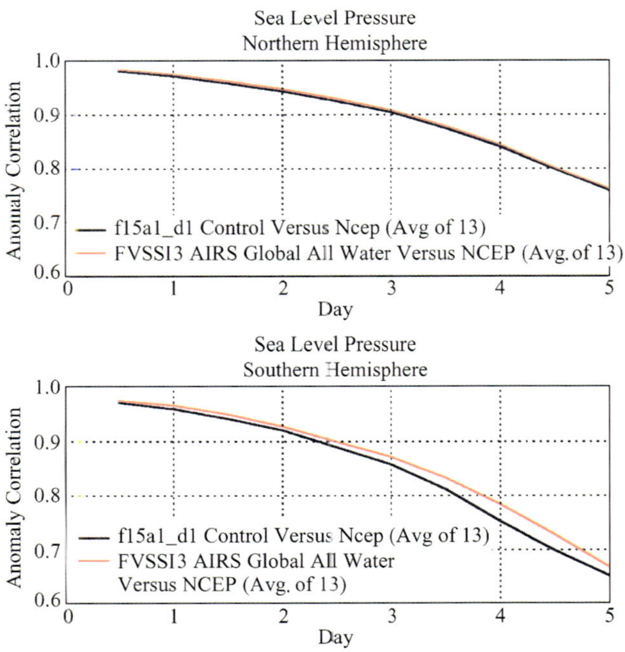

Figure 14.9 Sea level pressure forecast anomaly correlation coefficients with the NCEP analysis averaged over 13 forecasts using AIRS temperature soundings (red) and not using AIRS temperature soundings (black). Higher anomaly correlation coefficients indicate improved forecast skill

resulted in an improvement in average forecast skill of the order of 1 hour or less, but an improvement in average forecast skill in the Southern Hemisphere on the order of 6 hours results from assimilation of AIRS soundings, that is, equivalent forecast skill occurs 6 hours later when AIRS data is used. Assimilation of AIRS soundings only under essentially clear conditions (not shown), resulted in somewhat poorer forecasts than using all AIRS soundings. It should be noted that the Aqua orbit (1:30 ascending) is almost identical to that of NOAA 16 carrying HIRS3, AMSU-A and AMSU-B, so AIRS/AMSU/HSB soundings are providing additional information to that contained in the AMSU-A/AMSU-B radiances on NOAA 16 in the same orbit.

Figure 14.10 shows the RMS position error (km) and magnitude error (hPa) for 5-day forecasts of extra tropical cyclones in the three experiments. It is apparent that addition of AIRS soundings improved RMS forecast skill for both the position and magnitude of extra-tropical cyclones globally, and addition of AIRS soundings in partially cloudy areas further improved forecast skill as compared to use of soundings only in essentially clear conditions.

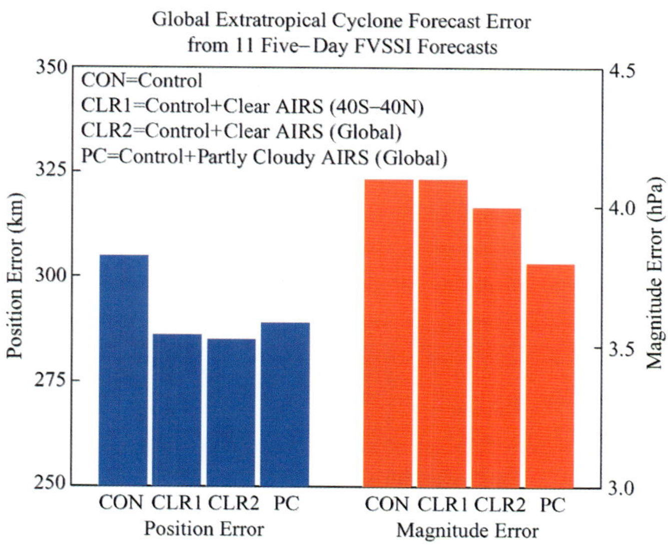

Figure 14.10 Global extratropical cyclone position and intensity RMS errors from 11 5 – day forecasts using differing amounts of AIRS temperature soundings

Several thousand cyclones verifications are included in these statistics. Addition of AIRS data did not improve forecasted cyclone position and intensity for each cyclone. Some were improved substantially however. Figure 14.11 shows the impact of AIRS data on the 24-hour forecast of position and intensity of tropical storm Beni, which was centered roughly 4° east of New Caledonia on January 31, 2003 with a central pressure of 990 mb (see Fig. 14.11(d)). The control forecast (see Fig. 14.11(a)) produced a relatively weak cyclone (1,007 mb) displaced considerably to the northwest, while the 24-hour forecast using AIRS data (Fig. 14.11(b)) was much more accurate in both position and intensity (995 mb). It is significant to note that the GSFC forecast using AIRS data was more accurate in both position and intensity than the NCEP operational forecast (see Fig. 14.11(c)) in this case, which, even though it used a higher resolution model and analysis system, did not have the benefit of AIRS data. The results shown indicate the potential of AIRS soundings to improve operational forecast skill.

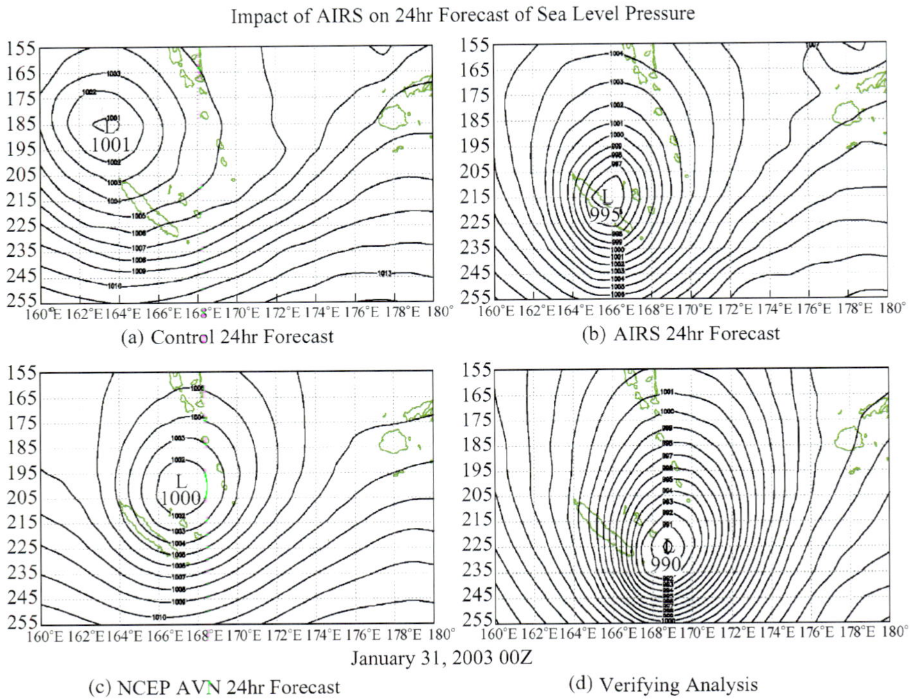

Figure 14.11 24-hour forecast of sea level pressure in the vicinity of tropical storm Beni based on forecasts with and without the benefit of AIRS temperature soundings, as well as the verifying analysis

ECMWF has been conducting research directly assimilating AIRS radiances, rather than AIRS retrieved temperature profiles, and has shown positive impact of AIRS on their forecast skill, but significantly less than found at GSFC. Based on this finding, assimilation of AIRS radiances is now operational at ECMWF. There are at least two reasons why the positive impact of AIRS data on ECMWF forecast skill is smaller than that obtained at GSFC. First, ECMWF forecast skill without AIRS data is even better than that of NCEP, and the ECMWF forecast may be harder to improve upon. Second, ECMWF only uses radiances thought to be unaffected by clouds, which are a small subset of the cloud cleared radiances used to produce the AIRS temperature soundings assimilated at GSFC. Use of cloud cleared radiances by ECMWF, with appropriate quality control, might further enhance the positive impact of AIRS data on their forecast skill. Dr. Atlas is working with NCEP to arrange an experiment to assimilate AIRS temperature soundings in the NCEP analysis as run on the NCEP computing system to see the extent, if any, that NCEP operational forecast skill can be improved upon.

14.5 Comparison of CrIS and AIRS

CrIS was designed to be a follow on to AIRS and has generally similar characteristics. AIRS is a multidetector array grating spectrometer and CrIS is an interferometer. These are two hardware approaches toward achieving roughly the same spectral and noise characteristics, and from the theoretical or practical perspectives, the difference in hardware approach poses only second order effects on the data.

The raw data product from an interferometer, the interferogram, is a cosine transform of the incoming radiance. One obtains the channel i radiance R_i by taking the cosine transform of the product of the interferogram $I(\delta)$ with an apodization function, where δ is the optical path difference. The cosine transform has finite limits, since the interferometer has a finite maximum optical-path difference L:

$$R_i = \frac{\int_{\delta=-\infty}^{\infty} A(\delta)\cdot I(\delta)\cdot\cos(2\pi(v-v_i)\cdot\delta)\cdot \mathrm{d}\delta}{\int_{\delta=-\infty}^{\infty} A(\delta)\cdot \mathrm{d}\delta} \tag{14.8}$$

In the interferogram domain, the unapodized (or boxcar) apodization function is defined as

$$A^u(\delta)=1 \ \text{ for} |\delta| \leqslant L = 0 \ \text{ for} |\delta| > L \tag{14.9}$$

The channel spectral response function, $f_i(v)$, is the cosine transform of the apodization function. For an unapodized interferometer, $f_i^u(v)$ is equal to

$$f_i^u(v)=\frac{\sin(y)}{y}+\frac{\sin(z)}{z}=\sin c(y)+\sin c(z) \tag{14.10}$$

where $y=2\pi L\cdot(v-v_i)$, $z=2\pi L\cdot(v+v_i)$, v_i is the channel center, and v is the frequency. Typically, the second term in Equation (14.10) is ignored.

The Nyquist sampling theorem states that the optimal sampling of channels is such that the channel spacing is $\delta v=1/(2L)$ in the frequency domain (Brigham, 1988). No additional information is gained by sampling the interferogram at a higher rate (although oversampling is used to reduce out-of-band aliasing). However, information is lost if the interferogram is sampled at a lower rate.

The sinc(y) function is shown as the red curve in Fig. 14.12 for $L = 0.8\ \text{cm}^{-1}$, the value of L in Band 1 of CrIS. The sinc(y) function has large side-lobes that alternate in sign between the zeros of the function spaced at $y = \pm n\pi$. FWHM of the sinc(y) function is equal to

$$\text{FWHM}^{u} = \frac{0.603355}{L} \approx \frac{1.2}{2L} \tag{14.11}$$

Only 45% of the area of the unapodized spectral response function comes from the central lobe.

The use of nonlocalized, unapodized radiances can produce complications in the retrieval of geophysical parameters. For multispectral retrievals (e.g. combining microwave and IR radiances) it is convenient, but not necessary, to represent radiances as brightness temperatures (i.e., the temperature of a blackbody with the same radiance). For unapodized spectra, brightness temperature is a meaningless concept due to the distortion caused by negative side-lobes that can produce negative-channel radiances.

Barnet et al. show that a Hamming apodization function is optimal for minimizing the side lobes of the interferometric spectrum response function, while keeping the FWHM as narrow as possible (Barnet at al., 2000). This spectral response function for the Hamming apodized interferogram is shown as the blue curve in Fig. 14.12. In this figure, $L = 0.8$ cm, corresponding to the CrIS value in this spectral region. Figure 14.12 also shows in black the spectral line shape for an AIRS channel at 720.5 cm^{-1}.

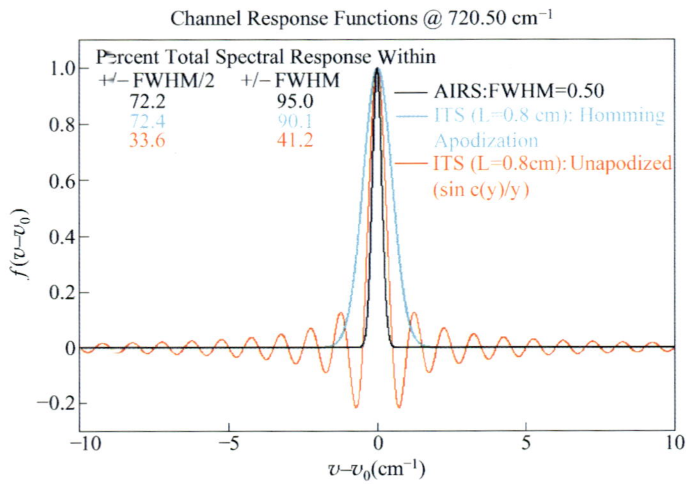

Figure 14.12 Channel response functions at 720.5 cm^{-1} for AIRS (black), CrIS without apodization (red), and CrIS with Hamming apodization (blue)

The unapodized CrIS spectral response function (red) has a narrow central lobe (0.6/*L* cm^{-1}, where *L* is the maximum optical displacement for a band) but side lobes that extend well beyond what is actually shown in the figure. Only 41.2% of the unapodized spectral response function lies within ±1 FWHM. The width of the central lobe of the Hamming apodized spectral response function (0.9/*L* cm^{-1}) is 50% larger than that of the unapodized spectral response function. CrIS spectral sampling (0.5/*L* cm^{-1}) is slightly larger than 1/2 of the Hamming apodized FWHM. The AIRS spectral response function at this frequency (it is narrower at lower frequencies and broader at higher frequencies) is roughly half the width of the Hamming apodized CrIS response function. 95% of the AIRS spectral response function lies within ± 1 FWHM, which is similar to the analogous statistic for the Hamming apodized CrIS. AIRS is sampled at two points per half width, or roughly twice the spectral density of CrIS. The information content of an unapodized spectrum and Hamming apodized spectrum is identical provided all channels in the band are used (Barnet et al., 2000). Because the Hamming apodized function is highly localized, 0.9/*L* is a better indication of the effective spectral resolution of an interferometer than is 0.6/*L*.

Figure 14.13 shows a sample spectrum observed by AIRS and CrIS. The spectral coverage of AIRS and CrIS is roughly the same. CrIS has 3 bands with the number of channels in each band indicated in the figure, with the sampling, Δv, of each band also indicated on the figure. AIRS is comprised of a number of

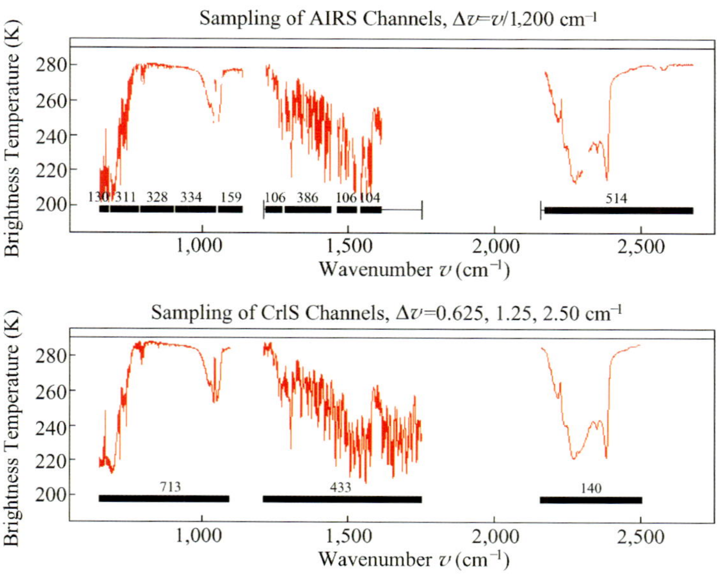

Figure 14.13 Simulated AIRS and CrIS Hamming apodized spectra for a sample scene. Gaps in AIRS spectral coverage relative to CrIS are indicated in the figure

mini-arrays of detectors with the number of channels in each array indicated. Some very small gaps exist between some of the arrays, as indicated in the AIRS figure. Spectral intervals covered by CrIS that are outside the AIRS spectral coverage are also indicated. In addition, AIRS has some "dead" channels, whose radiances are not included in the AIRS spectrum, but whose locations are not indicated by gaps in the black bars. Figure 14.14 shows a blow up of this spectrum from 650 cm^{-1} to 780 cm^{-1}. It is apparent that the spectral lines of the CO_2 absorption band in this spectral region are much better resolved by AIRS than by CrIS.

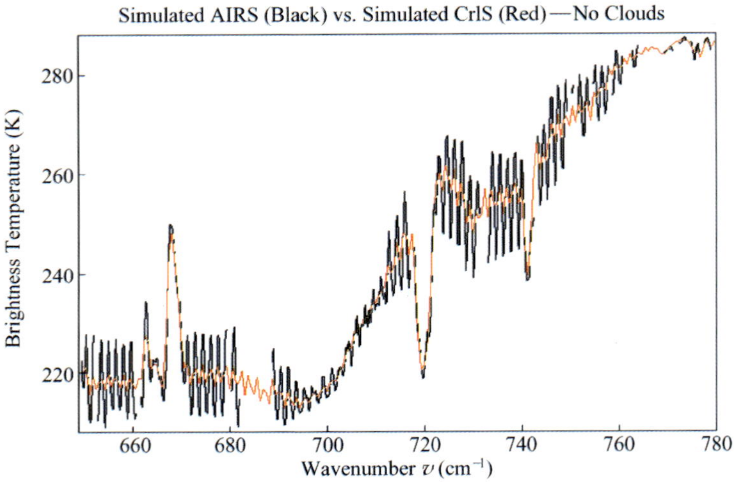

Figure 14.14 Comparison of AIRS and CrIS spectra between 650 cm^{-1} and 780 cm^{-1}

CrIS is comprised of three bands, with $L = 0.8$ cm^{-1}, 0.4 cm^{-1}, and 0.2 cm^{-1}. The spectral intervals of these bands are shown by the black bars in the lower panel of Fig. 14.13, as well as the number of channels in each band. The effective spectral resolution of CrIS, as well as the spectral sampling, is roughly a factor of 2 larger (poorer) than of AIRS in all bands. In band 2, there is little difference in the appearance of spectra of AIRS and CrIS because the water vapor lines are well resolved even with the poorer spectral resolution of CrIS. In band 3, there is also little difference, because neither spectral resolution is sufficient to resolve the lines.

Simulation studies have shown that products of similar accuracy can be obtained from both AIRS and CrIS, even with the poorer spectral resolution of CrIS. This is because CrIS is expected to have better signal to noise than AIRS. Figure 14.15 shows the AIRS flight model channel noise as well as the noise predicted for Hamming apodized CrIS channels (given by 0.6 times the unapodized CrIS channel noise). Figure 14.15 shows that CrIS is expected to have considerably less noise than AIRS. It is the lower noise of CrIS compared to AIRS that compensates for its poorer spectral resolution and spectral sampling, as compared to AIRS, and allows it to achieve soundings of similar accuracy compared to AIRS.

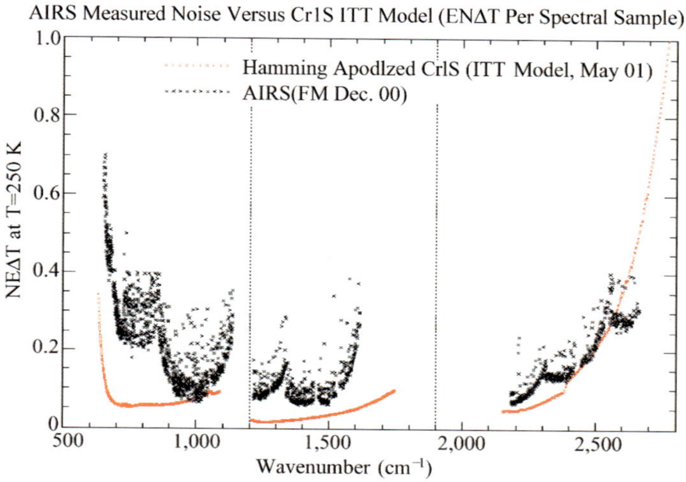

Figure 14.15 AIRS flight model noise and predicted CrIS Hamming apodized noise

14.6 Summary

The physics giving rise to thermal infra-red observations of the earth has been described, as well of the benefits of having a spectral resolving power on the order of 1,000, which resolves spectral features in the 15 μm CO_2 absorption band and 6.7 μm water vapor band, and also allows for very clean atmospheric windows. Both AIRS, which is flying on EOS Aqua, and CrIS, which will fly on NPP in 2006, were designed to take advantage of this high spectral resolution. Previous IR atmospheric sounders, such as HIRS, had resolving powers of the order of 100, and could not resolve individual absorption features.

Most IR observations are affected by clouds in the field of view. In order to best account for the cloud effects on the IR observations, it is optimal to have a microwave temperature profile sounder, such as AMSU-A, flying on the same platform. We have shown that accepted atmospheric soundings produced from AIRS under partially cloudy conditions (about 46% of cases) are only of slightly poorer quality than those produced under clear conditions (about 3% of the cases). Assimilation of AIRS temperature soundings has been shown to improve forecast skill, and a greater improvement results when all accepted soundings are assimilated than when only soundings under clear conditions are assimilated. Similar results should be obtained from CrIS/ATMS when it flies on NPP in 2006.

References

Barnet CD, Blaisdell JM, Susskind J (2000) Practical methods for rapid and accurate computation of interferometric spectra for remote sensing applications. IEEE Trans Geosci Remote Sensing 38: 169 – 183

Brigham EO (1988) The fast fourier transform and its applications. Englewood Cliffs, Prentice-Hall, NJ

Goldberg MD, Qu Y, McMillin LM, Wolff W, Zhou L, Divarkla M (2003) AIRS near-real-time products and algorithms in support of operational numerical weather prediction. IEEE Trans Geosci Remote Sensing 41: 379 – 389

Kaplan LD, Chahine MT, Susskind J, Searl JE (1977) Spectral band passes for a high precision satellite sounder. Appl Opt 16: 322 – 325

Lin SJ, Atlas R, Yeh KS (2004) Global weather prediction and high end computing at NASA. Computing in Science and Engineering: 29 – 35

Mehta A, Susskind J (1999) Outgoing longwave radiation from the TOVS Pathfinder Path A data set. J Geophys Res 104: 12,193 – 12,212

Pagano TS, Aumann HH, Hagan DE, Overoye K (2003) Prelaunch and in-flight radiometric calibration of the Atmospheric Infrared Sounder (AIRS). IEEE Trans Geosci Remote Sensing 41: 265 – 273

Rosenkranz PW (2000) Retrieval of temperature and moisture profiles from AMSU-A and AMSU-B measurements. In Proc IGARSS

Susskind J, Barnet CD, Blaisdell JM (2003) Retrieval of Atmospheric and Surface Parameters from AIRS/AMSU/HSB Data in the Presence of Clouds. IEEE Trans Geosc Remote Sensing 41: 390 – 409

15 The Ozone Mapping and Profiler Suite

Lawrence E. Flynn, Colin J. Seftor, Jack C. Larsen and Philippe Xu

15.1 Introduction

The Ozone Mapping and Profiler Suite (OMPS) is the next-generation US ozone monitoring system, designed for the National Polar-orbiting Operational Environmental Satellite System (NPOESS). The first flight of an OMPS is scheduled for late 2009 on the NPOESS Preparatory Project (NPP) satellite. The OMPS is designed to replace both the NASA Total Ozone Mapping Spectrometer (TOMS) and NOAA Solar Backscatter Ultraviolet Spectrometer/2 (SBUV/2) systems. The OMPS has two instrument modules: a combined Nadir Mapper and Nadir Profiler, and a separate Limb Profiler.

The OMPS was designed to meet the stringent set of performance requirements for atmospheric ozone products detailed in the original NPOESS system specifications. The instruments and algorithms were developed to give performance at least as good as the threshold requirements in Section 4.1.6.2.4 of the Integrated Operations Requirements Document (IORD). A copy of the IORD II is available at http://eic.ipo.noaa.gov/IPO archive/MAN/IORDII_011402.pdf.

The specifications for the ozone Environmental Data Records (EDRs) from OMPS are given in Tables 15.1(a) and 15.1(b). A Dobson Unit (DU) is equivalent to a milli-atmosphere centimeter. This means, for example, that if all the ozone in an atmosphere with a 300 DU column was collected at the Earth's surface at standard temperature and pressure, then it would form a gaseous layer 0.3 centimeters thick. The abbreviation ppmv stands for parts per million by volume. The 7-year stability requirement corresponds to the suite lifetime.

Table 15.1(a) Total Column Ozone EDR Performance

Measurement Parameter	Specification
Horizontal Cell Size	50 KM @nadir
Range	50 DU to 650 DU
Accuracy	15 DU
Precision	3 DU + 0.5%
Long-term Stability	1% over 7 years

Table 15.1(b) ozone profile EDR performance

Measurement Parameter	Specification
Vertical Cell Size	3 km
Vertical Coverage	Tropopause to 60 km
Horizontal Cell Size	250 km
Range	0.1 to 15 ppmv
Accuracy	
Below 15 km	Greater of 20% or 0.1 ppmv
Above 15 km	Greater of 10% or 0.1 ppmv
Precision	
Below 15 km	Greater of 10% or 0.1 ppmv
15 to 50 km	Greater of 3% or 0.05 ppmv
50 to 60 km	Greater of 10% or 0.1 ppmv
Long-Term Stability	2% over 7 years

15.2 Nadir Sensors

Ozone has four main absorption bands in the ultraviolet, visible and near-infrared as follows: the Hartley bands from 200 nm to 310 nm, the Huggins bands from 310 nm to 380 nm, the Chappuis bands from 400 nm to 650 nm, and the Wulf bands from 600 nm to 1,100 nm. The nadir sensor has two spectrometers, one with a wide, cross-track field-of-view (FOV) and spectral coverage in the Huggins ozone absorption bands, and the other with a smaller, nadir FOV and spectral coverage in the Hartley ozone absorption bands. Figures 15.1(a) and 15.1(b) show the ozone absorption cross-sections at a nominal atmospheric temperature for parts of these bands. These cross-sections are from work described in Bass and Paur (1984) and Paur and Bass (1984). (Figure 15.1(c) will be discussed later in the Limb Profiler Sections 15.4 and 15.5.)

Both sensors are designed to make measurements of the ultraviolet radiance backscattered by the Earth's atmosphere and surface (BUV) and of the extra-terrestrial solar irradiance. The radiance/irradiance ratios are the principal quantities used in the ozone retrieval algorithms. The sensor detectors are 2-dimensional Charge-Coupled Device arrays (CCDs) with spatial (cross-track) and spectral dimensions. The detectors are actively cooled to reduce dark currents and radiation damage. Each of the nadir spectrometers samples the spectrum at 0.4 nm with 1-nm Full-Width-Half-Maximum (FWHM) end-to-end resolution. Long-term calibration stability is maintained by periodic solar observations using a working and reference reflective diffuser system similar to the system successfully deployed on the recent TOMS sensors.

The total column sensor has a 110° cross-track FOV and 0.27° along-track slit width. The pixels are summed into 35 cross-track bins. These are 3.35° (50 km) at nadir and 2.84° at ± 55° (120 km). The resolution is 50 km along-track at

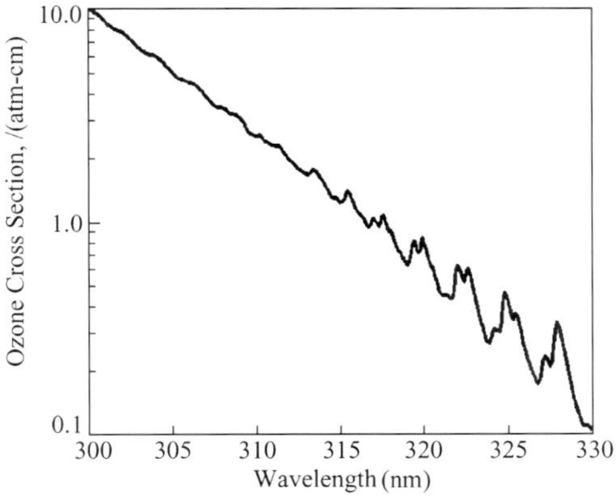

(a)

Figure 15.1(a) Ozone absorption cross-sections in the Huggins bands

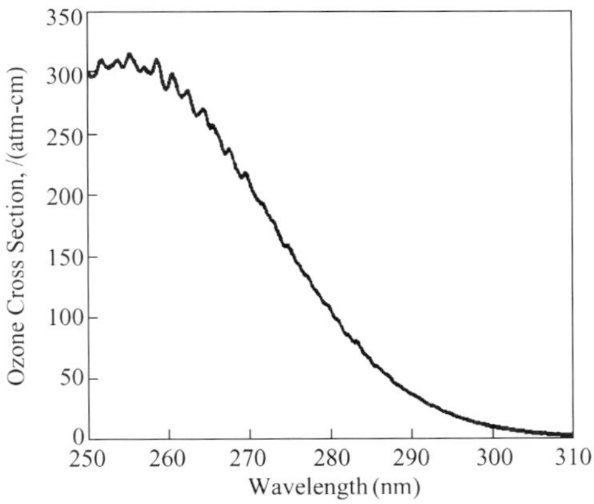

(b)

Figure 15.1(b) Ozone absorption cross-sections in the Hartley bands

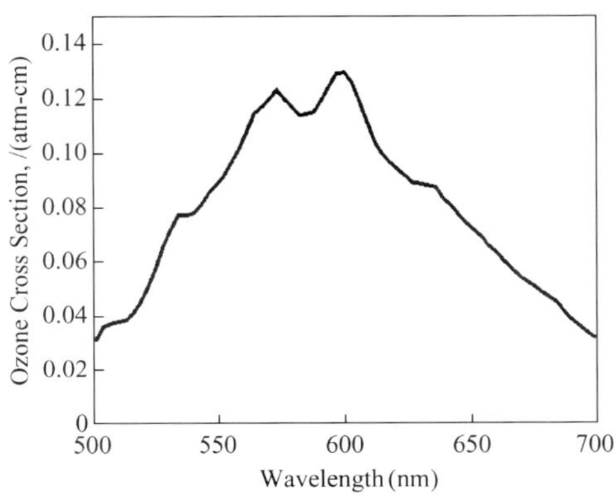

(c)

Figure 15.1(c) Ozone absorption cross-sections in the Chappuis and Wulf bands nadir with a 7.6 second reporting period. The spectral coverage is from 300 nm to 400 nm, with the signal shared with the nadir profiler between 300 nm and 310 nm through the use of a dichroic beam splitter. The nadir profile sensor has a 16.6° cross-track FOV and 0.26° along-track slit width. The pixels are summed into single cross-track bins forming 250 km by 250 km size cells collocated with five central total column cells. The reporting period is 38 seconds synchronized with five reporting periods for the nadir mapper. The spectral coverage is from 250 nm to 310 nm.

15.3 Nadir Retrieval Algorithms

The nadir total column and vertical profile ozone retrieval algorithms build on the algorithms from the heritage systems, TOMS Version 7 (McPeters et al., 1996) and SBUV/2 Version 6 (Bhartia et al., 1996). Information on the nadir and limb algorithms beyond that given here is in the OMPS Algorithm Theoretical Basis Documents (ATBDs) available from links in the IPO Web site at http://npoesslib.ipo.noaa.gov/index.php.

15.3.1 Total Column Ozone Algorithm

Extensive studies and simulations using the TOMS Version 7 algorithm indicated that its performance was close to that needed to meet the NPOESS specifications for accuracy, and that it would provide a strong foundation for meeting NPOESS specifications for precision. The OMPS total column algorithm was therefore based on Version 7 of the TOMS algorithm with enhancements designed to improve its performance to meet the NPOESS requirements.

The algorithm uses triplets of wavelengths to obtain estimates of the total column ozone. Table values computed for a set of standard profiles, cloud heights, latitudes, and solar zenith angles are interpolated and compared to the measured top-of-atmosphere albedos. The triplets combine an ozone insensitive wavelength channel (at 364, 367, 372 or 377 nm) to obtain cloud fraction and reflectivity information, with a pair of measurements at shorter wavelengths. The pairs are selected to have one "weak" and one "strong" ozone absorption channel. The hyperspectral capabilities of the sensor are used to select multiple sets of triplets to balance ozone sensitivity across the range of expected ozone column amounts and solar zenith angles. The "strong" ozone channels are placed at 308.5, 310.5, 312.0, 312.5, 314.0, 315.0, 316.0, 317.0, 318.0, 320.0, 322.5, 325.0, 328.0, or 331.0 nm. They are paired with a longer "weak" channel at 321.0, 329.0, 332.0, or 336.0 nm. Notice from Fig. 15.1(a) that the ozone cross-sections decrease from 3.0 $(\text{atm.cm})^{-1}$ to 0.3 $(\text{atm.cm})^{-1}$ over the range of "strong" wavelengths. Typical ozone columns range from 100 DU or 0.1 atm-cm to 600 DU or 0.6 atm-cm. At the earth's surface, the attenuation of incoming solar flux due to an atmosphere with a total ozone column Ω is approximately given by $exp[-s(\alpha\Omega+\beta p)]$ where s is the path length (approximately the secant of the solar zenith angle), α is the ozone cross section, β is the Rayleigh scattering coefficient, and p is the pressure at the surface. The quantity $exp(-s\beta p)$ is an approximation to the additional attenuation of the direct flux due to Rayleigh scattering. For p equal to 1 atm., βp varies from 1.13 at 305 nm to 0.71 at 340 nm. If the quantity $s(\alpha\Omega+\beta p)$ is large, then the backscattered radiance is mainly composed of photons scattered from high in the atmosphere, and the corresponding channel has little information about the full ozone column. If the quantity $s(\alpha\Omega+\beta p)$ is small, then the radiance has low sensitivity to any ozone changes. By varying the channels as viewing conditions and ozone amounts change, the algorithm is able to maintain good sensitivity to the full ozone column.

The hyperspectral capabilities of the sensor are further exploited through the selection of wavelengths with temperature-insensitive ozone cross-sections for use in the retrievals. Additional atmospheric temperature and ozone profile shape corrections are applied through the use of external temperature information obtained from NPOESS Cross-track Infrared Sounder (CrIS) measurements and external ozone profile information obtained from OMPS's own limb profile retrieval. The corrections are applied based on this external information through the use of first order expansions and retrieval sensitivity tables. If better tropospheric estimates than those in the standard ozone profile climatology are available, e.g., from more current local information or a forecast assimilation model, then a tropospheric ozone correction is applied to account for the inefficiency of the retrievals in sensing tropospheric ozone variations.

Simulated retrievals using synthetic radiances calculated from Stratospheric Aerosols and Gas Experiment (SAGE) ozone profiles and balloonsonde ozone and temperature profiles were generated to demonstrate the improved performance from retrievals using the OMPS algorithm over retrievals using the TOMS V7 algorithm. These simulations have shown that the OMPS system performance is within the NPOESS EDR specifications for total column ozone.

The retrieval algorithm includes calculations to identify atmospheres with high levels of UV-absorbing aerosols, smoke, volcanic ash, and dust. A quantity called the aerosol index is computed based on the differences between the measured versus modeled residuals for the 331 nm and 376 nm channels. This index is used to compute an adjustment to the ozone estimate to correct for wavelength-dependent effects of aerosols on the assumed reflectivity. See Torres et al. (1998) for more information on this correction.

The algorithm also identifies and flags atmospheres with large amounts of SO_2. The 310.5-nm, 312.0-nm, and 329.0-nm wavelength channels are chosen to compute an estimate of atmospheric SO_2. These wavelengths occur either at local maxima or minima with respect to the SO_2/O_3 cross-section ratios and were selected based on the conclusions in Gurevich and Krueger (1997) that such a choice optimizes the retrieval of SO_2. Since determination of a column SO_2 amount requires knowledge of its height distribution, this estimate is just used as an index to identify regions with elevated levels of SO_2, which could result in errors in the ozone estimates.

15.3.2 Nadir Profile Ozone Algorithm

The nadir profiler ozone algorithm is based on the heritage SBUV/2 maximum likelihood retrieval algorithm (Bhartia et al., 1996). The OMPS Nadir Profiler and Nadir Mapper make over 200 measurements across the spectral range covered by the twelve discrete measurements reported by the SBUV/2 instruments. The maximum likelihood algorithm can be adapted to use more measurements to provide a less noisy retrieval. The OMPS Nadir Profile sensor is designed with a double monochromator to give individual channel signal-to-noise ratio (SNR) performance at the SBUV/2 requirement levels, so the algorithm output even with just twelve wavelengths will be as good as or better than the current operations.

The single-scattering, normalized weighting functions are shown in Fig. 15.2. The curves, from top to bottom, are for eight SBUV/2 wavelengths

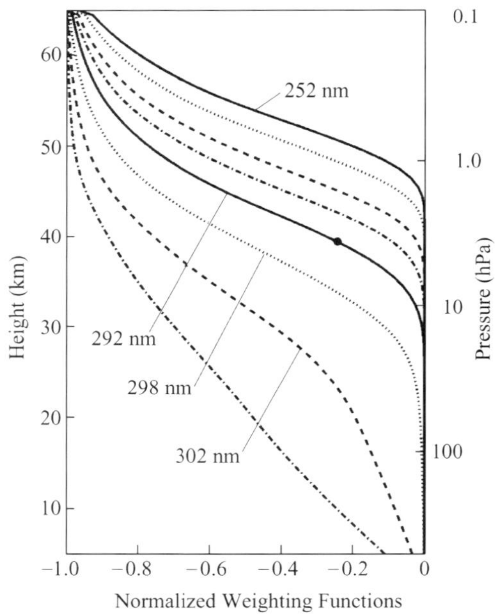

Figure 15.2 Weighting functions for the eight shortest SBUV/2 profile wavelengths—252, 273, 283, 288, 292, 298, 302 and 306 nm—for a standard 325-DU mid-latitude ozone profile at 30° SZA. The curves show the relative sensitivity of each channel to ozone changes as a function of height. Each curve is normalized by the channels absolute sensitivity at the top of the atmosphere, so the curves here give the relative change in the backscattered radiances for a one-unit change in the ozone at the corresponding altitude/pressure

from 252 nm to 306 nm. The curves show the relative penetration and backscatter of BUV photons into the atmosphere as functions of altitude/pressure, and thus their sensitivity to absolute ozone changes at different altitudes/pressures. The topmost dotted curve shows the results for 252 nm. The bottommost dash-dot curve shows the results for 306 nm. As the wavelengths increase, the ozone absorption cross-sections decrease (See Fig. 15.1(b)) and the channels "see" deeper into the atmosphere. For example, the solid dot on the curve at (0.25, 40 km) shows that the 292-nm channel will have 1/4 the response to absolute ozone changes at this height as it does to changes at the top of the atmosphere, i.e., 75% of the backscattered photons measured by a nadir viewing instrument do not make it down to the 40-km level. The weighting functions for the Nadir Profiler measurements produce broad contribution functions for ozone retrievals that cannot give adequate vertical resolution to meet the IORD ozone profile requirements, especially in the lower stratosphere. The Nadir Profiler will provide valuable data for the continuation of the SBUV/2 data record and validation and altitude registration for the Limb Profiler retrievals.

15.4 Limb Profiler Sensor

The limb sensor is based on the limb scattering technique developed for the Shuttle Ozone Limb Scattering Experiment/Limb Ozone Retrieval Experiment (SOLSE/ LORE) (McPeters et al., 2000). Similar measurements are made by the Optical, Spectroscopic and Infrared Remote Imaging System (OSIRIS) and by the Scanning Imaging Absorption Spectrometer for Atmospheric Chartography (SCIAMACHY) and the SAGE III in some of their operating modes. Plate 15.1 shows a view of the

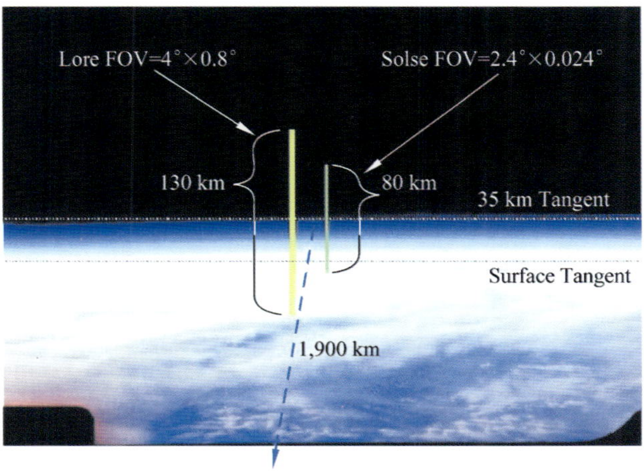

Plate 15.1 View of the Earth's limb from the space shuttle. The OMPS Limb profiler will fly in a higher orbit, so its 1.95° vertical FOV will give altitude coverage midway between SOLSE's and LORE's. This picture is a copy of one available at http://code916.gsfc.nasa.gov/Public/Space_based/solse/results.html

Earth's limb from the space shuttle. The OMPS Limb Profiler (LP) FOV will be similar to the LORE FOV with two additional FOVs spaced 250 km on either side.

Figure 15.3 gives a very simplified representation of some of the types of atmospheric interactions present in limb scattering observations. The dashed line goes perpendicularly from the surface to the limb tangent point. This distance equals the tangent height. The thick solid line represents the field-of-view of a satellite instrument for a specific tangent height. The thin solid lines show some sample paths that photons in incoming solar irradiance may take before being scattered toward the satellite within the line of sight (LOS). In addition to the redistribution by scattering and reflection, photons are lost through absorption, either by ozone or by other constituents. Further attenuation (absorption and scattering) takes place as photons travel along the LOS. The long geometric path of the LOS through the layer above the tangent point is a fundamental component of remote sensing for limb observations. The exponential increase in atmospheric pressure with depth in the atmosphere is another fundamental component, as the density of scatters at the tangent height is larger than that at other points along the LOS.

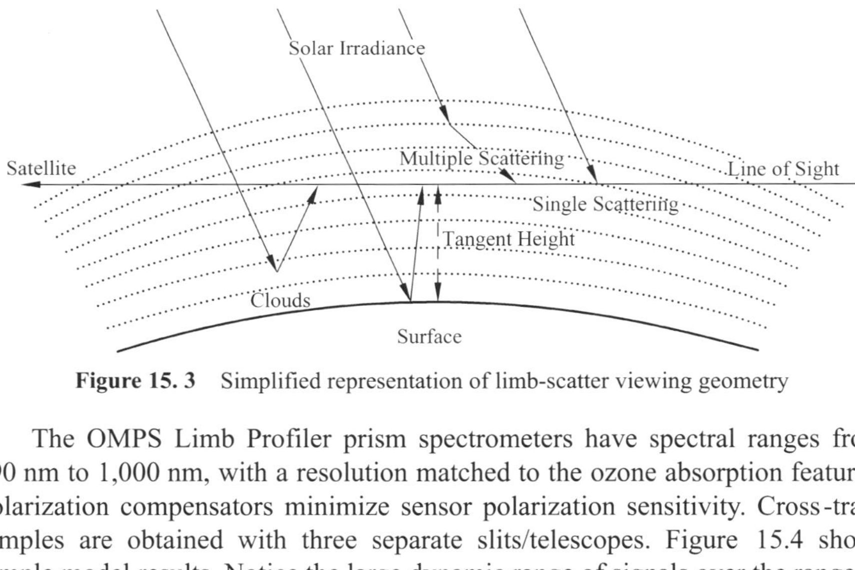

Figure 15. 3 Simplified representation of limb-scatter viewing geometry

The OMPS Limb Profiler prism spectrometers have spectral ranges from 290 nm to 1,000 nm, with a resolution matched to the ozone absorption features. Polarization compensators minimize sensor polarization sensitivity. Cross-track samples are obtained with three separate slits/telescopes. Figure 15.4 shows sample model results. Notice the large dynamic range of signals over the range of tangent heights and wavelengths. The large scene dynamic range is covered by using two separate apertures in each telescope, producing two optical gains, and by using two interleaved integration times, producing two electronic gains. All six spectra (resulting from three slits viewed through two apertures) are captured on a single CCD focal plane. The window above the detector is coated with different filters for the ultraviolet and visible regions of the spectra to reduce stray light.

The Limb Profiler slits are separated by 4.25° (250 km at the tangent points) and have a 38-second reporting period that equates to 250-km along-track motion.

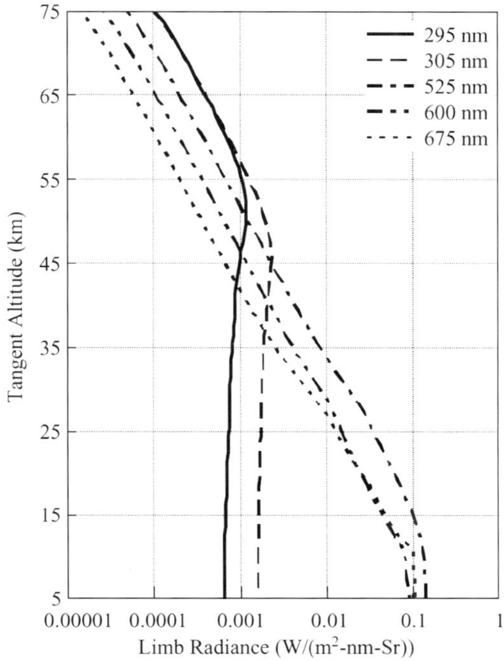

Figure 15.4 Tangent-height dependence of simulated radiances for selected channels for limb scattering observations. The curves give the radiances for different wavelength channels as functions of tangent height in km. The model atmosphere uses a standard, mid-latitude, 325 DU total column ozone profile and a 40° SZA

The slits have 1.95° vertical FOV that equate to 112 km at the limb giving 0 to 60 km atmospheric coverage plus offset allowances for pointing, orbital variation, and Earth oblateness. Individual pixels on the CCD are spaced every 1.1 km of vertical image with a spatial resolution of 2.1 km.

15.5 Limb Profiler Ozone Algorithm

The Limb Profiler retrieval algorithm is based on the limb scattering retrieval technique developed for the SOLSE/LORE measurements described in Flittner et al.(2000). Variations of this algorithm are currently used for some of the limb-scattering measurements made by the OSIRIS (described in von Savigny et al.(2003) and Haley et al.(2004)), SAGE III (described in Rault et al.(2003)), and the SCIAMACHY instruments.

For the OMPS Limb Profiler, the forward model uses the radiative transfer code developed at the University of Arizona (introduced in Herman et al.(1994) and Herman et al.(1995) and described in more detail by Loughman et al.(2004). It generates a multiple scattering solution in a spherical atmosphere, including molecular and aerosol scattering, ozone absorption, and polarization. The retrieval algorithm generates an optimal estimation solution (Rodgers, 1990) using both the Hartley-Huggins bands in the ultraviolet and the Chappuis bands in the visible wavelengths to sense ozone variations. The algorithm uses height- normalized radiances and wavelength pairs and triplets to reduce the effects of reflectivity contributions to the limb scattered radiance and to lessen the impact of instrument calibration and throughput changes. For example, the algorithm normalizes the visible channels by their radiances at 42 km, and then forms "triplets" of these normalized radiances, e.g., one set of triplets is the ratios of the 600-nm channel values to the geometric means of the 525-nm and 675-nm channel values. Notice in Fig. 15.1(c) that the ozone absorption cross-section has a local maximum for the 600 nm channel. Data in Fig. 15.1(c) comes from Burkholder and Talukdar (1994).

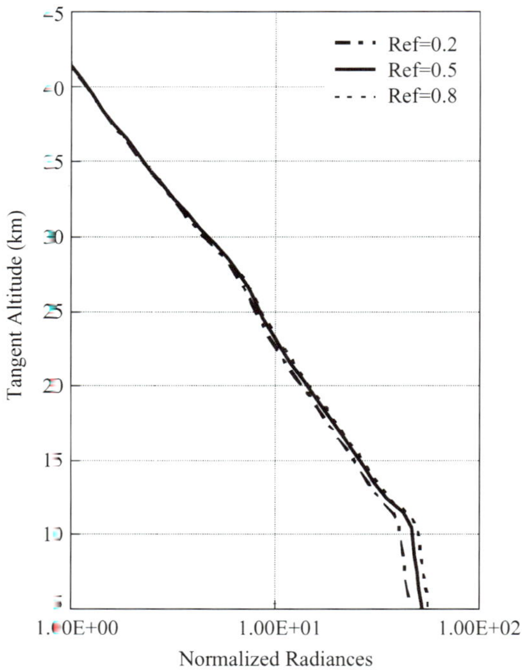

(a)

Figure 15.5(a) Normalized limb radiances as a function of tangent height. The curves show the 600 nm model radiances for a 40° SZA and mid-latitude 325 DU profile for three surface reflectivities cases (20%, 50% and 80%) normalized to their values at 42 km. The radiances at 42 km for the 20% and 80% cases differ from the 50% case by approximately – 25% and + 25%, respectively, prior to normalization

Simulated results are given in Figs. 15.5(a) and 15.5(c). The curves in Fig. 15.5(a) show the 600 nm channel model radiances for a 40° SZA and mid-latitude 325 DU profile for three surface reflectivity cases (20%, 50% and 80%) normalized to their values at 42 km. The radiances at 42 km for the 20% and 80% cases differ from the 50% case by approximately – 25% and + 25%, respectively, prior to normalization. The curves in Fig. 15.5(b) show the triplet results for the same model cases. Figure 15.5(c). shows how well the curves for the three reflectivity cases compare to each other. The values for the 20% and 80% reflectivity cases are divided by the values for the 50% reflectivity case. The thicker curves show the ratios of the triplets for 20% and 80% to the triplets for 50%. Notice that all three triplets stay within 3% of each other, meaning that an error in a modeled reflectivity should have little impact on the accuracy of a modeled triplet.

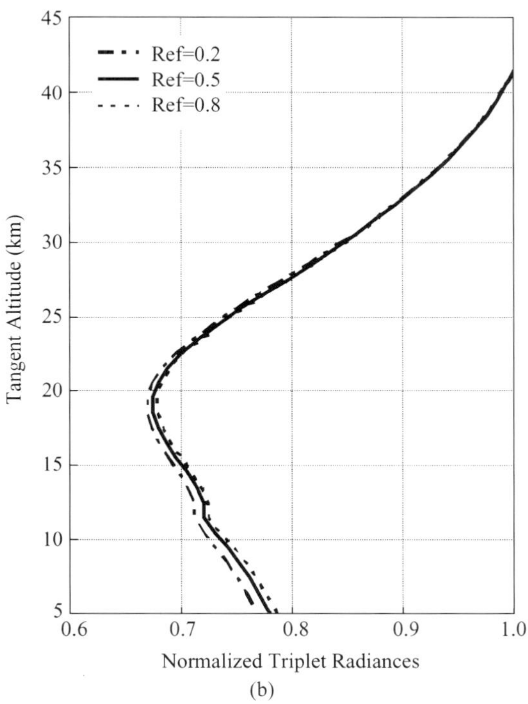

Figure 15.5(b) Visible triplet radiance ratios as a function of tangent height. The curves show the ratios of the 600-nm model radiances for a 40° SZA and mid-latitude 325-DU profile for three surface reflectivity cases (20%, 50% and 80%) to the geometric mean of the similarly computed 675-nm and 525-nm channels' radiances. All three results are further normalized to their values at 42 km

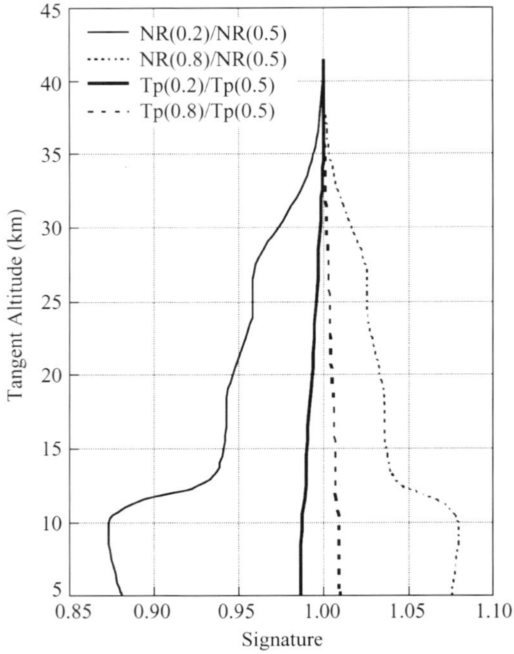

(c)

Figure 15.5(c) Deviations in normalized radiances for reflectivity cases. The curves show the ratios of the 600-nm model radiances for a 40° SZA and mid-latitude 325-DU profile for 20% (thin solid line) and 80% (thin dotted line) surface reflectivity compared to the model results for a 50% reflectivity case versus tangent height in km. The thicker curves show the ratios for the triplet values for 20% and 80% to the triplet for 50%. The triplet ratios are the ratios of the 600-nm channel radiances to the geometric mean of the 675-nm and 525-nm channels' radiances. All the results are further normalized to their values at 42 km

Plates 15.2(a) and 15.2(b) show the sensitivity of limb radiances at two selected channels and a range of tangent heights to ozone changes. Each curve shows how the radiance at a given observation tangent height would respond to relative changes in ozone in 1 km layers over a range of altitudes. The curves are given for tangent heights spaced every two kilometers. The exponential increase in pressure (scatterers) and the geometric increase in the path through the atmospheric layer at the tangent height, combine to give good vertical resolution, with the radiances often most sensitive to ozone changes within the layer at the tangent height. For example, the peak of the dark green curves shows that the radiance for the 49-km tangent height observation changes by -1% in response to a 10% ozone increase in the 1 km layer from 49 km to 50 km. Because of the large attenuation along the LOS, due to both Rayleigh scattering and ozone absorption, the ultraviolet channels lose sensitivity as the tangent height moves down in the atmosphere. Rayleigh scattering increases inversely with the fourth power of the wavelength.

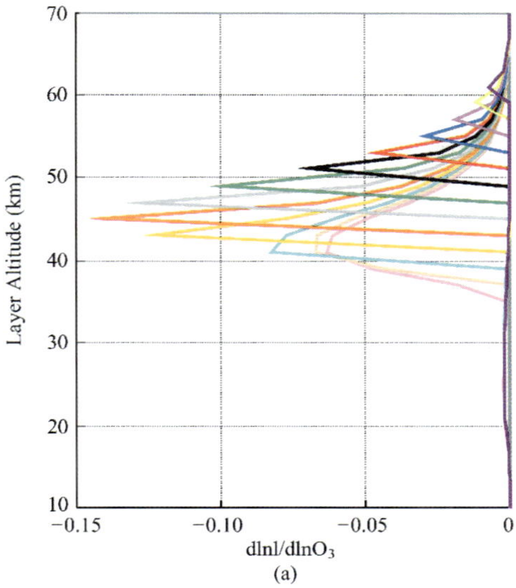

Plate 15.2(a) Sensitivity of the 305-nm channel limb radiances to layer ozone changes for 1 km layers for a 40° SZA and mid-latitude 325-DU profile. Each curve gives the ratio of changes in the limb radiance for a given tangent height to changes in the ozone amounts as a function of altitude. The changes in the natural log of both quantities are used to give a % radiance change / % ozone change interpretation to the results. The curves give radiances for observation tangent heights spaced out every 2 km. The orange curve with the highest peak is for the 45 km tangent height case

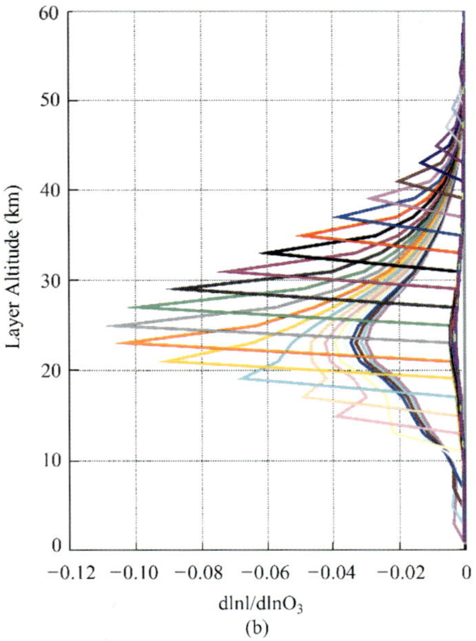

Plate 15.2(b) Same as Plate 15.2(a) except for the 600 nm channel. The grey curve with the highest peak is for the 25-km tangent height case

The information for the ozone profile in the lower stratosphere is obtained from visible channels. The ozone absorption cross-sections for parts of the Chappuis and Wulf bands for the visible channels were shown in Fig. 15.1(c).

The OMPS algorithm has been modified from the SOLSE/LORE basis to include additional ozone channels and a retrieval of the aerosol extinction profile by using wavelengths with little absorption by ozone. The Nadir Profiler ozone estimates are used as first guesses in the retrieval algorithm and to estimate the altitude registration of the limb profile measurements. The result is an iterative, maximum-likelihood retrieval algorithm that will use 26 spectral channels to provide NPOESS ozone profile EDR estimates with vertical coverage from the tropopause to 60 km. Plate 15.3 shows the channel locations and their uses.

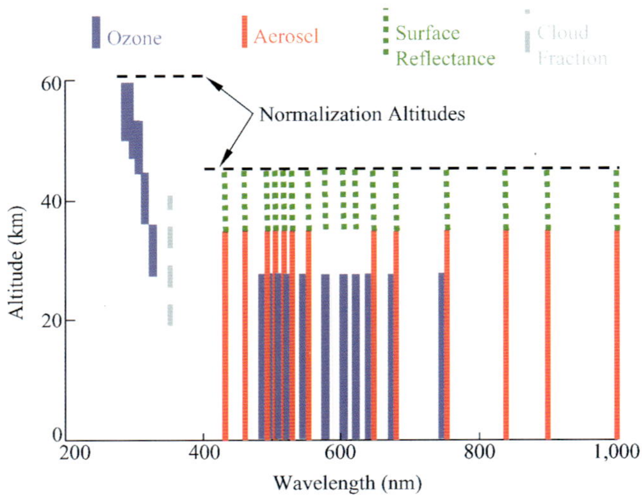

Plate 15.3 Channel positions and uses for the Limb Profiler retrieval algorithm

15.6 Limb Retrieval Challenges

The Limb Profiler retrieval faces several challenges. Three of the most important ones are discussed here. The first major source of error discussed here for the limb retrievals is altitude registration of the measurements. The long distance from the satellite to the limb poses a critical problem for pointing accuracy and precision. Small uncertainties and variations in the spacecraft attitude or alignment are leveraged by the 3,300 km LOS from the spacecraft to the atmospheric limb tangent point. A 10 arc second error in pitch becomes a 160 m error at the limb tangent. The algorithm attempts to estimate the height registration by matching limb radiances calculated by using ozone profile information from the nadir sensor with radiances measured by the limb sensor. Simulations indicate that, even with this correction, there is a 300-m uncertainty in the altitude registration, and ozone retrieval precision errors from 15 km to 28 km are still above the NPOESS requirements. This is primarily due to the errors in regions with large vertical ozone gradients. If the orbital variations are repeatable, it should be possible to reduce the height registration errors even farther through statistical analysis. Methods using characteristics of the UV limb measurements with tangent height, e.g., the Rayleigh Scattering Altitude Sensor (RSAS) technique described in Janz et al. (1996), to determine the height registration are under investigation.

A second source of error occurs due to effects of variations in reflectivity. Unlike occultation measurements where almost all of the photons for an observation at a given tangent height follow the same path through the atmosphere along the entire LOS, the photons in a limb scatter measurement are scattered toward the instrument from different locations along the LOS. The height normalization model computations assume that the measurements at different tangent heights are the result of interactions with spatially homogenous surface and/or cloud reflectivities. There are two main problems with this assumption. One arises from changes in the fields of influence below the LOS paths for different tangent heights. Consider a LOS for a 20-km tangent height and one for a 40-km tangent height with the same subtangent point. The 40-km case will have a wider cross track field of influence than the 20-km case. For example, if there is a single bright cloud 40 km wide directly below the LOS, then this will occupy an angle of ± 45° below the 20-km case, but only ± 30° for the 40-km case. If we compare the photons scattered into a LOS from below for the 40-km tangent height with those for the 20-km tangent height, then the effective reflectivity will differ between the two cases.

The other problem arises from reflectivity differences below the LOS as the distance from the satellite varies. For example, consider the scene depicted in Fig. 15.6, where the line-of-sight the passes from over open ocean to over an ice sheet at the subtangent point. For definiteness, let's take a case where there is a transition from 20% to 80% surface reflectivity at the sub-tangent point.

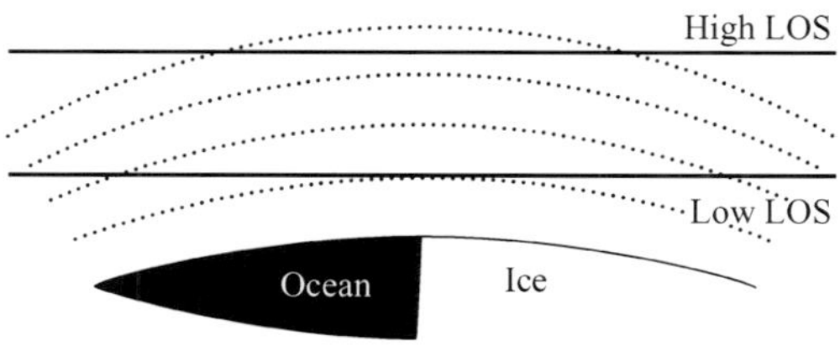

Figure 15.6 Simplified representation of reflectivity transition

As the tangent height decreases, the optical depth from the satellite to the tangent point increases. This means that a greater portion of the limb scattered photons for a low LOS are from scattering in the near part of the LOS than for a high LOS. That is, the measurement for a low LOS is less sensitive to the reflectivity farther from the satellite than the measurement for a high LOS is. Since the two LOS will have different weightings of the 20% and 80% reflectivity regions, a single average reflectivity cannot properly model both heights. The increase in optical depth to a point along the LOS as the tangent height decreases is not the same for different wavelength channels as it depends on absorption and scattering properties. Thus, the real measurements from the atmosphere might not behave with as nice reductions in sensitivity to reflectivity variations as shown in the Fig. 15.5 model results. The normalized radiances for sharp contrast and other varying reflectivity cases can deviate from the homogeneous case as functions of height and wavelength, and without additional modeling, could be interpreted erroneously as ozone or aerosol profile variations. Forward models with varying underlying reflectivity are under development to address these problems.

A third source of error is related to the simplified representation of the 3-D ozone field. The photons scattered toward the satellite within the line of sight travel different distances through the atmosphere along the LOS. This means that they are differentially sensitive to ozone variations along the LOS. The basic retrieval algorithm assumes that the ozone field has local spherical symmetry, that is, that the ozone in a given vertical layer is the same for the paths for the different tangent heights. In reality, for example, the LOS for a 30 km tangent height passes through the 1 km layer between 40 and 41 km over 350 km away from the tangent point (once on the near side and again on the far side). The assumption of spherical symmetry may not give accurate results. The limb scattering arrangement does not even have the balancing of linear ozone gradients present for occultation observations. For occultation measurements, if there is a linear ozone gradient in a layer, then the same photons will pass through both the elevated and lowered concentration regions relative to the layer amount at the tangent. For limb scattering, more photons will pass through the nearer layer than the farther one, adding to the discrepancy between a spherically symmetric model and reality. Multi-pass retrievals and tomographic approaches using 3-D ozone fields have been developed and tested and offer some promise for reducing the errors from this complication.

Measurements from the OMPS Limb Profiler on NPP will be used to research new science and to develop and test solutions for these and other complications. For more information on the current status of the OMPS and NPOESS, please visit the NPOESS Web site at www.ipo.noaa.gov.

Acknowledgements and Disclaimer

The authors wish to formally recognize that the progress on the NPOESS OMPS is the result of extensive research, development, fabrication and programming work performed by dozens of engineers, scientists, programmers, administrators and support staff at the Integrated Program Office, Ball Aerospace & Technologies Corporation, Northrop Grumman Corporation, and Raytheon Company, and by their subcontractors and consultants. This chapter contents are the opinions of the authors and do not constitute a statement of policy, decision, or position on behalf of NOAA or the US Government.

References

Bass AM, Paur RJ (1984) The ultraviolet cross-sections of ozone: I. The measurements in Atmospheric Ozone (Eds Zerefos CS, Ghazi A). Reidel, Dordrecht, Boston, pp 606 – 610

Bhartia PK, McPeters RD, Mateer CL, Flynn LE, Wellemeyer C (1996) Algorithm for the estimation of vertical ozone profiles from the backscattered ultraviolet technique. J Geophys Res 101: 18,793 – 18,806

Burkholder JB, Talukdar JR (1994) Temperature dependence of the ozone absorption spectrum over the wavelength range 410 to 760 nm. Geophys Res Lett 21(7): 581 – 584

Flittner DE, Herman B, Bhartia PK (2000) The retrieval of ozone profiles from limb scatter measurements: Theory. Geophys Res Lett 27: 2,601 – 2,604

Gurevich GS, Krueger AJ (1997) Optimization of TOMS wavelength channels for ozone and sulfur dioxide retrievals. Geophys Res Lett 24: 2,187 – 2,190

Haley CS, von Savigny C, Brohede S, Sioris CE, McDade IC, Llewellyn EJ, Murtagh DP (2004) A comparison of methods for retrieving stratospheric ozone profiles from OSIRIS limb-scatter measurements. Adv Space Res 34, doi: 10.1016/j.asr.2003.08.058

Herman BM, Ben-David A, Thome KJ (1994) Numerical techniques for solving the radiative transfer equation for a spherical shell atmosphere. Applied Optics 33: 1,760 – 1,770

Herman BM, Caudill TR, Flittner DE, Thome KJ, Ben-David A (1995) Comparison of the Gauss-Seidel spherical polarized radiative transfer code with other radiative transfer codes. Applied Optics 34: 4,563 – 4,572

Janz, SJ, Hilsenrath E, Flittner D, Heath D (1996) Rayleigh scattering altitude sensor. Proc SPIE Vol. 2831-45, pp 146 – 153

Loughman RP, Griffioen E, Oikarinen L, Postylyakov OV, Rozanov A, Flittner DE, Rault DF (2004) Comparison of radiative transfer models for limb-viewing scattered sunlight measurements. J Geophys Res 109(D6), doi: 10.1029/2003JD003854

McPeters RD, et al. (1996) Nimbus-7 TOMS Data Products User's Guide. NASA Reference Publication #1384

McPeters RD, Janz S, Hilsenrath E, Brown T, Flittner D, Heath D (2000) The retrieval of ozone profiles from limb scatter measurements: Results from the Shuttle Ozone Limb Sounding Experiment. Geophys Res Lett 27: 2,597 – 2,600

Paur RJ, Bass AM (1984) The ultraviolet cross-sections of ozone: II. Results and temperature dependence, in Atmospheric Ozone (Eds Zerefos CS and Ghazi A). Reidel, Dordrecht, Boston, pp 611 – 616

Rault DF, Loughman RP, Sioris CE (2003) Retrieval of atmospheric ozone and nitrogen dioxide vertical distribution from SAGE III limb scattering measurements. SPIE 10th International Symposium Proceedings, Barcelona, Spain

Rodgers CD (1990) Characterization and error analysis of profiles retrieved from remote sounding measurements. J Geophys Res 95: 5,587 – 5,595

Torres O, Bhartia PK, Herman JR, Ahmad Z, Gleason J (1998) Derivation of aerosol properties from satellite measurements of backscattered ultraviolet radiation: Theoretical basis. J Geophys Res 103: 17,099 – 17,110

Von Savigny C, et al. (2003) Stratospheric O3 Profiles retrieved from limb scattered sunlight radiance spectra measured by the OSIRIS instrument on the Odin satellite, Geophys Res Lett. 30(14), 1,755 – 1,758

The IORD Version II is available in a PDF at http://eic.ipo.noaa.gov/IPOarchive/MAN/IORDII_011402.pdf

Information on the NPP mission is available at http://jointmission.gsfc.nasa.gov/

16 Estimating Solar UV-B Irradiance at the Earth's Surface Using Multi-Satellite Remote Sensing Measurements

John J. Qu

16.1 Introduction

With the continued observation of stratospheric ozone depletion in both the northern and southern hemispheres (Bojkov et al., 1995; Farman et al., 1985; and Soloman et al., 1986), there is mounting interest in the solar ultraviolet radiation (UVR) received at the Earth's surface, especially biologically active solar ultraviolet-B (UV-B). During the last decade, some Earth surface UVR monitoring has been carried out (Scotto et al., 1988; Lubin et al., 1989; Webb, 1992; Booth et al., 1994; Qu et al., 1996; and Bigelow et al., 1998); however, the observations are still limited in duration and areal coverage. Specifically, these measurements have not been sufficient to substantiate or eliminate the possibility of an increase in the solar UVR in the long-term trend of the solar UVR observed on the Earth's surface, except in the Antarctic (Scotto et al., 1988). There are four main weaknesses in the existing data on UV-B: (1) the reliance of the data set on the use of limited broadband UV-B radiometers and spectroradiometers before 1980, (2) most solar UVR measurements have been made in urban areas where column ozone levels were impacted by significant local trends, (3) air pollution; comparisons are lacking among the different UV monitoring networks, and (4) some of the surface UV measurement instruments have not been properly calibrated. Data quality remains poor except for newer programs.

One response to the surface UVR data quality problem is the use of satellite estimation of spectral UV irradiance from remote sensing data, such as the National Aeronautics and Space Administration (NASA) Total Ozone Mapping Spectrometer (TOMS) and UV reflectivity, cloud and aerosol optical depths derived from other satellite instruments (Qu, 1997). Remote sensing of UV-B irradiance from satellite has the potential advantage of giving full daily coverage of spectral UV-B estimates (Eck, et al., 1995). Satellite-based derivations of the Earth surface UVR irradiance have been performed by different scientific groups (Lubin et al., 1998). The budget of spectral UVR between the top reflectance and the Earth surface was estimated using total ozone abundances and the solar backscatter ultraviolet (SBUV) measurement (Frederick and Lubin, 1988). The surface UV irradiances were derived directly from TOMS measurements (Eck, et al., 1995; Herman et al., 1996 and Krotkov et al., 1998). The global surface UVR flux climatology was also estimated using TOMS and NASA Earth Radiation Budget Experiment (ERBE) (Lubin and Jensen, 1995; Lubin et al., 1998).

The alternative method of this study was to use the satellite measurements at the top of the atmosphere, model the radiation transfer through the atmosphere, and test and calibrate the radiation transfer model with surface observations. Thus, the objectives of this study are: (1) to design a simple two-stream atmosphere radiation transfer model to simulate UV-B irradiances at the Earth's surface using remote sensing data from different satellite platforms, (2) to simulate the UV-B radiation attenuation profiles from the top of the atmosphere to the Earth surface (through atmospheric ozone, clouds and aerosols), (3) to examine the UV-B spectral and temporal characteristics of the Bureau of Land Management (BLM) and National Science Foundation (NSF) UV-B surface measurement network sites, (4) to examine the model input parameters that impact the output UV-B irradiances and calibrate the model output using the surface measurements.

In this study, successful simulation of surface UV-B radiation using satellite remote sensing data was demonstrated. Which factor (SZA, cloud, total ozone, Rayleigh scattering or aerosol) affecting the surface UV-B irradiance was most important was determined. Those factors in order of importance in surface UVR attenuation were cloud cover, ozone, Rayleigh scattering and aerosol depth. The relation between the surface UV-B and atmospheric ozone was not a simple linear rule, such as, for 1% decrease in total ozone, a 2% increase in surface UV-B irradiance. Model evaluation revealed that overall, the two-stream model simulated monthly UV-B radiation at the Earth's surface within 6% of the measured values 95% of the time. Good agreements between simulation and the well-calibrated surface UV-B observations, particularly for clear skies, and monthly total values, suggested that the two-stream atmospheric radiative transfer model is a useful tool to simulate surface UV-B variation. There are still conditions where model and observations vary significantly, especially for cloudy days, with high aerosol concentrations and high SZA, such as at high latitudes during winter.

16.2 Satellite Remote Sensing Measurements

16.2.1 Satellite TOMS Ozone and Backscatter Ultraviolet Measurements

The TOMS instrument measures atmospheric normalized radiances through backscatter ultraviolet (BUV) techniques. Dave and Mateer (1967) first demonstrated the feasibility of determining atmospheric ozone from measurements of the backscattered solar ultraviolet radiation at wavelengths near 310 nm. This was the basis of the algorithm later developed by Mateer et al. (1971), and used to obtain total ozone column from the measurements of the BUV experiment on the Nimbus-4 satellite for the initial months of operation. TOMS and SBUV measure ozone using two wavelengths in the ultraviolet, one is ozone sensitive and the other is not. The TOMS sensor has six channels, four of which have little absorption by ozone. The TOMS V7 algorithm took ratios of albedo from pairs of channels and found the ozone value by comparison with output of a multiple scattering radiative transfer calculation (lookup table) (McPeters et al., 1996).

In the mid-1970's NASA embarked on a long term program to monitor ozone and solar UV irradiance at the top of atmosphere (TOA) from space using the backscattering ultraviolet technique using the first TOMS carried by the Nimbus-7 satellite. Nimbus-7 SBUV/TOMS satellite instruments collected total ozone and BUV data from October 1978 to May 1993. In August 1991, the former Soviet Union launched the Meteor-3 satellite carrying a TOMS instrument provided by NASA. The Meteor-3 satellite TOMS instrument ensured continuity of the data when Nimbus-7/TOMS ceased operating in May 1993.

16.2.2 Shuttle Solar Backscatter Ultraviolet Measurements

The Shuttle Solar Backscatter Ultraviolet Experiment (SSBUV) uses a highly-calibrated instrument developed at NASA for periodic flight aboard the space shuttle. SSBUV can measure ultraviolet light from the sun and the amount of that light scattered back by the Earth's atmosphere. The change in radiation between the two measurements is used to calculate ozone levels. The Atmospheric Laboratory for Applications and Science (ATLAS) carries instruments to measure ozone and the other chemicals in the upper atmosphere and to measure the extraterrestrial solar spectrum. The SSBUV has flown 8 times since 1989 with the most recent mission completed with the ATLAS on Space Transportation System-66 (STS-66). Ozone comparisons were also conducted with the ATLAS instruments.

The ATLAS payload and companion instruments made measurements of solar irradiance and middle atmospheric temperatures and trace gas concentrations. The solar irradiance measurements included total and spectrally resolved solar irradiance. The atmospheric measurements include ultraviolet lamp sounding, nadir ultraviolet backscatter, and solar occultation techniques (Kaye and Miller, 1996). The SSBUV spectrometer simultaneously measured the solar spectral irradiance during the ATLAS-1 mission flown on board the Space Shuttle Atlantis in March 1992. The SSBUV data set was reported from 200 to 350 nm at 1.1 nm resolution. The fact that the calibrations of the instruments were based on standards provided confidence that the absolute solar spectral irradiance in the range 200 – 350 nm was now known with accuracy better than ± 5%. SSBUV supports the long-term global stratospheric ozone and solar UV monitoring programs by providing respected checks on the calibrations of the UV ozone and solar monitoring instruments flying on US and international satellites (Cebula et al., 1996). The SSBUV wavelength range is 200 to 405 nm and the bandpass is 1.1 nm. The instrument is identical to the SBUV/2 satellite instruments, except that SSBUV used a transmission rather that the reflection solar diffuser used on the satellite instruments. Each SSBUV solar observation consisted of 6 to 8 complete spectral scans of the sun. The SSBUV solar data were corrected for a small amount of degradation which occurred during flight.

16.2.3 Satellite Cloud Observations

The International Satellite Cloud Climatology Project (ISCCP) was established in 1982 as part of the World Climate Research Program (WCRP) to collect and analyse satellite radiation measurements to infer the global distribution of clouds, their properties, and their diurnal, seasonal, and interannual variations. Data collection began on July 1, 1983. The resulting datasets and analysis products are being used to improve understanding and modelling the role of clouds in climate, with the primary focus being the assessment of the effects of clouds on the radiation balance (Rossow and Schiffer, 1991). The ISCCP cloud D-2 (monthly three-hour average) data sets were used in this study.

Cloud type and cloud amount vary greatly in time and space, and there is no simple measure of the variability. Good cloud observations are representative for a large volume of space for air mass analysis over flat terrain and ocean areas. Interest is growing because of the effect of cloudiness on attenuation of UV radiation. Since cloud influences on radiative transfer of UV radiation is very complex, most recent studies are based on clear skies. Concerns over negative biological responses to attenuated atmospheric ozone are certainly valid, but we should not ignore the potential role of localized changes in cloud cover and cloud optical properties.

The ISCCP data processing group has only processed the data collected before the end of 1993; therefore, the ISCCP D-2 data of 1993 was used in this study. Data include cloud optical depth, cloud amount and surface reflectance. The cloud optical depth is based on the ISCCP D-2 monthly mean of three-hour average cloud optical depth.

16.2.4 Satellite Aerosol Observations

Atmospheric aerosols can affect the UV-B irradiances reaching the Earth's surface by scattering and absorbing. Aerosol optical depth data are from the operational satellite aerosol optical thickness at the National Oceanic and Atmospheric

Administration (NOAA)/National Environmental Satellite Data and Information Service (NESDIS, Stowe, 1991). A validation of the NOAA/NESDIS satellite aerosol product over the North Atlantic in 1989 has been performed (Ignatov et al., 1995). Comparing cloud optical depth ranges from 0.01 to 7.0 in July 1992, the aerosol optical depth ranges from 0.02 to 0.5 in July 1992.

The global monthly mean aerosol optical depth values used in this study were derived from AVHRR measurements from 1989 to 1992. Most aerosol optical depths are less than 0.4 over oceans and aerosol optical ranges are between 0.02 and 0.5 over land. The highest aerosol areas are located in Europe and Asia in the northern hemisphere. The volcanic aerosols from Mt. Pinatubo eruption (Philippines, 1991) may also affect the Earth's surface UV irradiance. SAGE II has operated aboard the ERBE since October 1984. It uses a limb-viewing solar occulation technique (McCormick et al., 1992). SAGE II aerosol profiles are also used in this study.

16.3 Ultraviolet Radiative Transfer Models

UV-B radiation transfer in the atmosphere is very complex. The UV-B reaching the Earth surface depends on many factors, such as atmospheric ozone, aerosols, clouds, and atmospheric absorption and scattering properties. The radiation transfer model chosen for this study can predict upward and downward irradiances at various levels in the atmosphere from the knowledge of the scattering and absorption characteristics of the atmosphere and the extraterrestrial spectrum.

16.3.1 Scheme of UV-B Radiation Model

A two-stream atmospheric radiative transfer model based on a delta-Eddington approximation was designed and used and its performance was evaluated. The model designed in this study takes input parameters of single scattering albedo, cosine solar zenith angle, ozone absorption cross section in inverse Dobson units (DU), and cloud and aerosol optical depths. The model radiation scheme is shown in Fig. 16.1 and the primary input parameters are described below.

Solar Zenith Angle: Solar zenith angle (SZA) is the most important input parameter associated with time (day, season), and geographic location (latitude and longitude). The SZA was measured from 10 to 90 degrees at San Diego during 1993 directly from the SUV-100A spectradiometer used for the surface measurements in the study.

Ozone: Total ozone column is the parameter best known to affect UV-B radiation in the atmosphere. The TOMS total ozone data are from satellite Nimbus-7 (from 1978 to May 1993) and satellite Metro-3 (1991 – 1994). The vertical distribution of ozone in the atmosphere and ozone absorption cross section with wavelength play an important role in the simulation model. For example, Tsay and Stamnes (1992) showed that a redistribution of stratospheric ozone to tropospheric ozone tends to decrease surface level UV irradiance, except for low solar elevations, when an increase of the UV-B irradiances can occur when total column ozone decreases.

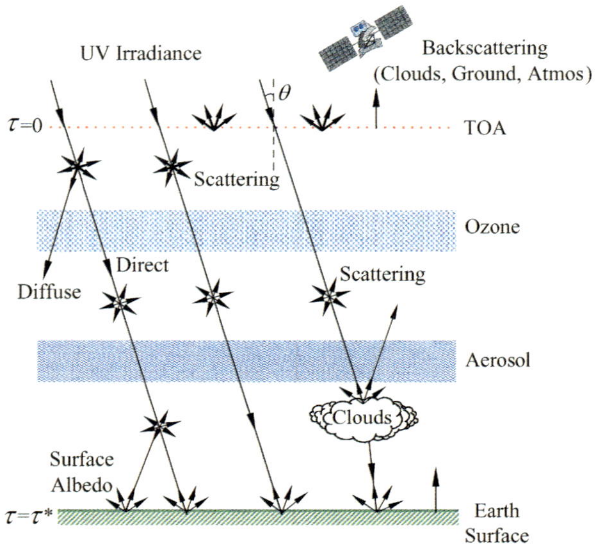

Figure 16.1 UV radiative transfer model scheme (after Qu and Stenphens, 1996)

Rayleigh Scattering: Rayleigh scattering due to gas molecules in the atmosphere is very important in the UV simulations because of the comparable size to wavelength. Scattering cross-sections with 5 nm resolution in the UV bands (WMO, 1986) were used for the model. Single scattering albedo and the asymmetry parameter of cloud particles were based on the use of ADT theory (Stephens and Tsay, 1990).

Surface Albedo: By introducing or altering multiple scattering paths, changing the surface albedo will alter the amount of scattering and absorption taking place in the atmosphere. Thus, not only is there reflection from the surface, but increased backscatter from the layer near the surface (Forster et al., 1995). The Earth's surface albedos in urban areas are between 0.03 and 0.10 (Eck et al., 1987). The surface albedo values were chosen as 0.05 and 0.10 in this study.

The Extraterrestrial Spectrum: The extraterrestrial spectrum data used are for the average sun-earth distance of one astronomical unit. These are extraterrestrial values, rather than top of the atmosphere values, since they do not include reflections from the Earth's atmosphere and the Earth's surface (Madronich, 1992). The high resolution solar irradiance data (from Dr. Cebula consulting, NASA) from the space shuttle ATLAS-1 and ATLAS-2 show the same variations.

The estimated accuracy range is between 3% (400 nm) to 10 % (280 nm) (WMO, 1986).

Clouds and Aerosols: A simplified form of atmosphere aerosols and clouds is currently included in the model. While the actual role of clouds and aerosols in atmospheric transfer is very complex, the simulations used an assumed stratospheric aerosol optical depth of 0.20. This value was based on (Tsay and Stamnes, 1992) which suggested using a simple layer and constraining optical depth values for modelling to values between 0.15 and 0.25. This probably constrains the model results to be applicable to large areas and time periods, rather than points and instaneneous calculations. The cloud optical depth will initially be based on the ISCCP-D2 monthly mean cloud optical depth. The AVHRR background aerosol data were obtained from NASA Langley Research Center. Standard atmosphere vertical temperature, water mixing ratio and ozone mixing ratio profiles were used in this study (NOAA/NASA, 1976).

16.3.2 Two-Stream UV-B Radiation Transfer Models

Two-stream models are simple to implement and can be run rapidly even on a moderate sized personal computer. The accuracy is sufficient for most UV radiation simulation, although the limitation imposed by the assumptions used should be kept in mind (Madronich, 1993). Five kinds of approximations are used in most empirical two-stream models. They are Eddington and modified Eddington, qundrature and modified quadrature methods, hemispheric constant method, delta function methods, and hybrid modified delta-Eddington function approximations (Meador and Weaver, 1980). A two-stream radiative transfer model was developed for the UVR (280 – 380 nm) and infrared (380 – 200 nm) wavelength ranges (Stackhouse and Stephens, 1991). In this study, some input parameters of this two-stream model have been changed, such as, wavelength, and extraterrestral spectrum. Cloud and aerosol information were also added to the model.

The two-stream equations define the energy balance of this thin slab of thickness z. The "+" superscript refers to quantities associated with flowing in an upward direction and "–" superscript refers to quantities relevant to downward flow. The two-stream equations are defined as the energy balance of a small volume of the atmosphere layer. This model is based on the simple two-stream version of the radiative transfer equation and can be written in the common form (Stephens and Tsay, 1990) :

$$\pm\frac{\mathrm{d}F^{\pm}}{\mathrm{d}T} = -[D(1-\bar{\omega}_0)+\bar{\omega}_0 b]F^{\mp} + S^{\mp} \qquad (16.1)$$

for upward F^{+} and downward F^{-} fluxes where D is a measure of the diffusivity of the radiation, $\bar{\omega}_0 b$ is the amount of unit flux along a direction which is backscattered in the opposite direction per unit depth of cloud. Here, $S^{\mp}$ is the source term which represents the diffuse radiation source created within the layer (Stephens, 1995). $\bar{\omega}_0$ is the single scattering albedo (amount of scattering/total extinction). This model simulates of upward and downward irradiances at any level in the atmosphere given different absorption and scattering conditions. The atmosphere is divided into sixty two layers in this study.

16.4 Sensitivity Study

A sensitivity study is important to determine the source of the largest uncertainties associated with the simulation of the Earth's surface UV irradiance. Few studies have examined the sensitivity of the UV simulation in regards to output in comparison to the assumption of existing radiative transfer models. For example, (Frederick and Lubin, 1988; Madronich, 1992; Eck et al., 1995; and Tsay and Stamnes, 1992) did not show the results of sensitivity analyses. The sensitivity of ultraviolet irradiances to changes in atmospheric elements are available (Foster et al., 1995). In studying the sensitivity of simulated surface UV-B irradiance to our model input variables, total ozone, SZA, signal scattering albedo, asymmetric factor, and cloud and aerosol optical depth were considered.

16.4.1 Sensitivity to Solar Zenith Angle

Solar zenith angle (SZA) is the angle the result of the relationship between the location of the Earth and Sun, and the rotation of the Earth. It can, thus, be calculated from the date, time, and geographic location. When the SZA is high (near sunrise or sunset and near midwinter), the optical path through the atmosphere is longest, absorption and scattering are greatest, and surface UVR is lowest. When SZA is at its minimum (near noon during summer), the largest amounts of solar UV radiation can arrive at the Earth's surface. In Fig. 16.2 the ratio of UV-B irradiances to total column ozone at the equator. It can be seen that the UV-B irradiance decreases as the SZA increases. The decrements depend on the total ozone column depths of the atmosphere. If the SZA increases by 10°, the UV-B irradiance decreases 10% for ozone at 325 DU.

16.4.2 Sensitivity to Atmospheric Ozone

The total ozone was varied from 250 DU to 350 DU in the model to test the sensitivity of UV-B simulation to ozone depletion. A ratio was developed for the model of surface UV-B irradiance with different total ozone to the UV-B irradiance with 350 DU total ozone. The results shown in Fig. 16.3 mirror those found by Booth and Madronich (1994). The relationship between ozone depletion and UV-B irradiance is not a linear rule such as "for 1% decrease in ozone, a 2% increase in UV expected" (Booth and Madronich, 1994).

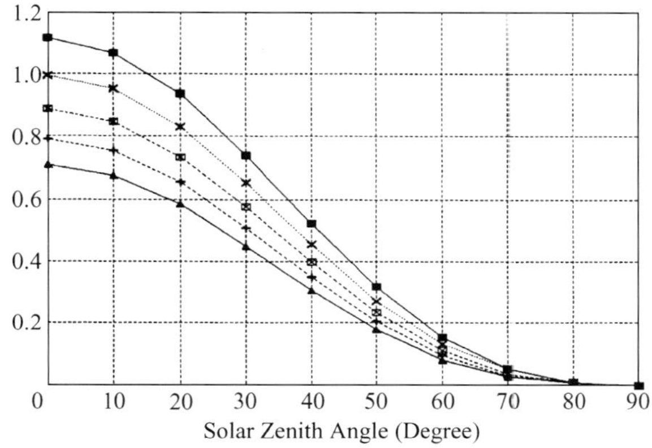

Figure 16.2 The ratio of UV-B irradiances as a function of SZA, model simulation results

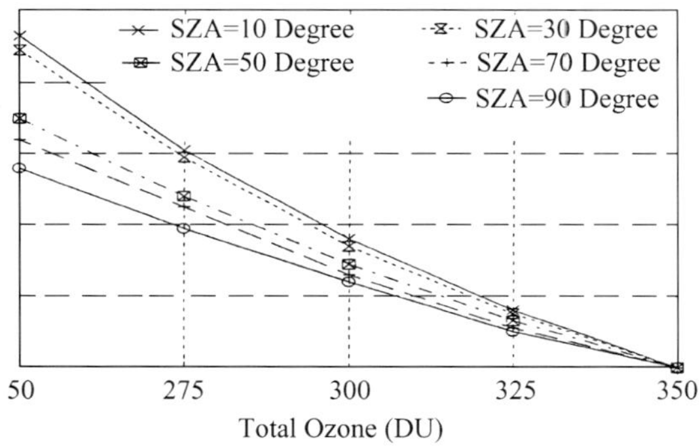

Figure 16.3 The ratio of surface UV-B irradiance as a function of atmospheric total ozone column (model simulation results)

16.4.3 Sensitivity to Surface Reflectivity

This part of the sensitivity study was done letting the surface UV reflectivity range from 0 to 1.0, with fixed 300 DU total ozone column and 30-degree SZA. As the surface albedo is increased, the UV-B irradiance increases. Figure 16.4 depicts both upward and downward UV-B irradiance ratios at the Earth's surface, where if surface albedo is 0, all UV-B is absorbed by the surface and the upward irradiance is 0. Whereas if the surface albedo is 1.0, no UV-B is absorbed by the surface, then upward UV-B irradiance equals downward UV-B irradiance at the surface. The importance of surface albedo can be seen as with increasing surface albedo, the atmospheric scattering layer above the surface will backscatter some of the surface reflected UV, increasing the downward UV-B irradiance.

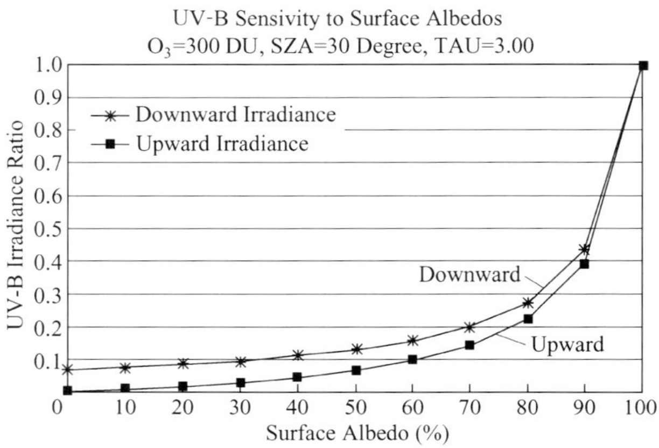

Figure 16.4 The ratio of surface UV-B irradiance bas a function of surface reflectivity (model silumation results), the atmospheric total ozone column is 300 DU, the SZA is 30 degree and the surface air pressure is 1,013 mb

16.4.4 Sensitivity to Cloud Optical Depth

It is well known that clouds can attenuate the surface UV-B irradiance, but the effect is not a simple process. The only cloud case included in this study is that for homogeneous cloud layers (Madronich, 1993). Broken clouds still pose a near insurmountable problem for atmospheric radiative transfer modelling. Cloud optical depth is, however, one of the important input parameters to the model used. Cloud optical depths ranging from 0 – 10 for 10, 30, and 60 degrees SZA were chosen for analysing the surface UV-B irradiance. The values chosen were based on monthly cloud optical depth measurements of ISCCP D2. The monthly average cloud optical depth range for the western United States was between 0.4 and 4.0 in July of 1992 (Qu et al., 1997).

Simulation results show the ratio of the surface cloudy UV-B irradiance to clear sky UV-B irradiance as a function of cloud optical depth with 325 DU atmospheric total ozone, 0.05 surface UV reflectivity, 0.999999 single scattering albedo, and a 0.85 asymmetric factor used. It can be seen that the surface UV-B irradiance decreases with increasing atmospheric cloud optical depth. Surface

UV-B irradiance ratios decrease with increasing SZA. When the SZA is 10° and the cloud optical depth is 1, the ratio of surface UV-B irradiance is approximately 0.4. This means that about 60% of the UV-B is attenuated by clouds and does not reach the surface. This relationship indicates that UV-B irradiance penetrates uniform clouds during the low SZA (or high solar elevation), but there is a lot of reflection with high SZA (or low solar elevations). Since real world cloud patterns are complex, not homogenous, with different types of clouds prevailing at different times and locations, this may not be a representative condition.

16.4.5 Sensitivity to Atmospheric Aerosols

Surface UV-B irradiance can be reduced by local, near surface atmospheric pollution. Absorbing pollutant gases, such as NO_2 and SO_2 can decrease UV-B radiation by a few percent in heavily polluted urban areas. Qu J (1997) determined the daily DNA dose sensitivity reduction due to aerosols. They concluded that the daily DNA dose decreases as the ground-level visual range decreases. To show the sensitivity of the simulated surface UV-B irradiance to the atmosphere aerosol concentration, aerosol values from 0 to 0.4 were chosen. The ratios of the surface UV-B irradiance with aerosols to the irradiance without aerosols as a function of aerosol optical depth were plotted. For this simulation the atmospheric total ozone column was 325 DU, the surface UV albedo was 0.05, and cloud optical depth was 0.0 (clear skies). The probability density distributions of the differences of the model simulation and measured total UV-B irradiances are plotted in Fig. 16.5. The differences between surface UV-B irradiances estimated from the two-stream radiative model with satellite derived inputs and the surface measurements are within about $\pm 6\%$ for 95% of the time. In Fig. 16.5, the reduction of the surface UV-B irradiance was related to both aerosol optical depth and solar zenith angles. Surface UV-B radiance decreases rapidly with aerosol during low SZA or high sun elevations. If the cloud optical depths were added, the sensitivity would have been different, because of the interactions between clouds and aerosols.

To estimate the effects of clouds, ozone, Rayleigh scattering and aerosols on surface UV-B radiation, a surface monthly average UV-B attenuation index (UVBAI) is introduced. The index is defined as:

$$\text{UVBAI}=\frac{\Delta \text{UVB} \bullet \tau}{\Delta \tau} \bullet \frac{1}{1{,}000} (\text{J/m}^2/\text{day}) \tag{16.2}$$

where Δ UVB is the net monthly UV-B at the surface. τ is optical depth and $\Delta\tau$ is the optical depth change. 0.2 is used as $\Delta\tau$ in this study. Table 16.1 shows the results of monthly net surface attenuation indexes as a function of optical depths of clouds, ozone, Rayleigh scattering and aerosols. Cloud UVBAI ranges from 17.60 to 50.00. Ozone UVBAI ranges from 4.5 to 15.5. Rayleigh scattering UVBAI ranges from 2.05 to 8.7.

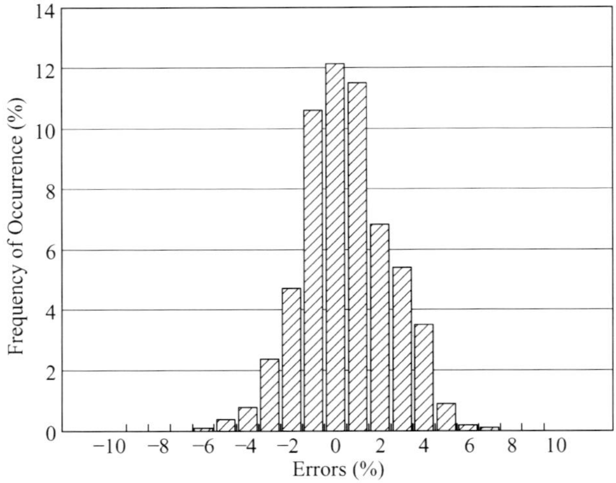

Figure 16.5 Histogram of the percentages of the errors in the surface UV-B irradiance. Results between the two-stream radiative transfer model and the surface measurements

Table 16.1 Net surface monthly UV-B attenuation indexes as functions of optical depths of clouds, total ozone, Rayleigh scattering and aerosols and SZA ($\Delta\tau$ =0.2)

SZA (Degree)	Cloud τ = 2	Cloud τ = 5	Ozone τ = 0.8	Ozone τ = 1.6	Rayleigh τ = 0.5	Rayleigh τ = 1	Aerosol τ = 0.2	Aerosol τ = 0.4
20	17.62	49.76	4.58	13.08	2.06	6.55	0.38	1.39
40	18.53	49.93	5.18	14.02	2.40	7.29	0.46	1.63
60	19.63	50.00	6.38	15.35	3.16	8.65	0.66	2.20

Aerosol UVBAI ranges between 0.35 and 2.21. These results show that cloud, ozone, Rayleigh scattering and aerosol have different attenuating effects on surface UV-B radiation. Clouds play the most important role in attenuating UV-B radiation. Ozone is the second most important factor, Rayleigh scattering is third and aerosol is the fourth, in terms of the magnitude of effects on surface UV-B radiation.

To determine the overall uncertainties that might be expected in estimating the surface UV-B irradiance, the two-stream simulations were conducted for a variety of conditions within the range of observations of a particular observation site. This involved about 80 combinations of model input variables with three

TOMS ozone types, three SZA types, three cloud optical depth types and three aerosol depth types. The simulations made to correspond to monthly total UV-B measurements made at Bear Trap, Montana, for July 1994. The monthly average total ozone and cloud optical depths were used as standard imputs. Then measurement errors, such as SZA with ± 1%, TOMS total ozone and ozone cross section with ± 3%, cloud and aerosol with ± 3%, were added to model input parameters and variables to run the model again.

16.5 The Effects of Clouds and Aerosols on UV-B Irradiance

Clouds cause complex changes in the UV radiation energy exchanges at the TOA, within atmosphere and at the surface because of large variations of their properties over a wide range of space and time scales. A change in each cloud property affects all the UV components. Cloud properties affect the radiation balance differently, so that understanding the link between cloud formation-decay processes and the radiation balance depends on examining variations of the individual radiative flux components together with a change in each of the key cloud properties. Lubin and Jensen (1995) used TOMS ozone abundance and ERBE cloud opacity to retrieve a preliminary estimate of the Earth's surface UVR budget.

16.5.1 The Effects of Cloud on the Surface UV-B Irradiance

The ISCCP cloud data are needed for the global UV-B radiation simulations. The two-stream atmospheric UV radiative transfer model described in Section 16.5 was employed to estimate UV-B irradiance at the Earth surface using ISCCP cloud data, TOMS total ozone, and satellite aerosol as input data. The relationship among the surface UV-B irradiances, atmospheric total ozone column, clouds and aerosols were already discussed in Section 16.3. Clouds and tropospheric aerosol attenuate the UV-B irradiance, with that effect depending little on wavelength. However, different types of clouds and aerosols affect the surface UV-B irradiance in different ways. Yet, very few studies that consider the UV radiative properties of clouds exist. The role of clouds on UV erythemal irradiance using UV data from the R-B meter network and a simple radiative transfer model was studied (Frederick and Snell, 1990). They found that if local fractional cloud cover during June and July varied by 10% from the monthly average value, erythemal irradiance at different sites was changed (in the oposite direction) by amounts ranging from 1.2% to 6.4%. An atmospheric radiation transfer model was used to study the combined effects of ozone depletion/ redistribution and particulate clouds on atmospheric heating rates and ultraviolet radiation reaching the biosphere (Tsay and Stamnes, 1992). Four types of particulate clouds prevalent in the summertime arctic were considered: Stratospheric aerosols, troposphere aerosols (arctic haze), cirrus clouds, and stratus clouds. They found that clouds affect both the UV and visible spectral regions. Stratus clouds provided a substantial shield from UV exposure; while stratospheric aerosols increased UV radiation exposure. Troposphere aerosol (arctic haze) resulted in a decrease in UVR.

16.5.2 The Effects of Aerosol on the Surface UV-B Irradiances

Atmospheric aerosols can affect the UV-B irradiances reaching the Earth's surface scattering and absorbing. The aerosol optical depth data are from the operational satellite aerosol optical thickness at the National Oceanic and Atmospheric Administration (NOAA)/National Environmental Satellite Data and Information Service (NESDIS) (Stowe, 1991). A validation of the NOAA/NESDIS satellite aerosol product over the North Atlantic in 1989 was done by Ignatov et al. (1995). Spectral aerosol optical depth retrieval can be determined using directly transmitted solar radiation as demonstrated (Stephens, 1994; Dutton, 1994). Total aerosol optical depth can be defined in several ways, but it may be most illustrative to refer to Eq. (16.3) which describes the total attenuation of a beam of direct radiation.

$$T_A(\lambda) = \ln\left[\frac{I_2(\lambda)}{I_1(\lambda)}\right] \tag{16.3}$$

Here $T_A(\lambda)$ is total aerosol optical depth. $I_1(\lambda)$ is the incident monochromatic intensity and $I_2(\lambda)$ is the exiting monochromatic intensity from a vertical path through aerosol only (Dutton, 1995). It is possible to find the $T_A(\lambda)$ of clear skies from observations of direct solar beam using photometric techniques (Dutton, 1994). Compared to the cloud optical depth range (from 0.01 to 7.0 in July 1992), the monthly aerosol optical depth range was from 0.02 to 0.4.

16.5.3 Model Calibration

Because the accuracy of UV-B measurements on the Earth's surface is limited, existing surface UV-B monitoring networks cannot provide nation-wide UV surface irradiance values. The two-stream solar UV-B radiation transfer model was used to simulate the Earth's surface monthly average daily total UV-B using a standard ozone vertical profile from TOMS total ozone data. Cloud optical property data were from Stephens' work (1979).

Table 16.2 shows the Rayleigh scattering, ozone, cloud and aerosol optical depths from satellite measurements or calculations for July 1992 in the western

United States (Qu et al., 1997). The ozone optical depth is a function of total ozone amount. The average total ozone amount (325 DU) was used for the simulation. Data (Table 16.2) suggest that cloud cover is the most important property in UV transfer, second is the total ozone column, and the atmospheric aerosol is the least important factor in UV transfer.

Table 16.2 Optical depths of cloud, Rayleigh scattering, ozone and aerosol of July, 1992

	Low	High	Mean
Ozone (325 DU)	0.2	1.6	0.9
Rayleigh Scattering	0.4	1.2	0.8
Cloud	0.4	4.0	2.2
Aerosol	0.02	0.5	0.26

To calculate surface UVR for cloudy days with aerosol, three levels in the atmosphere were used in the model simulation. Ozone was put in high level, cloud in middle and aerosol in lower level. Cloud height was estimated as 5 km. The global monthly cloud and aerosol data were also used for these simulations. The ISCCP D2 cloud optical depth data for 1992 were used because the 1993 data were not available. Monthly average climatic aerosol optical depths were also used in these simulations.

Table 16.3 shows the comparison test results between the simulation using satellite and remote sensing data and surface UV-B observations at sites in New Mexico, Wyoming, and Idaho. There is good agreement between the simulation and the surface observations for monthly total UV-B radiation. There were still some differences during summer (low SZA), but larger differences occurred during winter (high SZA). It may indicate that there are more errors in both measured and simulated data during high SZA, such as winter at Yakutsk, Russia and Fairbanks, Alaska (not shown). The Earth surface high-resolution UV-B observations in the spectral 280 – 320 band were compared with the two-stream atmospheric radiative model simulation results at four BLM sites (Bear Trap, MT; Fairbanks, AK; El Malpais, NM; Yakutsk; Russia) and NSF's sites (San Diego, CA; Barrow, AK) for both clear days and cloudy days with aerosol and cloud

Table 16.3 Simulation vs. Observation of July 1993 Monthly Total UV-B (kJ/m^2/month)

Station	Model	Observation
Snowy Range, WY	2,119.7	2,052.0
Little Wood River, ID	1,791.1	1,789.1
El Malpais, NM	1,900.4	1,881.8
Trapper Creek, WY	1,891.1	1,867.4

information deduced from satellite measurements. A study whether the calculated the Earth's surface UVR using TOMS, ISCCP, and SAGE II remote sensing data for BLM different remote sites shows the same trend with that measured by NSF's UV Spectroradiometer. These results confirm that a simple two-stream radiation transfer model can be used to simulate the climatological UV-B radiation at ground level.

16.6 Summary and Conclusions

The clear sky UV-B irradiances are simulated in this study as functions of SZA and total ozone. To understand the relationship between ozone depletion and UV-B radiation as a function of solar elevation, integrated UV-B irradiances were used to compare the model results and measurements (Fig. 16.6). For detecting the UV-B irradiance variations with SZA, values for the whole year of 1993 were chosen. Totalled together there are 5,638 observational data points for 1993 at San Diego.

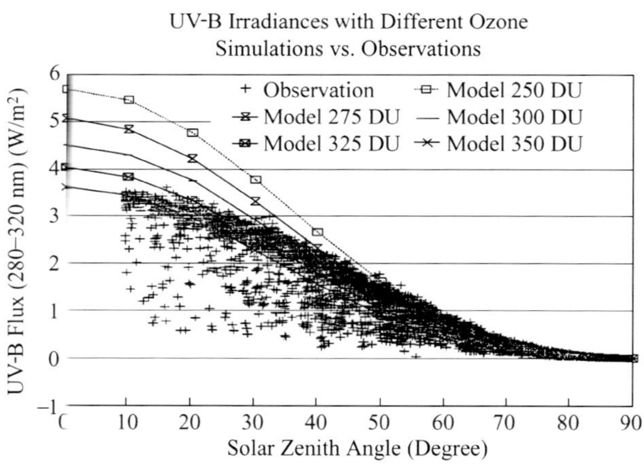

Figure 16.6 The Earth surface UV-B irrandiance with different ozone. The comparison between model simulations and surface observations. The surface measurements were taken at San Diego, CA, 1993

This study has presented results from a wide-ranging model simulation and comparison studies between modelling calculations and well-calibrated surface measurements. Successful simulations of surface UV-B radiation using satellite remote sensing data were demonstrated. The most important factors (e.g. SZA, cloud, total ozone, Rayleigh scattering and aerosol) affecting the surface UV-B irradiance were also evaluated. The most important internal factor in surface UVR attenuation is cloud cover, the second is ozone, the third is Rayleigh scattering and the fourth is aerosol. The relation between the surface UV-B and the atmospheric ozone is not a simple linear rule, such as, for 1% decrease in total ozone, a 2% increase in surface UV-B irradiance. The Earth's surface UV-B irradiance can be estimated using satellite measurements. Model evaluation revealed that overall, the two-stream model simulated monthly UV-B radiation at the Earth's surface within ± 6% of the measured values 95% of the time. As expected, model performance was best for clear days and poorer for cloudy days. Nevertheless, the research demonstrated that the two-stream model could reliably simulate UV-B radiation at the Earth surface, provided with satellite remote sensing data and other model input parameters. This research concluded that further improvements in simulating UV-B radiation at the Earth's surface will require greatly improved remote sensing data for clouds and aerosols.

Acknowledgements

This research is supported by the United States Department of the Interior, USGS. ISCCP cloud data and SAGE II aerosol data used in this study research was obtained from the Langley Research Center Distributor Active Center (DAAC). TOMS total ozone data was retrieved from the Goddard Space Flight Center. The UV radiative spectral and SZA data at San Diego were obtained from NSF Polar Program UV Spectroradiometer network. We would like to express sincere gratitude to Professor Graeme Stephens for providing his radiative transfer model code.

References

Bigelow DS, Slusser JR, Beaubien AF, Gibson JH (1998) The USDA ultraviolet radiation monitoring program. Bulletin of the American Meteor Soc 79(4): 601 – 625

Bojkov RD, Fioletov VE, Balis DS, Zerefos CS, Kadygrova TV, Shalamjansky AM (1995) Further ozone decline during the northern hemisphere winter-spring of 1994 – 1995 and the new record low ozone over Siberia. Geophys Res Lett 22: 2,729

Booth CR, Madronich S (1994) Radiation amplification factors: improved formulation accounts for large increases in ultraviolet radiation associated with Antarctic ozone depletion. Antarctic Research Series. CS Weiler and PA Penhale (eds) 62: 39 – 42

Cebula RP, et al. (1996) Observation of the solar irradiance in the 200 – 350 nm interval during the ATLAS-1 mission: a comparison among three sets of measurements-SSBUV, SOLSPEC and USIM. Geophys Res Lett 23(17): 2,289 – 2,292

Correll DL et al. (1992) Spectral ultraviolet-B radiation fluxes at the Earth's surface. J Geophys Res 97: 7,579 – 7,591

Dave JV Mateer CL (1967) A preliminary study on the possibility of estimating total atmospheric ozone from satellite measurements. J Atmos Sci 24: 414 – 427

Dutton EG (1995) Troposheric radiative forcing from El Chichon and Mt. Pinatubo: Theory and observations. PhD thesis, Colorado State University: 1 – 208

Dutton EG (1994) Features and effects of aerosol optical depth observed at Mauna Loa, Hawaii: 1982 – 1992. J Geophys Res 99: 8,295 – 8,306

Eck TF, Bhartia PK and Kerr JB (1995) Satellite estimation of spectral UVB irradiance using TOMS derived total ozone UV reflectivity. Geophy Res Lett 22(5): 611 – 614

Eck TF et al. (1987) Reflectivity of Earth's surface in ultraviolet observation. J of Geophys Res 92: 4,287 – 4,296

Farman JC, Gardinar BG, Shaklin (1985) Large losses of total ozone in Antarctica reveal season CLO_x/NO_x interation. Nature 315: 207 – 210

Frederick JE Lubin D (1988) The budget of biologically active ultraviolet radiation in the earth-atmosphere system J Geophys Res 93: 3,825 – 3,832

Frederick JE Snell HE (1990) Troposphere influence on solar ultraviolet radiation: The role of clouds. J Climate 3: 373 – 381

Forster M, Piers DeF (1995) Modeling Ultraviolet Radiation at the Earth Surface, Part I: The Sensitivity of Ultraviolet Irradiances to Atmospheric Changes. Journal of Applied Meteorology 34: 2,412 – 2,425

Forster M, Piers DeF (1995) Modeling Ultraviolet Radiation at the Earth Surface, Part II: Model and Instrument Comparison. Journal of Applied Meteorology 34: 2,426 – 2,439

Gleason J, Bhatia P, Herman J, McPeters R, Newman P, Stolarski R, Flynn L, Labow G, Larko D, Seftor C, Wellemeyer C, Komhyr W, Miller A, Planet W (1993) Record Low Global Ozone in 1992. Science 260: 523

Herman RJ, Bhartia PK, Ziemke J, Ahmad Z, Larko D (1996) UV-B increases (1979 – 1992) from decreases in total ozone. Geophy Res Lett 23: 2,117 – 2,120

Ignatov MA, et al. (1995) Validation of the NOAA/NESDIS satellite product over the North Atlantic in 1989. J Geophys Res 100(D3): 5,123 – 5,132

Josefsson WAP (1993) Monitoring Ultraviolet Radiation. In: Young AR, Bjorn LO, Moan J, Nultsch, W (eds) Environmental UV Photobiology, Plenum Press, New York, pp 73 – 88

Kaye JA, Miller TL (1996) The ATLAS series of shuttle missions. Geophy Res Lett 23: 2,285 – 2,288

Krotkov NA, Bhartia PK, Herman JR, Fioletov V Kerr J (1998) Satellite estimation of spectral surface UV irradiance in the presence of tropospheric aerosols. J Geophys Res (Atmos) 103(D8)8: 779 – 8,793

Lubin D, et al. (1989) Measurements of enhanced springtime ultraviolet radiation on Palmer station, Antarctica. Geophys Res Lett 16: 783 – 785

Lubin D, Jensen EH (1995) Effect of clouds and stratospheric ozone depletion on ultraviolet radiation trends. Nature 377: 710 – 713

Lubin D, Jensen EH, Gies HP (1998) Global surface ultraviolet radiation climatology from TOMS and ERBE data. J Geophys Res (Atmos) 103(26): 61 – 91

Madronich S (1992) Implication of recent total atmospheric for biologically active ultraviolet radiation reaching the Earth's surface. Geophys Res Lett 19: 37 – 40

Madronich S (1993) Atmosphere and UV-B radiation at ground level. In: Young AR, Bjorn LO, Moan J, Nultsch W (eds) Environmental UV Photobiology, Plenum Press, New York, pp 1 – 40

Mateer CL, Heath DF, Krueger AJ (1971) Estimation of total ozone from satellite measurements of backscattered ultraviolet earth radiance. J Atmos Sci 28: 1,307 – 1,311

McCormick MP, Veiga RE, Chu WP (1992) Stratospheric ozone profile and total ozone trends derived from the SAGE I and SAGE II data. Geophys Res Lett 19: 269 – 272

McPeters RD, et al. (1996) Nimbus-7 total ozone mapping spectrometer (TOMS) data products user's guide. NASA Ref Publ pp 1,384

Meador WE, Weaver WR (1980) Two-stream approximations to radiative transfer in planetary atmospheres: A unified description of existing methods and a new improvement. J Atmos Sci 37: 630 – 643

Molina LT, Molina MJ (1986) Absolute absorption cross sections of ozone in the 185 to 350 nm wavelength range. J Geophys Res (Atmos) 91(14): 501 – 508

NOAA/NASA/United State Air Force (1976) US Standard Atmosphere, US Government Printing Office 1976

Qu J, Smith F, Riebau A, Sestak M (1997) Surface solar ultraviolet-B radiation change detection in western United States using remote sensing, the preprint of the ninth radiation conference. Preprint for the Ninth Conference on Atmospheric Radiation, AMS 77th annual meeting, Long Beach, California, Feb 2 – 7, 1997

Qu J (1997) Golbal ozone and solar ultraviolet-B radiation detection using remote sensing. PhD Dissertation, Colorado State Universirty, pp 163

Qu J, Stephens G (1996) Estimation of the Earth's surface ultraviolet-B radiation using two-stream atmospheric radiation transfer model. Current Problems of Solar Radiation, Deepak Publisher, 962 – 965

Qu J, Riebau A, Sestake M, Smith L, Smith F, Coloff SG (1996) Global solar ultraviolet-B radiation measurements and simulation in remote areas. Current Problems of Solar Radiation, IRS's 96, Deepak Publisher 942 – 945

Rossow WB, Schiffer RA (1991) ISCCP cloud products. Bulletin of the American Meteor Soci 72: 2 – 20

Scotto, et al. (1988) Biological effective ultraviolet radiation, Science, 239: 762

Soloman S, Garcia RR, Rowland FS, and Wuebbles DJ (1986) On the depletion of Antarctic ozone. Nature, 321: 755 – 758

Stamnes K, et al. (1990) Biologically effective ultraviolet radiation, total ozone aboundance, and cloud optical depth at McMurdo station, Antarctica September 15 1988 through April 15, 1989. Geophys Res Lett 17: 2,181 – 2,184

Stackhouse P W, Stephens GL (1991) A Theoretical and Observational Study of the Radiative Properties of Cirrus: Results from FIRE 1986, Journal of the Atmospheric Sciences: Vol. 48, No. 18, pp. 2,044 – 2,059

Stephens GL (1995) Review of atmospheric radiation. Rev Geophys Supplement 785 – 794

Stephens GL (1994) Remote sensing of the lower atmosphere: An introduction. Oxford University Press, 118 – 317

Stephens GL, Tsay SC (1990) On the cloud absorption anomaly. QJR Meteorol Sco, 116: 670 – 704

Stephens GL, Webster PJ (1984) Cloud decoupling of the surface and planetary radiative budgets. J Atmos Sci 41(4): 681 – 686

Stephens GL (1979) Optical proprtties of eight water vloud types. Division of Atmospheric Physics Technical Paper No 36, Commonweath Scientific and Industrial Research Oranization, Australia, 1 – 35

Stowe LL (1991) Cloud and aerosol products at NOAA/NESDIS. Paleogeogr Paleoecol 90: 25 – 32

Tsay S, Stamnes K (1992) Ultraviolet radiation in the Arctic: The effect of potential ozone deplections and cloud effects. J of Geophy Res 97(D8): 7,829 – 7,840

Webb AR (1992) Spectral measurements of solar ultraviolet-B radiation in southeast England. J Appl Meteor 31: 212 – 216

WMO (1986) Atmospheric Ozone 1985. Global Ozone Research and Monitoring Project Report No. 16, World Meteorological Organization, Geneva, 349 – 392

17 Surface Rain Rates from Tropical Rainfall Measuring Mission Satellite Algorithms

Long S. Chiu, Dong-Bin Shin and John Kwiatkowski

17.1 Introduction

The Tropical Rainfall Measuring Mission (TRMM), jointly sponsored by the National Aeronautics and Space Administration (NASA) and Japan Aerospace Exploration Agency (JAXA, previously known as National Space Development Agency, or NASDA), is the first coordinated international satellite mission to monitor and study tropical and subtropical rain systems (Simpson et al., 1988). A detail description of the TRMM sensor package and a preliminary assessment of the sensor performance are given by Kummerow et al. (1998). The TRMM rain sensor package includes the first space-borne Precipitation Radar (TPR), a TRMM Microwave Imager (TMI) and a Visible and Infrared Scanner (VIRS). Rainfall estimates provided by the TRMM have found applications in climate analysis, data assimilation, water resource management, and decision support to agriculture and health issues.

The TRMM data have gone through major reprocessing cycles, as improved knowledge of the sensors and algorithms leads to improved sensor calibration and algorithms. The version 5 (V5) algorithms have shown significant improvement over the previous version of data (Kummerow et al., 2000). Over land, Shin et al. (2001) showed that TMI rainfall estimates are higher than TPR in the TRMM domain (35°N – 35°S) for the first two years of TRMM data. Adler et al. (2003a) pointed out that there is a distinct reversal of the algorithm bias between the tropics and the subtropics for the first three years of TRMM data. While global differences among the TRMM satellite estimates are of the order of 20%, there are large regional differences (Kummerow et al., 2000; Adler et al., 2003a). Nesbitt et al. (2004) compared V5 TMI and TPR with the Global Precipitation Climatology Center gauge analyses and TMI and TPR rain features using one year of data. Comparisons of TRMM satellite and ground validation radars have been carried out (Schumacher and Houze, 2000; Liao et al., 2001; Kim et al., 2004).

As the TRMM data are being used in research and applications, it is necessary to constantly evaluate the performance of the TRMM algorithm and quantify their relative biases for hypothesis testing, model validations, and operational decision-making, such as water management and crop yield monitoring (Teng et al., 2005).

A brief description of the data, the V5 algorithms, and improvements made in the V6 algorithms are given in Section 17.2. The data we used are the 0.5 degree binned Level-2 data and 1 degree Level-3 products. Paired-*t* tests were used to quantify the difference between the algorithms. The results for the algorithm intercomparison and comparison between V5 and V6 algorithm are presented in Section 17.3. The summary and discussions are contained in Section 17.4.

17.2 Satellite Algorithms and Data

The TRMM rain sensors and preliminary algorithm performance has been discussed by Kummerow et al. (2000), Smith and Hollis (2003) and Adler et al. (2003a). Figure 17.1 shows the TRMM satellite algorithm data flow, which is reproduced here for completeness (see also Chiu et al., 2005a, Fig. 13.3). The

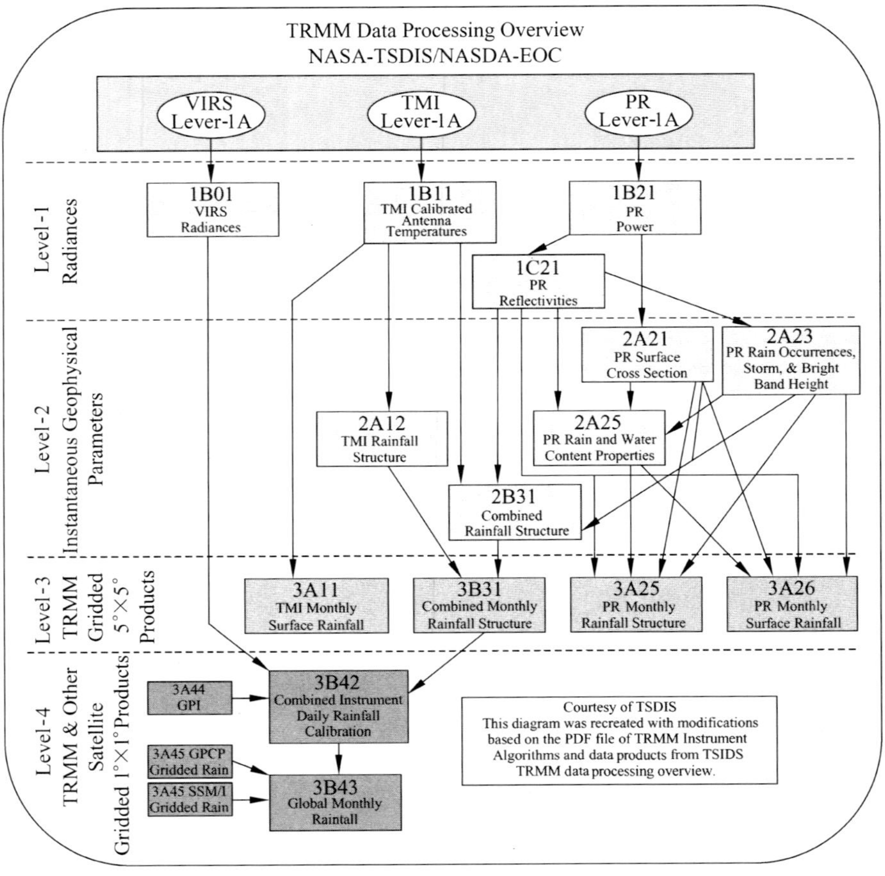

Figure 17.1 TRMM satellite algorithm data processing flow

TRMM level-1 products include VIRS radiance, TMI calibrated antenna temperature, TPR power and reflectivity. Level-2 products are geophysical parameters in satellite orbit coordinate and Level-3 and higher products represent space/time average products (Asrar and Greenstone, 1995). The TRMM level-2 products included TMI rain profile (2A12), TPR surface cross section (2A21), rain and bright band height (2A23), TPR rain profile (2A25) and TRMM Combined Instrument (2B31). There are three TRMM level-2 surface rain rate products (2A12, 2A25 and 2B31) and five Level-3 products derived from TRMM sensors. To facilitate analysis, the TRMM Science Data and Information System (TSDIS) produces product 3G68, which contains ASCII files of Level-2 rain products binned at 0.5 degree grids (Chiu et al., 2005a, this issue). This product is used in our analyses.

For operational decision support systems, high spatial and temporal resolution rainfall data in near-real time is of particular interest. Hence it is useful to examine the TRMM merged satellite product (3B42) which is available at 0.25 degree (1 degree) and 3 hourly (daily) resolution for V6 (V5) data. The level-2 satellite products provided input to V5 3B42. We will compare the level-2 satellite products and the level-3 TRMM and other satellites and merge TRMM and gauge products at the monthly scale.

Descriptions of the TRMM algorithms and algorithm updates can be found in the TRMM data and description page on line at the TRMM Web site (URL: http://trmm.gsfc.nasa.gov/data_dir/ProductStatus.html). These surface rain rate algorithms are briefly described below.

17.2.1 V5 Algorithms

The TMI Hydrometeor Profile (TMI or 2A12) rain rate is derived from the brightness temperature data observed by the TMI channels. The algorithm finds possible rainfall profiles from a database consistent with the TMI channel data and selects the most probable rain profile in a Bayesian formalism. The database of rainfall profiles is based on outputs of cloud resolving models, such as the Goddard cloud ensemble model (Kummerow et al., 2001).

The TPR profile algorithm (2A25) calculates the rain profile based on the measured radar reflectivity via a reflectivity-rain rate (Z-R) relation (Iguchi et al., 2000). It is a hybrid of the Hitschfeld-Bordan method and the surface reference method (Iguchi and Meneghini, 1994). The attenuation correction is based on the surface reference method which assumes that the decrease in the apparent surface cross section is caused by the propagation loss in rain only. The coefficient a in the attenuation *k-Z* relation ($k=a Z^b$) is adjusted in such a way that the path-integrated attenuation (PIA) estimated from the measured reflectivity-profile matches the reduction of the apparent surface cross section. The attenuation correction of Z is carried out by the Hitschfeld-Bordan method with the modified coefficient *a*.

A non-uniformity parameter is introduced to correct the bias in the surface reference arising from the horizontal non-uniformity of rain field within the beam. Radar echoes from near the surface are easily contaminated by the mainlobe clutter, hence only a near-surface rain rate, which is the rain estimate at the lowest point in the clutter-free region is given for V5.

The TRMM combined algorithm (TCA or 2B31) provides the vertical structure of rainfall (rates and drop-size-distribution parameters) based upon the TMI and the TPR within the TPR swath. The algorithm is based on a Bayesian approach. The radar measurements for every likely value of the drop-size-distribution (DSD) shape parameters are first inverted. The resulting rainfall estimates are used to produce the corresponding expected brightness temperatures, which are then compared to the actual passive measurements to select the most probable DSD shape parameter value (Haddard et al., 1997).

The V5 algorithm 3B42 produces TRMM-adjusted merged-infrared (IR) precipitation on a 1-day temporal resolution and a $1° \times 1°$ spatial resolution. Monthly IR calibration parameters are computed by comparing coincident VIRS data (1B01) and the TMI data (2A12) orbit by orbit. After a calendar month, the monthly gridded averages of coincident VIRS and TCA data are computed. For each gridbox, the monthly average coincident TMI data are converted to the corresponding TCA data using the TMI/TCA monthly calibration parameter. Using the monthly average VIRS and TCA data, the IR calibration parameters are computed. These IR calibration parameters are then applied to the merged-IR data derived from geosynchronous satellite measurements to produce the TRMM-adjusted merged-IR precipitation product.

Algorithm 3B43 is a monthly product at a $1° \times 1°$ spatial resolution. It is producesd by weighting the monthly 3B42 total and the monthly accumulated Climate Assessment and Monitoring System (CAMS) or Global Precipitation Climatology Centre (GPCC) monitoring rain gauge analysis by their error fields. 3B43 is produced after the end of the month and the selection of the CAMS or GPCC data are dependent on their availability at the time of production. For reprocessing, the GPCC monitoring product compiled from GTS observations is used.

17.2.2 V6 Algorithms

V6 algorithms are used for processing beginning in April 2004. Shortly afterwards, the V6 reprocessing began from the start of the TRMM data stream, Dec. 8, 1997. As of March 2005, four full years of reprocessed data are available. The three Level-2 surface rain retrieval algorithms (2A12, 2A25, and 2B31) were improved based on physical principles. An overview of the significant changes to the algorithms and products for V6 is discussed below. For complete V6 documentation on formats and science content, the reader is referred to the TSDIS Web page http://tsdis.gsfc.nasa.gov for File Specifications and Algorithm User's Guides written by the TRMM Algorithm Developers.

The V6 TMI algorithm (2A12) consists of three distinct algorithms depending on the surface type, ocean/land/coast. A key change has been an improved surface type mask from 25 km grid (V5) to 1/6 degree (V6) land mask.

The Precipitation Radar (PR) algorithms have undergone significant changes for V6, starting with the Level-1 TPR power (1B21) which converts instrument counts to echo power. In 1B21 the calibration was changed resulting in an effective increase in echo strength of 0.35 dBm, with a corresponding increase in TPR reflectivity (1C21). This change alone would increase PR rainfall. A hybrid algorithm (Meneghini et al., 2004) has been adopted over the ocean where the along track data is used as a reference to fit an idealized curve for the surface cross section as a function of incidence angle (cross-track direction). Deviations from the idealized curve are interpreted as attenuation from precipitation. This method allows for much smoother PIA fields over the ocean avoiding some discontinuities that were present in V5. The PR rain classification algorithm (2A23) has changed its classification scheme, resulting in a relative change in population of the three main rain classes: stratiform, convective and others. A preliminary calculation based on February 1998 data shows that the percent of rain are 78%, 12%, and 10% for V5 stratiform, convective and other rain types, respectively, and 80%, 17%, and 3%, for the corresponding V6 products. The Convective category in V6 includes a larger portion of shallow precipitation and the percentage of the "others" rain type is much reduced.

The V6 TPR algorithm has undergone the largest change. Attenuation from water vapor, cloud liquid water and molecular oxygen are now included in the attenuation correction algorithm. The most significant change in TPR is the estimation of attenuation below the clutter level using prescribed reflectivity slopes. This results in the addition of an estimated surface rainfall rate (at the Earth's Ellipsoid) in V6 compared to the near surface rainfall rate (above the clutter) in V5. This distinction is important since the height at which the near surface rainfall is reported is a function of incidence angle, generally closer to the Earth's Ellipsoid near nadir and up to 2 km above the Ellipsoid at the scan edges. The reflectivity slopes used for getting below the clutter depend on the surface and precipitation type. Averaging over all these cases, scan angles, and time shows that the estimated surface rainfall rate is about 6% lower than the near surface rainfall rate. The 2A25 algorithm developer suggested that all future comparisons of TPR to TMI rainfall use the TPR surface rainfall rates.

The V6 3B42 is a 3 hourly 0.25 degree product based on multi-satellite precipitation analysis (MPA, Huffman et al., 1995, 2001, 2004). First, all available 3-hourly microwave-IR combination estimates are put into the appropriate space/time bin. These high-resolution data are summed over a calendar month to create a monthly multi-satellite (MS) product. Due to the lack of global gauge data on a daily basis, the V6 3B42 product is scaled so that the short-period estimates sum to a monthly total that include monthly gauge analyses, as is done with the GPCP daily product (Huffman et al., 1997). The MS and gauge analysis are merged optimally as done in Huffman et al. (1997) to create a post-real-time monthly satellite-gauge (SG) product, which is the TRMM product 3B43. 3B42 is then scaled as the ratio of monthly MS to SG (the scale factor being limited to a range [0.2, 2]). The gauge analyses employed are presented on a 2.5° grid, so that the 0.25° 3B42 and 3B43 data do not have gauge information at their native resolution of 3 hourly and 0.25 degree.

17.3 Results

The TSDIS product 3G68 contains Level-2 TMI, TPR and TCA rain rates binned at a 0.5°×0.5° latitude-longitude grids. These data are available via ftp from the TRMM Web site or from the DAAC. 3B42 and monthly 3B43 at 1°×1° resolution are available through the DAAC. In order to compare the five algorithms, swath rain rates for TMI, TPR and TCA from 3G68 at 0.5° resolution are averaged to monthly rain rate at 1°×1° grid resolution.

17.3.1 Annual Means and Paired *t*-Tests

Figure 17.2 (left column) shows six-year (1998 – 2003) annual average rain rates for the V5 algorithms. The annual average TPR rain map is grainy, compared to the TMI, which has about three times the samples (220 km swath for TPR vs. 720 km swath for TMI). There are some interesting features at this resolution which are not easily discernible from previous analyses. There is a clear separation of the Intertropical Convergence Zone (ITCZ) and the South Pacific Convergence Zone (SPCZ). The existence of a double ITCZ in the eastern Pacific, the primay one being north of the equator and a much weaker secondary peak to the south, is discernable. This feature is very prominent in March to May in non-El-Nino years. There are also distinct dry zones between Borneo and the Philippine Islands, and between Sumartra and New Guinea, reflecting the importance of topography, land sea contrast and prevailing wind in controlling rainfall distribution. The contrast between land and ocean precipitation in the maritime continents is also noted in all algorithms.

While the major rain features are well represented in all algorithms, there are distinct differences in their intensities. TMI shows the highest domain average rain rate (2.99 mm/day), followed by 3B42, 3B43, TCA and TPR at 2.66, 2.60, 2.43, and 2.30 mm/day, respectively.

Figure 17.2 (right column) also shows the annual average rain rate for the

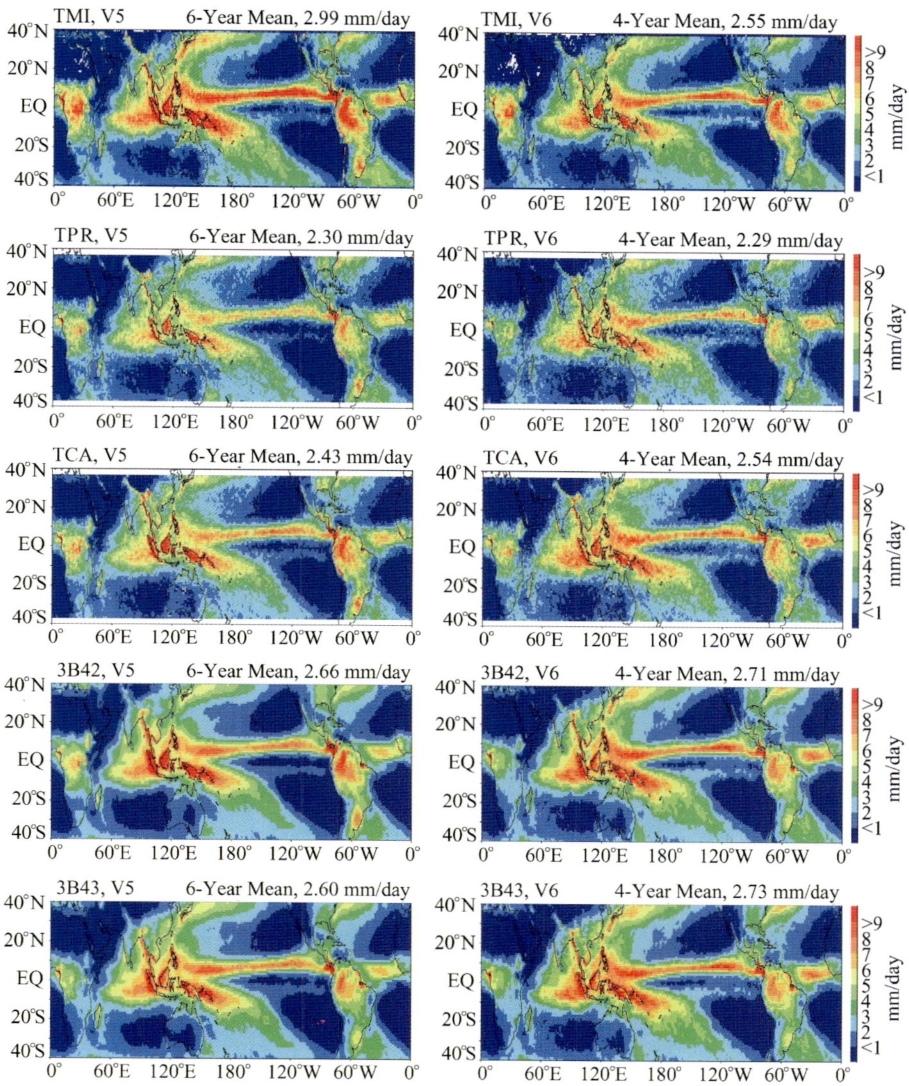

Figure 17.2 Annual average rain rates for the V5 (left column) and V6 (right column) TRMM algorithms: 6-year (1998 – 2003) averages for V5 and 4-year (1998 – 2001) averages for V6

V6 algorithms for 1998 – 2001. The decrease in V6 TMI from V5 is clearly discernable, both over land and oceans. All other V6 algorithms (TPR, TCA, 3B42 and 3B43) increase over oceans, but decrease over land, particularly over the Amazon and the African rain belts. 3B43 and 3B42 shows the two highest domain average rain rate (2.73 mm/day and 2.71 mm/day). TMI and TCA rain rates are smaller, at 2.55 mm/day and 2.54 mm/day, respectively. TPR is the lowest at 2.29 mm/day.

Figure 17.3 shows the annual zonal average rain rates for V5 (1998 – 2003) and V6 algorithms (1998 – 2001). The 4-year V6 algorithm zonal climatologies show much better agreement among themselves than the V5 algorithms. For V6 algorithms, TCA, 3B42 and 3B43 are in good agreement and are in general higher than TPR and TMI over the oceans. TMI is higher than TPR in the tropical zonal rainfall maximum, but is slightly lower for latitudes between 10°N – 20°N. Over land, TMI rain rate is highest and TPR lowest. 3B42, 3B43 and TCA are in between and track each other quite well. The zonal peaks of the V6 algorithms are lower than the V5.

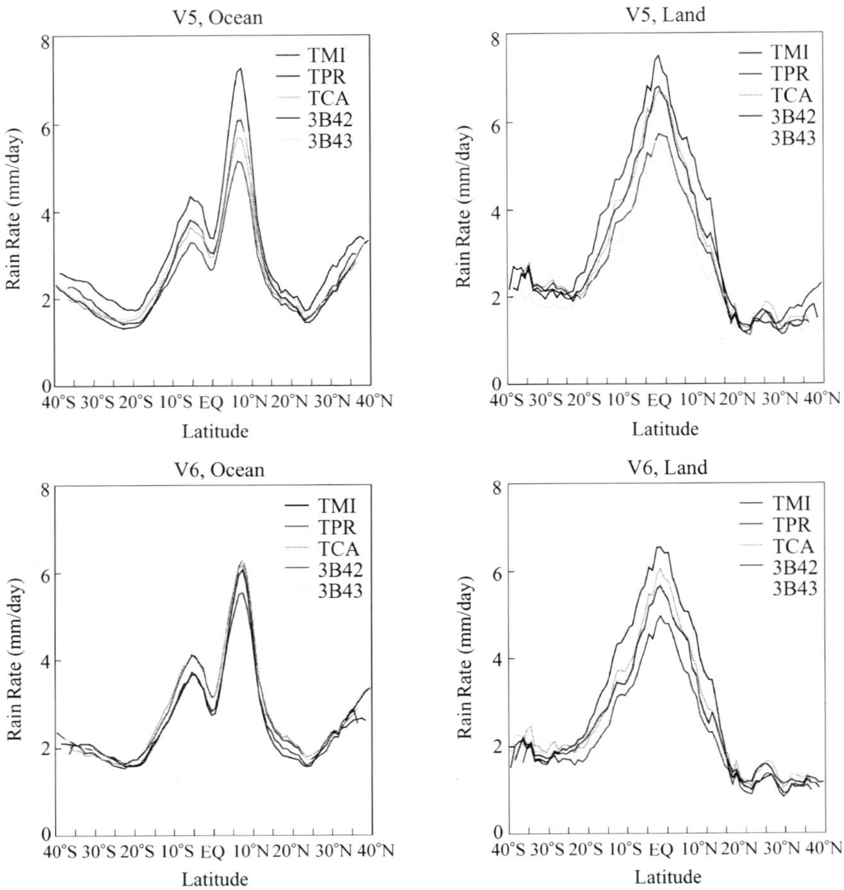

Figure 17.3 Zonal mean rain rates over ocean and land for V5 (upper panels) and V6 (lower panels) algorithms

To quantify the difference between the algorithms, a paired t-test,

$$t = [z]/\sigma_z / n^{1/2} \tag{17.1}$$

where $z = x - y$ is the difference between the two samples, x and y, respectively, and $[z]$ and σ_z are the sample mean and standard variations of z. n, the number of x and y pairs, is used (Chang et al., 1999; Shin et al., 2001). The x and y series represents the time series of the two algorithms under consideration at a grid point. The validity of the paired t-tests relies on the normality of rain rate distributions. Yu and Chiu (2005) showed that in the heavy rain areas, the assumption of normality of monthly rain rates is accepted, however, in the oceanic dry zones and over some land areas, this assumption is violated. They proposed the use of a non-parametric test, which is a more strigent test than the paired t-test.

Both the annual total and seasonal statistics from these algorithms are computed. For the annual case for V5, $n = 72$, and the null hypothesis that no difference exists between x and y (difference is zero) can be rejected at the 95% level if $|t| > 1.96$. For the seasonal case, $n = 18$ (3 months per season for 6 years), $|t| > 2.10$ in order to reject the null hypothesis. The $|t|$ values are slightly higher for four years of data used in V6.

The paired-t statistics between the various V5 algorithms using all six years (72 months) of data are shown in the left coumn of Fig. 17.4. The first panel shows the t-statistics of TMI—TPR. Grids with t values higher than 2 are colored red and regions with $t < -2$ are colored blue. It can be seen that in the oceanic rain belts, such as ITCZ, the SPCZ, over the storm tracks in the western Pacific and Atlantic ocean, and in the Amazon, West Africa and in the southern slopes of the Himalayas, TMI is significantly higher than TPR. However, in the oceanic dry region, and in the mountainous regions, such as the Andes and the Sierra Nevada, and in the Himalayas, TPR is significantly higher.

The second panel shows the t-statistics between TPR and TCA. There is no significant difference between TPR and TCA, except in some limited area over the maritime continent. This is consistent with earlier results (Shin et al., 2001).

The third and fourth panels show the t-statistics between TMI and 3B42 and between TPR and 3B42. 3B42, the merged satellite algorithm, uses the TCA rain rates to calibrate the rain rates estimated from IR thresholding technique from the more frequent geosynchronous and low Earth orbiting satellite observations. Over the oceans, TMI is higher than 3B42 in the major rain belts, but lower than TPR in oceanic dry regions, hence 3B42 is intermediate between TMI and TPR. However, over a major part of Asia and eastern US, and in the southern slope of the Himalayas, 3B42 is higher than both TPR and TMI.

The fifth panel shows the t-statistics between 3B42 and 3B43. There is no difference between 3B42 and 3B43 over a major part of the oceans, except in the oceanic dry regions. Over land, however, 3B42 is significantly higher than 3B43 except in eastern China, over the Tibetian plateau, in the coastal inland of Brazil and in northern and eastern highland of Africa. Since 3B43 is a combination of gauge analyses and 3B42, this result reflects the fact that the TRMM based satellite estimates (3B42) are in general higher than gauge analyses over land.

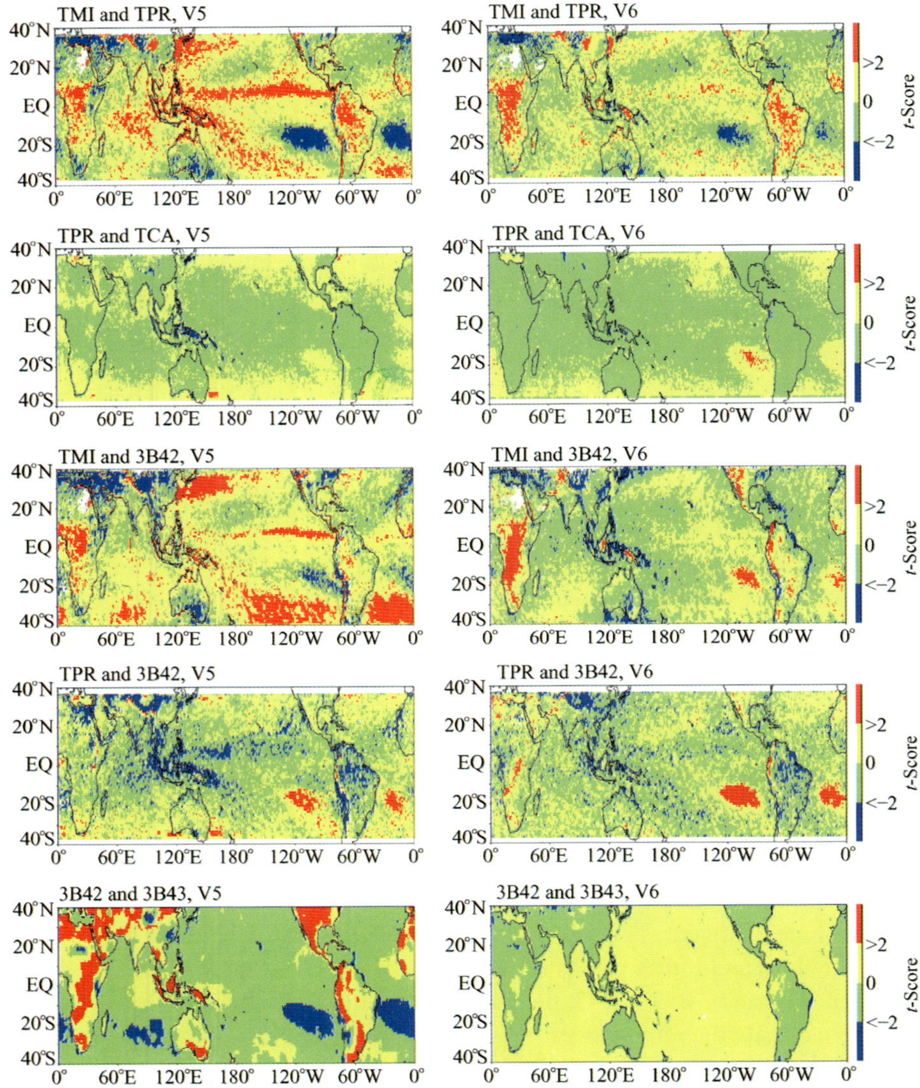

Figure 17.4 Paired t-tests between the TRMM algorithms: the left and right columns indicate V5 and V6, respectively

Figure 17.4 (right column) also shows the similar t-statistics for the V6 algorithms. The difference between TMI and TPR over the oceans are much reduced as evidenced in the major rain belts (ITCZ, and SPCZ), the mid-latitude storm tracks and in the oceanic dry zones. The difference over land is comparable to that for V5, however. There are reversals in the sign of the difference, notably in some coastal regions in the maritime continent, presumably due to the development of a V6 TMI coastal algorithm. The difference between TPR and TCA continues to be insignificant, except in the oceanic dry regions. The difference between TMI and

3B42 is much reduced over oceans as compared to V5. In the oceanic dry zones, there is a reversal, as TMI is now larger than 3B42. TMI is also larger than 3B42 in the western part of north America and the west coast of south America. The area of signficance in Africa between 0°S – 20°S is also increased. The paired t- statistics between TPR and 3B42 show little difference between these two versions. The lowest panel shows no significant difference between 3B42 and 3B43 for the V6 data, which is consistent with the design of the V6 3B2 algorithm.

17.3.2 Seasonal Differences

Figure 17.5 shows the seasonal pattern of the paired t-statistics between V5 TMI and TPR. In general, the seasonal patterns are similar to the annual case.

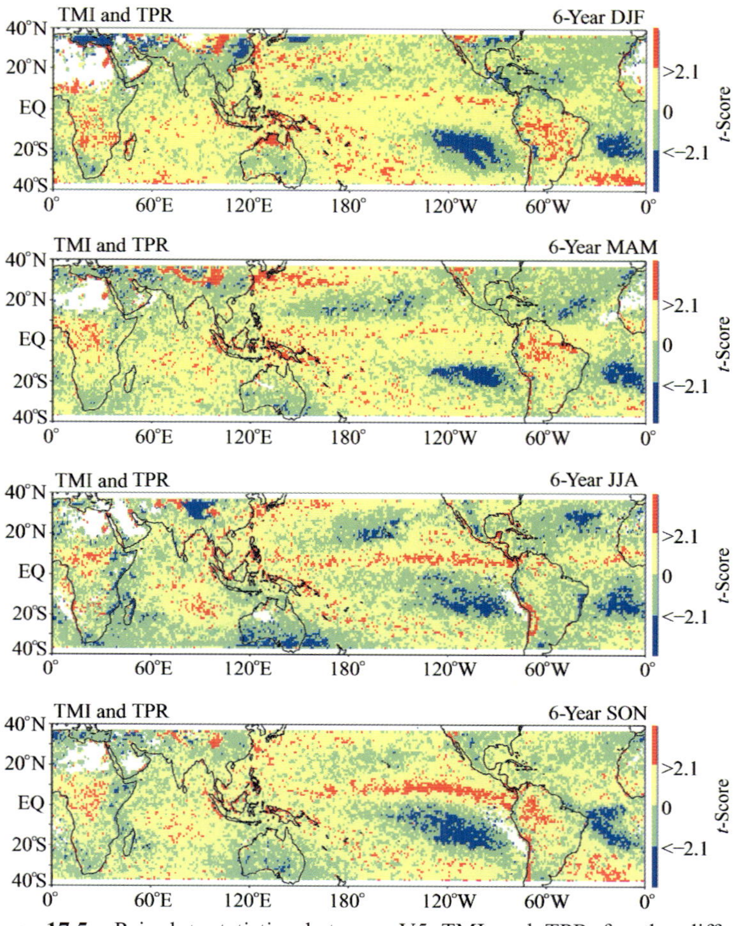

Figure 17.5 Paired t- statistics between V5 TMI and TPR for the different seasons (DJF, MAM, JJA, and SON). Areas shaded white indicate no data

However, during the northern winter (DJF), TMI is significantly higher over the the western US, but lower in the eastern US, in Central America, in the eastern part of China, and in the Mediterrean sea. Examinations of these regions show significant seasonal trends in the paired t-statistics. In particular, in eastern China, the difference reverses sign during the course of the year. In other regions, the seasonal trends are less pronounced but with no sign of reversal. In the northern summer (JJA), TMI is significantly lower in southwest Australia. The seasonal difference between the V6 TPR and TMI is also computed. In general, the sign of the difference remains but the significance is much reduced.

Seasonal patterns of the difference between 3B42 and 3B43 have been computed. Figure 17.6 shows the percent bias of V5 3B42 relative to 3B43. The seasonal patterns of the difference are in general similar to the annual case. During DJF, there is a belt of negative bias (3B42 < 3B43) of more than 80%

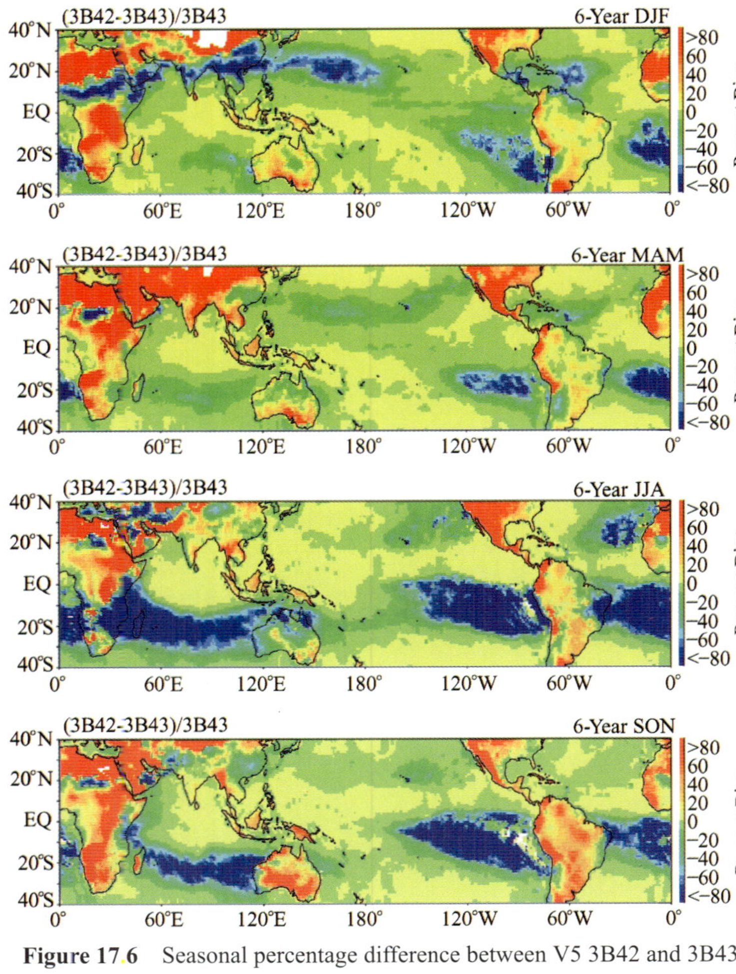

Figure 17.6 Seasonal percentage difference between V5 3B42 and 3B43

(blue area) in the latitude belt extending from around 20 degrees north in the middle Pacific to China, IndoChina, India, the Arabia sea and into Africa.

However, over the other land areas, the biases are positive, and are typically > 80% (shown in red). This belt of negative bias disappears in the northern spring (MAM) and the 3B42 is larger than 3B43 by about 80% over land. The belt of negative bias reappears in the southern hemisphere oceans between the equator and around 20 degrees south in JJA. This negative belt is not confined only to the oceans, but extends to Magadesdar and inland of Africa. The negative bias over Indochina and India found in DJF also becomes positive. The biases reverse sign in monsoon Asia, India and in the SW American monsoon area.

For the V6 data, there is no significant difference between 3B42 and 3B43, which is consistent with the design of the V6 3B42 algorithm. Note that daily 3B42 are scaled from fine resolution MPA rain rates, hence non-detection and false detection rates are unchanged on the daily or 3 hourly scale.

17.3.3 Interannual Variations

Figure 17.7 (upper panels) shows the time series of the domain average V5 monthly rain rate computed separately over ocean and land for all algorithms. TMI

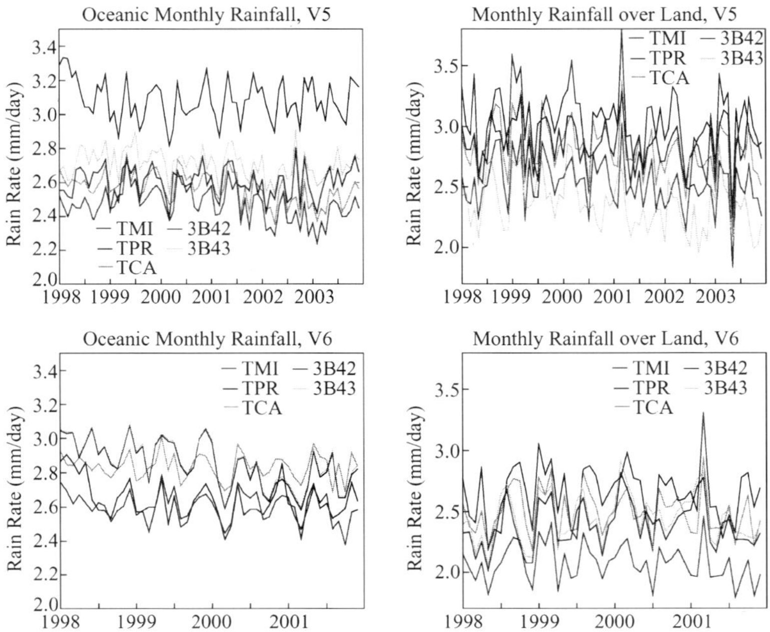

Figure 17.7 Time series of the domain (37°N – 37°S) average rain rates from V5 (upper panels) and V6 (lower panels) TRMM algorithms computed separately over ocean and land

shows the highest global (ocean+land) rain rate, followed by 3B42, 3B43, TCA, and TPR (see Fig.17.3). Over the oceans, the order of these algorithms are the same, except TMI is substantially higher than 3B42. There is a reversal of the domain bias between 3B43 and 3B42 over land. 3B42 is slightly lower than 3B43 over the oceans, but it is substantially higher over land.

Similar time series for V6 are shown in the lower panels of Fig.17.7. As seen previously, the absolute value of TMI is much reduced over the oceans and now tracks TPR fairly closely except in early 1998. There is no difference between 3B42 and 3B43, both of which are higher than TMI and TPR and track the TCA quite well. Over land, TMI continues to show the largest rain rate, and TPR the lowest. There is a slight difference between 3B42 and 3B43. All level-2 monthly rain rates show similar peaks and dips.

Figure 17.8 (upper panels) shows the domain average bias of V5 TMI relative to TPR and of 3B42 relative 3B43. Over the oceans, TMI bias over TPR is a maximum (30% – 40%) in 1998, which decreases to less than 20% in 1999, and seems to increase after 2000 to about 20% – 30% in 2002 – 2003. Over land, the bias is about 25% with a range of 15% – 40%. The oceanic difference between 3B42 and 3B43 ranges from 0% – 10%, with a mean of 2% – 3%. There is a very distinct semiannual cycle.

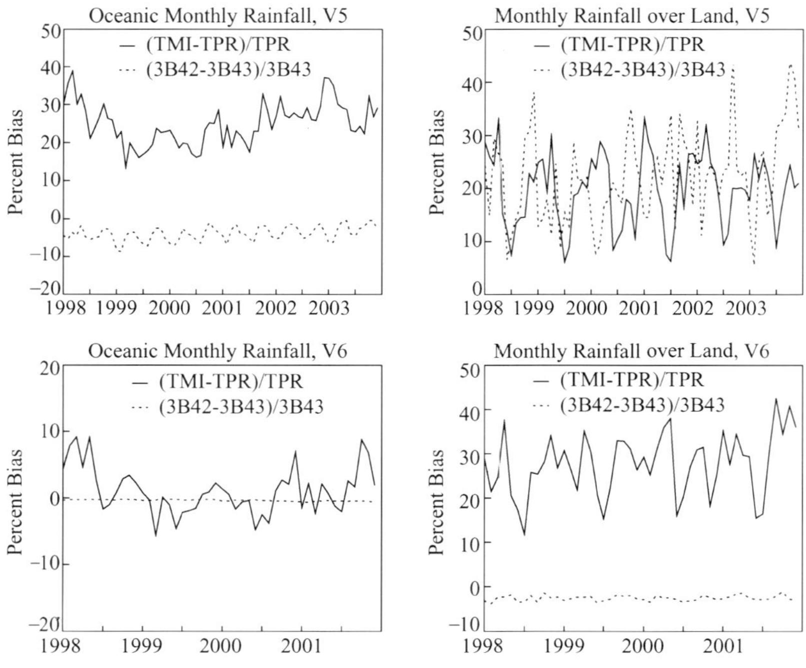

Figure 17.8 Domain average bias of TMI relative to TPR and 3B42 relative to 3B43 over ocean and land for V5 (upper panels) and V6 (lower panels) TRMM algorithms

The month-to-month variations of the domain bias over land is much larger than that over the oceans. The relative biases of TMI to TPR and of 3B42 to 3B43 are about 20%, with a range of 10% – 30%.

Figure 17.8 (lower panels) shows the similar time series for V6 data. There is no difference between monthly 3B42 and 3B43. Over the oceans, the percent bias of TMI relative to TPR is much reduced, but the general trend is similar to that of V5. Over land, TMI is higher than TPR by about 20%, and the V5 and V6 curves are quite similar.

We next examine the seasonal anomalies of each algorithm from the six (four for V6)-year means. Figure 17.9 shows the time series of the non-seasonal rain rates for V5 and V6, respectively, for all five algorithms. For non-seasonal data, the monthly climatology (1998 – 2003 for V5 and 1998 – 2001 for V6) have been subtracted. The non-seasonal departures of the algorithms track each other quite well over oceans (except in 1998) and over land in general. However, there are periods when the algorithms are quite different over oceans such as in early 1998 and between late 2000 – early 2001. While the absolute values of these rain estimate differ considerably, their anomalies from climatology track each other quite well. Hence, it may be advantageous to examine the relative anomalies for nonseasonal analyses.

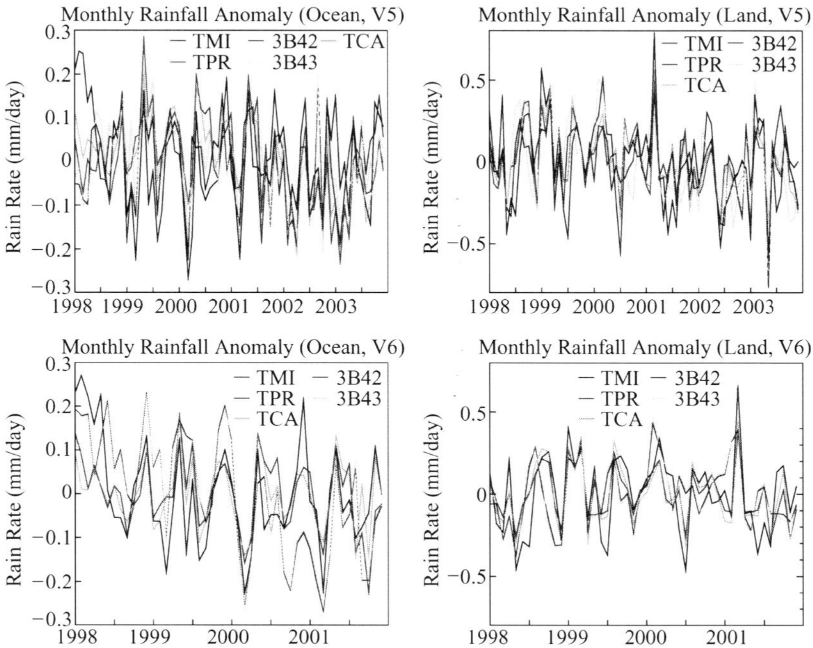

Figure 17.9 Time series of TRMM rain rate departures from climatology (6-year average for V5, 4-year average for V6) computed over ocean and land

17.4 Summary and Discussion

We compared the V5 and V6 rain rates produced by five TRMM algorithms. For V5, the TMI shows the largest rain rate, followed by 3B43, 3B42, TCA and TPR over the TRMM domain. Over the oceans, this order is followed, except the TMI rain rate is much higher than the rest. Over land, however, 3B43 shows the lowest rain rate. The V6 TMI and TPR retrievals are more internally and physically consistent. Over land, a V6 TPR surface rain rate is introduced and is lower than the near surface rain rate in V5. For V6 algorithms, 3B43 is highest, followed by 3B42, TCA, TMI and TPR, averaged globally for 4 years of data.

Table 17.1 summarizes the domain average rain rate for V5 and V6 algorithms, computed separately over land and oceans. The V6 TMI decreases by about 15% over both land and oceans. The other algroithms increase by about 5% (TPR and 3B43) to 10% (TCA and 3B42) over oceans and decrease by 15% (TCA) to 20% (TPR and 3B42) over land. 3B43 remains relatively unchanged between V5 and V6. If we use 3B43 to gauge the difference among the algorithms and between the versions, our results suggest that the V6 TCA and 3B42 are now on par with 3B43 over the oceansand land. While the oceanic difference between TPR and TMI are much reduced for V6, they are now biased low compared to 3B43 and the TAO buoyed gauges. Over land, TMI remains high and TPR low compared to 3B43.

Table 17.1 Comparison of V5 and V6 average rain rates over land and oceans for 1998–2001

	Ocean			Land		
	V5	V6	%Diff	V5	V6	%Diff
TMI	3.09	2.63	– 15	3.06	2.64	– 13
TPR	2.50	2.61	+ 4	2.57	2.07	– 19
TCA	2.60	2.84	+ 9	2.86	2.49	– 13
3B42	2.61	2.85	+ 9	2.88	2.30	– 20
3B43	2.73	2.86	+ 4	2.40	2.45	+ 2

Units: mm/day. The percent difference (%Diff) = (V6 – V5)/V5 × 100%.

Paired *t*-tests are performed to evaulate quantitatively the difference between the algorithms. There is no signficant difference between TCA and TPR for both V5 and V6. However, substantial difference exists between TMI and TPR. The bias between V5 TMI and TPR has been noted (Adler et al., 2003a). Adler et al. (2003b) compared three years of TMI and TPR with GPCC analysis (Rudolf et al., 1994). They found large TMI biases with respect to the GPCP analysis, but the bias becomes smaller poleward of 15°N – 15°S. TMI is higher in the oceanic rain belts. However, TPR is higher in the dry oceanic regions. The sampling afforded by TMI is better than TPR due to its larger swath. Inadequate sampling tends to underestimate, hence the small TMI rain rates cannot be attributed to sampling.

TMI shows lower sensitivity at the low rain rates and a rain-no rain threshold of 0.4 mm/hr is used. The rain rates in these oceanic dry regions are, in general, low. Monthly histograms of the TPR rain rates show the existence of a large fraction of TPR rain rates below 1 mm/hr (Robertson et al., 2003). The TMI rain rate threshold introduces a large non-detection rate in these regions and contributes to the low TMI monthly rain rates there. This pattern is carried over to V6 algorithms, although the difference is much reduced.

Our results show that V5 TPR is significantly lower than TMI in the oceanic rain belts. Serra and McPhaden (2003) compared V5 TRMM rain rates with the Tropical Atmosphere Ocean (TAO) array and the Pilot Research Moored Array in the Tropical Atlantic array and found that there is almost no bias between the buoy gauges and TMI rain rate, however, the TPR rain rates are smaller by 30% – 40%. A similar conclusion is also reached by Bowman et al. (2003). These buoyed arrays of gauges are located in the heavy rain belts of the ITCZ in the Pacific and Atlantic, hence our results are consistent with theirs. For the V6 data, the TMI is reduced by about 15% while TPR increased by 4%, overall. Hence both TPR and TMI are now biased low compared to the TAO gauges.

Chiu et al. (2005b) compared V5 TRMM products with gauge analysis in New Mexico, USA and found the satellite algorithms, especially TPR, significantly overestimated gauge analyses in the summer, which they attributed to evaporation of hydrometeors before reaching the surface. The V6 TPR surface rain rate accounts for decreases below cloud base, and hence better agreement with gauge analyses can be anticipated.

The discrepancy between TPR and the other TRMM estimates for 1998 ENSO has been noted. Robertson et al. (2003) showed large interannual variability of the TPR rain rate histogram, especially in the low rain rate categories, and suggested the assumption of drop size and associated path attenuation may be a source of uncertainty in interannual variability of TPR rain rate. The incorporation of water vapor, cloud liquid water and molecular oxygen attenuation in V6 TPR results in higher rain rate, and the increases in early 1998 are now noted in both V6 TMI and TPR (Fig. 17.9). However, this increase in V6 TPR (and decrease in TMI) seems insufficient to reconcile the difference between TPR and TMI rain rates in the El-Nino event of 1998.

Our analysis showed that the bias of V5 3B42 (AGPI) relative to the merged satellite product (3B43) is, in general, small (~10%) in JJA in western Africa. Nicholson et al. (2003) used a dense rain gauge network situated between the equator and 20°N over Africa for the period May-September 1998 to validate the GPCC and V5 TRMM products. They found almost no bias for TRMM merge (3B43) and a small bias (0.2 mm/hr) for AGPI (3B42) at the seasonal scale. However, large biases of TPR, TMI and TCA are found. Our results are therefore consistent with the analysis of Nicholson et al. Better agreement can be anticipated for the V6 analyses since there is little change in 3B43 and 3B42 is now scaled to match the 3B43 monthly results.

In general, the seasonal pattern of the different fields between V5 3B42 and 3B43 (Fig. 17.6) are similar to the annual pattern and shows a seasonal trend in most oceanic regions. However, there is a reversal of the bias over land areas, in particular in the southern part of China, Indochina, India and in Central America. This is consistent with the results of Adler et al. (2003b) who noted a reverse of the bias from low to mid-latitude belts. The absence of high-level clouds may lower the estimate of 3B42 since it is dependent on the fraction of cold cloudiness. Lau and Wu (2003) noted a large fraction of TRMM rainfall are due to warm rain process.

The TRMM project provided a prototype near realtime product 3B42RT for operational use. For V6, 3B42RT is MPA. Because of its high temporal and spatial resolution, 3B42RT is the most popular data set for applications (see (Chiu et al., 2005a), this issue). It is therefore important to characterize the seasonal biases of 3B42RT and to continue to monitor the algorithm performance as the algorithms improve at each reprocessing.

In 2003 the International Precipitation Working Group began a project to verify and intercompare operational and semi-operational satellite rainfall estimates on a routine basis over Australia, US and Europe (URL: http://www.bom.gov.au/bmrc/SatRainVal/IPWG_precip_archive.html) (Ebert, 2002). These sites provide rain maps and validation statistics over land for some of the TRMM algorithms. The validation activities will provide useful insights for the interpretation of algorithm errors and improvement on rain algorithm physics.

Acknowledgements

We thanked Drs. Roberrt Adler, George Huffman, and an anomynous reviewer for useful comments, Dr. Yimin Ji and Mr. John Stout for computing statistics for February 1998 data for V6. The authors acknowledge support from the NASA TRMM program. LSC is also partially supported by the NASA REASoN program and NASA grant NNG04GB02G. The TRMM data is processed by TSDIS and distributed by the NASA GSFC DAAC.

References

Adler RF, Kummerow C, Bolvin D, Curtis S, Kidd C (2003a) Status of TRMM monthly estimates of tropical precipitation, in Cloud systems, Hurricanes and the Tropical Rainfall Measuring Mission (TRMM)—A tribute to Dr. Joanne Simpson, W-K. Tao and R. Adler, (Editors), American Meteorological Society, Boston, MA

Adler RF, Huffman GJ, Chang A, et al. (2003b) The version-2 global precipitation climatology project (GPCP) monthly precipitation analysis (1979-present). J Hydrometeor 4 (6): 1,147 – 1,167

Asrar G, Greenstone R, (Eds) (1995) MTPE/EOS Reference Handbook, National Aeronautics and Space Administration, Washington D.C., NP-215, pp 276

Bowman KP, Phillips AB, North GR (2003) Comparison of TRMM rainfall retrievals with rain gauge data from the TAO/TRITON buoy array. Geophys Res Lett 30(12): 1,757

Chang ATC, Chiu LS, Kummerow C, Meng J (1999) First results of the TRMM microwave image (TMI) monthly oceanic rain rate: Comparison with SSM/I. Geophys Res Lett 26(12): 2,379 – 2,382

Chiu L, Liu Z, Rui H, Teng W (2005a) Tropical Rainfall Measuring Mission (TRMM) data and access tools, Earth System Science Remote Sensing, J. Qu et al., (Editors), Springer (this issue)

Chiu L, Liu Z, Vongsaard J, Morain S, Budge A (2005b) Comparison of TRMM and Water Division rain rates over New Mexico. Advances in Atmospheric Sciences (accepted)

Ebert EE, (2002) Verifying satellite precipitation estimates for weather and hydrological applications. 1st Intl Precipitation Working Group (IPWG) Workshop, Madrid, Spain, 23 – 27 September 2002

Haddad ZS, Smith EA, Kummerow CD, Iguchi T, Farrar MR, Durden SL, Alves M, Olson WS (1997) The TRMM "day-1" radar/radiometer combined rain-profiling algorithm. J Meteor Soc Japan, 75: 799 – 809

Huffman GJ, Adler RF, Rudolph B, Schneider U, Keehn P (1995) Global Precipitation Estimates Based on a Technique for Combining Satellite-Based Estimates, Rain Gauge Analysis, and NWP Model Precipitation Information. J Clim 8: 1,284 – 1,295

Huffman GJ, Adler RF, Morrissey MM, et al. (2001) Global precipitation at one degree daily resolution from multisatellite observations. J Hydrometeor 2(1): 36 – 50

Huffman GJ, Adler RF, Arkin P, et al. (1997) The Global Precipitation Climatology Project (GPCP) Combined Precipitation Dataset, Bull Amer Meteor Soc 78 (1): 5 – 20

Huffman GJ, Adler R, Bolvin D, Nelkin E (2004) Uncertainty in fine-scale MPA precipitation estimates and implications for hydrometeorological analysis and forecasting, 18th Conf. On Hydrology, 11 – 18 January, 2004, Seattle, WA.

Iguchi T, Kozu T, Meneghini R, Awaka J, Okamoto K (2000) Rain-Profiling Algorithm for the TRMM Precipitation Radar. J Appl Meteor 39(12): 2,038 – 2,052

Iguchi T, Meneghini R (1994) Intercomparison of Single Frequency Methods for Retrieving a Vertical Rain Profile from Airborne or Spaceborne Data. J Atmos and Ocean Tech 11: 1,507 – 1,516

Kim MJ, Weinman JA, Houze RA (2004) Validation of maritime rainfall retrievals from the TRMM microwave radiometer. J Appl Meteor 43(6): 847 – 859

Kummerow C, Barnes W, Kozu T, Shiue J, Simpson J (1998) The Tropical Rainfall Measuring Mission (TRMM) Sensor Package. J Atmos and Ocean Tech 15: 808 – 816

Kummerow C, coauthors (2000) The status of the Tropical Rainfall Measuring Mission (TRMM) after two years in orbit. J Appl Meteor 39: 1,965 – 1,982

Kummerow C, coauthors (2001) The Evolution of the Goddard Profiling Algorithm (GPROF) for Rainfall Estimation from Passive Microwave Sensors. J Appl Meteor 40: 1,801 – 1,820

Lau KM, Wu HT (2003) Warm rain processes over tropical oceans and climate implications. Geophys Res Lett 30(24): 2,290

Liao L, Meneghini R, Iguchi T (2001) Comparisons of Rain Rate and Reflectivity Factor Derived from the TRMM Precipitation Radar and the WSR-88D over the Melbourne, Florida, Site. J Atmos and Oceanic Tech 18(12): 1,959 – 1,974

Meneghini R, Jones JA, Iguchi T, Okamoto K, Kwiatkowski J (2004) A Hybrid Surface Reference Technique and Its Application to the TRMM Precipitation Radar. J Atmos & Ocean Tech 21: 1,645 – 1,658

Nesbitt SW, Zipser EJ, Kummerow CD (2004) An examination of the version-5 rainfall estimates from the TRMM Microwave Imiager, Precipitation radar and rain gauges on global, regional and storm scales. J Appl Meteor 43(7): 1,016 – 1,036

Nicholson SE, coauthors (2003) Validation of TRMM and other rainfall estimates with a high-density gauge dataset for west Africa. Part II: Validation of TRMM rainfall products. J Appl Meteor 42(10): 1,355 – 1,368

Robertson F, Fitzjarrald D, Kummerow C (2003) Effects of uncertainty in TRMM precipitation radar path integrated attenuation on interannual variation of tropical oceanic rainfall. Geophy Res Lett 30(4): 1,180

Rudolf B, Hauschild H, Ruth W, Schneider U (1994) Terrestrial precipitation analysis: operational method and required density of point measurements. Global Precipitation and climate change. M. Dubois and M. Desalmand, Eds Springer Verlag, pp 173 – 186

Schumacher C, Houze RA (2000) Comparison of radar data from the TRMM satellite and Kwajalein ocean validation site. J Appl Meteor 39: 2,151 – 2,164

Serra YL, McPhaden MJ (2003) Multiple time-and space-scale comparison of ATLAS buoy rain gauge meausrments with TRMM satellite precipitation measurements. J Appl Meteor 42: 1,045 – 1,059

Shin D-B, Chiu LS, Kafatos M (2001) Comparison of the monthly precipitaiton derived from the TRMM satellite. Geophys Res Lett 28(5): 795 – 798

Simpson J, Adler RF, North GR (1988) A Proposed Tropical Rainfall Measuring Mission (TRMM) Satellite. Bull Am Meteor Soc 69: 278 – 295

Smith EA, Hollis TD (2003) Performance Evaluation of Level-2 TRMM Rain Profile Algorithms by Intercomparison and Hypothesis Testing. Meteorological Monographs 29: 207

Teng W, Chiu L, Doraiswamy P, Kempler S, Liu Z, Pham L, Rui H (2005) An interoperable agricultural information system based on satellite remote sensing data. ASPRS annual conference, March 7 – 11, 2005, Baltimore, MD

Yu C, Chiu L (2005) Comparison of TRMM rain rates using non-parametric statistical methods. Preprint, Conference on Hydrology, AMS Conference, January, 2005, San Diego, CA

18 Use of Satellite Remote Sensing Data for Modeling Carbon Emissions from Fires: A Perspective in North America

Zhanqing Li, Ji-Zhong Jin, Peng Gong and Ruiliang Pu

18.1 Introduction

Accurate accounting of carbon cycling is paramount to understanding and modeling global climate change. At present, a considerable amount of global carbon uptake (~2 Gt/year) remains unaccounted for in the carbon budget. It has been argued that the missing carbon may be absorbed in the terrestrial biomes of the Northern Hemisphere (Tans et al., 1990), in particular the temperate and boreal forests in North America (NA), which could account for the bulk (1.7 Gt/year) of the missing carbon (Fan et al., 1998). Fire is a driving factor controlling the carbon dynamics in NA, which affects both the sign and magnitude of the carbon budget (Stocks et al., 1996; Conard and Ivanova, 1997; Kasischke, 2000). According to the modeling results of Chen et al. (2000), boreal forests in NA have undergone tremendous fluctuations in its carbon budget over the last 200 years (Fig. 18.1). Around 1880 when widespread severe fires released a huge amount of carbon into the atmosphere, the forests were a very large source of carbon (~140 GtC/year) while around 1940, the forests became a very large sink of carbon (~200 GtC/year) due to fast forest regeneration that absorbed a large quantity of atmospheric carbon. The net carbon exchange is now so small that it is being debated whether the boreal forest is currently a sink (Chen et al., 2000) or a source (Kurz et al., 1995). The close correlation between the carbon budget and fire activity demonstrates the importance of the accurate estimation of carbon emissions from fires.

So far, the continental-scale estimates of carbon emission were made mainly for fires occurring over the forest ecosystem using ground-based fire datasets (French et al., 2000). Few attempts have been made employing remote sensing data from coast to coast. While ground-based data are valuable, they have certain limitations that can be overcome by remote sensing. Ground-based fire data are mainly restricted to total burned area with their quality and completeness varying from year to year and region to region. Remote sensing is capable of providing additional spatial and temporal fire information to improve fire emission estimations.

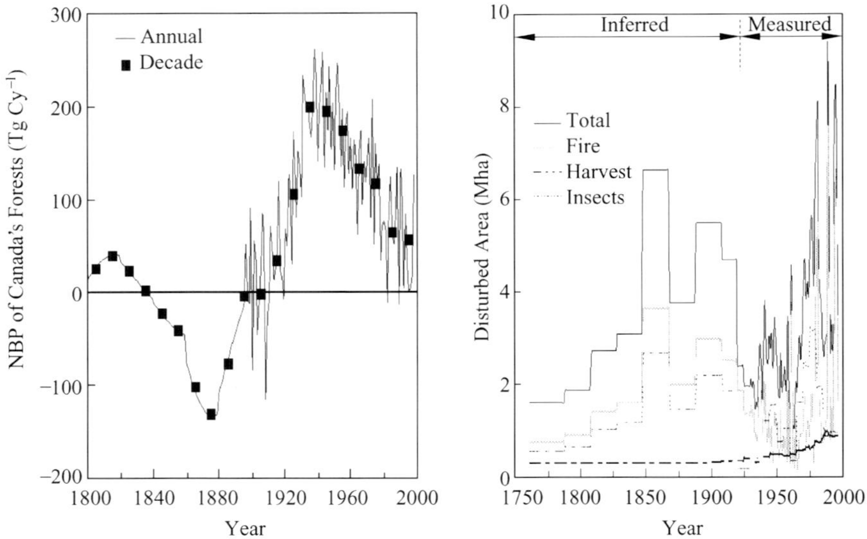

Figure 18.1 Model simulated time series of net carbon source/sink from boreal forests in the North America (left) and major disturbance factors (Chen et al., 2000)

In addition to burned area, remote sensing can provide "snapshots" of fire dynamics information (starting and ending dates and daily spread), and spatial heterogeneity (the degree of burning, fragmentation of burned scars, fuel type, biomass amount, etc.). On the other hand, remote sensing has its own limitations that do not allow providing all emission related parameters such as emission factor. Therefore, the best strategy is to combine conventional and satellite data to maximize their utility for fire emission estimation.

18.2 Carbon Emission Estimation

There is a simple mathematical formula to compute the emission of any chemical gas or particle species as originally proposed by Seiler and Crutzen (1980):

$$E = BA \times FL \times FF \times EF \tag{18.1}$$

where E is the emission of a gas (x) or particulate matter from fire (g)—(here, mainly for CO_2); BA is burned area (ha); FL fuel loading (or density) (kg/ha); FF = fraction of fuel consumed (%); EF emission factor for gas species (x) or particulate matter (g/kg of fuel consumed). It has been, however, a daunting task to obtain any of these variables from ground, air-borne, or space-borne observations. Nearly none of them are trivial to derive from any observation platform, nor from modeling.

To effectively make use of a variety of spatial data in raster or vector formats, GIS-based emission modeling systems as is shown in Fig. 18.2 have been developed. For fire emission estimation, major fire emission attributes can be obtained from satellite and other sources as model inputs. They include burned area, degree of burning, fuel loading, above-ground and ground layer biomass amount, burning conditions and emission factors, etc.. Before running the system, the input data need to be acquired, edited and transformed as different attribute data layers. The system consists of several modular subsystems that simulate the burning processes. Some of the systems are based on the algorithms from the Forest Service First Order Fire Effects Model (FOFEM, Reinhardt, 1997) coded into Avenue (the ArcView scripting language) for implementation in the GIS.

The system needs to be able to use remote sensing data in combination with conventional data in order to enhance the estimation of carbon emissions and cycling.

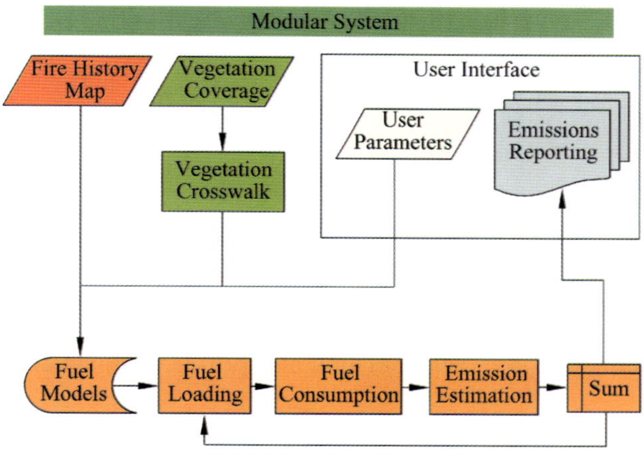

Figure 18.2 Flowchart represents the Emission Estimation System (EES) modules. Boxes represent module components. Shading distinguishes modules from each other

18.3 Fire Emission Parameters and Modeling

18.3.1 Burned Area

There have been a large number of studies using satellite data to monitor and map fires around the globe. Recent reviews on fire detection, burned area mapping and fire observation systems/products may be found in Li et al. (2001a), Arino et al. (2001), and Grégoire et al. (2001), respectively.

Among all fire emission related factors, burned area can be inferred most accurately by satellite. This is because fires usually leave a distinct "footprint" that can be captured by satellite. Two types of "fire footprints" have been traced for fire remote sensing, namely, the long-lived fire scars and short-lived fire hot spots. Fire scars may be measured by the drastic decrease in vegetation index such as the Normalized Difference Vegetation Index (NDVI) computed from reflectance in visible (VIS) and near infrared (NIR) channels:

$$\text{NDVI} = \frac{(\text{NIR} - \text{VIS})}{(\text{NIR} + \text{VIS})} \tag{18.2}$$

Numerous attempts have been reported to estimate burned areas based on changes in NDVI (Kasischke and French, 1995; Martin and Chuvieco, 1995; Li et al., 1997, 2000b). However, use of the NDVI alone tends to cause significant commission errors, since the NDVI decrease may be unrelated to fire and more related to drought, seasonal vegetation senescence, timber harvesting, image mis-registration, and cloud contamination. A further difficulty lies in the selection of effective thresholds for separating burns that are spatially and temporally variable (Fernandez et al., 1997).

One might composite all fire hot spots to obtain the burned area. As optical remote sensing of fires is only feasible under clear-sky conditions (Li et al., 2000a), an accumulation of fire hot spots may be substantially less than the actual area of burning, depending on cloud cover and frequency (Li et al., 2000b). To overcome these limitations, methods have been developed that combine synergistic information on fire hot spots and vegetation damage indicated by a vegetation index (Roy et al., 1999; Fraser et al., 2000a). A method named Hotspot and NDVI Differencing Synergy (HANDS) proposed by Fraser et al. (2000a) has a high degree of automation and self-adaptation(see Fig. 18.3). The

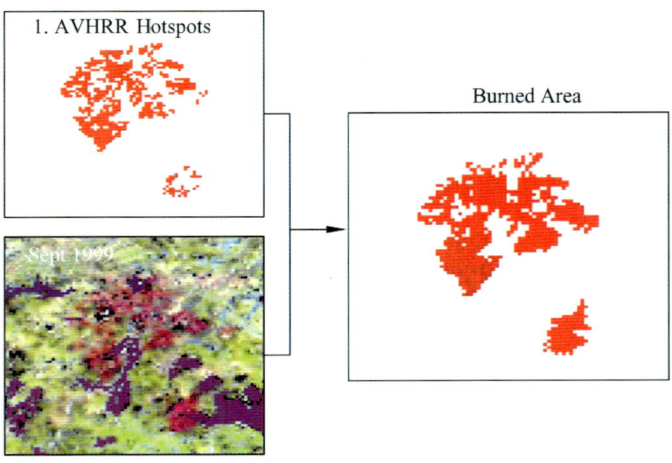

Figure 18.3 A simplified schematic of the HANDS method proposed by Fraser et al. (2000a) in generating burned area (right) by synergetic use of fire hot spot (left top) detected by the Li et al. (2000a) algorithm and vegetation index (left bottom)

principle of this method is to use hotspot locations to train spatially variable NDVI difference threshold, while changes in NDVI between two periods (before and after a fire, or on two anniversaries in two consecutive years) allow to eliminate false fire hot spots.

Using the fire detection algorithm of Li et al. (2000a) and the mapping algorithm of Fraser et al. (2000a), a long-term (1989 – 2000) daily 1-km fire product has been generated across the NA from the historical AVHRR archive (Pu et al., 2006, c.f. Fig. 18.4). Validations of the satellite mapped burned areas against fire polygons generated by air-borne surveillance showed a very close match (Li et al., 2003). Moreover, the satellite mapping method can pick up fires that were missed by the conventional method. This is especially the case over remote regions where fires are usually allowed to develop in their own natural course and so manual mapping is less complete. This and other validations (Fraser et al., 2000b; Fraser and Li, 2002) have revealed consistent high accuracy in mapping burned areas over forests.

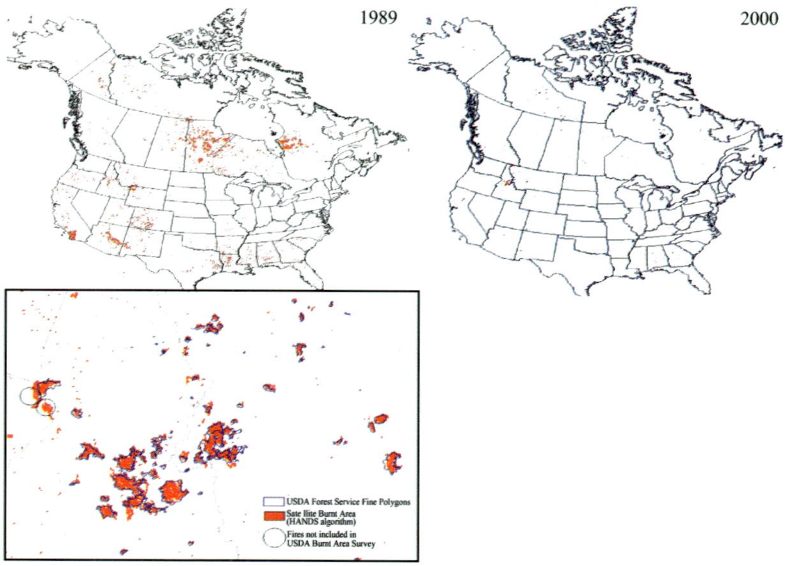

Figure 18.4 Upper panels: nation-wide fire burned scars mapped from AVHRR data in 1989 and 2000. Such fire maps are available on a daily basis across NA continent. Lower left panel: comparison of fire burn scars mapped from satellite and the USDA Forest Service for the fires occurred in the western states in 2000 (Li et al., 2003)

Large errors are found in mapping fires over non-forest land (Csiszar et al., 2003). This has been a major problem for all satellite-based fire products using the mid-infrared channel of AVHRR around 3.7 μm, which is also a key channel for fire detection. The problem arises from channel saturation and the contribution of solar reflection (Li et al., 2001a). Note that the main purpose of this channel was originally designed for mapping sea surface temperature during night overpass. The maximum detection limit is usually set around 320 K with a range of variation (Csiszar and Sullivan, 2002). This does not incur any problem unless exceptionally hot targets are detected such as fires or volcanoes that can readily saturate the channel. Were the saturation caused by hot targets alone, it would not be a problem so far as fire detection is concerned. Unfortunately, saturation may be caused by unusually large thermal emissions or solar reflection, or both. As illustrated in Li et al. (2001a), a scene of albedo over 20% in this wavelength can reflect enough solar radiation to saturate the channel. Although channel 3 is known to have a significant contribution of solar reflection in the mid-IR, none of the existing fire detection algorithms have taken it into consideration (Li et al., 2001a). This is partially because much attention has been paid to fires occurring over forests where reflectivity in that channel is very small (< 5%). For other natural scene types, surface reflectivity may be large, variable and difficult to obtain (Salisbury et al., 1991; Salisbury and D'Aria, 1994; Snyder et al., 1997), as shown in Fig. 18.5.

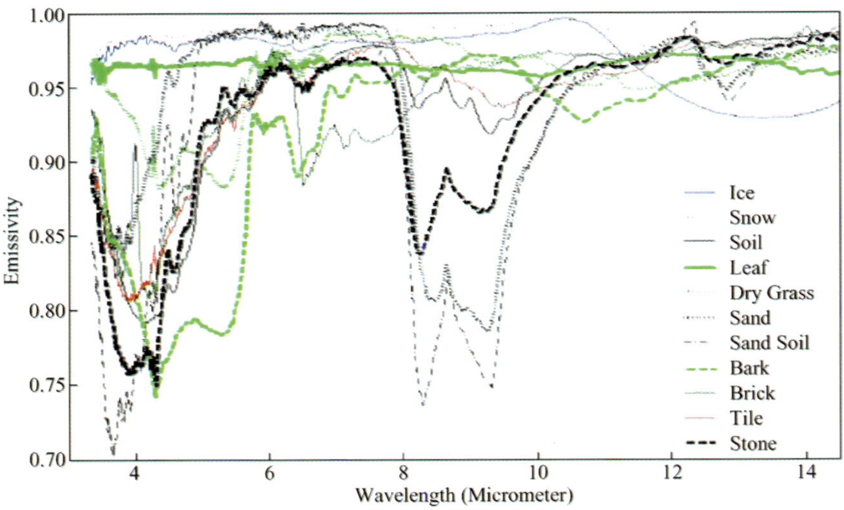

Figure 18.5 The emissivity of some natural materials (Salisbury et al., 1991). Note that emissivity is equal to 1 minus albedo

This inherent problem can be resolved/lessened by the MODIS sensor. MODIS has several advantages over the AVHRR for fire monitoring (Kaufman et al., 1998a; Justice et al., 2002; Roy et al., 2002; Ichoku et al., 2003; Kaufman et al., 2003; Li et al., 2004). First, the saturation limit for the MODIS 3.7 μm channel is much higher. Second, the MODIS products include estimates of surface emissivity at this and other IR channels. Use of the emissivity data and solar radiative transfer calculation, one can determine and subtract the contribution of solar reflection from the total radiance measured at this channel so that the remaining signal is a true measure of thermal emission. Third, MODIS includes a longer mid-IR channel near 4 μm where incoming solar radiation is much reduced. Fourth, MODIS has several additional channels that can greatly facilitate fire detection and mapping. One most useful channel is the shortwave (SW) IR channel around 1.6 μm.

While NDVI has been widely used for mapping burned area, its signal may fade out quickly for regions where understory vegetation grows rapidly after fire. A vegetation index derived from a combination of SWIR channel around 1.6 μm and NIR as was proposed by Kaufman and Remer (1994) has a long-lasting "memory" of burning:

$$\mathrm{SWVI} = \frac{\mathrm{NIR} - \mathrm{SWIR}}{\mathrm{NIR} + \mathrm{SWIR}} \tag{18.3}$$

This is demonstrated in Fig. 18.6 showing a comparison of NDVI and SWVI of an old fire scar inside the white polygon. The scar was generated by a big forest fire that occurred in 1995, but the vegetation indices were computed from the SPOT/VGT (a French satellite) image obtained in 1998. It is seen that the NDVI shows almost no trace of burning, but there is a distinct fire boundary markedly discernible from the SWVI. Another advantage of using SWVI over the NDVI lies in its lack of influence by smoke plume. Note that smoke consists of fine-mode particles whose transmittance to solar radiation increases with wavelength. Taking advantage of these properties, Li et al. (2004) proposed a method for near real-time mapping of burned area using multiple MODIS channels from SWIR to mid-IR (also see Fig. 18.9).

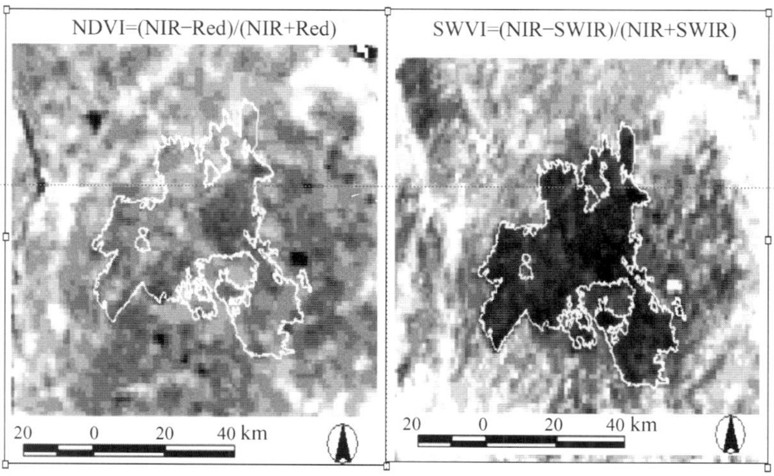

Figure 18.6 Comparison of NDVI and SWVI image derived from SPOT/VGT image in August 1998 over a burned scar created in 1995 in Alberta, Canada

18.3.2 Spatial Fragmentation and Temporal Expansion of Burned Area

Burn fragmentation due to varying degrees of burning is next to total burned area in determining fire emissions. The methodology described based on AVHRR and MODIS sensors is adequate to delineate fire boundaries. However, a burning field generally shows very inhomogeneous distribution due to varying degrees of burning severity. Burn fragmentation is very important for estimating fire emissions. High-resolution data such as that from LANDSAT-7 TM is valuable for assessing fire severity (Michalek et al., 2000) and calibrating burned areas mapped by the coarse resolution data (Fraser et al., 2004).

Figure 18.7 shows a comparison of burned areas extracted from LANDSAT-5 TM and SPOT/VGT data for fires that occurred during May 1998 in Alberta, Canada. The outer burn perimeter derived using VGT data corresponds quite well to the TM boundary. However, as a result of the coarser resolution of VGT imagery (1 km), most interior small-scale unburned islands are mapped as being burnt. This leads to a systematic over-prediction of burnt area, and thus also the emission estimation. By double sampling a representative selection of fires with both TM and VGT data, a function of the two burned areas was derived that can be used to calibrate the coarse resolution burned area at a continental scale. The calibration coefficient may be parameterized by the change in a vegetation index before and after burning.

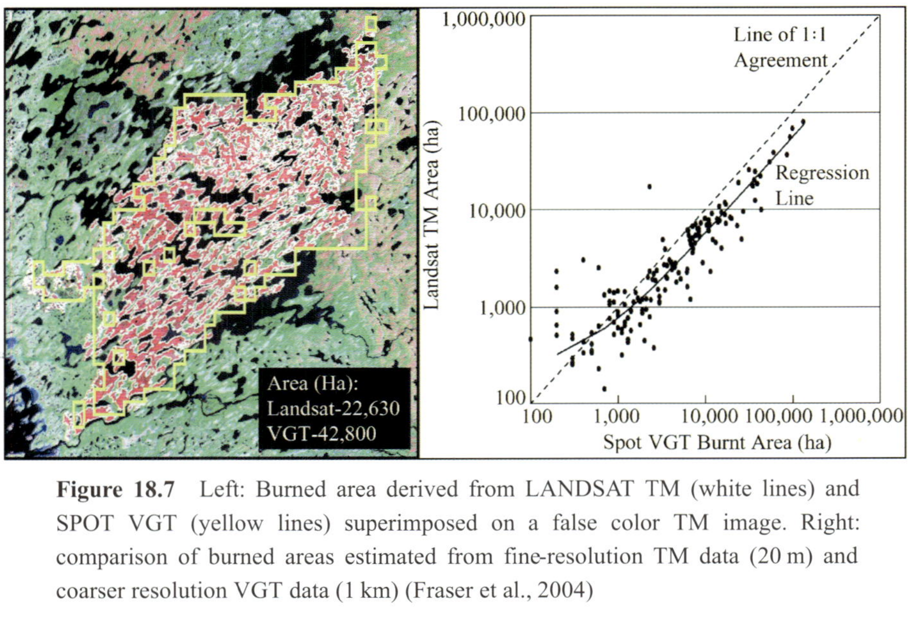

Figure 18.7 Left: Burned area derived from LANDSAT TM (white lines) and SPOT VGT (yellow lines) superimposed on a false color TM image. Right: comparison of burned areas estimated from fine-resolution TM data (20 m) and coarser resolution VGT data (1 km) (Fraser et al., 2004)

For fire emission modeling, it is also necessary to have dynamic information on the progress of burned scars. Such fire-spread information can be coupled with daily fire weather data into the emission model (discussed later) to simulate fuel consumption from crown and ground fires. At a minimum, one may use daily fire hotspots to roughly represent a fire episode. While this temporal accumulation of fire hot spots provides valuable dynamic information that is in addition to the total burned area, it is often interrupted by any presence of cloud cover. To generate a more continuous fire development data with the traditional AVHRR imagery, we developed a burn growth algorithm using current hotspots as seeds. Burned pixels are iteratively "grown" from hotspot locations if they satisfy three conditions:

(1) Joined to a hotspot or previously identified burned pixel;

(2) Not cloudy, based on a thermal infrared (channel 4) threshold; and

(3) Elevated mid-infrared (channel 3) signal with respect to background.

Tests #2 and #3 are specifically designed to permit mapping of burned areas that are covered by smoke plumes.

This near real-time burned area growing algorithm was applied to daily AVHRR images corresponding to a big fire occurred in Virginia Hill in Alberta, Canada from May 3 to 20, 1998. Figure 18.8 shows the resulting cumulative burned area for each day. Burned pixels identified on a given day are then linked to current fire weather data in order to estimate fuel consumption.

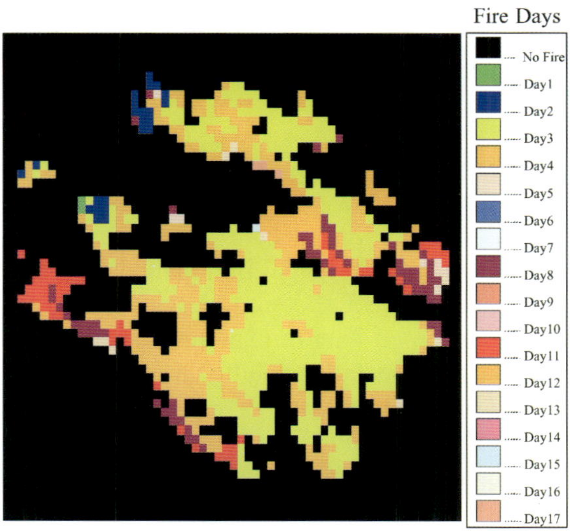

Figure 18.8 Development of daily burned area for a fire in Canada

For MODIS, this complicated procedure may be avoided by using the multiple SW IR channels at 1.24, 1.64 and 2.13 μm, following the new method of mapping fire scars proposed by Li et al. (2004). Fresh fire scars are clearly visible on a false color image composited by data from the three SWIR channels even under a heavy smoke. In contrast, the scar is completely marked by the smoke for a real color image using data from red, blue and green channels (see Fig. 18.9). This technique allows for mapping of real-time burned areas as long as no cloud is present.

Figure 18.9 Left: a true color AVIRIS image of a fire smoke scene composited from red, green and blue channels. Right: a false color image composited from three AVIRIS SWIR channels that are equivalent to MODIS channels at 1.24, 1.64 and 2.13 μm (Li et al., 2004)

18.3.3 Fuel Loading

Fuel loading varies considerably with fuel type, tree density, species composition, age, etc. Due to a lack of available spatial information on fuel loading, previous estimates of fire emissions (e.g., Cahoon et al., 1994) have assumed total fuel loading and consumption based on fire experimental data (e.g. 2.5 kg/m^2). To improve estimates of the large spatial variability in fuel loading, a recent emission study by French et al. (2000) characterized average fuel loading for nine forested ecozones in Canada. They employed the Canadian Forest Service's forestry inventory (CanFI) data and allometric equations to convert stand information to aboveground biomass. Below ground carbon was estimated from a national soil map, which is available in GIS format from CanSIS. CanFI provides general information pertinent to spatial changes in fuel loading. They do not provide spatial details, as they are given on the scale of ~10,000 km^2. The area of an ecoregion is usually much larger than 100 km^2, while fuel loading within an ecoregion could be highly variable at much finer spatial scales. Since burn scars are usually patchy and non-uniform, more spatially resolved fuel loading information would help improve estimates of fire emissions.

In order to reduce the uncertainties in emission estimation, it is highly desirable to obtain dynamic information to characterize the spatial and temporal variations in fuel loading. This is an extremely challenging task and thus no large-scale fine-resolution data on fuel loading exist at present. While no attempt has been made to extract fuel loading information from remote sensing, certain qualitative measures of biomass content could be derived from satellites (Arseneault et al., 1997). The most promising approach would be to use space-borne lidar to measure the height of vegetation. This concept has been demonstrated with a proposed space-borne vegetation lidar (Dubayah and Drake, 2000). Satellite imagery, particularly at SW infrared channels, conveys certain information pertinent to vegetation biomass. A preliminary analysis indicates that there is a correlation between SWVI and total forest biomass and the vegetation index can explain between 60% – 66% of the variation in post-fire forest regrowth age, an indirect measure of biomass content (Fraser and Li, 2002)(see Fig. 18.10). Although both correlations are weak, they may still be valuable for regions where there does not exist any biomass data, which is the case over the vast majority of forest land anywhere in the world. So, any information is better than none.

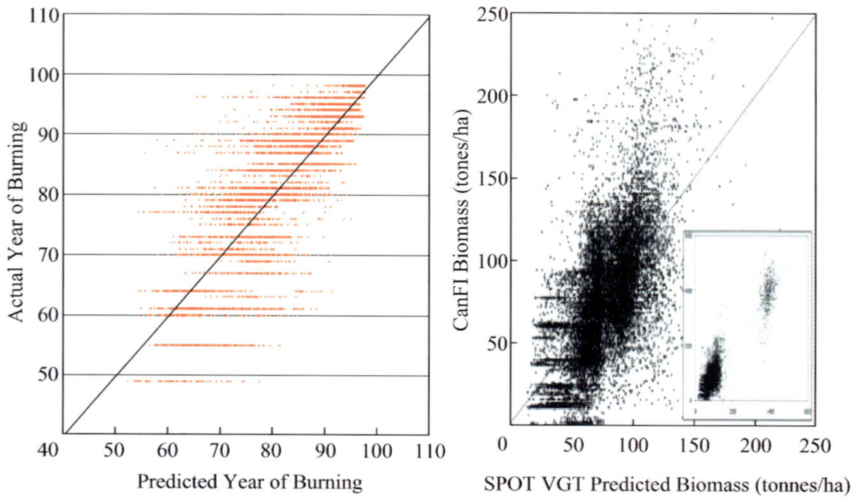

Figure 18.10 Left: Relationship between post-fire regeneration ages obtained from historical fire record and predicted from a SWVI. Right: same as left but for above-ground biomass content (Fraser and Li, 2002)

Combining the forest regrowth age with another proxy of fuel loading would provide a much improved estimate of fuel loading; that is the tree density or the fraction of forest cover inside a satellite pixel. Global distribution of the percentage of tree cover has been derived from 500 m resolution MODIS data (Hansen et al., 2002).

Use of the tree density information alone has proven valuable in estimating carbon emissions from the tropical deforestation and regrowth from 1980s to 1990s (DeFries et al., 2002). More accurate estimation of fuel loading would require a field survey to determine the average tree densities and diameters of different forest stands (Michalek et al., 2000) in order to better estimate above-ground biomass from the tree density data for different forest cover types. Allometric equations can be used to determine average tree biomass.

Despite these promising attributes from satellite, our ability to map fuel loads with satellite imagery is still very limited, as they are all concerned mainly with above-ground biomass. Below-ground biomass is even more difficult to estimate by any means, but it is very important for estimating the total emission from fires. Variations in the burning of organic soils account for a large portion of uncertainty in the estimates of emissions from forest fires, but not so for non-forest fires.

Average pre-burn ground layer biomass estimates are usually determined by collecting a number of ground layer profiles in each tree density class. The depths of different strata within the ground layer (e.g., litter, live moss, dead moss, fibric and humic soil) are measured at each profile. Samples of each stratum can be analyzed in the laboratory to determine bulk density and biomass.

While satellite remote sensing cannot be used to infer underground biomass directly, satellite observations of smoke loading integrated over the lifetime of burning could provide certain qualitative information of the biomass burned above and below ground, should the mode (smoldering or flaming) and temperature of burning be known (Kaufman et al., 1990). It is worthwhile to explore the relationship between burned biomass and smoke emissions. There are many ways to identify smoke from analysis of satellite imagery such as the threshold method, neural network method, pattern recognition method, etc.. Figure 18.11 presents an example of classification of smoke, cloud and clear land by applying the methods proposed by Li et al. (2001b) to AVHRR data. For MODIS, smoke can be much more easily identified from clouds whose reflectivity remains high for all solar channels using the combination of channels at short and longer wavelengths as illustrated in Fig. 18.9. Smoke emissions are generally proportional to smoke optical depths.

Aerosol optical depth has been retrieved from both sensors (Mishchenko et al., 1999; Kaufman et al., 2002). However, large uncertainties exist for smoke aerosol whose retrieval depend critically on aerosol absorption properties (Wong and Li, 2002). For smoke, absorption depends on fuel type and burning conditions. While this is a promising approach, little effort has been made, for it demands sizable resources to establish the relationship between the two quantities, as it requires fire-by-fire analyses of total, above- and below-ground biomass amounts measured before, during and after burning.

Figure 18.11 A smoke image classified by a method of Li et al. (2001b) applied to the AVHRR data

18.3.4 Fuel Type

Fuel type is important for characterization of fire behavior, fuel loading and emission efficiency (Anderson, 1982). In NA, continental-scale fuel type data have been generated both in the US and Canada from land cover type (Eyre, 1980), forest inventory, above-ground and ground layer biomass survey, ecoregions, drainage classes, topography, and soil types (Bailey, 1998). In the US, fuel model and types are available from the National Fire Danger Rating System (NFDRS, http://www.fs.fed.us/land/ wfas/nfdr_map.htm) and the Forest Service Wildland Fire Assessment System (Burgan et al., 1998; Reinhardt et al., 1997). The conventional fuel type data are static and have coarse spatial resolution relative to other fire attributes extracted from satellite as elaborated above. As more refined and more reliable land cover classification data are now available from numerous satellite sensors (AVHRR, MODIS, TM/ETM +) (DeFries et al., 1998; Cihlar et al., 1999; Hansen et al., 2000), it is possible to generate continental-scale fuel type at high resolutions, together with low-resolution ground-based survey data on forest inventory, soil drainage class, and ecozone. An ecosystem map may help confirm the classified fuel types by examining whether they fall into a sound ecozone, such as the Terrestrial Ecozones and Ecoregions of Canada produced by the Agriculture and Agri-Food Canada, Ecological Stratification Working Group (1996). A drainage map provides additional information that may be used to discriminate similar fuel types. After the various data sets are synthesized, a set of rules need to be established and applied by an expert system to classify the fuel types with multi-layer information. The criteria for the classification depend on the characteristics of fire behavior for a particular fuel type. The classification may follow a hierarchical structure in order to establish the relations between various input data sets and to organize the classification rules. Use of satellite data allows updating of the fuel type change due to biomass burning and other land cover and land use events.

18.3.5 Fraction of Fuels Consumed

The fraction of fuels consumed (FFC) by fire depends on fuel type, fuel loading, fuel moisture dictated by fire weather conditions, as well as the phase of combustion, i.e., smoldering or flaming. Studies on the estimation of fire emissions have either assumed a constant FF (Cahoon et al., 1994) or used temporally (monthly or seasonally) and spatially (regionally or provincially) averaged values (Vose et al., 1996). Cairns et al. (2000) assigned constants of above-ground biomass burned to four categories of forest. French et al. (2000) used a simple model to estimate a weighted fraction of biomass consumed for each year and ecozone based on annual area burned.

The spatial and temporal variations of FFC may be resolved by integrating various data sets, in particular the remote sensing data and the forest fire danger estimates provided by forest services on a daily basis. In Canada, such data are available from the Fire Weather Index (FWI) System (Van Wagner, 1987) and Fire Behavior Prediction (FBP) System (Forestry Canada Fire Danger Group, 1992). The systems have been operated during the fire season. The FWI system calculates fuel moisture codes and fire behavior indices based on daily weather conditions. The FBP system simulates fire behavior for each of the 16 fuel types based on the FWI indices, topography, and fuel type. The primary outputs of the FBP system are rate of spread, fuel consumption, and fire intensity. Total fuel consumption (TFC) includes surface fuel consumption (SFC) and crown fuel consumption (CFC). The moisture codes and indices of FWI system, fuel types, and topography determine how much surface fuel is consumed, whether a crown fire occurs, and what fraction of the crown fuel is consumed.

Figure 18.12 presents the flowchart of a fuel consumption and fire emissions modeling system developed based on the Canadian FWI and FBP modules. The system can make direct use of satellite-based products such as daily burned area and fuel type data. Therefore, the output can depict the spatial and temporal variation of fire emission. The system was run for estimating emissions from a big fire in Canada with the results presented in Fig. 18.13 (Li et al., 2000c). This approach is better than the polygon-based model simulation that assumes homogenous fuel distribution within a fire polygon.

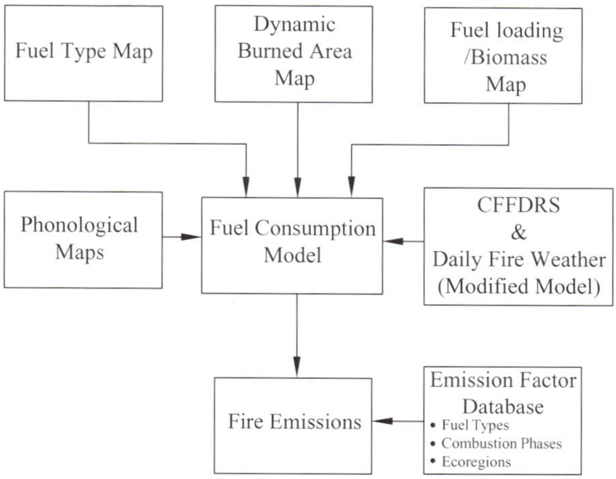

Figure 18.12 A flowchart of a fuel consumption model for computing fraction of fuel consumed

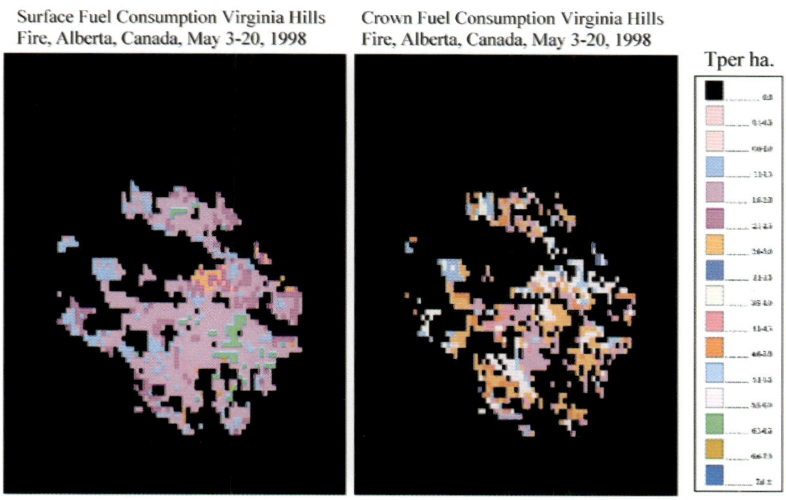

Figure 18.13 Estimated fuel consumptions from surface (left) and tree crown fires for a big fires in Virginia fires in Alberta, Canada from May 3 – 20, 1995 (Li et al., 2000c)

This approach is similar to that of Cairns et al. (2000), but uses improved estimates of several key parameters from satellite: (1) more precise fire starting date, (2) fire ending date, (3) daily fire spread, and (4) burn severity and fragmentation. For fire emissions modeling, fire weather is the primary factor dictating fuel consumption. Precipitation amount is an input of the system for determining fuel moisture. Information on vegetation moisture may also be extracted from NDVI or other vegetation indices from MODIS or AVHRR (Illera et al., 1996; Gonzalez-Alonso et al., 1997).

Kaufman et al. (1998a) proposed a "short-cut" approach to estimate fuel consumption by using the MODIS thermal radiance at IR channels around 4 and 11 μm. It was assumed that the rate of biomass consumption and emission of trace gases and aerosol particles are proportional to radiation emitted at the two channels. The assumption may only be valid if the channels are not saturated. Unlike the low saturation level (~320 K) of AVHRR ch. 3, the 4 and 11 μm MODIS channels are sensitive to temperatures up to 450 and 400 K respectively. It is worth noting that fire temperature varies considerably within a burning field in general due to inhomogeneous fuel loading, moisture, heat-induced turbulence, wind, etc. This is clearly illustrated in Fig. 18.14, obtained by an air-borne thermal imager during a fire experiment conducted over a Canadian boreal forest. While the hottest portion of the burning field is around 700 K, hot burning is confined to small areas. The bulk of the area is less than 400 K. Besides, the brightness temperature as measured from space is always lower than surface temperature due

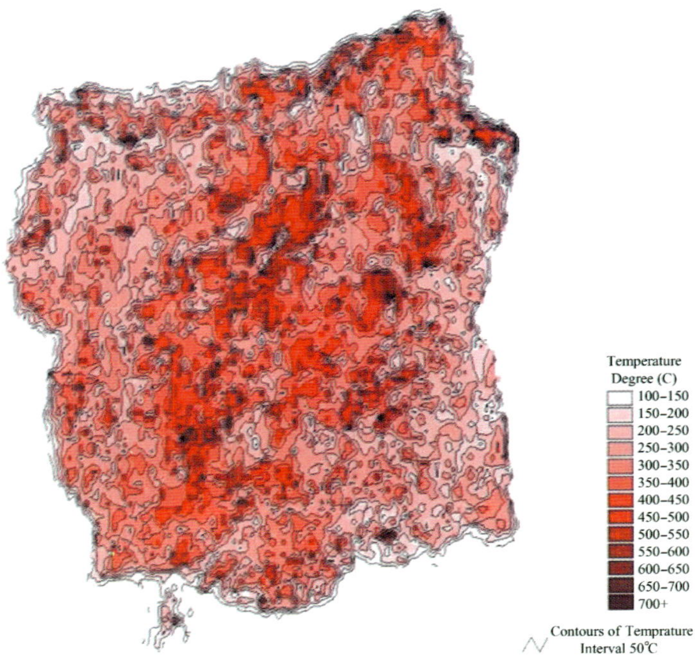

Figure 18.14 Temperature distribution over a burning field over a Siberia forest in Yartsevo, Russia on July 18, 2000 (Courtesy of Doug McRae at the Canadian Forest Services)

to atmospheric attenuation and contribution of the atmospheric emission at much colder temperatures. As a result, the spatially integrated temperatures as registered by the MODIS sensor over a field of view of 1 km rarely exceed the saturation limits for ordinary wild fires. The relationship between the brightness temperatures at the two channels can serve as an indicator of relative contribution from smoldering and flaming fires to the total biomass consumption and evaluate their effect on the combustion efficiency and emissions (Kaufman et al., 1998a, 1998b).

Combustion efficiency may be measured as the fraction of carbon released from the biomass in the form of CO_2 over the sum of CO_2 and CO (Ward et al., 1996). To make this approach work, it requires extensive in-situ measurements of fire thermal energy, burned area and rate of emission of gases and particle matters in order to establish their relationships (Kaufman et al., 1998b). Yet, the relationship may be complex, as it is likely to change with vegetation types.

18.3.6 Emission Factor

Emission factors (EFs) are required to convert fuel consumption (kg) to gas and particulate emission yields (g). For large-scale fire emissions modeling, the EFs have normally been compiled from biomass burning experiments (Cofer et al., 1988; FIRESCAN Science Team, 1996; Goode et al., 2000). Comprehensive measurements have been made in some of the experiments that document the proportion of biomass burned, the combustion efficiency and the emission factors of CO_2 and other trace gases. EFs were measured during the International Crown Fire Modeling Experiment for a jack pine with spruce understory fuel type (Cofer et al., 1998). The Boreal Forest Island Fire Experiment measured EFs for Scots Pine in Siberia (FIRESCAN Science Team, 1996). The experiments include reports on weather condition and vegetation characteristics (species, composition, fuel loading, fuel moisture content, etc.). Relationships between fire emission factors and burning and environmental conditions established from these experiments are valuable to better quantify the variation of emission factors (Hao et al., 1998; Susott and Ward, 1999; Yokelson et al., 1999).

A comprehensive database of EFs was assembled from the literature based on results from fire experiments conducted in Canada and the United States. Table 18.1 lists the emission factors that we have compiled and used in the current version of our fire emission estimation system. A table in the FOFEM literature (Reinhardt et al., 1997) shows different combinations of combustion efficiencies for flaming and smoldering phases of combustion under varying moisture regimes. Ward and Hardy (1991) reported empirically derived equations for CO_2 and CO as functions of combustion efficiency. Reinhardt et al. (1997) used the CO equation with the table of combustion efficiencies to create CO emission factors for different moisture regimes. These emission factors vary with moisture regime due to different ratios of flaming to smoldering combustion.

Table 18.1 Table of emission factors used in the EES for different moisture conditions. Emission factors for CO, PM_{10}, and $PM_{2.5}$ were obtained from FOFEM literature. All other emission factors were derived by the Center for the Assessment and Monitoring of Forest and Environmental Resources (CAMFER), University of California, Berkeley for use in the EES

Pollutant	Moisture Regime	Litter, Wood 0 – 1 Inch	Wood 1 – 3 Inches	Wood 3 + Inches	Herb, Shrub, Regen	Duff	Canopy Fuels
		emission factor in pounds of emissions per ton of fuel consumed					
PM_{10}	Wet	9.30	14.00	26.60	25.10	28.20	25.10
PM_{10}	Moderate	9.30	14.00	21.60	25.10	30.40	25.10
PM_{10}	Dry	9.30	14.00	19.10	25.10	30.40	25.10
PM_{25}	Wet	7.90	11.90	22.50	21.30	23.90	21.30
PM_{25}	Moderate	7.90	11.90	18.30	21.30	25.80	21.30
PM_{25}	Dry	7.90	11.90	16.20	21.30	25.80	21.30
CO	Wet	52.40	111.40	268.90	249.20	288.60	249.20
CO	Moderate	52.40	111.40	205.80	249.20	316.10	249.20
CO	Dry	52.40	111.40	174.40	249.20	316.10	249.20
CH_4	Wet	2.10	4.46	10.76	9.97	11.54	9.97
CH_4	Moderate	2.10	4.46	8.23	9.97	12.64	9.97
CH_4	Dry	2.10	4.46	6.98	9.97	12.64	9.97
TNMHC	Wet	3.67	7.80	18.82	17.44	20.20	17.44
TNMHC	Moderate	3.67	7.80	14.41	17.44	22.13	17.44
TNMHC	Dry	3.67	7.80	12.21	17.44	22.13	17.44
NH_3	Wet	0.52	1.11	2.69	2.49	2.89	2.49
NH_3	Moderate	0.52	1.11	2.06	2.49	3.16	2.49
NH_3	Dry	0.52	1.11	1.74	2.49	3.16	2.49
N_2O	Wet	0.49	0.47	0.43	0.43	0.42	0.43
N_2O	Moderate	0.49	0.47	0.45	0.43	0.42	0.43
N_2O	Dry	0.49	0.47	0.45	0.43	0.42	0.43
NO_x	Wet	8.23	7.97	7.27	7.36	7.19	7.36
NO_x	Moderate	8.23	7.97	7.55	7.36	7.07	7.36
NO_x	Dry	8.23	7.97	7.69	7.36	7.07	7.36
SO_2	Wet	2.53	2.45	2.24	2.27	2.21	2.27
SO_2	Moderate	2.53	2.45	2.33	2.27	2.18	2.27
SO_2	Dry	2.53	2.45	2.37	2.27	2.18	2.27

For other chemical species, e.g. CH_4, Total Non-Methane Hydro-Carbons (TNMHC), and NH_3, one can use their emission ratios to CO based on field experiments to create the emission factors in Table 18.2.

Table 18.2 Total emission of carbon species from the Virginia Hill fire computed the fire emission model coupled with various satellite derived parameters

Type of Burning	Pixels	Avg. Fuel Consumption (t/ha)	Emissions From Fire (megatons)		
			CO_2	CO	CH_4
Surface	994	20.4	2.7743	0.2873	0.0103
Crown	716	3.5	0.3716	0.0272	0.0007
Total	994	23.9	3.1459	0.3145	0.011

* Note that such ratios are much less variable than their emission magnitudes (Hao, private communication).

18.3.7 Fuel Moisture Content

It should be stated that several of the emission parameters as discussed above are affected by fuel moisture content (FMC). FMC is difficult to obtain over large scales. Limited success has been reported to estimate FMC by means of remote sensing using shortwave (SW) reflective, thermal and microwave data (Bowman, 1989; Carter, 1991; Gogineni et al., 1991; Chuvieco et al., 2004). The basic information content for optical remote sensing of MFC comes from the SW infrared (SWIR) reflectance around 1.6 μm (Tucker, 1980; Hunt and Rock, 1989), that is available from many common sensors such as MODIS, AVHRR and VGT (Fraser and Li, 2002). Because of water absorption around this wavelength, SWIR is negatively correlated with FMC. While SW NIR measurements may convey certain information on FMC, it is so weak that the investigations are far from being conclusive (Hunt and Rock, 1989; Carter, 1991).

Thermal emission is related to FMC by altering the latent heat release due to evapotranspiration that is proportional to FMC. For plants of high FMC, an increase in latent heat release leads to a decrease in sensible heat and thus lowers the air temperature. Temperature differences between the ground and air may thus serve as an alternative measure of FMC. Based on this principle, several indices have been proposed including the Stress Degree Day (SDD) (Jackson, 1986), the Crop Water Stress Index (CWSI) (Jackson et al., 1981), and the Water Deficit Index (WDI) (Moran et al., 1994). WDI has been successfully tested as a predictor of fire danger with NOAA-AVHRR and Landsat-TM data (Vidal et al., 1994; Vidal and Devaux-Ros, 1995).

18.4 Summary

Biomass burning emits huge amount of gases and particles in various forms of carbon compounds and thus play a key role in global carbon cycling. To reach a closure in carbon balance, we need a full and accurate accounting of carbon emissions due to fire activities. Remote sensing is the only feasible means of monitoring fires around the globe. Maximum usage of satellite data is thus highly desired to achieve this goal. However, none of the emission related fire attributes are measured directly by any satellite sensors. Inversion algorithm and modeling are required to estimate fire emissions using satellite data.

This chapter provides an extensive review of remote sensing data and methods that could be brought to bear on fire emission estimation in terms of their information content, extraction method, strengths and limitations. These parameters include burned area, burning fragmentation and spreading, fuel loading, fraction of fuel consumed by fire, and emission factors for different gases. In general, satellites can provide good information on burned area by combined use of hot spot data together with changes in vegetation indices. Fire fragmentation and/or severity depend critically on satellite sensor resolution. For moderately coarse resolution data like MODIS and AVHRR, unburned fire islands inside the fire polygons provided by forest agencies may be singled out, but little can be gained concerning inhomogeneous degree of burning. Limited information may be extracted on fuel loading in terms of forest regrowth age, fraction of tree coverage by using a combination of measurements from several passive channels, especially NIR and SWIR data. Vegetation height detected by space-borne lidar may be linked to biomass content. Satellite-based land cover classification on continental scale may help refine fuel type classification. Determination of the fraction of fuel consumption (crown and surface) usually requires modeling, except for a short-cut approach that links radiation emission with fuel consumption.

References

Air Quality Policy on Agricultural Burning, Recommendation from the Agricultural Air Quality Task Force (AAQTF) to US Dept. of Agriculture, November 10, 1999

Anderson HE (1982) Aids to Determining Fuel Models for Estimating Fire Behavior. USDA Forest Service. General Technical Report INT – 122

Arino O, Piccolini I, Kasischke E, Siegert F, Chuvieco E, Martin P, Li Z, Fraser HE, Stropplana D, Pereira J, Silva JMN, Roy D, Barbosa P (2001) Mapping of burned surfaces in vegetation fires, in Global and Regional Vegetation Fire Monitoring from Space, Planning and Coordinated International Effort (Eds. Ahern F, Goldammer JG, Justice C). pp 227 – 255

Arseneault D, Villeneuve N, Boismenu C, Leblanc Y, Deshaye J (1997) Estimating Lichen Biomass and Caribou Grazing on Wintering Grounds of Northern Quebec: an Application of Fire History and Landsat Data. Journal of Applied Ecology 34 (1): 65 – 78

Bailey RG (1998) Ecoregions: The Ecosystem Geography of the Oceans and Continents. Springer-Verlag, New York, 192 P

Bowman WD (1989) The relationship between leaf water status, gas exchange, and spectral reflectance in cotton leaves, Remote Sensing of Environ 30: 249 – 255

Burgan RE, Klaver RW, Klaver JM (1998) Fuel models and fire potential from satellite and surface observations. International Journal of Wildland Fire 8 (3): 159 – 170

Cahoon DR, Stocks BJ, Levine JS, Cofer WR, Pierson JM (1994) Satellite analysis of the severe 1987 forest fires in northern China and southeastern Siberia. Journal of Geophysical Research 99: 18,627 – 18,638

Cairns MA, Hao WM, Alvarado E, Haggerty PK (2000) Carbon emission from Spring 1998 fires in tropical Mexico. Proceedings of the International Conference on Tropical Forests and Climate Change: Status, Issues and Challenges. October 19 – 22, 1998, Manila, Philippines

Carter GA (1991) Primary and secondary effects of water content on the spectral reflectance of leaves. American J Botany 78: 916 – 924

Chen JM, Chen WJ, Liu J, Cihlar J, Gray S (2000) Annual carbon balance of Canada's forest during 1895 – 1996. Global Biogeochemical Cycles 14: 839 – 850

Chuvieco E, Conalton RG (1988) Mapping and inventory of forest fires from digital processing of TM data. Geocarto International 4: 41 – 53

Chuvieco E, Vaughan P, Riano D, Cocero D (2004) Fire danger and fuel moisture content estimation from remotely sensed data. Proceedings of the Joint Fire Science Conference and Workshop. Boise, Idaho, June 15 – 17, 1999

Cihlar J, Beaubien J, Latifovic R, Simard G (1999) Land Cover of Canada 1995 Version 1.1. Digital data set documentation. Natural Resources Canada, Ottawa, Ontario

Clinton N, Gong P, Pu R (2003) Evaluation of wildfire mapping with NOAA/AVHRR data by land cover types and eco-regions in California. Submitted to Int. J. Wildland Fire, 2003

Cofer III WR, Winstead EL, Stocks BJ, Goldammer JG, Cahoon DR (1998) Crown fire emissions of CO_2, CO, H_2, CH_4, and TNMHC from a dense jack pine boreal forest fire. Geophysical Research Letters 25: 919 – 922

Cofer III WR, Levine JS, Riggan PJ, Sebacher DI, Winstead EL, Shaw EF, Jr. Brass JA, Ambrosia VG (1988) Trace gas emission from a mid-latitude prescribed chaparral fire. Geophysical Research 93: 1,653 – 1,658

Conard SG, Ivanova GA (1997) Wildfire in Russian boreal forests—Potential impact f fire regime characteristics on emissions and global carbon balance estimates. Environ Pollut 98: 305 – 313

Csiszar I, Abdelgadir A, Li Z, Jin J, Fraser R, Hao W (2003) Interannual changes of active fire detectability in North America from long-term records of the Advanced Very High Resolution Radiometer. J Geophy Res 108(D2): 4075, doi: 10.1029/2001JD001373

Csiszar I, Sullivan J (2002) Recalculated pre-launch saturation temperatures of the AVHRR 3.7 micrometer sensors on board the TIROS-N to NOAA-14 satellites. International Journal of Remote Sensing 23: 5,271 – 5,276

DeFries RS, Hansen M, Townshend JRG, Sohlberg R (1998) Global land cover classification at 8 km spatial resolution. The use of training data derived from Landsat imagery in decision-tree classifiers. International Journal of Remote Sensing 19: 3,141 – 3,168

DeFries RS, Houghton RA, Hansen M, Field C, Skole DL, Townshend J (2002) for the 1980s and 90s. Proceedings of the National Academy of Sciences, 99 (22), Carbon emissions from tropical deforestation and regrowth based on satellite observations 14,256 – 14,261

Dubayah R, Drake J (2000) Lidar remote sensing for forestry applications. J Forestry 98: 44 – 46

Eyre FH (ed) (1980) Forest Cover Types of the United States and Canada. Society of American Foresters

Fan S, Gloor M, Mahlman J, Pacala S, Sarmieno J, Takahashi T, Tans P (1998) A large terrestrial carbon sink in North America implied by atmospheric and oceanic carbon dioxide data and models. Science 282: 442 – 446

Fernandez A, Illera P, Casanova JL (1997) Automatic mapping of surfaces affected by forest fires in Spain using AVHRR NDVI composite image data. Remote Sens Environ 60: 153 – 162

FIRESCAN Science Team (1996) Fire in ecosystems of boreal Eurasia: The Bor Forest Island Fire Experiment fire research campaign, Asia-North (FIRESCAN). In Biomass Burning and Global Change Vol 2, pp 848 – 873. Edited by Levine JS, MIT Press, Cambridge, Massachusetts

Forestry Canada Fire Danger Group (1992) Development and structure of the Canadian Forest Fire Behavior Prediction System. Information Report ST-X-3, Forestry Canada

Fraser R, Li Z (2002) Estimating fire-related parameters in the boreal forest using SPOT VEGETATION. Rem Sens Environ 82: 95 – 110

Fraser R, Li Z, Cihlar J (2000a) Hotspot and NDVI differencing synergy: A new technique for burned area mapping over boreal forest. Rem Sens Environ 74: 362 – 375

Fraser R, Li Z, Landry R (2000b) SPOT VEGETATION for characterizing boreal forest fires. Int J Rem Sens 21: 3,525 – 3,532

Fraser RH, Hall RJ, Landry R, Landry T, Lynham T, Raymond D, Lee B, Li Z (2004) Validation and calibration of Canada-wide coarse-resolution satellite burned maps. Photogrammetric Engineering and Remote Sensing 70: 451 – 460

French NHF, Kasischke ES, Stocks BJ, Mudd JP, Martell DL, Lee BS (2000) Carbon release from fires in the North American boreal forest, pp 377 – 388. In: Fire, Climate Change and Carbon Cycling in the Boreal Forest (Eds. Kasischke ES, Stocks BJ). Ecological Studies Series, Springer-Verlag, New York

Garcia MJ, Caselles V (1991) Mapping burns and natural reforestation using thematic mapper data. Geocarto International 1: 31 – 37

Gogineni S, Ampe J, Budihardjo A (1991) Radar estimates of soil moisture over the Konza Prairie. Int J Rem Sens 12: 2,425 – 2,432

Gong P, Pu R et al. (2004a) Spatial and Temporal Distribution of Forest Fires in North America as Determined with NOAA/AVHRR Data. Being prepared for submitted

Gong P, Pu R et al. (2004b) A Report on Development of a Long-term (1989 – 2000) Inventory of Fire Burned Areas and Emissions of North America's Boreal and Temperate Forests. Being prepared for submitted

Gonzalez-Alonso F, Cuevas JM, Casanova JL, Calle A, Illera P (1997) A forest fire risk assessment using NOAA AVHRR images in the Valencia area, eastern Spain. Int J Remote Sensing 18(10): 2,201 – 2,207

Goode JG, Yokelson RJ, Ward DE, Susott RA, Babbitt RE, Davies MA, Hao WM (2000) Measurements of excess O_3, CO_2, CO, CH_4, C_2H_4, C_2H_2, HCN, NO, NH_3, HCOOH, CH_3COOH, HCHO, and CH_3OH in 1997 Alaskan biomass burning plumes by airborne Fourier transform infrared spectroscopy (AFTIR). Journal of Geophysical Research 105(22): 147 – 166

Grégoire J-M, Cahoon DR, Stroppiana D, Li Z, Pinnock S, Eva H, Arino O, Rosaz JM, Csiszar I (2001) Forest fire monitoring and mapping for GOFC: current products and information networks based on NOAA-AVHRR, ERS-ATSR, and SPOT-VGT, in Global and Regional Vegetation Fire Monitoring from Space. Planning and Coordinated International Effort (Eds. Ahern F, Goldammer JG, Justice C), pp 105 – 124

Hansen MC, DeFries RS, Townshend JRG, Sohlberg R (2000) Global land cover classification at 1 km spatial resolution using a classification tree approach. International Journal of Remote Sensing 21: 1,331 – 1,364

Hansen MC, DeFries RS, Townshend J, Carroll M, Dimiceli C, Sohlberg R (2003) Global percent tree cover at a spatial resolution of 500 meters: First results of the MODIS Vegetation Continuous Fields algorithm. Earth Interactions Vol 7

Hao WM, Liu M.-H (1994) Spatial and temporal distribution of tropical biomass burning. Global Biogeochemical Cycling 8: 495 – 503

Hao WM, Ward DE, Olbu G, Baker SP (1996) Emissions of CO_2, CO, and hydrocarbons from fires in diverse African savanna ecosystems. J Geophys Res 101: 23,577 – 23,584

Hao WM, DE Ward, Babbitt RE, Susott RA, Wang JS, Davies M, Baker SP (1998) Influences of weather and vegetation on biomass burning. Paper presented in the Joint International Symposium on Global Chemistry, Seattle, WA, August 19 – 25

Hunt ER, Rock BN (1989) Detection of changes in water content using near and middle-infrared reflectances. Rem Sens Environ 30: 43 – 54

Ichoku C, Kaufman Y, Giglio L, Li Z, Fraser RH, Jin J, Park B (2003) Comparative analysis of daytime fire detection algorithms using AVHRR data for the 1995 fire season in Canada: Perspective for MODIS. Int J Rem Sens 24: 1,669 – 1,690

Illera P, Fernandez A, Delgado JA (1996) Temporal evolution of the NDVI as an indicator of forest fire danger. Int J Remote Sensing 17(6): 1,093 – 1,105

Jackson RD, Idso SB, Reginato RJ, Pinter PJ (1981) Canopy temperature as a crop water stress indicator. Water Resources Res 17: 1,133 – 1,138

Jackson RD (1986) Remote sensing of biotic and abiotic plant stress. Ann Rev Phytopathology 24: 265 – 287

Justice CO, Giglio L, Korontzi S, Owens J, Morisette JT, Roy D, Descloitres J, Alleaume S, Petitcolin F, Kaufman Y (2002) The MODIS fire products. Remote Sensing of Environment 83: 244 – 262

Kasischke ES, French NHF (1995) Locating and estimating the areal extent of wildfires in Alaskan boreal forest using multiple-season AVHRR NDVI. Remote Sensing of Environment 51: 263 – 275

Kasischke ES, French NHF, Bourgeau-Chavez LL, Christensen NL (1995) Estimating release of carbon from 1990 and 1991 forest fires in Alaska. Journal of Geophysical Research 100: 2,491 – 2,951

Kasischke ES (2000) Boreal ecosystems in the global carbon cycle. pp 19 – 30. In: Fire, Climate Change and Carbon Cycling in the Boreal Forest (Eds. Kasischke ES, Stocks BJ). Ecological Studies Series, Springer-Verlag, New York

Kaufman YJ, Tucker CJ, Fung I (1990) Remote sensing of biomass burning in the tropics. J Geophys Res 95: 9,927 – 9,939

Kaufman YJ, Remer LA (1994) Detection of forests using mid-IR reflectance: an application for aerosol studies. IEEE Transactions on Geoscience and Remote Sensing 32: 672 – 683

Kaufman YJ, Justice C, Flynn L, Kandall J, Prins E, Ward DE, Menzel P, Setzer A (1998a) Monitoring Global Fires from EOS-MODIS. J Geophys Res 103 (32): 215 – 239

Kaufman YJ, Kleidman RG, King MD (1998b) SCAR-B Fires in the Tropics: properties and their remote sensing from EOS-MODIS. J Geophys Res 103 (31): 955 – 969

Kaufman YJ, Tanré D, Boucher O (2002) A satellite view of aerosols in the climate system. Nature 419: 215 – 223

Kaufman YJ, Ichoku C, Giglio L, Korontzi S, Chu DA, Hoa WM, Li R-R, Justice CO (2003) Fires and Smoke Observed from the Earth Observing System MODIS Instrument-Products, Validation, and Operational Use. International Journal of Remote Sensing 20: 1,031 – 1,038

Kurz WA, Apps MJ, Bekema SJ, Lekstrum T (1995) 20th century carbon budget of Canadian forests. Tellus (B) 47: 170 – 177

Li Z, Cihlar J, Moreau L, Huang F, Lee B (1997) Monitoring fire activities in the boreal ecosystem. J Geophy Res 102 (29): 611 – 624

Li Z, Nadon S, Cihlar J (2000a) Satellite detection of Canadian boreal forest fires: Development and application of an algorithm. Int J Rem Sens 21: 3,057 – 3,069

Li Z, Nadon S, Cihlar J, Stocks BJ (2000b) Satellite mapping of Canadian boreal forest fires: Evaluation and comparison and comparisons. Int J Rem Sens 21: 3,071 – 3,082

Li Z, Jin J-Z, Fraser RH (2000c) Mapping burning areas and estimating fire emissions over the Canadian boreal forest. Proceedings of the Joint Fire Science Conference and Workshop, Idaho, USA, June 15 – 17, 1999, Vol 2, pp 247 – 251

Li Z, Kaufman YJ, Ithoku C, Fraser R, Trishchenko A, Giglio L, Jin J, Yu X (2001a) A review of AVHRR-based active fire detection algorithms: Principles, limitations, and recommendations, in Global and Regional Vegetation Fire Monitoring from Space. Planning and Coordinated International Effort (Eds. Ahern F, Goldammer JG, Justice C), pp 199 – 225

Li Z, Khananian A, Fraser R, Cihlar J (2001b) Detecting smoke from boreal forest fires using neural network and threshold approaches applied to AVHRR imagery, IEEE Tran Geosci & Rem Sen 39: 1,859 – 1,870

Li Z, Fraser R, Jin J, Abuelgasim AA, Csiszar I, Gong P, Pu R, Hao W (2003) Evaluation of algorithms for fire detection and mapping across North America from satellite, J Geophys Res 108(D2): 4 076, doi: 10.1029/2001JD001377

Li R, Kaufman YJ, Hao WM, Salmon JM, Nordgren BL, Gao BC (2004) A technique for detecting burn scars using MODIS data, IEEE Trans Geosci Rem Sens, 42(6): 1,300 – 1,308

Martin MP, Chuvieco E (1995) Mapping and evaluation of burned land from multitemporal analysis of AVHRR NDVI images. EARSEL Adv Remote Sens 4: 7 – 13

Michalek JL, French NHF, Kasischke ES, Johnson RD, Colwell JE (2000) Using Landsat TM data to estimate carbon release from burned biomass in an Alaskan spruce forest complex. Int J Remote Sensing 21(2): 2,323 – 2,338

Mishchenko MI, Geogdzhayev IV, Cairns B, Rossow WB, Lacis AA (1999) Aerosol retrievals over the ocean by use of channels 1 and 2 AVHRR data: sensitivity analysis and preliminary results. Appl Optics 38: 7,325 – 7,341

Moran MS, Clarke TR, Inoue Y, Vidal A (1994) Estimating crop water deficit using the relation between surface-air temperature and spectral vegetation index. Rem Sens Environ 49: 246 – 263

Pu R, Gong P (2003) Determination of Burnt Scars Using Logistic Regression and Neural Network Techniques from a Single Post-Fire Landsat-7 ETM + Image. Photogrammetric Engineering and Remote Sensing, 70(7): 841 – 850

Pu R, Li Z, Gong P, Fraser R, Csiszar I, Hao W, Kondragunta S, Weng F (2006) Development and Analysis of a 12-year Daily 1-km Forest Fire Dataset across North America from NOAA/AVHRR Data, Rem. Sens. Environ., in press

Reinhardt ED, Keane RE, Brown JK (1997) First Order Fire Effects Model: FOFEM 4.0, User's Guide. USDA Forest Service. General Technical Report INT-GTR-344

Roy DP, Giglio L, Kendall JD, Justice CO (1999) Multi-temporal active-fire based scar detection algorithm. International Journal of Remote Sensing 20: 1,031 – 1,038

Roy DP, Lewis PE, Kendall J, Justice CO (2002) Burned area mapping using multi-temporal moderate spatial resolution data—a bi-direction reflection model-based expectation approach. Rem Sens Environ 83: 263 – 286

Salisbury JW, D'Aria DM (1994) Emissivity of terrestrial materials in the 3 – 5 mm atmospheric window. Rem Sens Environ 47: 345 – 361

Salisbury JW, Walter LS, Vergo N, D'Aria DM (1991) Infrared (2.1 – 25 μm) spectra of minerals, The Johns Hopkins University Press, Baltimore, MD

Seiler W, Crutzen PJ (1980) Estimates of gross and net fluxes of carbon between the biosphere and atmosphere. Climate Change 2: 207 – 247

Snyder W, Wan Z, Zhang Y, Feng YZ (1997) Thermal infrared (3 – 14 μm) bi-directional reflectance measurements of sands and soils. Rem Sens Environ 60: 101 – 109

Still CJ, Collatz GJ, Berry JA, DeFries RS (2003) The global distribution of C3 and C4 plants: Carbon Cycle Implications. Global Biogeochemical Cycles 17(1):1006, doi: 10.1029/2001GB001807

Stocks BJ, Lee BS, Martell DL (1996) Some potential carbon budget implications of fire management in the boreal forest. In: Forest ecosystems, Forest management and the global carbon cycle (eds Apps MJ, Price DJ). NATO ASI Series, Springer-Verlag, Berlin, pp 90 – 96

Susott RA, Ward DE (1999) Ward, Smoke emission from ponderosa pine fuels exposed to a variety of fire histories and site preparation treatments. Final Report submitted to Arizona National Forests

Tans PP, Fung IY, Takahashi T (1990) Observational constrains of the global atmospheric CO_2 budget. Science 247: 1,431 – 1,438

Tucker CJ (1980) Remote sensing of leaf water content in the near infrared. Rem Sens Environ 10: 23 – 32

Van der Werf GR, Randerson JT, Collatz GJ, Giglio L, et al. (2003) Continental-scale partitioning of fire emissions during the 1997 to 2001 El Nino/La Nina period. Science 303: 73 – 76

Van Wagner CE (1987) Development and structure of the Canadian forest fire weather index system. Canadian Forestry Service, Forestry Technical Report 35, Ottawa

Vidal A, Pinglo F, Durand H, Devaux-Ros C, Maillet A (1994) Evaluation of a temporal fire risk index in Mediterranean forest from NOAA thermal IR. Rem Sens Environ 49: 296 – 303

Vidal A, Devaux-Ros C (1995) Evaluating forest fire hazard with a Landsat TM derived water stress index. Agricul Forest Meteor 77: 207 – 224

Vose JM, Swank WT, Geron CD, Major AE (1996) Emissions from forest burning in the Southern United States: application of a model determining spatial and temporal fire variation. Biomass Burning and Global Change, Vol 2 edited by Levine JS, The MIT Press, Cambridge, Massachusetts, London, England, 1996

Wan Z, Li Z (1997) A physics-based algorithm for retrieving land surface emissivity and temperature from EOS/MODIS Data. IEEE Trans Geosci Rem Sens 35: 980 – 996

Ward DE, Hardy CC (1991) Smoke Emissions from Wildland Fires. Environment International 17: 117 – 134

Ward DE, Hao WM, Susott RA, Babbitt R, Shea RW, Kauffman JB, Justice CO (1996) Effect of fuel composition on combustion efficiency and emission factors for African savanna ecosystems. J Geophy Res 101(23): 569 – 576

White JD, Ryan KC, Key CC, Running SW (1996) Remote sensing of forest fire severity and vegetation recovery. Int J Wildland Fire 6: 125 – 136

Wong J, Li Z (2002) Retrieval of optical depth for heavy smoke aerosol plumes: uncertainties and sensitivities to the optical properties. J Atmos Sci 59: 250 – 261

Yokelson RJ, Goode JG, Bertschi I, Susott RA, Babbitt RE, Ward DE, Hao WM, Wade DD, Griffith DWT (1999) Emissions of formaldehyde, acetic acid, methanol, and other trace gases from biomass fires in North Carolina measured by airborne fourier transform infrared spectroscopy (AFTIR). J Geophy Res, in press

19 TRMM Fire Algorithm, Product and Applications

Yimin Ji and Erich Stocker

19.1 Introduction

Land fires are frequent menaces to human lives and property. They also change the state of the vegetation and contribute to the climate forcing by releasing large amounts of aerosols and greenhouse gases into the atmosphere. It is estimated that fires may contribute about 30% to the total amount of tropospheric ozone, global carbon monoxide and carbon dioxide (Crutzen and Andreae, 1990; Levine, 1991). Therefore, fires are significant and continuous contributors not only to the earth's deforestation and ecology but also the physical and chemical processes taking place in the atmosphere. The release of aerosols during fires may also lead to significant changes of cloud microphysics and radiative properties. A recent study of TRMM data in Indonesia showed evidence that smoke from sustained fire may suppress regional rainfall completely for certain rain types and therefore create an even more favorable environment for fire to occur (Rosenfeld, 1999).

The purpose of this chapter is to review fire algorithms, products, and applications that have been developed at the TRMM Science Data and Information System (TSDIS). This section provides an overview of the existing and future satellite fire and aerosol products. TSDIS fire algorithms are described in Section 19.2. Section 19.3 describes the TSDIS fire products and provides inter-comparisons between selected TSDIS fire data and AVHRR fire products. Parallel analyses of fire and aerosol data on seasonal and interannual variability are presented in Section 19.4. Section 19.5 provides statistical results of diurnal cycle and intraseasonal variations of land fires derived from the TSDIS data. Relations between fire and rainfall variability are discussed in both Section 19.5 and Section 19.6.

19.1.1 Satellite Fire Products

During the last several decades, more and more satellite fire products have been emerging and becoming available to the community. The Visible Infrared Spin Scan Radiometer and Atmospheric Sounder (VAS) on board the Geosynchronous Operational Environmental Satellites (GOES) has been used to monitor biomass burnings in many selected regions since 1980 (Dozier, 1981; Matson and Dozier,

1981). The GOES/VAS provides unique high temporal resolution products that are useful in determining the diurnal variations of intense fires. But the coarse spatial resolution (4 – 13.8 km at nadir) of the products can significantly affect the accuracy of locations and extents of the fire.

The Advanced Very High Resolution Radiometer (AVHRR) of the National Oceanic and Atmospheric Administrator (NOAA) satellite series has been able to provide fire monitoring at a higher spatial resolution (1.1 km at nadir) for various areas (Matson and Dozier, 1981; Flannigan and Vonder Haar, 1986; Kaufman et al., 1990; Justice et al., 1996). Unfortunately, the full resolution AVHRR data (LAC) are only recorded for selected regions. There have been efforts (Pinnock and Gregoire, 1999) to integrate the AVHRR high-resolution data into a global 10-day composite. However, the 10-day composite only covers a small portion of the earth's surface. Further, since the coverage of LAC varies significantly from one region to another, the composite count does not have absolute quantitative representation. The available daily AVHRR data is the reduced resolution GAC (Global Area Coverage) for which every third scan of the full resolution orbit data is processed and four out of every five pixels along the third scan are averaged. Therefore, the GAC has a resolution of 1.1 km$\times$4.4 km at nadir with a 3 km gap between pixels across the scan. While the long history and global coverage of GAC data are very attractive for the study of interannual fire variability, the GAC resampling scheme would result in unreliable characterization as well as bias in the detection of fire. Since there is no long-term global GAC fire product readily available, substantial efforts may be needed to further analyze the quality of GAC fire.

Pioneering studies based upon GOES/VAS and NOAA/AVHRR data have provided scientific fundamentals on fire retrieving using remotely sensed data. New network techniques have demonstrated the potential for providing real-time and long-term fire products using operational satellites. The satellite fire detection has been further advanced due to the recent launch of Terra and Aqua. The sophisticated MODIS is on board both Terra (EOS AM-1) and Aqua (EOS PM-1). The MODIS 1 km 4 μm high gain channel with a saturation level of 500 K is specifically designed to improve the fire detection (Kaufman and Justice, 1998). MODIS data have been available since early 2000. The MODIS Thermal Anomalies product provides fire occurrence (day/night), fire location, the logical criteria used for the fire selection, and an energy calculation for each fire. The product also includes composite 8-day-and-night fire occurrence (full resolution), composite monthly day-and-night fire occurrence (full resolution), gridded 10-km summary per fire class (daily/8-day/monthly), and a gridded 0.5° summary of fire counts per class (daily/8-day/ monthly). The Level-2 product provides various fire related parameters including the occurrence of day and night thermal anomalies, flagged and grouped into different temperature classes with emitted energy from the fire. These parameters are retrieved daily at 1-km resolution. Further, the coincident fire observation from high resolution ASTER on board Terra can provide instantaneous validation for MODIS.

It is expected that better global fire products may also emerge in the near future from the Visible Infrared Imager Radiometer Suite (VIIRS) of National Polar-Orbiting Operational Environmental Satellite System (NPOESS) and the NPOESS Preparatory Project (NPP). Fire is one of the major parameters in both EOS and NPOESS missions. Other satellite sensors being used for fire detection include Defense Mapping Satellite Program (DMSP), Landsat Thematic Mapper, the Advance Along Track Scanning Radiometer (AATSR) onboard the European Space Agency's ENVISAT-1.

While the main objective of TRMM (Kummerow et al., 1998) is to improve observations and understanding of the tropical rainfall variability, the VIRS (Barnes et al., 1996), one of the three primary TRMM sensors, is similar to the NOAA/AVHRR. The VIRS data are well calibrated and recorded at VIRS full resolution (2.11 km at nadir, 3.02 km at edge of scan) globally. Therefore, TRMM/VIRS provides the capability of producing a continuous global fire data set over tropics and subtropics.

19.1.2 Satellite Aerosol Product

Both TOMS daily and monthly products provide mean aerosol index at $1.25^\circ \times 1.0^\circ$ resolution. The TOMS aerosol algorithm is a modified spectral contrast method, or a residue method. It utilizes the contrast between TOMS 340 nm reflectance and 380 nm reflectance. One of the unique strengths of this technique is that the presence of sub-pixel clouds does not affect the aerosol detection as clouds produce near-zero residues. For UV-absorbing aerosol, the residues are positive. However, because the residue has strong altitude dependence due to the effects of aerosol absorption on molecular scattering, the TOMS aerosol index has a decreasing sensitivity at low altitudes. The UV-absorbing aerosol within the boundary layer near the ground is hardly detectable by TOMS. However, for most sustained fires, the related aerosol transports occur in mid-troposphere and can be readily detected by TOMS. For weak fires and in cold seasons, the smoke is shallow and less detectable by TOMS. The detail of the TOMS aerosol algorithm can be found in Herman et al. (1997). Fires are important sources of atmospheric aerosols and greenhouse gases.

This research uses TOMS index data (Herman et al., 1997) to compare with the TSDIS fire products. This effort sought to investigate whether the TRMM observed spatial and temporal variations of biomass burnings were consistent with the variations of atmospheric aerosols from TOMS observations. TOMS measures the variations of various atmospheric gases such as ozone and atmospheric aerosols. These variations are associated with global fires (Levine, 1991; Andreae, 1991; Hsu et al., 1996). Since the distribution of gases and aerosols released by fires relates not only to the position and strength of fires but also to the atmospheric circulation, comparisons at either the pixel or grid level were inappropriate. Rather, comparisons are made with emphasis on visual patterns on a continental scale. By comparing with TOMS daily map, strong thermal anomalies that are not associated with fires may be identified. These anomalies, such as the open shrub land at daytime, may have similar thermal spectral properties as fire pixels, but does not release smoke.

19.2 TSDIS Fire Algorithms

Using measurements of VIRS 3.75 μm and 11 μm bands, the TSDIS fire product algorithm was built on heritage algorithms of NOAA/AVHRR (Kaufman et al., 1990; Flasse and Ceccato, 1996). The TRMM/VIRS and NOAA/AVHRR have five channels with similar center wavelengths and bandwidths across visible and infrared spectrums (0.63 μm, 1.61 μm, 3.75 μm, 10.8 μm and 12 μm). VIRS scans a 45° swath with a 2.11 km Instantaneous Field of View (IFOV) at nadir and 3.02 km IFOV at the edge of scan from the non-sun-synchronous 350 km TRMM orbit. The VIRS geometric registration has an uncertainty about 0.5 ~ 0.8 of the pixel size. VIRS provides twice daily observations over most of the tropical and subtropical areas (180°W – 180°E and 40°S – 40°N).

The fundamental physics of the fire algorithm using VIRS 3.75-μm and 11-μm bands is the Wien displacement law:

$$\lambda_{\max} = \frac{C}{T} \tag{19.1}$$

where $\lambda_{\max}$ is the wavelength at which the radiation is at a maximum if the radiative temperature is at T. $C = 2{,}898$ μm·K is a constant. Thus, the vegetated surface over tropics and sub-tropics, with a radiative temperature of about 300 K, has a peak around the 11 μm band. Fire pixels, with radiative temperatures about 800 K, have a radiative peak around the 3.75 μm band. Therefore, if fires occur in a portion of a pixel, the radiant energy of 3.75 μm band increases much more rapidly than that of 11 μm band, resulting in a larger than normal difference of brightness temperatures between the two bands.

If a large part of a pixel is filled with fires, the 3.75 μm band could be saturated because the saturation temperature of VIRS 3.75 μm band is only around 322 K. We have not observed saturation of the VIRS 11 μm although the saturation temperature of this band is similar to that of 3.75 μm band. The 11 μm band is not as sensitive as the 3.75 μm band to the thermal anomalies happening within a pixel. The saturation temperature is the maximum output brightness temperature for a thermal infrared channel of the sensor. The NOAA/AVHRR sensors have saturation temperatures between 320 K to 330 K. ATSR, which is designed to measure sea surface temperatures, has a saturation temperature of

312 K only. However, the recent MODIS has a saturation temperature of 500 K for 4 μm band.

19.2.1 Nighttime Algorithm

The TSDIS nighttime algorithm is basically a traditional threshold method using only the VIRS thermal band brightness temperatures (*Tb*s). In this chapter, we chose the East-Kalimantan region of the Indonesia as an example to demonstrate the day/night thermal properties of VIRS IR bands.

The nighttime images (see Figs. 19.1(a) and 19.1(b)) were from VIRS orbit 1509 of March 3, 1998. The environmental surface temperature of the area was about 300 K. In clear sky conditions, the channel 3 (3.75 μm) brightness temperature (*Tb*3) should range between 290 K to about 300 K as indicated in the image (see Fig. 19.1(a)). However, as fire occurred, either the channel 3 was saturated or the *Tb*3 reached a level much higher than 300 K. Of a total of 9,853 pixels, 75 pixels were saturated and 63 pixels had *Tb*3 greater than 315 K. These fire pixels covered an area of approximately 1,000 km^2. There were also hundreds of pixels with temperatures around 310 K. March of 1998 was at the peak of the devastating 1998 Indonesian fires due to the dry and hot weather in the El-Nino event and also the feedback of the smoke released by the fires (Schindler, 1998; Rosenfeld, 1999). As a comparison, the 11 μm image (see Fig. 19.1(b)) did not show any thermal anomalies over the region. Therefore, for background pixels, the differences between *Tb*3 and *Tb*4 (brightness temperature of 11 μm band, or channel 4) were usually a few degrees as expected, while for fire pixels, the differences between *Tb*3 and *Tb*4 were much larger than a few degrees (see Fig. 19.1(c)).

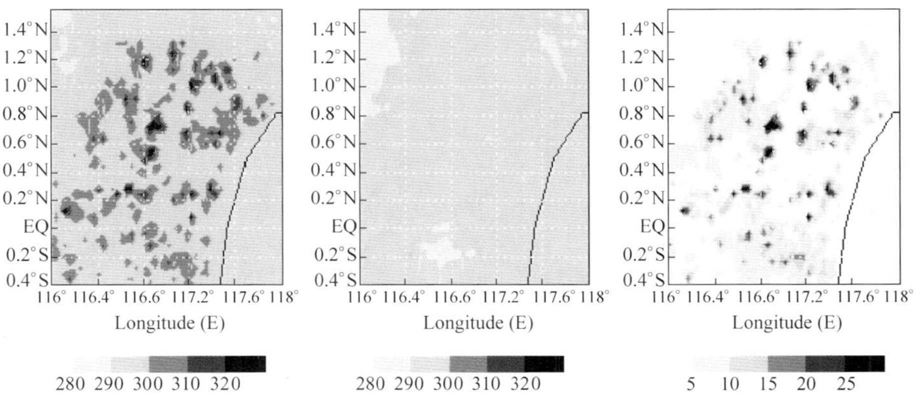

(a) Ch3 BT (b) Ch4 BT (c) Ch3 BT−Ch4 BT

Figure 19.1 VIRS night time images over East-Kalimantan on March 3, 1998. (a) 3.75 μm band, (b) 11 μm band, and (c) the difference

Based on careful studies over various regions globally using VIRS data as well as examining the results of AVHRR experiments (Matson and Dozier, 1981; Kaufman et al., 1990), the TSDIS fire algorithm uses the following criteria to detect nighttime fire pixels:

- Channel 3 (3.75 μm) is saturated or
- $Tb3 > 315$ K and $Tb3 - Tb4 > 15$ K.

In general, only a small portion of a pixel may be occupied by the fire when a fire occurs. Because the environmental *Tb*3 (*Tb*3env) is about 300 K and the fire *Tb*3 is about 800 K, the criteria presented above may reject a pixel that has less than 2% of its area occupied by fire. Compared to the current existing algorithms (see Table 19.1), the TSDIS algorithm is reasonable. Table 19.1 shows typical threshold of various sensors. For some of these sensors, multi-algorithms may exist.

Table 19.1 Nighttime fire algorithm comparison

Sensor	*Tb*3	*Tb*3 – *Tb*4	*Tb*3 Saturation	IFOV
AVHRR	> 316 K	> 10 K	322 K	1 km
ATSR	> 312 K		312 K	1 km
MODIS	> 315 K	> 10 K	500 K	0.5 km
VIRS	> 315 K	> 15 K	322 K	2 km

19.2.2 Daytime Algorithm

The daytime algorithm uses a contextual approach described by Flasse and Ceccato (1996). The contextual approach uses a certain threshold to determine a hot spot pixel and then uses various ancillary data and a series of tests to exclude false alarms. The TSDIS daytime algorithm first uses the thermal band Tbs to choose a candidate of fire pixels. After that, additional tests are performed to make final choices. These tests use VIRS visible/near-infrared data as well as ancillary data sources.

Compared to the nighttime retrieval, the daytime fire detection is much more difficult and complicated as the 3.75 μm band is affected by solar radiation in many aspects. The reflected solar radiation from the surface may significantly increase the 3.75 μm channel radiances. Studies over desert, open shrub and other bare ground area indicated that the VIRS 3.75 μm band might be saturated when solar radiation was at its maximum. In a vegetated area, the increase in Tb3 due to solar radiation was not as significant. As a result, the TSDIS daytime fire algorithm uses a 1 km resolution (Townshend et al., 1994) global surface type data to exclude bare ground pixels and other thermal anomalies related to the solar radiation and surface properties. The land type data set was generated by the UMD using near nadir AVHRR data. The uncertainty of geometric registration of this data is about 0.8 km. Therefore, a significant number of false fire pixels may exist over areas with mixed surface types in summer or spring season.

The daytime images (see Figs. 19.2(a) and (b)) were from VIRS orbit 1470 of March 1, 1998. The daytime background *Tb*3 was higher than the nighttime *Tb*3 across the area, but still lower than 310 K. Of the 9,512 pixels within this image, 51 pixels were saturated and 10 pixels had *Tb*3 greater than 320 K. The daytime *Tb*4 was significantly lower than the *Tb*3 and there were no thermal anomalies. Unlike the nighttime instance, the difference of *Tb*3 and *Tb*4 in daytime (see Fig. 19.2(c)) failed to indicate the position of fire pixels. The difference over much of the area was greater than 30 K. This was because the smoke released by fires attenuated the surface radiance of 11 μm band. The TOMS aerosol data showed that the area was covered by significant smoke during this day. While the surface radiation of 3.75 μm band may also be attenuated by the smoke, there is increased solar radiation reflected and scattered by the smoke in 3.75 μm channel.

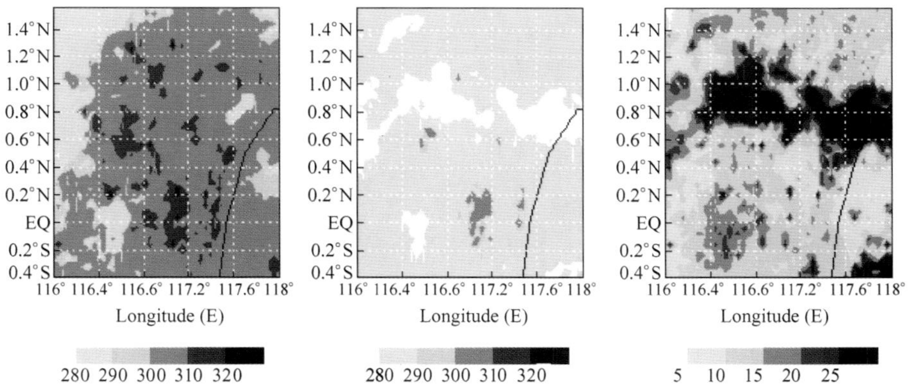

(a) Ch3 BT (b) Ch4 BT (c) Ch3 BT−Ch4 BT

Figure 19.2 VIRS day time images over East-Kalimantan on March 3, 1998 (a) 3.75-μm band, (b) 11-μm band, and (c) the difference

Another problem for daytime fire retrieval is the occurrence of sun glint. When sun glint occurs, the 3.75 μm pixels become saturated or have *Tb*s similar to that of fire pixels. The sun glint pixels have been observed from VIRS measurements in the oceanic area. The sun glint can be detected using variations of reflectance of VIRS channel 1 and channel 2 (Flasse and Ceccato, 1996; Kaufman and Justice, 1998). Sun glint may also occur in the cloud surface and cloud edges. Discussion of such problems is beyond the scope of this chapter. In daytime detection, the algorithm also calculates background *Tb*3 and *Tb*4 by scanning an area of 25 pixels and 25 scans with the target pixel as the center. The purpose of this calculation is to further exclude false fire pixels. In the final product, a fire pixel may be rejected if the difference between *Tb*3 and the background *Tb*3 is less than 8 degree.

Based on these analyses, the TSDIS daytime fire algorithm uses the following criteria to identify daytime fire pixels:

- Channel 3 (3.75 μm) is saturated and $Tb4 > 285$ K or $Tb3 > 320$ K and $Tb3 - Tb4 > 20$ K and $Tb4 > 285$ K;
- Pixel is not a bare ground, or open shrub pixel;
- $Tb3 - Tb3\mathrm{env} > 8$ K.

The *Tb*4 requirement is used to exclude cloud and heavy aerosol pixels. Under these conditions, the algorithm is unable to detect fire pixels. Both criteria 2 and 3 may reject real fire pixels. However, our day/night fire product comparison showed that the fire occurrence under these two conditions was rare while a large number of false fire pixels might be generated without these daytime constraints. Table 19.2 shows typical thresholds of various sensors in daytime. However, the most important thing for the daytime algorithm is to exclude false fire pixels. Except the contextual threshold, manual methods are often used for regional retrieval. Further discussion of daytime false alarm removal will be given in Section 19.5.1.

Table 19.2 Daytime fire algorithm comparison

Sensor	*Tb*3	*Tb*3 – *Tb*4	*Tb*3 Saturation	IFOV
AVHRR	> 320 K	> 15 K	322 K	1 km
ATSR				
MODIS	> 320 K	> 20 K	500 K	0.5 km
VIRS	> 320 K	> 20 K	322 K	2 km

19.3 TSDIS Fire Products

TSDIS fire algorithm routinely creates global daily and monthly fire products. The daily product is a hot spot map with VIRS pixel resolution (2.1 km at nadir). The processing of daily products starts as soon as the production of the VIRS L1B files for the current day is completed. There are typically 16 orbits, or L1B files daily. The operational L1B creation lags real-time by about 12 hours. Therefore, the global fire product has a lag time of approximately 14 hours. The daily products include two files. The first file, the basic file of all products, is a text file that details necessary information for all fire pixels. The information includes date, orbit number, pixel number, solar zenith angle, latitude, longitude, UTC time, reflectance of channel 1 and channel 2, brightness temperatures and background brightness temperatures of channel 3 and channel 4. This text file is then used to create the second file, the daily global hot spots.

Figure 19.3 shows a daily fire product of November 26, 2003. The upper panel displays the distribution of global fire pixels detected by the TSDIS fire algorithm. The three lower panels show enlarged area maps over Northern America, Southern America, and the Indonesian area. On this particular day, a large number of fire pixels were detected over Southern California, reflecting a devastating fire in this area in late November 2003.

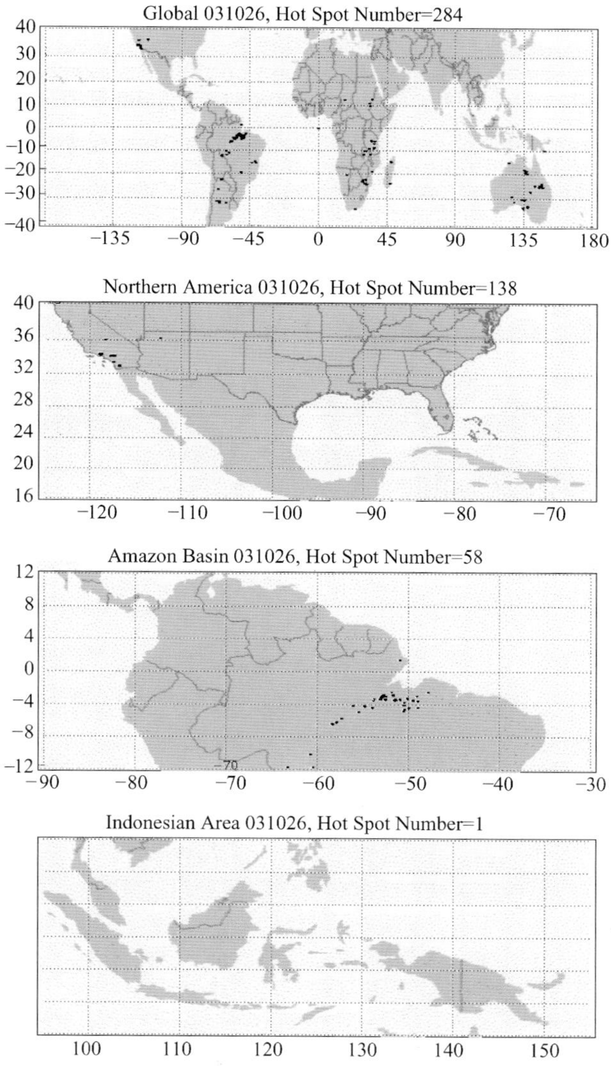

Figure 19.3 TSDIS daily fire image of November 26, 2003

The monthly image is a $0.5°\times0.5°$ resolution composite of fire counts for the month. The monthly text files provide information for all fire pixels observed within the month. Any information missed in daily text files due to operational delays is displayed in the monthly file. The monthly text files have been reprocessed recently from the beginning of the TRMM mission such that overpass time for each fire pixel may be retrieved. In general, the monthly mapping is able to show both intensity and duration. An intense fire usually lasts longer and covers a larger area. Both factors would add significantly to the fire count. The monthly product may be useful for understanding the effect of fires on the variations of aerosol (Hsu et al., 1996) ozone and other greenhouse gases (Levine, 1991). They may be used to simulate the physical and chemical processes in the climate models. Users may download or ftp the description of the algorithm, as well as the daily and monthly products (image and data) from TSDIS Home page or anonymous ftp site. However, as described in the above section, daytime hot spots contained in these files may contain a large number of false fire pixels. A typical way for a user to further exclude false alarms is day/night screening. Such screening may reject most false fires in non-fire season. The detailed description of the TRMM fire algorithm and products can be found in Ji and Stocker (2002a). The data are available at an anonymous ftp site (http: // ftp-tsdis.gsfc.nasa.gov) in "pub/yji/MONTHLY" and "pub/yji/DAILY".

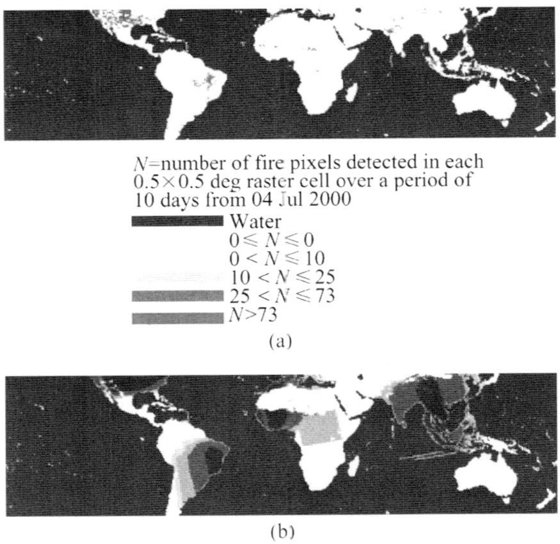

Figure 19.4 (a) EC/JRC AVHRR 10-day fire composite during 4 – 13 July 2000, and (b) EC/JRS AVHRR 10-day coverage of 4 – 13 July 2000

The 1 km AVHRR fire data are now available at EC/JRC World Fire Web (WFW) Home page (Pinnock and Gregoire, 1999). This data set provides a good opportunity to compare TSDIS fire products and AVHRR fire products. The spatial coverage of the 1-km AVHRR Local Area Coverage (LAC) data for a certain time period varies significantly. Therefore, the current EC/JRC fire count does not have absolute quantitative representations across the globe. Figure 19.4(a) shows a 10-day composite of the WFW AVHRR 1 km fire counts during July 4 to July 13, 2000 over the covered area (see Fig.19.4(b)). The data indicated that fires occurred in Northern America, Southern America and Sub-Saharan areas during this period. The TSDIS 10-day composite (see Fig. 19.5(a)) during the same period showed similar spatial distribution of fires in the WFW AVHRR covered area. However, the values of TSDIS counts were significantly smaller than the WFW AVHRR fire counts in Northern America and Southern America. The composite of TOMS aerosol index during the same period (see Fig. 19.5(b)) showed aerosol anomalies in Northern America and Central Africa.

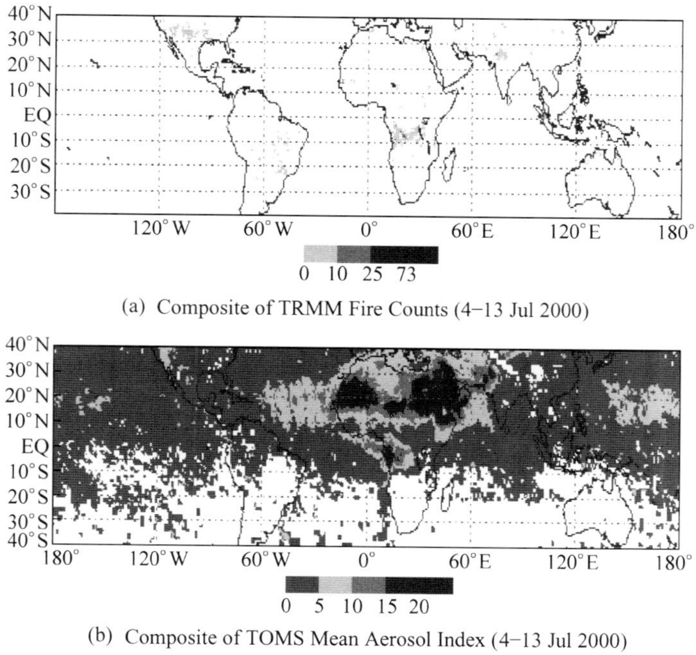

Figure 19.5 (a) TSDIS 10-day fire composite during 4 – 13 July 2000, and (b) TOMS 10-day aerosol index composite of 4 – 13 July 2000

19.4 Seasonal and Interannual Variability

19.4.1 Fire and Aerosol Comparison

Fires may occur randomly but are constrained by the regional climate and vegetation state. The TRMM yearly mean fire count of 1998 – 2003 years (see

Fig. 19.6(a)) showed intensive fires in South America, Africa, Australia, Southeast Asia, and Indonesia. Moderate fires occurred in China, Northern and Central America. In February, March, April, and May (FMAM) season, most fires occurred in the northern hemisphere (see Fig. 19.6(b)), especially in Southeast Asia, sub-Saharan Africa and Central America. In June, July, August, and September (JJAS), fires were observed in the southern hemisphere and northern America (see Fig. 19.6(c)).

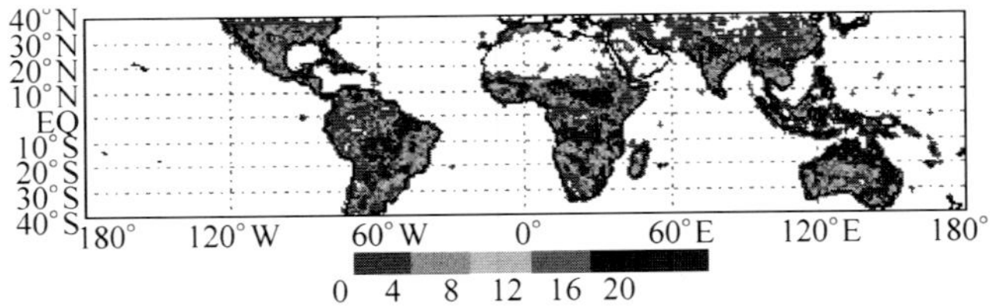

(a) 1998–2003 Mean Fire Count

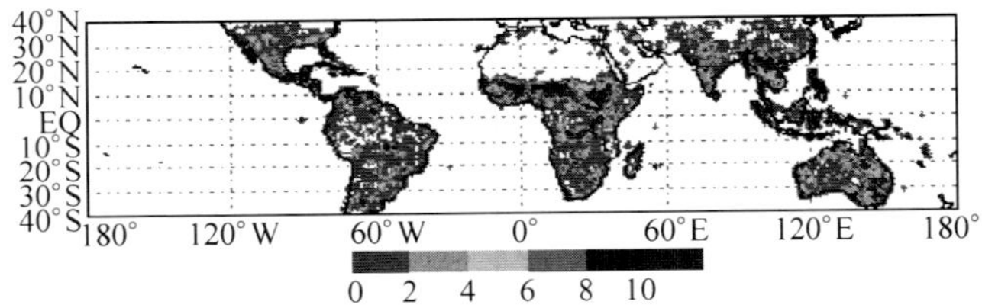

(b) 1998–2003 FMAM Mean Fire Count

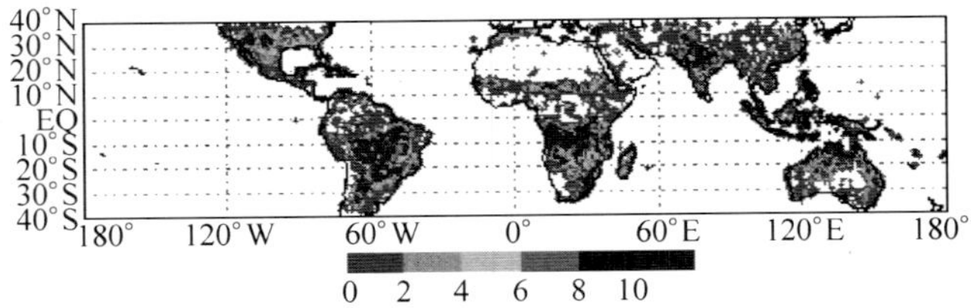

(c) 1998–2003 JJAS Mean Fire Count

Figure 19.6 Global distribution of fire count (resolution: $0.5° \times 0.5°$), (a) Annual mean from 1998 to 2003 (unit: count/year), (b) FMAM mean from 1998 to 2003 (unit: count/4-month), and (c) JJAS mean from 1998 to 2003 (unit: count/4-month)

The dominant pattern of aerosol index is the Sahara dust (see Fig. 19.7(a)). However, the contribution of intensive fires is evident. Since fires occur only in a certain season while the Sahara dust is almost a static feature, the magnitude of mean aerosol index in burning areas is significantly smaller than that of the Sahara dust. In FMAM season, the magnitude of aerosol index over fire centers of Southeast Asia and Indonesia is comparable to that over Sahara desert area (see Fig. 19.7(b)). Fires in these areas have been very serious in the recent decade (Malingreau, 1990; Giri and Shrestha, 2000). Almost no aerosols were observed over the southern hemisphere in this season. In JJAS season (see

Fig. 19.7(c)), the strength of aerosol index in southern Africa and southern America is quite strong but still considerably weaker than that of Sahara dust. There were moderate aerosols in northern America in this season.

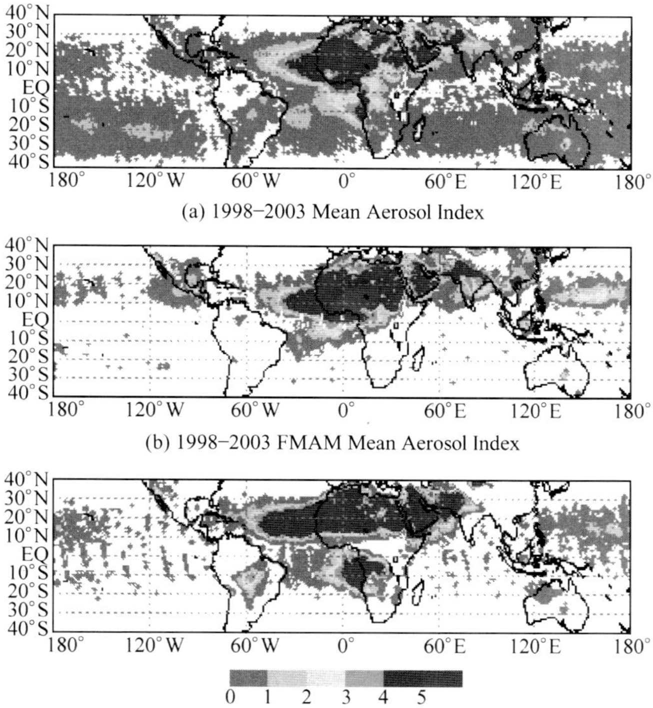

(a) 1998–2003 Mean Aerosol Index

(b) 1998–2003 FMAM Mean Aerosol Index

(c) 1998–2003 JJAS Mean Aerosol Index

Figure 19.7 Global distribution of aerosol index (resolution 1.25°× 1.0°), (a) Annual mean from 1998 to 2003, (b) FMAM mean from 1998 to 2003, and (c) JJAS mean from 1998 to 2003

The aerosol index maps are generally in agreement with the fire maps except the desert area. Since the smoke can be transported far beyond its fire origin, the unconditional correlation between fire count and aerosol index is only about 0.22. For conditional (only compare pixels where fire exists) cases, the correlation between aerosol index and fire count is as high as 0.55 if Australia is excluded. In Australia, there were virtually no aerosols observed while substantial fires were observed. This inconsistency may reflect false fire in TRMM fire algorithm for vegetation/desert mixed pixels. The false fire pixels have also been noticed in the sub-Saharan and other areas. However, the decreasing sensitivity of TOMS algorithm toward lower altitudes may also contribute to this difference. The depth of smoke in the Southwestern Australia fire during the winter season may be quite small.

Interannual variations were significant in Indonesia and Central America. For example, in FMAM 1998, an El-Nino season, intensive fires and aerosols in the two regions were observed by TRMM and TOMS, respectively (see Figs. 19.8(a) and 19.9(a)). The magnitude of aerosol index in Indonesia and Central America in 1998 FMAM season is comparable to that of Saharan dust. However, in FMAM 1999, a normal season, fires were moderate and no aerosols were observed (see Figs. 19.8(b) and 19.9(b)).

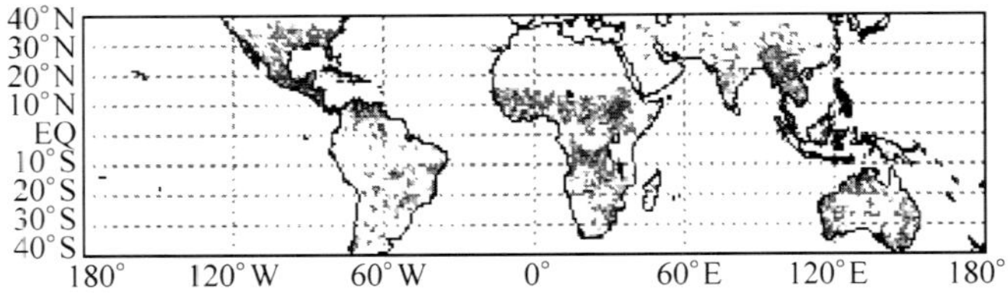

(a) 1998 FMAM Fire Count

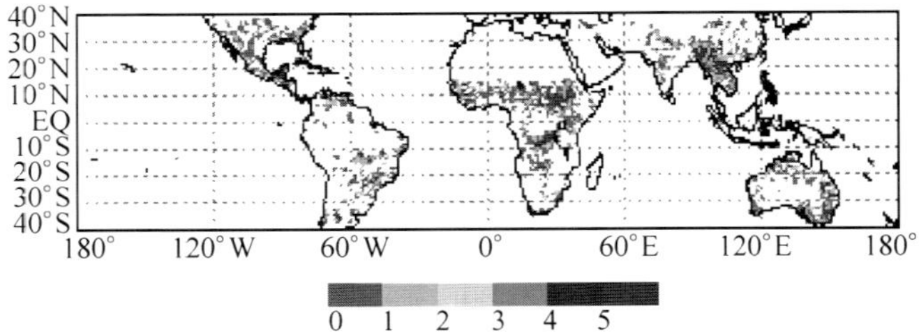

(b) 1999 FMAM Mean Fire Count

Figure 19.8 FMAM mean fire count (unit: count/4-month, resolution 0.5°×0.5°)

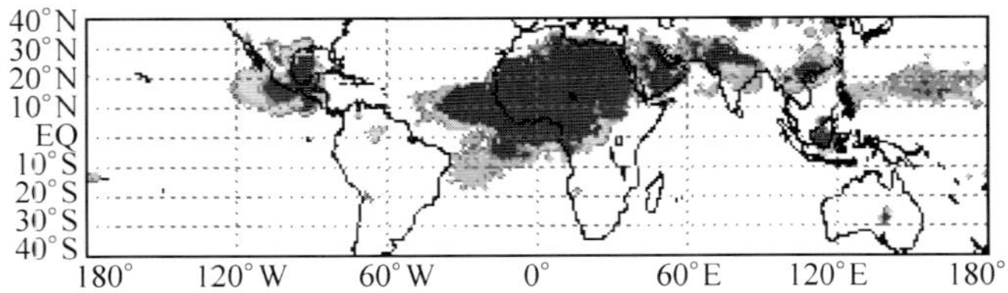

(a) 1998 FMAM Aerosol Index

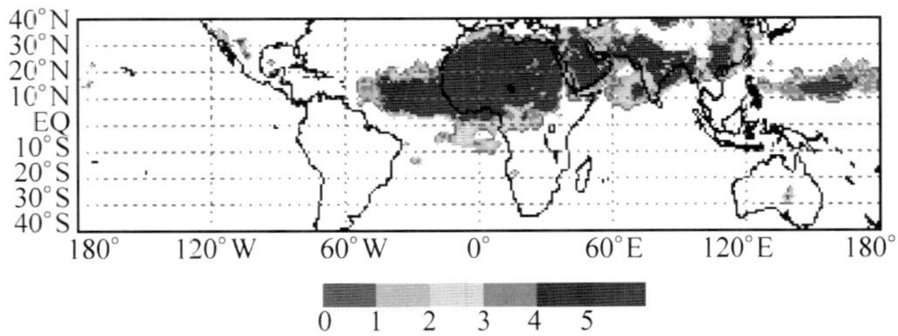

(b) 1999 FMAM Aerosol Index

Figure 19.9 FMAM mean aerosol index (resolution 1.25°×1.0°)

19.4.2 Statistical EOF Analysis

In order to further investigate the global fire variability, traditional Empirical Orthogonal Function (EOF) was used to extract qualitative information from fire and aerosol data. The EOF (Lorenz, 1956; Barnett, 1977; Broomhead and King, 1986; Ji and Stocker, 2002b) method involves the solution of the eigenvector equation:

$$(\boldsymbol{R} - \mu \boldsymbol{I})\boldsymbol{e}_m = 0 \tag{19.2}$$

where $\boldsymbol{R}$ is the data covariance matrix, $\boldsymbol{I}$ is a unit matrix, μ is eigenvalue and $\boldsymbol{e}_m$ is the resulting eigenvector (EV), and $m = 1, 2, \cdots, M$, where M is the dimension of the spatial field. Each of the eigenvalues explains a fraction of total variance and leads to an eigenvector solution that describes one mode of the spatial variability. The temporal behavior of data associated with the EV mode is represented by the principal components (PC) that are coefficients in reconstructing the data in eigenvector space. In normal cases, the dimension of the temporal vector (N) should be larger than the dimension of the spatial vector (M) to generate a normal covariance matrix $\boldsymbol{R}$ (M, M). This often limits the resolution of the spatial data.

The data used in EOF analysis are 5-day fire composites and 5-day mean aerosol index derived from the TRMM daily product and TOMS daily product, respectively. The temporal resolution may not go higher than the pentad because the data would be too noisy. The temporal resolution is also limited by the spatial coverage of these sensors. Both VIRS and TOMS take daily global coverage. However, there are gaps between orbits in daily coverage. Both sensors need about 2~3 days to get a full global coverage. The TRMM covers an area of 180°W – 180°E in longitude and 40°S – 40°N in latitude. For a $10° \times 10°$ resolution, the spatial dimension is about 288. However, at this latitude range, area from 160°E to 130°W is pure ocean area (except islands which should not affect the global fire pattern). Therefore, the EOF analyses actually excluded this area so that the dimension of the spatial vector is reduced to about 220. In this case, there are about 220 eigenvectors. However, only a few leading EVs are used in analyses to describe dominant patters of the variability.

If the annual cycle was not removed from the data, it was anticipated that the spatial patterns of the first EV of the EOF analysis would show seasonal mode. The first fire EV (see Fig. 19.10(a)) indeed showed contrast between North and South hemispheres. The corresponding PC (see Fig. 19.11(a)) indicated seasonal variations with contrast between spring and late summer and also 25 – 60-day oscillations that were superimposed on the seasonal cycle. This EV explained 24% of the total variance. The second EV (see Fig. 19.10(b)), which explained 11% of the total variance, showed inter-continental correlation. There are contrasts between sub-Sahara and southern Africa, and also between Central America and southern America. The PC of EV showed weak annual cycles with January as positive peak and July as negative peak. This contrast indicates a transition of fires from the sub-Saharan region to the south in Africa during a course from early spring to the summer. Similar patterns can be found in South America. The third EV (see Fig. 19.10(c)) does not indicate any contrast. The pattern is similar to the mean fire distribution of the four years. However, the PC (see Fig. 19.11(c)) indicates intra seasonal oscillations with a period of about 25 – 60 days. This eigenvector explains 7% of the total variance.

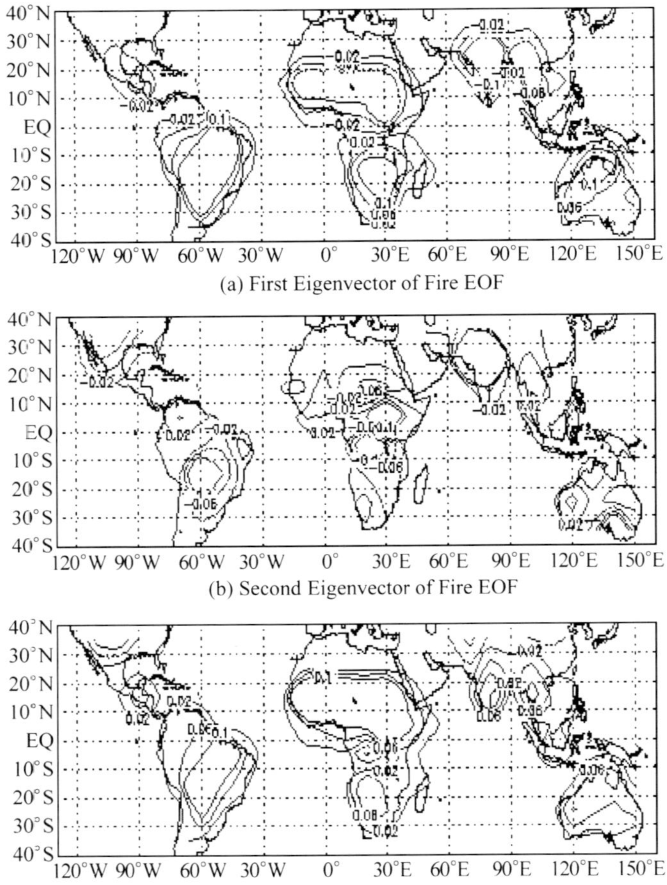

(a) First Eigenvector of Fire EOF

(b) Second Eigenvector of Fire EOF

(c) Third Eigenvector of Fire EOF

Figure 19.10 First (a), second (b), and third (c) eigenvectors of EOF analyses derived from TRMM global fire data

Since the dominant aerosol variation is the Saharan dust, the first EV (not shown), that explained 34% of the total variance of the TOMS aerosol EOF,

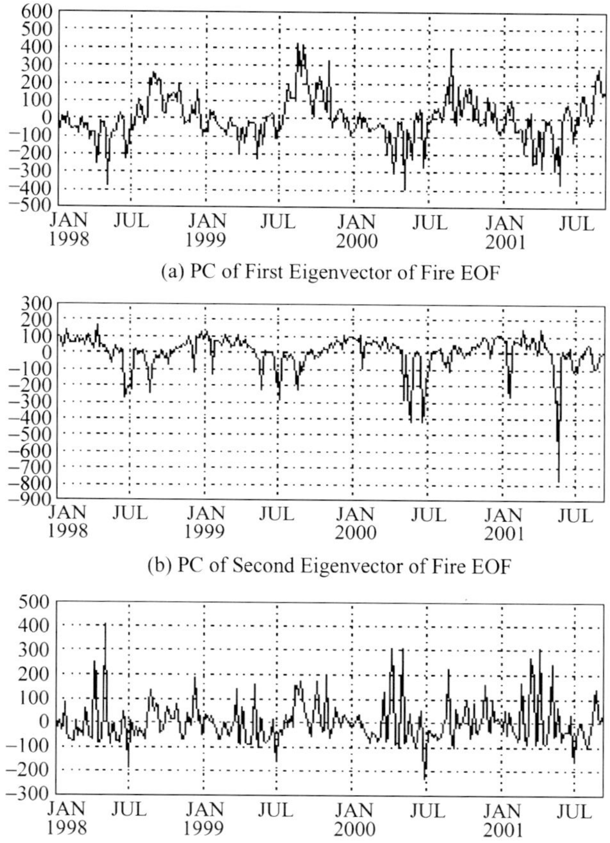

Figure 19.11 Principal components of first (a), second (b), and third (c) eigenvectors of TRMM fire data EOF analyses

showed static patterns of the desert dust. However, the second EV of TOMS aerosol EOF (see Fig. 19.12(a)) that explained 12% of the total variance, showed similar patterns to the first fire EV except in the Australia area where no aerosols were observed. The PC of this EV (see Fig. 19.13(a)) showed a strong annual cycle. Similar to the first fire PC, the second aerosol PC showed a contrast between spring and summer. However, the high frequency signal was weaker compared to the fire results. This may reflect a weak intra-seasonal variation of aerosol as compared to the fire. The third aerosol EV (see Fig. 19.12(b)) showed inter-continental transition. The contrast exists between sub-Sahara and Southern Africa, and also between Central America and South America. The PC of this EV also indicated a seasonal cycle (see Fig. 19.13(b)). This aerosol mode bears similarity of the second fire EOF mode. The fourth EV and PC showed intra-seasonal variation of aerosol with a cycle of about 25 – 60 days. The ENSO signal did not appear in both fire and aerosol EVs. The effect of the 97/98

El-Nino on fires was only evident in Indonesia and Central America and it ended in May 1998.

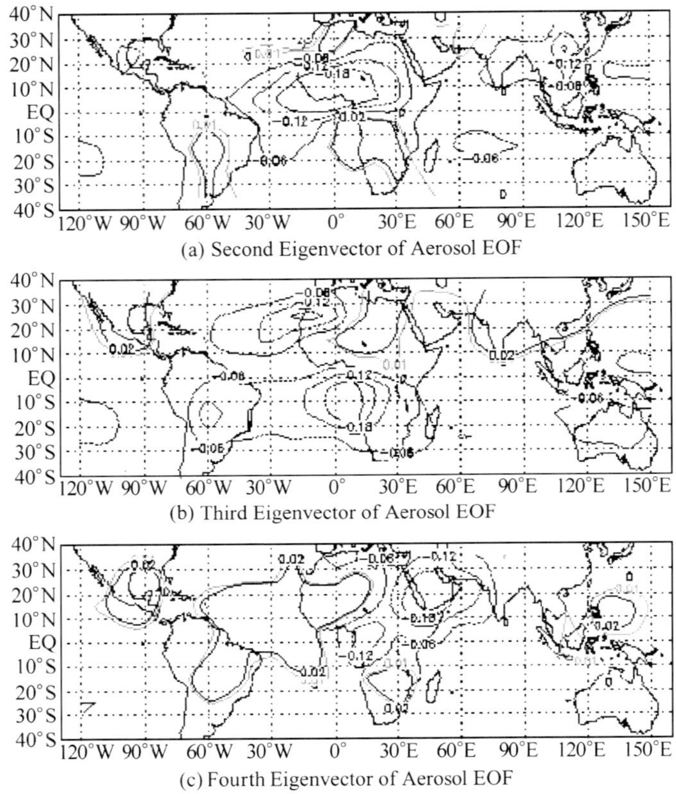

Figure 19.12 Second (a), third (b), and fourth (c) eigenvectors of EOF analyses derived from TOMS global aerosol data

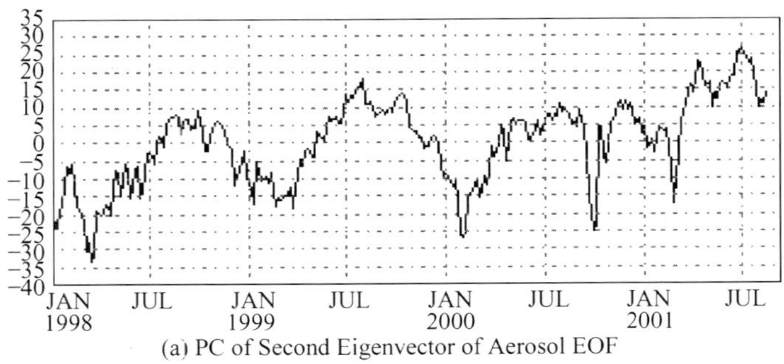

Figure 19.13 Principal components of second (a), third (b), and fourth (c) eigenvector of aerosol EOF analyses

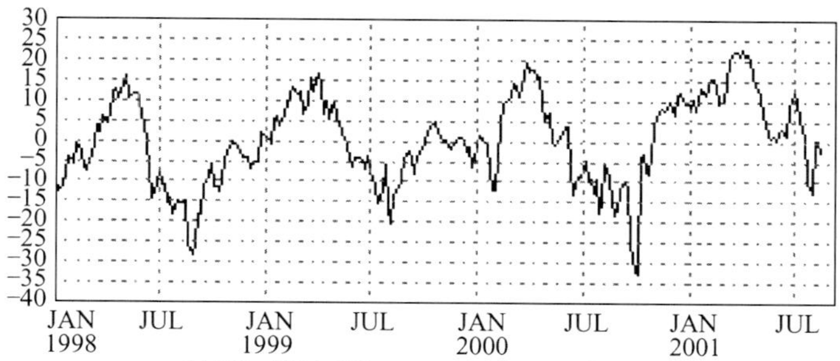

(b) PC of Third Eigenvector of Aerosol EOF

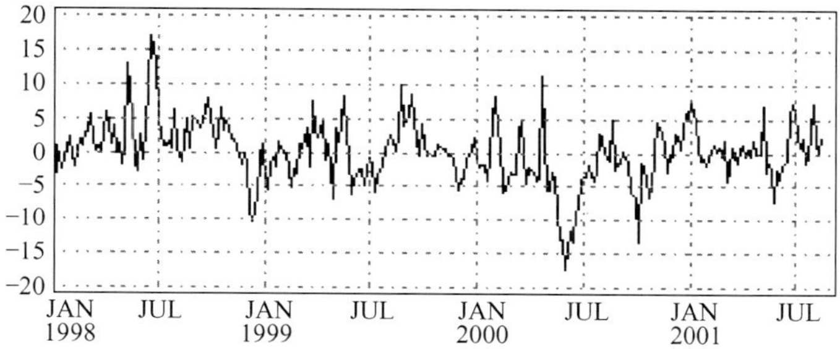

(c) PC of Fourth Eigenvector of Aerosol EOF

Figure 19.13 (Continued)

19.5 Diurnal Cycle and Intraseasonal Variability

Studies have indicated a rich spectrum of intraseasonal variations superimposed on the annual cycle and interannual variability (e.g. Ji and Stocker, 2002b). In order to further investigate the intraseasonal fire variability, the TRMM daily fire product must be properly averaged both specially and temporally. One particular problem for such averages is related to the specific TRMM orbit geometry. The TRMM is designed such that the local overpass time of the satellite normally drifts slightly each day completing a daily cycle in 46 days. Therefore, only a 46-day averaging for a particular location at pixel level would completely remove the aliasing of the diurnal burning cycle. However, the twice-daily observations in the lower tropics from descending and ascending TRMM orbits are usually more than several hours apart. For example, this time interval is about $10-12$ hours in Southeast Asia. The five-day time averaging and $10-20$ degree spatial averaging in the longitude direction contribute an additional three hours to the averaging window. Further, following the TRMM over flight calculation, the $10-20$ degree averaging in latitude direction may add about two hours to the moving window. Therefore, the five-day averaging of TRMM data for a $10°\times 10°$ box in Southeast Asia gives a moving average of about 15 hours. As a result, the effect of aliasing a strong diurnal burning cycle may have certain effects in determining the intraseasonal variability from the TRMM/VIRS product.

19.5.1 Diurnal Cycle Aliasing

In order to study the diurnal cycle and intraseasonal variability, the day/night hot spots must be carefully studied for reliable fire seasons and the daytime false fire pixels during non-fire seasons must be eliminated. Table 19.3 shows the day/night contrasts of fire counts in various areas.

The data used in Table 19.3 excluded the non-fire season observations; during these seasons, a large number of false fire pixels may occur in daytime due to the errors in the land type screening. As shown in Table 19.3, in Southeast Asia, South America, and Africa, the numbers of fire pixels in daytime and nighttime do not differ substantially. The ratios are about 1 – 1.5. In Indonesia, the nighttime fire pixels are outnumbered by the daytime fires pixels. Figure 19.14 shows a typical fire diurnal cycle in Southeast Asia from February – March 1998 – 1999 data and South America from July – August 1998 – 2001 data. The maximum fire counts appear between noons to almost mid-night. Since each pentad covers at least 15 hours of a daily cycle, the effect of aliasing on the intraseasonal variability is limited. The diurnal cycle study can also be used to exclude false alarms in the non-fire season (Ji and Stocker, 2004).

In order to further remove the effect of diurnal cycle aliasing, a simplified model is developed to transform the TRMM fire data. Only nighttime results are presented in this chapter to avoid discussions of issues such as false fire and day/night screening although the methods can be used for daytime too. In the

Table 19.3 Comparison of day/night fire observations

Region	Longitude	Latitude	Time Period	Day Count	Night Count
Indonesia	110°E – 120°E	10°S – 0°	03/01/1998 – 04/30/1998	152	157
Southeast Asia	90°E – 110°E	5°N – 25°N	02/01/1999 – 03/31/1999	717	612
Southeast Asia	90°E – 110°E	5°N – 25°N	02/01/1999 – 03/31/1999	744	640
South America	70°W – 50°W	25°S – 5°S	07/01/1998 – 08/31/1998	2,136	1,450
South America	70°W – 50°W	25°S – 5°S	07/01/1999 – 08/31/1999	2,149	2,078
South America	70°W – 50°W	25°S – 5°S	07/01/2000 – 08/31/2000	658	592
South America	70°W – 50°W	25°S – 5°S	07/01/2001 – 08/31/2001	790	630
Africa	25°E – 35°E	0° – 10°N	01/01/1998 – 03/31/1998	709	559
Africa	25°E – 35°E	0° – 10°N	01/01/1999 – 03/31/1999	850	540

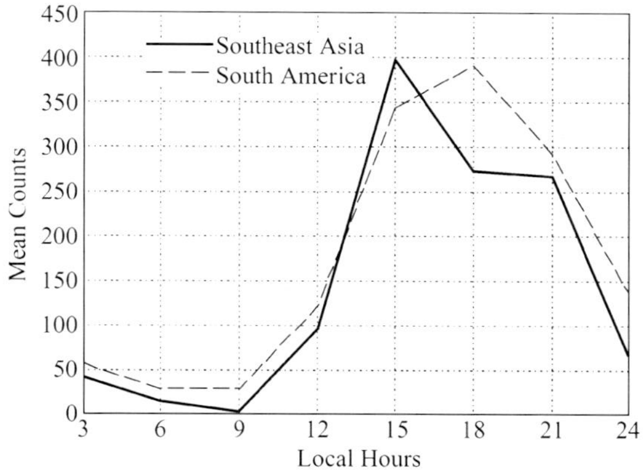

Figure 19.14 Typical diurnal cycles in Southeast Asia and South America

prototype, the nighttime period is divided into four particular time windows (see Table 19.4) and the model assumes one overpass for each window for all pentads. Multi-overpasses are normalized before processing. The model then calculates the average fire counts per overpass within each window using TRMM observations during major fire seasons. As an example, the ratios for each window and each fire season in Southeast Asia (90°E – 110°E, 5°N – 25°N) are listed in Table 19.4. Since each pentad of this region has observations that cover about 2/3 of the daily cycle, TRMM overpasses for at least one of the four windows in all pentads are guaranteed. Available observations within certain windows and pre-calculated ratio lookup tables are then used to extrapolate fire counts for windows with no observed overpasses.

Table 19.4 Count per overpass for time windows in Southeast Asia

Year (Time)	6 pm – 9 pm	9 pm – 12 pm	12 pm – 3 am	3am – 6 am
1998 (01/01 – 05/20)	5.10	1.64	1.09	1.11
1999 (01/01 – 04/05)	6.59	1.92	1.42	0.94
2000 (01/01 – 04/10)	2.52	0.26	0.35	0.13
2001 (01/01 – 04/15)	2.35	0.82	0.22	0.10

Time series of TRMM observed fire count (count/day) and transformed fire count (count/day) in Southeast Asia is displayed in Fig. 19.15. In the observed time series, the effect of satellite aliasing can be seen from a number of dips during certain fire episodes. Such dips are largely eliminated in the transformed time series. The fire time series in Southeast Asia was also compared with the Global Precipitation Climatology Project (Janowiak and Arkin, 1991) rainfall over land. The results (see Fig. 19.15) indicate that the fire intraseasonal variability is indeed closely related to the rainfall variations. The intraseasonal fire variability is dominated by fire episodes relative to the rainfall variability rather than a few dips relative to the aliasing. The transformation does not substantially change the pattern of time series. The comparison also indicates that the onset and duration of fire seasons are also related to the intraseasonal variability of rainfall. The 15 – 30-day and 30 – 60-day intraseasonal oscillations of tropical rainfall have been well defined. Further discussion about the diurnal cycle aliasing can be found in (Ji and Stocker, 2003) and (Giglio and Kendall, 2003).

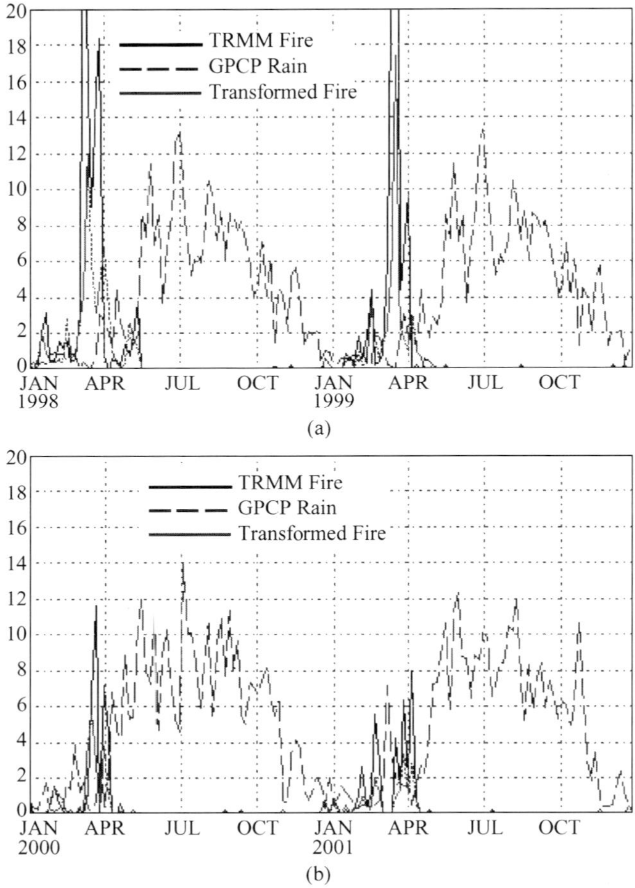

Figure 19.15 Time series of TRMM fire count(count/day, solid line), TRMM transformed fire count (count/day, dotted line), and GPCP rainfall (mm/day) in Southeast Asia (90°E – 110°E, 5°N – 25°N)

19.5.2 Single Spectrum Analysis

The Singular Spectrum Analysis (Ji and Stocker, 2002b) was used to study the intraseasonal variability. The SSA has been developed and used in atmospheric science quite recently (Vautars and Ghil, 1989; Rasmusson et al., 1990). Although the SSA involves the solution of the same eigenvector equation, it addresses the spectrum aspect of the chaotic data instead of the spatial patterns in EOF. The data used in SSA is a one dimensional time series. In SSA, the $\boldsymbol{R}$ in Eq. (19.2) is an auto-correlation matrix of the time series. The size of the auto- correlation matrix of the times series is determined by the number of lags (M). Normally, the number of lags must be at least an order of magnitude smaller than the size of the time series. This requires a time period of data be much longer than the spectrum of interest. In this study, the SSA is used to capture the intra-seasonal spectrum; the number of lags is about 20 (100 days) that is an order smaller than the length of the time series (four years). The eigenvectors of SSA capture the spectrum of the time series, while the principal components show the temporal behavior of the spectrum.

The singular spectrum analysis (Ji and Stocker, 2002b) from TRMM transformed nighttime fire counts for Southeast Asia during 1998 – 2002 (see Fig. 19.16)

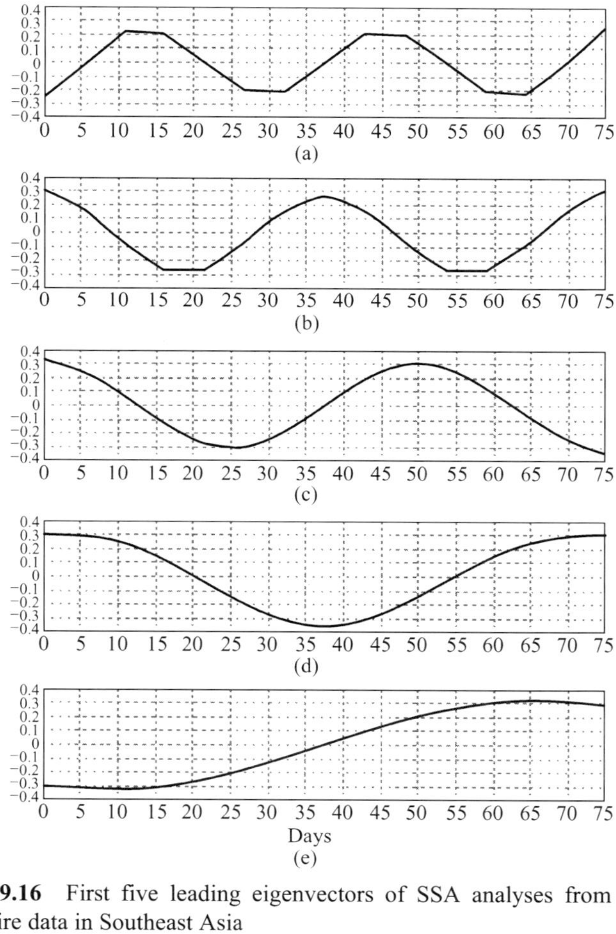

Figure 19.16 First five leading eigenvectors of SSA analyses from nighttime TRMM fire data in Southeast Asia

clearly shows the dominant modes of 30 – 60-day oscillations. These oscillations may be related to the Rossby wave and Madden-Julian oscillation (Madden and Julian, 1994) in tropics. The time series of these modes are shown in Fig. 19.17.

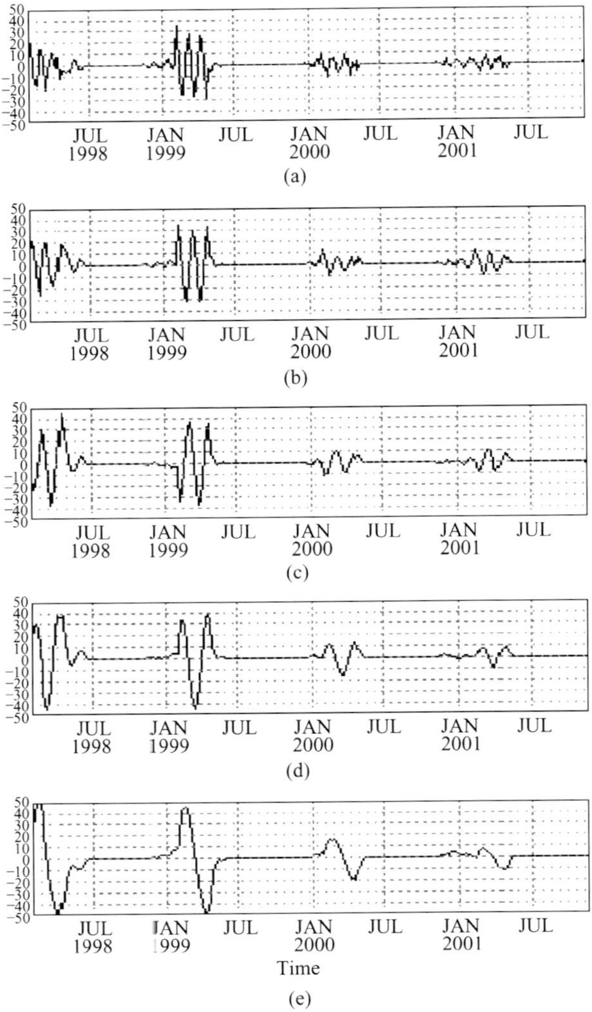

Figure 19.17 Principal components of the five leading eigenvectors of fire SSA in Southeast Asia derived from TRMM nighttime data

19.6 Interaction between Fire and Rainfall

Although the fire season and variability are dominated by global rainfall variations, the positive feedback from cloud microphysics on the land fire has been suspected for many years. The hypothesis is that the large concentration of small CCN in the smoke from fire nucleates many small cloud droplets that coalesce inefficiently into raindrops. Direct and satellite observations have shown strong support on this hypothesis (Kaufman and Fraser, 1997; Rosenfeld, 1999). It is now widely recognized that aerosols have a significant impact on the global climate and hydrological cycle (Ramanathan et al., 2001; Harshvardhan et al., 2002). There are three major sources of aerosols, i.e., desert dust, smoke from biomass burning, and anthropogenic air pollution. It is found that the major effects of the smoke from biomass burning on the clouds lead to the formation of a high concentration of small cloud droplets and little coalescence, and therefore to an increased cloud albedo (Kaufman and Nakajima, 1993) and suppressed precipitation (Kaufman and Fraser, 1997; Rosenfeld et al., 1998; Rosenfeld, 2000) due to the fact that these smoke aerosols are recognized as the sources of large concentration and small CCN (Rosenfeld et al., 2001). This is similar to the observed effect of desert dust and anthropogenic air pollution on clouds. These effects have been recently verified by TRMM observations (Rosenfeld, 1999).

A typical example is that during the FMAM 1998, the fire area in eastern Kalimantan seldom got any rainfall while the rain events occurred in the nearby areas for many times (Ji and Stocker, 2002a). Figure 19.18 shows maps of VIRS

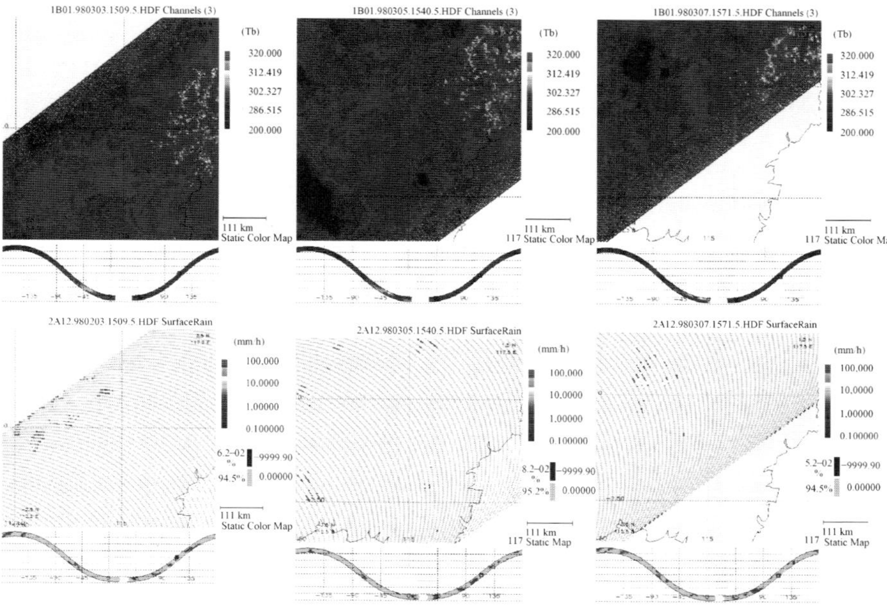

Figure 19.18 VIRS channel 3 T_b (upper) and coefficient TMI microwave rainfall (low) from TRMM 1509 orbit on March 03, 1998 (left), from TRMM 1540 orbit on March 05, 1998 (middle), and from TRMM 1571 orbit on March 07, 1998 (right)

channel 3 T_b and coincident TMI microwave rainfall from the TRMM March 03 – 07 1998 overpasses. These maps show clear distinction of cloud property between fire and non-fire areas. However, observation and analyses may not sufficiently explain these phenomena. Significant cloud modeling efforts combined with observation and analysis is needed to clearly describe the cause and effect. The central question is how to simulate and model the cloud droplet nucleation-growth process, smoke and cloud droplet collision-coalescence process, and seeding effect of smoke aerosols on clouds involving the fire smoke.

19.7 Summary

The TSDIS fire product has provided global fire information since January 1998 and will continue to do so for the TRMM lifetime. The results presented in this chapter indicate a strong seasonal cycle of fire occurrences over Southeast Asia with peaks in March and over South America and Africa with peaks in northern summers. The fire occurrences in the Indonesian region and Central America were dominated by the 1998/1999 ENSO cycle. Significant fires occurred over Indonesia with a peak in March and also occurred in Central America with peaks between March and May in 1998. Releases of smoke related to the global fires in the 1998 and 1999 years were observed from TOMS. The results of the two different satellite (TOMS and VIRS) observations were consistent.

The dominant modes of EOF showed contrasting of fire and aerosols between North and South hemispheres. The PCs of these modes indicating the peaks of fires and aerosol index in the northern hemisphere including Southeast Asia, sub-Sahara and Central America were during late spring to early summer. In southern Africa and southern America, the peaks appeared in late summer to early autumn. The intra-seasonal variability of fire and aerosols were also captured in the leading EOF modes. SSA analyses confirmed the intra-seasonal and interannual variability of fires in Southeast Asia and Indonesia. The dominant spectrum of intra-seasonal variability is about 25 – 60 days. The intra-seasonal spectrum derived from the fire data showed similarity of Madden-Julian (Madden, 1986) 30 – 60-day oscillation mode of meteorological parameters in tropics such as wind and precipitation.

Acknowledgements

The authors would like to acknowledge the support of many individuals at NASA/TSDIS, especially Tony Stocker for his efforts to create automated routines that display fire images daily in the TSDIS Home page and Michael Hensley who reprocesses all monthly fire data recently such that local times for all hot spot can be retrieved from the monthly text files. We sincerely appreciate the efforts of NASA/TOMS team, especially Dr. Richard McPeters, for the new version of the aerosol index data. We also wish to acknowledge the EC/JRC WFW group, especially Dr. Simon Pinnock, for allowing us to use the WFW products.

References

Andreae MO (1991) Biomass burning: its history, use, and distribution and its impact on environmental quality and global climate. In: Levine S (ed) Global Biomass Burning. The MIT Press

Barnes WL, Barnes RA, Holmes AW (1996) Characterization and calibration results from the Visible and Infrared Scanner (VIRS) for the Tropical Rainfall Measuring Mission (TRMM). Proceedings of the Advanced and Next-Generation Satellites II, Taormina, Italy, 23 to 26 September 1996, 2957: 266 – 276

Barnett TP (1977) The principal time and space scales of the Pacific trade wind fields. J Atmos Sci 34: 221 – 236

Broomhead DS, King GP (1986) Extracting qualitative information from experimental data. Physica D 20: 217 – 236

Crutzen PJ, Andreae MO (1990) Biomass burning in the tropics: impact on atmospheric chemistry and biogeochemical cycles. Science 250: 1,669 – 1,678

Dozier J (1981) A method for satellite identification of surface temperature fields of sub pixel resolution. Remote Sensing of Environment 11: 221 – 229

Flannigan MM, Vonder Haar TH (1986) Forest fire monitoring using the NOAA satellite AVHRR. Canadian J of Forest Research 16: 975 – 982

Flasse SP, Ceccato PS (1996) A contextual algorithm for AVHRR fire detection. Int J of Remote Sensing 17: 419 – 424

Giglio L, Kendall JD (2003) Comment on "Seasonal, intra-seasonal, and interannual Variability of Global Land Fire and Their Effects on Atmospheric Aerosol Distribution" by Y. Ji and E. Stocker. J of Geophys Res 108: doi: 101029/2003JD003548

Giri C, Shrestha S (2000) Forest fire mapping in Huay Kha Khaeng wildlife sanctuary. Int J of Remote Sensing 21: 2,023 – 2,030

Harshvardhan S, Schwartz E, Benkovitz CM, Guo G (2002) Aerosol influence on cloud microphysics examined by satellite measurements and chemical transport modeling. J Atmos Sci 59: 714 – 725

Herman JR, Bhartia PK, Torres O, Hsu C, Seftor C, Celarier E (1997) Global distribution of UV-absorbing aerosols from Nimbus 7/TOMS data. J of Geophys Res 102: 16,911-16,922

Hsu NC, Herman JR, Bhartia PK, Seftor CJ, Torres O, Thompson AM, Eck TF, Holben BN (1996) Detection of biomass burning smoke from TOMS measurements. Geophysical Research Letters 23: 745 – 748

Janowiak JE, Arkin PA (1991) Rainfall variations in the tropics during 1986– 1989, as estimated from observations of cloud-top temperature. J of Geophys Res 96: 3,359 – 3,373

Ji Y, Stocker E (2002a) An overview of the TRMM/TSDIS fire algorithm and products. Int J of Remote Sensing 23: 3,285 – 3,303

Ji Y, Stocker E (2002b) Seasonal, Intra-seasonal, and Interannual Variability of Land Fires and their effect on the Atmospheric Aerosols. J of Geophys Res 107: 4697, doi: 101029 /2002JD002331

Ji Y, Stocker E (2003) Reply to comment by Giglio et al. on "Seasonal, intra-seasonal, and interannual Variability of Global Land Fire and Their Effects on Atmospheric Aerosol Distribution". J of Geophys Res 108: 4697, doi: 10 1029/2003JD004115

Ji Y, Stocker E (2004) Comment on "Improving the Seasonal Cycle and Interannual Variations of Biomass Burning Aerosol Sources" by Generoso et al., Atmospheric Chemistry and Physics Discuss 4: 2,161 – 2,166

Justice CO, Kendell JD, Dowty PR, Scholes RJ (1996) Satellite remote sensing of fires during the SAFARI campaign using NOAA-AVHRR data. Journal of Geophysical Research 23: 851 – 863

Kaufman YJ, Justice CO (1998) MODIS fire products algorithm technical background document. EOS ID# 2741, USA

Kaufman YJ, Tucker CJ, Fung I (1990) Remote sensing of biomass burning in the tropics. J of Geophys Res 95: 9,927 – 9,939

Kaufman YJ, Fraser RS (1997) The effect of smoke particles on clouds and climate forcing. Science 277: 1,636 – 1,639

Kaufman YJ, Nakajima T (1993) Effect of Amazon on cloud microphysics and albedo-analysis from satellite imagery J Appl Meteor 32: 729 – 744

Kummerow C, Barnes W, Kozu T, Suiue J, Simpson J (1998) The tropical rainfall measuring mission (TRMM) sensor package. Journal of Atmospheric and Oceanic Technology 15: 808 – 816

Levine JS (1991) Global biomass burning: atmospheric, climatic, and biospheric implications. In: Levine JS (ed) Global Biomass Burning. The MIT Press

Lorenz EN (1956) Empirical orthogonal functions and statistical weather prediction. Department of Meteorology, MIT Science Report 1:49

Madden RA (1986) Seasonal variations of the 40 – 50 day oscillation in the tropics. J Atmos Sci 43: 3,138 – 3,158

Madden RA, Julian PR (1994) Observations of the 40 – 50-Day Tropical Oscillation—A Review. Mon Wea Rev 122: 814 – 837

Malingreau JP (1990) The contribution of remote sensing to the global monitoring of fires in tropical and sub tropical ecosystems. In: Goldammer JG (ed) Fires in Tropical Biota. Berlin: Springer

Matson M, Dozier J (1981) Identification of sub resolution high temperature sources using a thermal IR sensor. Photo Engr and Remote Sensing 47: 1,311 – 1,318

McKnight TL (1996) Physical Geography, 5rd edn (Upper Saddle River, New Jersey: Prentice Hall), pp 188 – 189

Pinnock S, Gregoire JM (1999) World Fire Web: A global fire observation system. Proceedings of the IUFRO Remote Sensing and Forest Monitoring Conference, Rogow, Poland, 1 to 3 June 1999

Ramanathan V, Crutzen PJ, Kiehl JT, Rosenfeld D (2001) Aerosol, climate, and hydrological cycle. Science 294: 2,119 – 2,123

Rasmusson EM, Wang X, Ropelewski CF (1990) The biennial component of ENSO variability. J Mar System 1: 71 – 96

Rosenfeld D (1999) TRMM observed first direct evidence of smoke from forest fire inhibiting rainfall. Geophysical Research Letters 26: 3,105 – 3,108

Rosenfeld D (2000) Suppression of rain and snow by urban and industrial air pollution. Science 287: 1,793 – 1,796

Rosenfeld D, Lensky IM (1998) Satellite-based insights into precipitation formation processes in continental and maritime convective clouds. Bull Am Meteorol Soc 79: 2,457 – 2,476

Rosenfeld D, Rudich Y, Lahav R (2001) Desert dust suppressing precipitation: A possible desertification feedback loop. PNAS, 98: 5,975 – 5,980

Schindler L (1998) The Indonesian fires and SE Asian haze 1997/1998 review, damages, causes and necessary steps. Proceedings of the Asia-Pacific Regional Workshop on Transboundary Atmospheric Pollution, Singapore, 27 to 28 May 1998

Townshend JR, Justice CO, Skole D, Malingreau JP, Cihlar J, Teillet PM, Sadowski F, Ruttenberg S (1994) The 1 km resolution global dataset: needs of the International Geosphere-Biosphere Programme. Int J of Remote Sensing 15: 3,417 – 3,441

Vautard R, Ghil M (1989) Singular spectrum analysis in nonlinear dynamics, with applications to paleoclimatic time series. Physica D 35: 395 – 424

20 China's Current and Future Meteorological Satellite Systems

Wenjian Zhang, Jianmin Xu, Chaohua Dong and Jun Yang

20.1 Introduction

Meteorological satellites have become an indispensable tool for weather and environment observations in China. These satellites, viewing the Earth from large scale to a global perspective, are integrated components of China's Earth observation system for meteorological operations, major natural disasters monitoring and improving the efficiency of many sectors of our national economy. Therefore, the meteorological satellite has been regarded as a kind of application satellite with notable social and economic benefits among man-made satellites. It is not feasible nowadays to ignore the space observation data in the field of meteorology, hydrology, agriculture, environment as well as disaster monitoring in China, such a geographically and economically big country with the largest population in the world. For this reason, China is making her unremitting efforts on building up the meteorological satellite systems and data application systems.

The meteorological satellite program of China consists of two series: the polar orbiting series and geostationary series. The main objectives of the program are to establish, with the combination of polar and geostationary orbits, a comprehensive operational meteorological satellite system as well as the ground monitoring and application data system, in order to meet the needs on various aspects in China, and enhance the ability of contribution to the international community. The China Meteorological Administration contracts China Aerospace Science and Technology Corporation (CASC) to develop the space segment including the launchers and the satellites, while National Satellite Meteorological Center, a sub-organization of China Meteorological Administration, is responsible for developing the satellite ground segment.

In China, meteorological satellites are named simply as Feng-Yun series, abbreviated as FY-series. The Chinese words Feng-Yun stand for "Winds and Clouds". The FY-odd numbers are used for the generations of the polar orbiting satellite series, i.e. FY-1 for first generation, FY-3 for second generation, etc., whereas the FY-even numbers are used for the generations of the geostationary series, i.e. FY-2 for first generation, and FY-4 for the second generation, etc..

20.2 The Polar Orbiting Meteorological Satellites of China

20.2.1 The First Generation of Polar Orbiting Operational Meteorological Satellites of China

1. The Experimental Satellites: FY-1A and FY-1B

China began her polar meteorological satellite program since early 1970. The first two experimental satellites, FY-1A and FY-1B, were launched in 1988 and 1990 respectively. These two satellites are controlled with three-axis stabilization, and the main payload is a five-channel imager and a Space Environment Monitor (SEM). FY-1A only operated for 39 days, with the main trouble on the satellite being attitude control. FY-1B operated discontinuously for more than one year, but a lot of images had been sent back to the ground. The experiences of these two satellites laid solid foundations for the success of the successor operational polar satellites. Based upon the lessons learned during the development and operation of FY-1A and FY-1B, the final design and configuration were decided for the two operational polar satellites, i.e., FY-1C and FY-1D, with considerable improvements.

2. The Improvements and Specifications of FY-1C and FY-1D

The major improvements are summarized as follows:

(1) To improve the satellite attitude control and the overall quality by improving the reliability of the mechanical moving components, to ensure the design life-span of the satellites.

(2) To greatly improve the imaging payload. The Multi-channel Visible and Infrared Radiometers (MVISR) onboard FY-1A and FY-1B were improved into a ten-channel imager (see Table 20.1), which enables the imager to have more powerful capabilities to observe the clouds, land and oceans. FY-1C satellite is equipped with two identical MVISR imagers for backup of each other for ensuring operational continuity (FY-1D satellite equipped with the same payload as FY-1C).

(3) For accommodating the MVIRS data transmission, the data rate of the FY-1C and FY-1D High Resolution Picture Transmission format (named as CHRPT) is increased to 1.3308 Mbps for the 10-channel data (NOAA and FY-1A/FY-1B data rate is of 0.6654 Mbps). The data format is compatible with NOAA/HRPT, which makes the worldwide NOAA/HRPT receiving and data processing systems can easily receive and process the FY-1C and FY-1D data with smallest modifications.

(4) With the three national ground acquisition stations located in Beijing, Guangzhou and Urumqi respectively, China can daily receive the real-time CHRPT data covering whole China territory. In addition, with the on board data

storage capacity of 300 minutes, China has the alternative of receiving global coverage data of four selected channels with reduced resolution (3.3-km resolution at sub-satellite point) once each day (defined as Delayed Global Picture Transmission, DGPT), or playbacking 20 minutes orbit observation data of ten channels with original resolution (1-km resolution at sub-satellite point) at any region of the world (defined as Delayed Local Picture Transmission, DLPT).

The channel features of the main payload on FY-1C and FY-1D is the Multi-channel Visible and IR Scan Radiometer (MVISR) with the channel characteristics indicated in Table 20.1.

Table 20.1 The channel characteristics of MVISR onboard FY-1C and FY-1D

Channel	Wavelength (m)	Primary Use
1	0.58 – 0.68	Daytime cloud, ice and snow, vegetation
2	0.84 – 0.89	Daytime cloud, vegetation
3	3.55 – 3.95	Heat source, night cloud
4	10.30 – 11.30	SST, day/night cloud
5	11.50 – 12.50	SST, day/night cloud
6	1.58 – 1.64	Soil moisture, ice/snow distinguishing
7	0.43 – 0.48	Ocean colour
8	0.48 – 0.53	Ocean colour
9	0.53 – 0.58	Ocean colour
10	0.90 – 0.985	Water vapour

3. The Current Status of FY-1C and FY-1D

FY-1C was launched on May 10, 1999, with two ten-channel radiometers (backup each other) as the primary sensing instruments and a Space Environmental Monitor (SEM). FY-1C operates in a sun-synchronous orbit with the following orbit parameters depicted in Table 20.2. Till the writing of this chapter, FY-1C has been operating in the orbit for more than five years, and still operating now with partial capacity of observations. FY-1C is the longest surviving earth observation satellite of China and many valuable success experiences can be drawn from it.

The FY-1D, the successor of FY-1C and last one of FY-1 series, was launched on May 15, 2002, with the same instruments and very similar orbital parameters (see Table 20.2) as FY-1C. Now FY-1D is in operation for more than 2 years and is still in good health.

Many operational products are generated daily based on FY-1C and FY-1D satellites data, such as sea surface temperature (SST), NDVI, cloud cover and classification, fog identification, land surface temperature, aerosol, ocean colour, snow cover, water vapour etc., as well as rapid response products such as forest fire, flooding area identification, air pollution, sea ice identification, etc.. These image data and products are used for the daily environmental monitoring and

Table 20.2 Orbital and CHRPT parameters of FY-1C and FY-1D satellites

Satellite	FY-1C, FY-1D
Launch Date	FY-1C: May 10, 1999; FY-1D: May 15,2002
Orbit	un-synchronous
Altitude (km, @launch)	FY-1C: 873; FY-1D: 866
Weight (kg)	950
Period (min)	102.3
Inclination (degrees)	98.80
Eccentricity	< 0.005
Descending Node (LST @launch)	FY-1C : 08: 15; FY-1D: 08: 45
Average Power Output (@launch)	256 Watts
Design Life	2 years
Attitude Control	Three-axis stabilized
CHRPT Transmission Frequency	1,708.0 MHz (1,704.5 MHz as backup)
DPT Transmission Frequency	1,700.0 MHz
EIRP	39.4 dbm
Polarization	Right hand circular
Modulation	PCM-PSK
Modulation Index	67.5° ± 7.5°
Bit Rate of CHRPT	1.3308 Mbps

weather forecasting services in China and around the globe for more than five years (Zhang, 1999). With the high quality of the imager data and ten-channel remote sensing capability, the FY-1C and FY-1D satellites data are widely accepted and received by USA, European countries, Australia and Asia countries.

20.2.2 The Second Generation of Polar Orbiting Operational Environmental Satellites of China: FY-3 Series

1. The Mission Objectives of FY-3

FY-3 series, the second generation and the successor of FY-1 series of Chinese polar orbiting environmental satellites, will be developed during 2000 to 2016 with two research and development (R&D) satellites and five operational satellites, according to present planning. With the planned first launch in late 2006, the FY-3 series will provide meteorological and environmental services during 2006 – 2020.

The primary mission objectives of the FY-3 series are:

- To sound the 3-dimensional thermal and moisture structures of the Earth's atmosphere globally, obtaining the cloud pattern and other key parameters measurements such as precipitation, ozone, etc., to support global numerical weather prediction and environmental services.
- To image the Earth's surfaces for monitoring large scale meteorological and/or hydrological disasters and biosphere environment evolvement.
- To establish long-term environmental data sets with retrieved important geophysical and space environmental parameters for climate monitoring, global predication estimation, space environmental monitoring and Earth sciences researches.
- To carry on data collection and re-transmission by DCPs.

The first two satellites FY-3A and FY-3B and the on-board instruments are underway, and the corresponding ground segment is being designed and developed.

2. The Major Specifications of the FY-3A Satellite

FY-3A is a sun-synchronous polar-orbiting environmental satellite. The satellite in general is a hexahedron of 4.4 m$\times$2.0 m$\times$2.0 m and the configuration is depicted in Fig. 20.1.

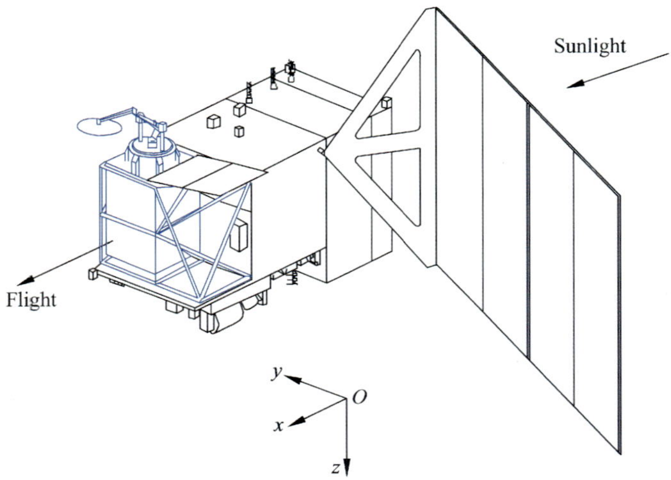

Figure 20.1 Diagram of the FY-3A satellite

The total weight is estimated 2,400 kg. The one solar panel is mounted on one side of the satellite main body which makes the span length of the satellite 10 meters in flight. The attitude control of the satellite is three-axis stabilized with a measuring precision of 50 meters with the measurement of a star sensor onboard the satellite. Table 20.3 depicts the major orbital parameters of the satellite.

Table 20.3 FY-3A satellite design specifications and major orbital parameters

Orbit	Sun-Synchronous
Altitude (km)	836.4
Power (W)	1,100
Weight (kg)	2,400 kg
Size	4.38 m×2.0 m×2.0 m (in stowed) 4.44 m×10.0 m×3.79 m (in flight)
Orbital Period (minutes)	101.603
Inclination (degrees)	98.753
Eccentricity	< 0.0025
Equatorial Crossing Time	10:10 (a.m.)
Orbital maintenances	10 minutes within two years
Onboard Data Storage	144 Gbits
Attitude Control	Three-axis stabilization
Launch Vehicle	LM-4B
Launch Plan	FY-3A:September, 2006
Design Lifetime	Three years

20.2.3 Payloads Onboard FY-3A

To achieve FY-3 objectives, the designing and manufacturing a core meteorological payload with imaging mission, sounding mission and complementary mission instruments are underway. The major payload characteristics are depicted in more detail as follows.

1. The Imaging Mission Payloads

There are 4 imaging instruments in FY-3A mission as following:

(1) Visible and InfraRed Radiometer (VIRR)

This is an instrument heritaged from Mulit-channel Visible and Infrared Scanning Radiometer (MVISR, 10 channels) onboard FY-1C and FY-1D satellites for ensuring the data continuity of the FY-1 series operational observations. For risk reduction purpose, the VIRR instrument will remain basically the same as the MVISR with one central wavelength change as specified in Table 20.4.

(2) Medium Resolution Spectral Imager (MERSI)

This is a 20-channel VIS/NIR instrument onboard FY-3A with the major purpose to observe the Earth's clouds and surfaces characteristics with multi-channel measurements. The MERSI channels are mainly located at VIS and NIR spectral region. With the combination of VIRR instrument which has the major important IR channels, these two instruments together will make a powerful observation to the Earth systems, including ocean, atmosphere and land. The main characteristics of MERSI is that there are five channels (4 VIS and 1 thermal IR) with the 250 m spatial resolution, which enables it to image the Earth with higher resolution and near true colour imagery during the day and get high resolution thermal IR image during the night. The major specifications and channel wavelength characteristics are depicted in Table 20.5 and Table 20.6, respectively. It is planned from the development of FY-3B, the MERSI and VIRR instruments will be merged into one imaging instrument suite.

Table 20.4 Visible and Infrared Radiometer (VIRR) channel characteristics onboard FY-3A

Channel No.	Spectral Range (μm)	NEDP ρ (%) NEDN (300 K)	Dynamic Range
1	0.58 – 0.680	0.1%	0 – 100%
2	0.84 – 0.890	0.1%	0 – 100%
3	3.55 – 3.930	0.3 K	180 – 350 K
4	10.3 – 11.30	0.2 K	180 – 330 K
5	11.5 – 12.50	0.2 K	180 – 330 K
6	1.55 – 1.640	0.15%	0 – 90%
7	0.43 – 0.480	0.05%	0 – 50%
8	0.48 – 0.530	0.05%	0 – 50%
9	0.53 – 0.580	0.05%	0 – 50%
10	1.325 – 1.395	0.19%	0 – 90%

Table 20.5 Major specifications of MERSI

Quantizing Levels	12 bits
Earth Scan Angle	± 55.4°
Pixels for each scan line	2,048 (for 1,000 m); 8,192 (for 250 m)
Pixel Registration	< 0.3 pixel
MTF	⩾ 0.27 (for 1.0 km channels); ⩾ 0.25 (for 250 m)
Onboard Calibration	Yes
Lab Calibration Accuracy	VIS/Near IR: 5% (ρ); IR: 1 K (270 K)

Table 20.6 MERSI channel characteristics

Channel	Wavelength (μm)	Bandwidth (μm)	Resolution (meter)	NED ρ (%) NEDT (300 K)*	Dynamic Range
1	0.470	0.05	1,000	0.10	100%
2	0.550	0.05	250	0.20	100%
3	0.650	0.05	250	0.30	100%
4	0.865	0.05	250	0.30	100%
5	1.640	0.05	250	0.10	90%

Continued

Channel	Wavelength (μm)	Bandwidth (μm)	Resolution (meter)	NED ρ (%) NEDT (300 K)*	Dynamic Range
6	2.130	0.05	1,000	0.15	90%
7	0.412	0.02	1,000	0.10	80%
8	0.443	0.02	1,000	0.10	80%
9	0.490	0.02	1,000	0.05	80%
10	0.520	0.02	1,000	0.05	80%
11	0.565	0.02	1,000	0.05	80%
12	0.650	0.02	1,000	0.05	80%
13	0.685	0.02	1,000	0.05	80%
14	0.765	0.02	1,000	0.05	80%
15	0.865	0.02	1,000	0.05	80%
16	0.905	0.02	1,000	0.10	90%
17	0.940	0.02	1,000	0.10	90%
18	0.980	0.02	1,000	0.10	90%
19	1.030	0.02	1,000	0.10	90%
20	11.500	2.00	250	0.3 K	330 K

* User required speficications.

(3) Microwave Radiation Imager (MWRI)

This is a conical scanning microwave Imager at 6 frequency points with dual polarizations (12 channels). This sensor measures thermal microwave missions from land and ocean surfaces, and can measure various forms of waters in the atmosphere, clouds and surfaces. For microwave wavelengths are much longer in the electromagnetic spectrum compared with visible and infrared, the microwave imager can penetrates clouds, and provides forecasters with all weather measurement capability. At higher frequency channels, such as 89 GHz and 150 GHz, the scattering signatures from the precipitation are also good indicators for detecting of rainfall over both land and ocean. The spatial resolutions are from 15 km to 80 km, depending on the wavelengths. Table 20.7 shows the major specifications of the instrument.

Table 20.7 Major specifications of MWRI

Frequency (GHz)	10.65	18.7	23.8	36.5	89	150
Polarization	H,V	H,V	H,V	H,V	H,V	H,V
Bandwidth (MHz)	180	200	400	1,000	2,000	2,000
Integral Time (ms)	12	6	6	4	2	2
Dynamic range (k)	3 – 350	3 – 350	3 – 350	3 – 350	3 – 350	3 – 350

(4) Total Ozone Mapping Unit (TOMU)

The Total Ozone Mapping Unit (TOMU) is a cross scanning instrument which uses UV spectral information. The main specification is given in Table 20.8 and Table 20.9, respectively.

Table 20.8 Major specifications of TOMU

Specifications	Value
Spectral Range (nm)	308 – 360
Sensitivity	$\leqslant$ 0.0014 μW/cm^2·sr·nm(*S*/*N*=1)
Spots of Each Scan Line	31 spots
Scan Period for Each Line (seconds)	8.16
Quantizing Levels	12 bits
Sub-Point Resolution (km)	50

Table 20.9 Spectral characteristics of TOMU

CH No.	Central Wavelength (nm)	Bandwidth (nm)
1	308.68 ± 0.15	1 + 0.3, − 0
2	312.59 ± 0.15	1 + 0.3, − 0
3	317.61 ± 0.15	1 + 0.3, − 0
4	322.40 ± 0.15	1 + 0.3, − 0
5	331.31 ± 0.15	1 + 0.3, − 0
6	360.11 ± 0.25	1 + 0.3, − 0

2. FY-3A Sounding Mission Payload

(1) Infrared Atmospheric Sounder (IRAS)

This is the primary sounder for FY-3A. It is a HIRS/3-like instrument and the required first 20-channel frequencies and specifications are the same as HIRS/3 (for channel information, please see NOAA-KLM user guide for details). However, there are additional 6 channels which will enable IRAS having capacity of measuring aerosols, stratosphere temperature, carbon dioxide content and cirrus. The instrument optical FOV is 0.97 degrees, which makes the ground IFOV of 14 km at nadir (with nominal altitude of 900 km). Table 20.10 depicts the major specifications for IRAS.

(2) Microwave Atmospheric Temperature Sounder (MWTS)

This is an 8-channel passive scanning microwave sounder with the purpose of temperature sounding in the cloudy region. There are four channels around 50 GHz and another four channels located at 19.35, 23.9, 31.0 GHz and 89.0 GHz. Table 20.11 shows the major specifications of MWAS. It is planned that the MWTS will continuing evolve during FY-3B development and will reach 15 – 20 channels for the FY-3C and the following satellites.

Table 20.10 IRAS spectral characteristics

CH. No.	Central Wave Number (cm^{-1})	Wave Length (μm)	Half Power Bandwidth (cm^{-1})	Max. Scene Temperature (K)	NEΔN(mW/ $m^2 \cdot sr \cdot cm^{-1}$)
1	669	14.95	3	280	4.00
2	680	14.71	10	265	0.80
3	690	14.49	12	250	0.60
4	703	14.22	16	260	0.35
5	716	13.97	16	275	0.32
6	733	13.84	16	290	0.36
7	749	13.35	16	300	0.30
8	802	12.47	30	330	0.20
9	900	11.11	35	330	0.15
10	1,030	9.71	25	280	0.20
11	1,345	7.43	50	330	0.23
12	1,365	7.33	40	285	0.30
13	1,533	6.52	55	275	0.30
14	2,188	4.57	23	310	0.009
15	2,210	4.52	23	290	0.004
16	2,235	4.47	23	280	0.006
17	2,245	4.45	23	266	0.006
18	2,388	4.19	25	320	0.003
19	2,515	3.98	35	340	0.003
20	2,660	3.76	100	340	0.002
21	14,500	0.69	1,000	100%A	0.10%A
22	11,299	0.885	385	100%A	0.10%A
23	10,638	0.94	550	100%A	0.10%A
24	10,638	0.94	200	100%A	0.10%A
25	8,065	1.24	650	100%A	0.10%A
26	6,098	1.64	450	100%A	0.10%A

(3) Microwave Atmospheric Humidity Sounder (MWHS)

The MWHS is humidity sounder, mainly sounding the atmosphere with high microwave frequencies. It is noted based on the AMSU-B experiences that these high frequency channels are not only useful for the moisture sounding, but also sensitive to the cloud liquid water, ice particles and precipitating clouds. The MWHS instrument and channel characteristics are shown in Table 20.12 and Table 20.13, respectively.

Table 20.11 Major specifications of MWTS

CH.	Frequency (GHz)	Absorber	Band width (MHz)	NEDT (K)	Calibration Accuracy (K)	Resolution (Nadir, km)
1	19.35	Window	220	0.3	1.5	50
2	23.90	H_2O	250	0.3	1.5	50
3	31.00	Window	600	0.25	1.5	50
4	50.31	Window	220	0.3	1.0	50
5	53.74	O_2	220	0.3	1.0	50
6	54.96	O_2	220	0.3	1.0	50
7	57.95	O_2	220	0.3	1.0	50
8	89.00	Window	6,000	0.8	1.5	50

Table 20.12 MWHS instrument characteristics

Characteristics	Value
Earth Scan Angle	$\pm 48.95°$
Earth Swath Width	2,200 km
Each Scan Line	90 spots
Spatial Resolution at Sub-Satellite Point	15 km
Calibration	Black body and deep space
Scan Period	8/3 s
Quantizing Levels	14 bits
Cal. Accuracy	1.5 K

Table 20.13 MWHS channel characteristics

CH No.	1	2	3	4	5
Center Freq. (GHz)	150(V)	150(H)	183.31 ± 1	183.31 ± 3	183.31 ± 7
Main Absorber	Window	Window	H_2O	H_2O	H_2O
Band Width (MHz)	$1,000 \times 2$	$1,000 \times 2$	500×2	$1,000 \times 2$	$2,000 \times 2$
NEΔT (k)	0.9	0.9	1.1	0.9	0.9
Freq. Stability (MHz)	50	50	30	30	30
Antenna Beam Efficiency	$\geqslant$95%	$\geqslant$95%	$\geqslant$95%	$\geqslant$95%	$\geqslant$95%
Dynamic Range (K)	3 – 340	3 – 340	3 – 340	3 – 340	3 – 340

(4) Solar Backscatter Ultraviolet Sounder (SBUS)

The purpose of the SBUS instrument is to measure the Solar irradiance and Earth radiance in the near ultraviolet spectrum. From these data, the global and vertical distribution of stratospheric ozone can be deduced.

The SBUS is a nadir pointing 12-channel spectrograph sensitive to radiation in the 250 nm to 400 nm ultraviolet spectrums. The overall radiometric resolution is approximately 1 nm in this spectral band. The spatial resolution of instrument is around 200 km at nadir. Two optical radiometers form the heart of the SBUS instrument: a mono-chromatograph and a small but very important Cloud Cover Radiometer (CCR). Table 20.14 shows the SBUS spectral characteristics.

Table 20.14 SBUS spectral characteristics

CH No.	Center Wavelength (nm)	Bandwidth (nm)
1	252.00 ± 0.05	$1 + 0.2, - 0$
2	273.62 ± 0.05	$1 + 0.2, - 0$
3	283.10 ± 0.05	$1 + 0.2, - 0$
4	287.70 ± 0.05	$1 + 0.2, - 0$
5	292.29 ± 0.05	$1 + 0.2, - 0$
6	297.59 ± 0.05	$1 + 0.2, - 0$
7	301.97 ± 0.05	$1 + 0.2, - 0$
8	305.87 ± 0.05	$1 + 0.2, - 0$
9	312.57 ± 0.05	$1 + 0.2, - 0$
10	317.56 ± 0.05	$1 + 0.2, - 0$
11	331.26 ± 0.05	$1 + 0.2, - 0$
12	339.89 ± 0.05	$1 + 0.2, - 0$
C. C. Radiometer	379.00 ± 1.00	$3 + 0.3$

20.2.4 Complementary Mission

- The Earth Radiation Budget Unit (ERBU)

This instrument is for measuring the radiation budget of the Earth-atmosphere system. There are wide-FOV and narrow-FOV observation units, separately, with two channels on each of the units. The broad-band channel covers the spectral range from 0.2 μm to 50 μm, and the narrow-band channel covers 0.2 μm to 3.5 μm, respectively. A separate solar constant monitoring instrument is included in the ERBU package.

- Space Environment Monitoring Unit (SEMU)

Space Environment Monitor Unit (SEMU) onboard FY-3 is the modified version of FY-1 Space Environment Monitor (SEM) instrument, with improved accuracy and measuring capacity for high-energy particles.

All together there will be more than 10 instruments on board FY-3A satellite. Table 20.15 summaries the main payload of FY-3A satellite with indications of major corresponding applications.

Table 20.15 FY-3A major remote sensing instruments

Instrument Name	Major Characteristics	Major Applications
Imaging Mission		
Visible and InfraRed Radiometer (VIRR), heritage from FY-1C/D	Spectral range: 0.43 – 12.5 μm Channel numbers: 10 Cross track scanning: ± 55.4° Spatial resolution: 1.1 km	Cloud, vegetation, snow and ice, SST, LST, water vapour, aerosol, ocean colour , etc.
Medium Resolution Spectral Imager (MERSI)	Spectral range: 0.43 – 12.5 μm Channel numbers: 20 Cross track scanning: ± 55.4° Spatial resolution: 0.25 – 1 km	True colour imagery, cloud, vegetation, snow and ice, ocean colour, aerosol, rapid response products(fires, flooding, etc.)
Microwave Radiation Imager (MWRI)	Frequency range: 23 – 89 GHz Channel numbers: 12 (6 frequencies with H,V polarization) Conical scanning: 110.8° Spatial resolution: 15 – 80 km	Rainfall, soil moisture, cloud liquid water, sea surface parameters
Total Ozone Mapping Unit (TOMU)	Spectral range: 0.2 – 3.8 μm Channel numbers: 6 Cross track scanning: ± 56.0° Spatial resolution: 50 km	Total ozone distribution
Sounding Mission		
Infrared Atmospheric Sounder (IRAS)	Spectral range: 0.69 – 15.5 μm Channel numbers: 26 Cross track scanning: ± 49.5° Spatial resolution: 14.0 km	Atmospheric temperature profile, atmospheric humidity profile, total ozone content, cirrus, aerosol, etc.
Microwave Atmospheric Temperature Sounder (MWTS)	Frequency range: 23 – 89 GHz Channel numbers: 8 Cross track scanning: ± 48.6° Spatial resolution: 50 km	Atmospheric temperature profile, rainfall, cloud Liquid water, surface parameters, etc.
Microwave Atmospheric Humidity Sounder (MWHS)	Frequency range: 150 – 183 GHz Channel numbers: 5 Cross track scanning: ± 48.6° Spatial resolution (SSP): 15 km	Atmospheric humidity profile, water vapour, rainfall, cloud liquid water, etc.

Continued

Instrument Name	Major Characteristics	Major Applications
Solar Backscatter Ultraviolet Sounder (SBUS)	Spectral range: 0.2 – 50 μm Channel numbers: 12 Cross track scanning: ± 56.0° Spatial resolution (SSP): 200 km	Ozone profile, total ozone amount.
Complementary Mission		
Earth Radiation Budget Unit (ERBU)	Spectral range: 0.2 – 50 μm Channel numbers: 12 Cross track scanning: ± 56.0° Spatial resolution (SSP): 200 km	Earth-atmosphere radiation budget, cloud radiation forcing.
Space Environment Monitoring Unit (SEMU)	X-ray	Space environmental parameters for space environmental monitoring and for safety protection of the satellite

* All the Spatial resolutions indicated in the table is for sub-satellite point.

FY-3 Data Format and Data Transmission Scheme

The FY-3 interface formats are consistent with the requirement of the Consultative Committee for Space Data Systems (CCSDS) standards. However, further definition may be required to establish the physical layers of the data formats. This document is structured to identify FY-3 selected options and requirements.

The spacecraft communications links are S-band and X-band. Commands are via S-band only. Command and telemetry links may be active simultaneously. The S-band section of the communications subsystem provides primary telemetry and command (T&C) service to and from FY-3 ground stations. The X-band section of the communication subsystem provides the science and engineering data downlink for the FY-3 common spacecraft. Three modes of data transmission are provided.

(1) Mission Picture Transmission (MPT) data format MPT is actually in the direct playback (DP) mode. All the global stored science and engineering data (except MERSI data is 20 minutes) are transmitted in high data rate at 110 Mbps, to the NSMC national ground playback stations (Beijing, Guangzhou, Urumuqi and Svarbld) when the satellite fly over the acquisition range of these stations. The transmission band frequency will be within 8,025 – 8,215 MHz or 8,215 – 8,140 MHz, with bandwidth of 140 MHz.

(2) X-band Data Transmission (XDT) data format XDT will be in the X-band Direct Broadcast (DB) mode. The main function of this data format is for real-time broadcasting of the science and engineering data of MERSI, with the data rate of 20 Mbps, to any receiving station with line of sight view. The

broadcasting band frequency is between 7,750 – 7,850 MHz, with the bandwidth of about 25 MHz.

(3) AHRPT data format This data format will observe WMO AHRPT data format. AHRPT transmission band frequency is within 1,698 – 1,710 MHz at the data rate of 4.2 Mbps. The data will be in real-time global broadcasting.

FY-3 satellite science data downlink design will follow CCSDS space data system standard, which makes the FY-3 data compatible with NPOESS and METOP satellite data transmission characteristics.

20.3 The First Generation Geostationary Meteorological Satellites of China

20.3.1 The FY-2A and FY-2B Satellites

The first generation geostationary meteorological satellite of China consists of five satellites. The first two satellites, i.e., FY-2A and FY-2B with the experimental purposes, were launched on June 10, 1997 and June 25, 2000 respectively.

1. Specifications of FY-2A and FY-2B Satellites

FY-2A/B meteorological satellites have the following functions:

- Obtaining visible, infrared and water vapour cloud images by a radiometer on board satellite. Sea surface temperature, cloud analysis chart, cloud parameters and wind vector can be derived from these data.
- Collecting and transmitting observed data from widely dispersed data collection platforms.
- Broadcasting S-VISSR data, WEFAX and S-FAX or processed cloud images.
- Monitoring space environmental from satellite.

The major satellite specifications are depicted in Table 20.16.

Table 20.16 FY-2A and FY-2B satellite specifications

Dimensions	Diameter	2.1 m
	Height	1.6 m (cylinder)
Mass	Launch	1,200 kg
	On Station	520 kg
Life Span	Designed	3 years
Orbit	Geostationary	Located at 105°E
Attitude	Spin-Stabilized, Spin Rate	(100 ± 1) rpm
Launch Vehicle	Long March-3	

2. Visible and Infrared Spin Scan Radiometer

The major payload of FY-2A and FY-2B meteorological satellites is the three-channel Visible and Infrared Spin Scan Radiometer (VISSR). The characteristics of the instrument are shown in Table 20.17.

Table 20.17 Major characteristics of VISSR onboard FY-2A and FY-2B

	Visible	Infrared	Water Vapour
Wavelength	0.5 – 1.05 μm	10.5 – 12.5 μm	6.2 – 7.6 μm
Resolution	1.25 km	5 km	5 km
FOV	35 μrad	140 μrad	140 μrad
Scan Line	2,500×4	2,500	2,500
Detector	Si-photo-diode	HgCdTe	HgCdTe
Noise Performance	S/N=6.5 (albedo=2.5%) S/N=43 (albedo=95%)	NEDT=0.5 – 0.65 K (300 K)	NEDT=1 K (300 K)
Quantification Precision	6 bits	8 bits	8 bits
Scan Step Angle	140 μrad (N-S scanning)		

3. The Operation History and Current Status of FY-2A

FY-2A, the first experimental Chinese geostationary meteorological satellite, was launched on 10 June 1997. On 17 June, 1997 FY-2A was located at 105°E. The satellite started operational imagery transmission on 1 January, 1998.

The normal operation schedule for FY-2A is that the operation of ground system is commanded and scheduled by SOCC (Satellite Operation and Control Center) to automatically acquire VIS, IR, and WV images. After being registered at the Image Acquisition System (IAS) of CDAS (Command and Data Acquisition Station), the S-VISSR images are generated and retransmitted to users through the satellite. Normally, the operational schedule is acquiring 28 earth images a day, among which 4 images are for wind observation. FY-2A broadcasts WEFAX images 16 times and takes ranging 4 times a day except when the satellite is performing orbit, or attitude control or equipment check.

The operation was disrupted on 8 April, 1998, due to the defect of satellite de-spin subsystem that is designed to drive the S-band antenna to spin in counteraction of the spinning of the satellite itself, by so doing to keep the antenna pointing toward the earth all the time. Since 6 July, 1998, FY-2A had to work discontinuously everyday in order to cool down the frictional heat built up between the antenna and satellite.

On 26 April, 2000, before the launch of FY-2B, FY-2A was moved to the backup position of 86.5°E. On July 27, 2000, the results of check out for FY-2A showed that after 3 years in the orbit the FY-2A satellite system remained a good condition except for the S-band antenna that cannot be allowed to work longer than 10 hours everyday. FY-2A satellite was switched from aboard system A to system B (redundancy) successfully in the process, then the system B was checked out thoroughly. The results showed all the engineering parameters of system B is quite similar to the system A.

After the launch of FY-2B, FY-2A has been in standby mode, mainly served as a contingent and backup satellite of FY-2B. The satellite manufacturer uses FY-2A as an engineering testing satellite for drawing useful information for satellite engineering. The ground system makes only the orbit control and the eclipse management.

During the interruption of FY-2B transmission, a contingent plan was implemented to let the FY-2A work 4 hours a day in order to replace FY-2B in case the latter cannot be recovered. A test was made on FY-2A, during which, FY-2A transmitted 6 full disc images, undertook the turn around ranging 3 times. Since the FY-2B is recovered, FY-2A is closed. It is closed as long as FY-2B works everyday.

4. The Operation History and Current Status of FY-2B

FY-2B is the second geostationary meteorological satellite of China with both purpose for operational service and engineering research and development (R&D) as well. FY-2B was launched on June 25, 2000 with Long-March 3 vehicle from Xichang Satellite Launch Center. The satellite was stationed at 105°E on July 10, 2000. On January 1, 2001, the FY-2B was declared operational and started broadcasting S-VISSR and WEFAX images.

On February 28, 2001, the first day after the satellite entered the spring eclipse, the up-converter of the transponder ceased working, leading to interruption of image transmission and anomaly of DCP subsystem. It turned out that the local oscillator of the up-converter was too much sensitive to temperature. During the whole eclipse period, the satellite temperature was carefully controlled. Several adjustments were made after the satellite came out of the earth shadow and that brought the DCP subsystem back to work again.

Through further implementation of temperature control, the transponder worked again, but the power output decreased. On June 18, 2001, image transmission recovered; however, the EIRP (Effective Isotropic Radiated Power) is 8 dBW below the normal level. Though it is possible for user to receive the data using a 2.4-meter antenna, the bit error rate of transmission signal is comparatively high and the quality of imagery is affected. Since its recovery, the up-converter of S-band transmitter has been working normally. The image acquiring, data transmission, data collection and turn around ranging have all recovered for operation.

The working state of FY-2B's transponder is susceptible to the change of temperature that must be kept precisely within a very narrow range around 8.4 C degrees. Its temperature depended much upon energy supply to maintain this condition. Therefore, during eclipse period when energy supply is less, FY-2B has to stop the image transmission completely to ensure enough energy for the safe management of satellite during eclipse period. Some equipment in the satellite must be switched off during autumn and spring eclipses (92 days per year in all) due to the limitation of energy. Therefore, the number of images was reduced to 25 and WEFAX broadcasting to14.

On June 8, 2003, the scan mirror of VISSR was stuck due to insufficient lubrication in the process of mirror retrace from scanning the south. The quality of the image was affected. The VISSR was then reset to get back the image quality. To prevent the problem from becoming deteriorated, hemispheric scanning mode has since then been implemented to let VISSR scan only the north hemisphere.

On August 30, 2004, FY-2B satellite was moving to the second operational position of 123.5°E. On September 12, 2004, it was repositioned on 123.5°E successfully. After engineering checking and ground segment adjustment, this satellite started and maintained hemisphere observation since October 27, 2004.

20.3.2 The First Generation of Chinese Geostationary Operational Satellite: FY-2C Series

1. The Improvement of FY-2C Series Upon the Basis of FY-2A and FY-2B

The current plan of FY-2C series (operational series) consists of three satellites, i.e., FY-2C, FY-2D and FY-2E. The improvements of FY-2C series benefit greatly from FY-2A and FY-2B satellites, and all the emerged problems have been extensively analysed and simulated, and effective measures have been taken to prevent these problems occurring again in the successor satellites. In addition, some improvements have been achieved for enhancing the performance of the FY-2C series of satellites, mainly the following aspects:

(1) A split window (a pair channels) has been formed within 10.5 – 12.5 μm on the radiometer, which allows the satellite having more capability to observe the clouds property, atmospheric moisture and surface temperatures of both land and oceans.

(2) A mid wave Infrared channel of 3.9 μm has been added. As this channel is less affected by water vapour, when it combines with thermal infrared window channels, more accurate surface temperature can be acquired. For this channel is sensitive to high temperature, therefore it is helpful for detecting warm targets on surface, such as forest and grassland fires. It is also used to obtain information of low-level cloud and fog. This channel is also helpful to distinguish between low-level clouds and snow/ice coverage. With this five-channel radiometer, the satellites have more capability to observe the clouds, atmosphere and the surface (both land and ocean).

(3) By decreasing the noise equivalent temperature difference (NEDT) , the instrument measurement accuracy has been improved, which allows the quantization levels of the radiometer output to be increased from 8 bits to 10 bits for the thermal infrared channels. The quantization level of the visible channel is also increased by improving the signal-noise ratio of the visible channel.

(4) Power supplies of the FY-2C series satellites have been increased to support the eclipse management.

(5) WEFAX has been replaced by LRIT.

2. Specifications of VISSR of FY-2C

S-VISSR format of FY-2C series is compatible with FY-2A and FY-2B. The spin rate of FY-2C series is changed from 100 rpm (of FY-2A and 2B) to 98 rpm in order to have enough time to transfer the raw data and S-VISSR in a cycle. The full disk observation takes 25.0 minutes. The channel characteristics of VISSR onboard FY-2C series are shown in Table 20.18, with comparison against the instrument onboard FY-2A and B.

Table 20.18 The comparison of the IR channels characteristics of VISSR

	FY-2 A,B		FY-2 C,D,E			
	IR	WV	IR1	IR2	IR3	WV
Wavelength (μm)	10.5 – 12.5	6.3 – 7.6	10.3 – 11.3	11.5 – 12.5	3.5 – 4.0	6.3 – 7.6
FOV (μrad)	160	160	140	140	140	140
Spatial resolution (km)	5.76	5.76	5	5	5	5
Dynamic range (K)	180 – 330	190 – 290	180 – 330			180 – 280
Temperature resolution (K)	0.6	1.0	0.4 – 0.2	0.4 – 0.2	0.5 – 0.3	0.6 – 0.5
Number of detectors	1(main) + 1 (alternate)	1(main) + 1 (alternate)	1(main) + 1 (alternate)	1(main) + 1 (alternate)	1(main) + 1 (alternate)	1(main) + 1 (alternate)
Quantization level	256	256	1024	1024	1024	256
Calibration	On board blackbody calibration, once every 3 disks		The ground calibration accuracy is 1K.Cool space and planet calibration is used for on-board calibration, once every 2 disks.			

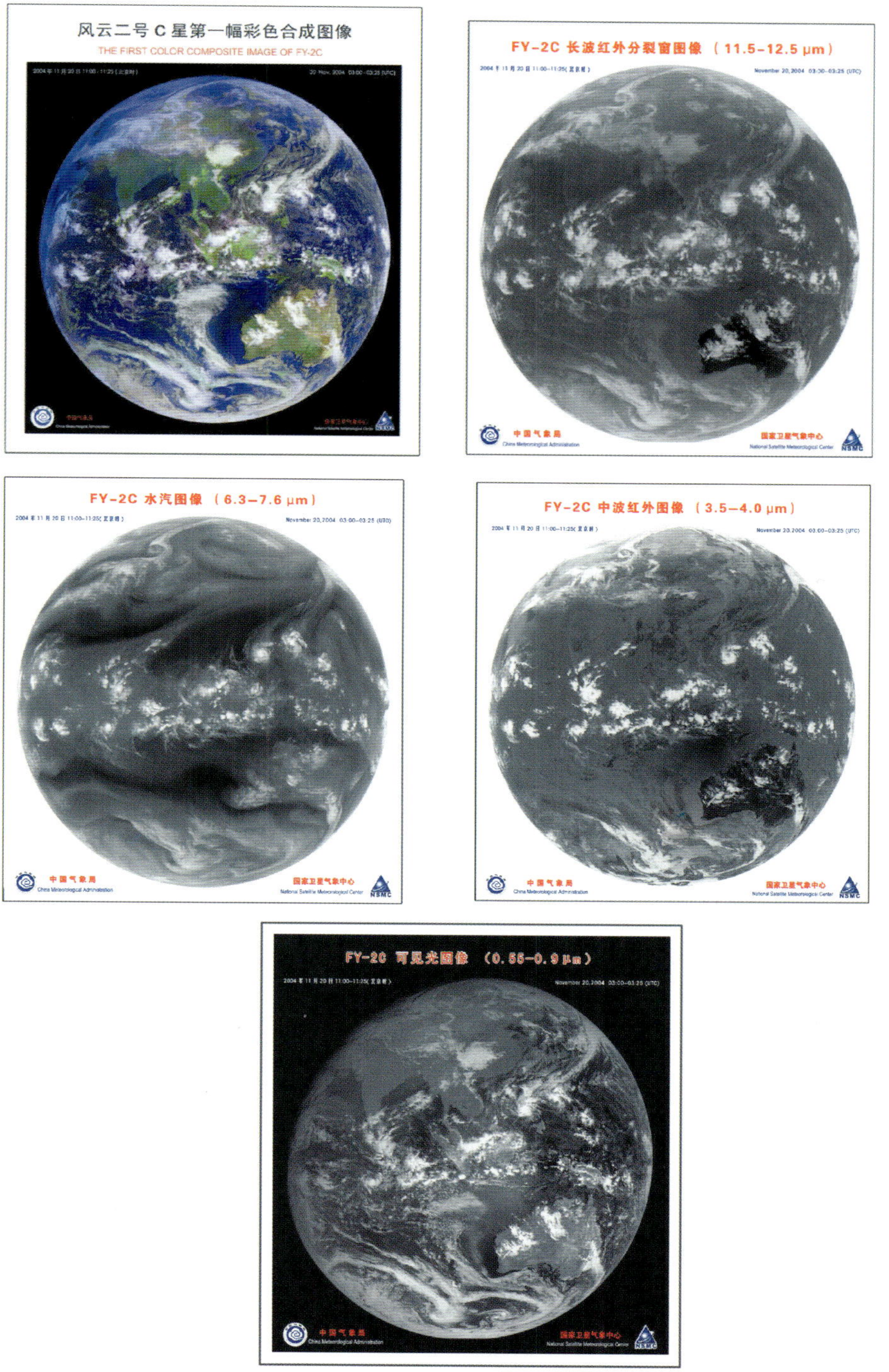

Figure 20.2 Five-channel images of FY-2C

3. The Launch and Commissioning of FY-2C

The FY-2C was successfully launched at 01:20 (UTC), October 19, 2004. The satellite positioned at 105°E on October 27, 2004. At 03:00 (UTC), October 29, 2004, the first visible channel image was obtained. At 03:00 (UTC), November 20, 2004, the remaining IR channel images were acquired. Soon after the turn-on full observation of the radiometer, the commissioning was carried out. The initial results showed that the performance of the FY-2C matched with the specified specifications. The Fig. 20.2 shows the five-channel images obtained on November 20, 2004. According to the schedule, the FY-2C will be put into full operation on January 1, 2005.

The FY-2D and FY-2E are planned to be launched in 2006 and 2009, respectively. It was decided that the observation scheme is also improved for the FY-2C series, i.e., during the flooding season (June to August each year), FY-2C series will make half-hourly observations for northern hemisphere to meet meteorological service requirements. For contingency strategy, the FY-2D will be ready by the end of 2005 and can meet on-demand launch requirement by then.

20.4 The Planning of the Second Generation Geostationary Meteorological Satellites of China: FY-4

China now is planning the second generation geostationary meteorological satellite program. This satellite series is the successor of the FY-2C series. The major user requirements come from the monitoring, now casting and very short range forecasting of the severe weather, as well as the surface environmental applications. The basic considerations are the following:

(1) The constellation will consist of two series, i.e., VIS/IR satellite (refer to "A series") with the early launch around 2012, and Microwave satellite (refer to "B series") with the early launch around 2015. All the satellites platform will be the three axis stabilization satellites.

(2) The major remote sensing instruments of the "A series" satellites will consist of a powerful imager with more than 12 channels, reference to SEVIRI and GOES-R imager; A hyper-spectral instrument is pursued on as the IR sounder with reasonable spatial resolution; A lighting mapper for locating the thunderstorm in the flooding season and for preventing the losses from the forest fires. The "A series" could remain two satellites constellation, and each satellite carry only one major payload for reducing the interference of different instruments.

(3) The instruments for the "B series" will mainly consist of a microwave instrument working on high frequencies for sounding the cloudy atmosphere, and a high spatial resolution CCD camera for frequent high resolution imagery observations for the meso-scale severe weather.

(4) The ground segment will co-design with the satellites and will have the enhanced control capability. The application segment will cover the weather, climate and environment, as well as rapid response fields.

Now FY-4 is at the definition and pre-configuration stage, some of the key technical feasibility studies are underway. The first satellite is scheduled to be developed during 2006 – 2012, and launch after the last satellite of the FY-2C series.

20.5 Summary

Stepping into the 21st century, we are now in an era that space technologies and applications are undergoing a fast development. China will continue her efforts to develop two types (LEO and GEO) of meteorological and environmental satellites to meet national requirements and modernizing the meteorological service of China and world meteorological community. Chinese meteorological satellite program is one of the components of space-based Global Observing System (GOS) of WMO. It is believed that the Chinese satellites are not only benefit the nation of China, but also a valuable contribution to the international meteorological, hydrological and environmental community.

References

The planning on Chinese Meteorological Satellites: from 1998 – 2010, China Meteorological Administration, April 1998 (In Chinese)

Zhang W (1999) Meteorological Satellite Program of China. Proceedings of Asia Conference on Remote Sensing, Hong Kong, China

Index

Index

Index

Index